AF905541

Introductory Nuclear Physics

Introductory Nuclear Physics

SECOND EDITION

SAMUEL S.M. WONG
University of Toronto

Wiley-VCH Verlag GmbH & Co. KGaA

All books published by Wiley-VCH are carefully produced.
Nevertheless, authors, editors, and publisher do not warrant the information contained in these books, including this book, to be free of errors.
Readers are advised to keep in mind that statements, data, illustrations, procedural details or other items may inadvertently be inaccurate.

Library of Congress Card No.:
Applied for

British Library Cataloging-in-Publication Data:
A catalogue record for this book is available from the British Library

Bibliographic information published by Die Deutsche Bibliothek
Die Deutsche Bibliothek lists this publication in the Deutsche Nationalbibliografie; detailed bibliographic data is available in the Internet at <http://dnb.ddb.de>.

© 1998 by John Wiley & Sons, Inc.
© 2004 WILEY-VCH Verlag GmbH & Co. KGaA, Weinheim

All rights reserved (including those of translation into other languages).
No part of this book may be reproduced in any form – nor transmitted or translated into machine language without written permission from the publishers.
Registered names, trademarks, etc. used in this book, even when not specifically marked as such, are not to be considered unprotected by law.

ISBN-13: 978-3-527-41445-1
ISBN-10: 3-527-41445-2

Printed and bound by CPI Group (UK) Ltd, Croydon, CR0 4YY
C9783527414451_220824

Contents

Useful Constants

Quantity	Symbol	Value			
Universal constants:					
fine structure constant	$\alpha = \frac{e^2}{[4\pi\epsilon_0]\hbar c}$	1/137.0359895(61)			
Planck's constant	h	$6.6260755(40) \times 10^{-34}$	J-s	4.135669×10^{-21}	MeV-s
(reduced)	$\hbar = \frac{h}{2\pi}$	$1.05457266(63) \times 10^{-34}$	J-s	$6.5821220 \times 10^{-22}$	MeV-s
	$\hbar c$			197.327053(59)	MeV-fm
speed of light	c	299792458	m/s		
unit of charge	e	$1.60217733(49) \times 10^{-19}$	C	$4.8032068 \times 10^{-10}$	esu
Conversion of units:					
area	barn	10^{-28}	m^2		
charge	C	2.99792458×10^9	esu		
energy	eV	$1.60217733(49) \times 10^{-19}$	J		
length	fm	10^{-15}	m		
mass	eV/c^2	$1.78266270(54) \times 10^{-36}$	kg		
	u	$1.6605402(10) \times 10^{-27}$	kg	931.49432(28)	MeV/c^2
Masses:					
electron	m_e	$9.1093897 \times 10^{-31}$	kg	0.51099906(15)	MeV/c^2
muon	m_μ	$1.8835327 \times 10^{-28}$	kg	105.65839(6)	MeV/c^2
pions	$\pi^\pm$	$2.4880187 \times 10^{-28}$	kg	139.56755(33)	MeV/c^2
	π^0	2.406120×10^{-28}	kg	134.9734(25)	MeV/c^2
proton	M_p	$1.6726231 \times 10^{-27}$	kg	938.27231(28)	MeV/c^2
				1.007276470(12)	u
neutron	M_n	$1.6749286 \times 10^{-27}$	kg	939.56563(28)	MeV/c^2
				1.008664898(12)	u
Lengths:					
Bohr radius	$a_0 = \frac{r_e}{\alpha^2}$	$5.29177249(24) \times 10^{-11}$	m		
classical electron radius	$r_e = \frac{\alpha\hbar}{m_e c}$	$2.81794092(38) \times 10^{-15}$	m		
Compton wavelength					
electron	$\lambda_{Ce} = \frac{h}{m_e c}$	$2.426310585(22) \times 10^{-12}$	m		
proton	$\lambda_{Cp} = \frac{h}{M_p c}$	1.32141×10^{-17}	m		
Others:					
Avogadro number	N_A			$6.0221367(36) \times 10^{26}$	mol^{-1}
Bohr magneton	$\mu_B = \frac{e\hbar[c]}{2m_e c}$			$5.78838263(52) \times 10^{-11}$	MeV/T
Boltzmann constant	k	1.380658×10^{-23}	J/K	$8.617385(73) \times 10^{-11}$	MeV/K
Fermi coupling constant	$\frac{G_F}{(\hbar c)^3}$	$1.43584(3) \times 10^{-62}$	$J\text{-}m^3$	$1.16637(2) \times 10^{-5}$	GeV^{-2}
Gamow-Teller to Fermi coupling constants	$\frac{G_A}{G_V}$	−1.259(4)			
magnetic dipole moment:					
electron	μ_e			1.001159652193(10)	μ_B
proton	μ_p			2.792847386(63)	μ_N
neutron	μ_n			−1.91304275(45)	μ_N
nuclear magneton	$\mu_N = \frac{e\hbar[c]}{2M_p c}$			$3.15245166(28) \times 10^{-14}$	MeV/T
permeability, free space	μ_0	$4\pi \times 10^{-7}$	N/A^2	$\epsilon_0\mu_0 = c^{-2}$	
permittivity, free space	ϵ_0	$8.854187817 \times 10^{-12}$	C^2/Nm^2		
Rydberg energy	$Ry = \frac{1}{2}m_e c^2\alpha^2$			13.6056981(40)	eV

Preface to the Second Edition

In the half dozen years or so since the first publication of *Introductory Nuclear Physics*, there have been several new developments and changes in the emphasis in the field. This, together with the enthusiastic feedback from colleagues and students, makes it imperative to publish a new edition.

For an active topic of research, a textbook cannot stay static. There are large areas that are basic and well established. These form the core of the first edition and they have stayed more or less the same. At the same time, the students should be made aware of certain new trends, such as superdeformation, relativistic heavy-ion reactions, nuclear astrophysics, and radioactive beams. At the same time, the preparation of students taking a course on nuclear physics is changing as well. Assumptions of a good working knowledge of angular momentum algebra and basic methods of quantum mechanics may no longer be correct for many. For this reason, some parts of the core of nuclear physics have been rewritten to make it more accessible.

The main changes in the second edition are the addition of two new chapters. Heavy-ion reactions, from high-spin states to ultra-relativistic collisions, are now in a totally new chapter. The same approach is also taken on nuclear astrophysics. To keep the book from getting too big, a few of the appendices in the first edition are either incorporated into the main text or taken out. In addition, some material that is no longer in the forefront of nuclear physics research is shortened or removed altogether.

The Internet has increasingly become the means of providing up-to-date information. From the latest description of major projects to comprehensive data bases, the World Wide Web is now the source of choice. For this reason, Uniform Source Locators (URL) are given as the "reference" for such topics as nuclear binding energies. Unfortunately, changes are frequently made to these electronic addresses and the reader may have to do some search to find the latest one if a particular URL is moved to a new location.

S. S. M. Wong

Preface to the First Edition

Nuclear physics is a subject basic to the curriculum of modern physics. There are several good reasons for this to be so. First and foremost is the intrinsic interest of the subject itself: The study of atomic nuclei has historically given us many of the first insights into modern physics. Furthermore, the potential of future discoveries remains very promising. Second, nuclear physics is closely associated with several other active branches of research: particle physics, in terms of the large overlap of interests in fundamental interactions and symmetries, and condensed matter physics, through the many-body nature of the problems involved. Third, nuclear physics may be usefully applied to other fields: chronology in geophysics and archaeology, tracer element techniques, and nuclear medicine, just to name a few.

The diversity of interest in nuclear physics also makes it very difficult to cover the entire subject in any satisfactory manner; some philosophy and guiding principles had to be adopted in selecting the material to be presented. The basic principle used for this book was to include what I believe every serious student of physics should know about the atomic nucleus. It was not always possible to live up to this principle. First, an appreciation of nuclear physics today will require not only a good knowledge of quantum mechanics and many-body theory but also quantum field theory. This, in general, is too much to expect for the average reader and some sacrifice must be made. Second, there are many interesting techniques, both experimental and theoretical, that form a part of the subject itself. Any reasonable coverage of these technical aspects will greatly expand the size of the book and make it useless in practice.

On the other hand, it is not possible to give a true flavor of nuclear physics without some background in quantum mechanics. In preparing this volume I have assumed that the student has the equivalent of a one-year undergraduate course in quantum mechanics or is taking concurrently an advanced quantum mechanics course at the level of one of the textbooks listed as general reference at the end. A basic knowledge of electromagnetic theory is also assumed; it is, however, unlikely that the background required here will be a problem to most students.

Some effort has been devoted to make the book as self-contained as possible. For this purpose, references to the literature are kept to a minimum. A specific paper published in scientific journals is mentioned only if a direct quotation is taken from it or if there is some historical interest associated with it. If references are needed, the first preference has been given to books that are readily available. However, this is not always possible. As a second choice, review articles are cited because a student starting out in the field may better comprehend this type of article than the original paper. Conference proceedings are used only as a last resort since it is difficult to expect standard libraries to be stocked with the multitude of proceedings published every year.

One result of adhering to this philosophy is that very few of the excellent papers of my colleagues have been cited. I have also had some difficulty in selecting standard textbooks for reference in subjects such as quantum mechanics, classical mechanics, electromagnetism, and statistical mechanics. Here, I have relied purely on my own biases without guidance from a general philosophy, as I have done with papers.

One decision that had to be made concerns the system of units used for equations involving electromagnetism. The Système International (SI) or meter-kilogram-second (mks) system would have been the more correct choice since essentially all students have been exposed to it and are more likely to be familiar with it. However, many of the advanced treatments on the subject, and nearly all the standard references on the topic in subatomic physics, are written using centimeter-gram-second (cgs) units. It is therefore more practical to use the latter system here so that it is easier for a reader to make use of other references. For the convenience of those who are more comfortable with SI units, most of the equations (except those in §V.2) have the necessary additional factor enclosed in large square brackets to convert the expressions to SI units. In most cases, it is possible to write the equations involving electromagnetism in a form independent of the system of units by making use of the fine structure constant α and by measuring charge in units of e, the absolute value of charge carried by an electron, and magnetic dipole moment in units of μ_N, the nuclear magneton.

The book is aimed at physics students in their final year of undergraduate or first year of graduate studies in nuclear physics. There is enough material for a one-year course though it could be used for a one-semester course by leaving out some of the detail and peripheral topics. The selection of material is guided in part by current interests in the field; no attempt has been made to give a complete account of everything that is known in nuclear physics. However, sufficient knowledge is provided here so that a student may then go to the library and obtain information on a particular nucleus or a special aspect of a topic.

S. S. M. Wong

Chapter 1

Introduction

Nuclear physics is the study of atomic nuclei. From deuteron to uranium, there are almost 1700 species that occur naturally on earth. In addition, large numbers of others are created in the laboratory and in the interior of stars. The main force responsible for nuclear properties comes from strong interaction. However, both weak and electromagnetic interactions also play important roles. For these reasons, nuclear physics serves as an important platform where basic properties of subatomic matter can be examined and fundamental laws of physics can be studied. We shall in this chapter give a brief history of the subject, its role in modern physics, and some of the general properties of nuclei we wish to study before going on into more detailed examinations in subsequent chapters.

1-1 Brief Early History of Nuclear Physics

The beginning of nuclear physics may be traced to the discovery of radioactivity in 1896 by Becquerel. Almost by accident, he noticed that well-wrapped photographic plates were blackened when placed near certain minerals. To appreciate the significance of this discovery, it is useful to recall that the time was before the era of quantum mechanics. The only known fundamental interactions were gravity and electromagnetism. In fact, just before the end of the nineteenth century, most of the observed physical phenomena were considered to be well understood in terms of what we now refer to as classical physics. Radioactivity was one of the few examples of unsolved problems. It was through the desire to understand these "exceptions" to otherwise well-established set of physical laws that gave birth to modern physics.

Two years after Bacquerel's discovery, Pierre and Marie Curie succeeded in separating a naturally occurring radioactive element, radium ($Z = 88$), from the ore (pitchblende). Soon afterward, it was realized that the chemical properties of an element were changed by such activities. When a source was placed in a magnetic field, it was found that there were three different possible types of activity, as the trajectories of some of the "rays" emitted were deflected to one direction, some to the opposite direction, and some not affected at all. These were named α-, β-, and γ-rays, as nothing more was known about them until much later. Subsequently, it was found that α-rays consist of positively charged ^{4}He nuclei, β-rays are made of electrons or positrons, and γ-rays are nothing but electromagnetic radiation that carries no net charge.

The existence of the nucleus as the small central part of an atom was first proposed by Rutherford in 1911. Later, in 1920, the radii of a few heavy nuclei were measured by Chadwick and were found to be of the order of 10^{-14} m, much smaller than the order of 10^{-10} m for atomic radii. The experiments involve scattering α-particles, obtained from radioactive elements, off such heavy elements as copper, silver, and gold, and the measured cross sections were found to be different from values expected of the Rutherford formula for Coulomb scattering off point charges.

The building blocks of nuclei are neutrons and protons, two aspects, or quantum states, of the same particle, the nucleon. Since a neutron does not carry any net electric charge and is unstable as an isolated particle, it was not discovered until 1932 by Chadwick, Curie, and Joliot. The only charged particles inside a nucleus are protons, each of which carries a positive charge of the same magnitude, but opposite in sign, as an electron. Since only positive charges are present, the electromagnetic force inside a nucleus is repulsive and the nucleons cannot be held together unless there is another source of force that is attractive and stronger than Coulomb. Here we have our first encounter with strong interaction.

Both gravitational and electromagnetic forces are infinite in range and their interaction strengths diminish with the square of the distance of separation. Clearly, nuclear force cannot follow the same radial dependence, else nucleons in one atom would have felt the attraction of those in nearby atoms. Being much stronger, it would have pulled the nucleons in different nuclei together into a single unit and destroy all the atomic structure we are familiar with. In fact, nuclear force has a very short range, not much beyond the confine of the nucleus itself, in marked contrast to the fundamental forces that were familiar at the time.

In 1935, Yukawa proposed that the force between nucleons arises from meson exchange. This was the start of the concept of field quantum as the mediator of fundamental forces. The reason that nuclear force has a finite range comes from the nonzero rest mass of the mesons exchanged. In contrast, the field quantum for electromagnetic force is the massless photon and, for gravitational force, the graviton. With the introduction of quantum chromodynamics, we come to realize that the Yukawa picture of meson exchange is only an effective theory for the force between nucleons. The fundamental force responsible for nuclear properties is the strong interaction between quarks. Most of this interaction is restricted to between the quarks inside a nucleon with gluons as the field quanta. However, some small "residue" goes outside and gives us the interaction between nucleons. This is very similar to chemical interactions. Even though atoms and molecules are electrically neutral, small remanents are found in the electromagnetic force between the atomic nucleus and its surrounding electrons, and these give rise to the wide diversity of chemistry around us.

For the nucleons inside a nucleus, nuclear force is far stronger than that due to electromagnetic interaction, as can be seen from the comparisons of the relative strengths, or coupling constants, made in Table 1-1. This presented some difficulties in understanding spontaneous α-particle decay of some heavy nuclei in the early part of the twentieth century. If the interaction is strong, how can α-decays have such long lifetimes? For example, nuclei such as ^{238}U ($\tau_{1/2} = 4.47 \times 10^9$ years) were created before the solar system was born and must have half-lives comparable to or longer than the age of the earth or else it cannot be found as ores today. The solution of the puzzle is

Table 1-1: Fundamental interactions.

Interaction	Field quantum	Range (m)	Relative strength	Typical cross section (m^2)	Typical time scale (s)
Strong	Gluon	10^{-15}	1	10^{-30}	10^{-23}
Weak	$W^{\pm}, Z^0$	10^{-18}	10^{-5}	10^{-44}	10^{-8}
Electromagnetic	Photon	∞	$\alpha = \frac{1}{137}$	10^{-33}	10^{-20}
Gravity	Graviton	∞	10^{-38}	–	–

quantum-mechanical tunneling, a direct evidence of the wave nature of particles, as we shall see later in §4-10.

Before the discovery of the neutron, it was assumed that a nucleus is made of protons and electrons. The presence of electrons inside the nucleus was made necessary for the following reason. The electric charge of a nucleus is, without exception, some integer multiple of e, the absolute value of the charge on an electron. Let us use Z to represent this integer. At the same time, the nuclear mass is essentially given by some integer A times the proton mass m_p. In the case of the hydrogen nucleus, we have $Z = A = 1$. For a nucleus made of A protons (as neutrons were not known), the charge should have been Ae. Instead, it is observed to be Ze, with $Z < A$ for all nuclei beyond hydrogen. To get around this difficulty, it was proposed to include $A - Z$ electrons in the nucleus to "neutralize" some of the proton charges.

This simple model fails when we include more data into our study. Nuclei with odd number of nucleons (A = odd) are known to have half-integer value spins, the total angular momentum and intrinsic spin of all the nucleons. On the other hand, nuclei with even A have integer value spins. Since particles with half-integer spins are fermions, particles that obey Fermi-Dirac statistics, an odd-A nucleus must be a fermion. Both electrons and protons are also fermions by virtue of the fact that their intrinsic spins are half integers. An electron and a proton may be combined to form an electrically neutral object, but their total spin is an integer and the combined object, as a result, cannot be a fermion. If there were no neutrons, the question of whether the spin of a nucleus takes on integer or half-integer values would have to be determined entirely by whether Z is even or odd. This is not found to be true in practice, and a model of the nucleus made of protons and electrons cannot be correct, as it violates the fundamental relationship between spin and quantum statistics.

The same quantum statistics consideration comes into play also in the discovery of the neutrino in β-decay. A free neutron is more massive than a proton and decays into a proton with a half-life of about 10 min. To conserve charge, an electron is emitted. However, this cannot be the complete picture, as all the particles involved are fermions. Furthermore, there are some difficulties with energy conservation as well. In nuclei, β-decay can transform one of the protons in the nucleus to a neutron with the emission of a positron and one of the neutrons to a proton by emitting an electron. The electrons and positrons are found to have a continuous spectrum of energy up to some maximum value known as the end-point energy. This seemed, on the surface, to violate energy conservation, as there is a definite energy difference ΔE between the parent and daughter nuclei. If the final state of the decay involves only two particles, an electron and the much more massive daughter nucleus, the kinetic energy of the electron is essentially fixed and completely specified by conservation of energy and momentum in the reaction. A continuous distribution of electron kinetic energy violates this simple argument. The neutrino was proposed by Pauli in 1931 and used by Fermi in 1933 to explain the puzzle. In addition to the electron or positron, a neutrino is also emitted in nuclear β-decay. It was not observed in the reaction because it carries no charge and very little, if any, rest mass. This "unobserved" fermion is even more elusive than the neutron: It hardly interacts with any other particles and is so light that even today we are still uncertain whether it is massless or not.

The concept of parity violation, the first one of a series of "broken" symmetries found in physics, was confirmed through nuclear β-decay. Both strong and electromagnetic interactions are known to conserve parity, i.e., experiments give the same results whether they are viewed in right-handed coordinate systems or left-handed coordinate systems. In the early 1950s, it was almost unthinkable to doubt that weak interaction should be any different from the other known ones, and certainly there were no reasons to suspect that parity needs to be treated any differently. However, there were baffling experimental data involving particles which seemed to be identical except for their decay modes. The concept of parity violation, proposed by Lee and Yang in 1957, was confirmed by a β-decay experiment using ^{60}Co in which it was observed that more electrons were emitted with momentum components opposite to the orientation of the nuclear spin than along it (for more details see §5-5). This is a clear violation of the invariance of operations under space inversion, i.e., a reflection through the origin of the coordinate system used. Violation of parity has led to a better understanding of the weak interaction itself, and the concept of broken symmetry opens a new horizon for us to view fundamental laws of physics.

1-2 What Is Nuclear Physics?

Since nuclei are involved in a wide variety of applied and pure research, nuclear physics overlaps with a number of other fields. In particular, it shares common interest with elementary particle physics in many respects. For example, the study of quark-gluon plasma in relativistic heavy-ion collisions involves both particle and nuclear physics. In astrophysics, stellar evolution and nucleosynthesis are intimately related to low-energy nuclear reaction rates, and the subject is of interest to nuclear physicists as well as astrophysicists. Many applications of nuclear properties, such as nuclear energy, nuclear medicine, tracer element techniques, involve a knowledge of nuclear physics, and nuclear physicists are often involved in the development of these areas. A broad definition of nuclear physics will therefore include far too much material than a single volume can reasonably cover. For our purpose, we shall only be concerned with the core of nuclear physics, its place as an integral part of modern physics, and its relation with some of the closed related disciplines.

The primary aim of nuclear physics is to understand the force between nucleons, the structure of nuclei, and how nuclei interact with each other and with other subatomic particles. These three questions are, to a large extent, related with each other. Furthermore, their interests are not necessarily confined to nuclear physics alone.

Nuclear force. One may argue that, since nuclear force is only one aspect of the strong interaction between quarks, all we need to do is to understand quantum chromodynamics (QCD), the theory for strong interaction. This is, however, not the complete picture. Nuclear interaction operates at the low-energy extreme of QCD where the interaction is strong and most complicated. This is one reason why studies in particle physics are often carried out at high energies where things are believed to be far simpler and we have a chance to unravel the mystery of the fundamental force between quarks. Needless to say, we do not yet understand strong interaction anywhere as well as, for example, electromagnetic interaction. In fact, studies made on nuclei constitute some of the best means to clarify certain aspects of QCD.

Even a thorough knowledge of QCD may not solve the problem of nuclear force. Again we can make an analogy with chemistry. All chemical interactions between atoms and molecules are electromagnetic in nature. However, this does not mean that we can calculate the structure of a DNA molecule starting from Maxwell's equations. The same is true between the fundamental strong interaction and nuclear force. We need QCD to provide us with an understanding of the foundation of nuclear force—any practical applications in nuclei must still come from a direct knowledge of the interaction among nucleons. It is also very likely that, from an operational point of view, strong interaction is too complicated to be applied directly to nuclei, and nuclear force derived from studies made on nuclei may be far more convenient to use in practice.

Nuclear structure. Nuclei are usually found in their individual ground states, by virtue of the fact that these are the lowest ones in energy. However, in the laboratory, and in the interior of stars, energy can be injected into nuclei to promote them to excited states. Besides energy, other properties for many of these states, such as electromagnetic moments and transition rates, can also be observed. In addition, β-decay, nucleon transfer, fission, and fusion transform one nuclear species to another. The study of these quantities supplies us with information on the structure of nuclei. In addition to its intrinsic values, nuclear structure can also provide us with the "data" on the nature of nuclei and the forces acting on the system.

From a quantum mechanics point of view, nuclear structure studies, for the most part, may be classified as bound state problems. Given an interaction, solution to the eigenvalue problem provides us with the energy level positions and wave functions. From the eigenfunctions, we can calculate matrix elements of operators corresponding to observables. The interaction of primary concern here is the strong force between nucleons. The effect of Coulomb force, in many cases, can be treated as perturbation to the predominant nuclear interaction. This comes, in part, because of the simple radial dependence of electromagnetic force, in contrast to that for strong interaction. On the other hand, weak interaction has extremely short range and, for all practical purposes in nuclear physics, may be treated as a zero-range, or "contact," interaction. Its presence is mainly felt in β-decay and related processes.

We are, however, faced with several difficulties here. The first is that nuclear interaction is not well known. In fact, the interaction between nucleons bound in nuclei can be somewhat different and, perhaps, even simpler than nucleon-nucleon interaction in general. For this reason, an important part of nuclear structure studies involves effective potentials between bound nucleons. A second difficulty is the Hilbert space that must be used to obtain a solution. In principle, the dimension is infinite. To reduce the problem to a tractable one, sever truncations are necessary. It is possible to compensate in part the errors introduced in making the calculation within a restricted space by adjusting, or "renormalizing," the interaction. We shall see in Chapter 7 that we have quite a bit of success in understanding nuclear structure by proceeding in this way.

Most of our information is obtained from studies made on stable nuclei for the simple reason that they are far easier to handle in the laboratory. Since this is a very special group among all the possible ones that can be formed, it is likely that our knowledge is biased. Furthermore, unstable nuclei form important intermediate steps in nucleosynthesis and are crucial in stellar evolution. With the advent of radioactive beams, large quantities of a variety of short-lived "exotic" nuclei will soon become available to enrich our data bank on nuclear structure.

Nuclear reaction. In nuclear reactions, we study the behavior of nuclei in the relation with other subatomic particles. From a quantum mechanics point view, it is primarily a scattering problem. There are several marked differences from nuclear structure studies.

First, it involves kinematics, and the results depend very much on the reaction energy as well. Besides elastic scattering, we can have inelastic processes that lead to different final states and create particles not present in the initial state. In addition, the reaction may also be sensitive to any momentum dependence of the interaction between particles.

Second, the probe itself is often a complex object and may be modified by the reaction. For example, when a light ion, such as ^{16}O, is used to scatter off a nuclear target, both the incident and target nuclei may be excited or transformed into other particles. This complicates the analysis as well as opens up new channels for nuclear studies.

A third aspect is that the scattering problem involving strong interaction is perhaps too complicated to be solved. In fact, for many purposes, the complete solution may not be of interest. The study of reaction theory is developed, to a large extent, because of such interests in strong interaction processes. Unfortunately, the topic can be rather formal at times. For our purpose, we shall only make very limited use of this vast resource in Chapter 8.

A good example among those of current interest is heavy-ion reactions. At low energies, the reaction creates a large number of exotic nuclear states that further enhance our knowledge of nuclear physics. At the other extreme of ultra-relativistic energies, it allows us to study the fundamental strong interaction itself.

Understanding nuclear structure and nuclear reaction is interesting and important by its own merits. However, the benefit goes beyond nuclear physics. We have already seen examples of new insight in terms of quantum-mechanical tunneling from nuclear α-decay, in confirmation of parity nonconservation using nuclear β-decay, and in using relativistic heavy-ion collision to create quark-gluon plasma. As an integral part of modern physics, nuclear phenomena can give and have given deep insight in understanding physics. The possibility is only limited by our imagination.

1-3 General Properties of Nuclei

The intense activity in the last century has resulted in a large body of knowledge on nuclear physics. We shall summarize in this section some of the general properties that are basic to the subject.

Valley of stability. Stable nuclei are found with proton number $Z = 1$ (hydrogen) to $Z = 82$ (lead). There are, however, a few minor exceptions, and we shall come back in Chapter 9 to see the significance of some of these in astrophysics. For each proton number, there are usually one or more stable or long-lived nuclei, or isotopes, each having a different number of neutrons. Since the chemistry of an element is determined by the electrons outside the nucleus and, hence, the number of protons inside, the chemical properties of different isotopes are fairly similar to each other. However, since they are made of different neutron numbers N, their nuclear properties are quite different.

The only unstable nuclei found naturally on earth are those with lifetimes comparable to or longer than the age of the solar system ($\sim$5 billion years) or as decay products of other long-lived species. However, in stars, unstable nuclei are being created continuously by nuclear reactions in an environment of high temperature and high density. Many short-lived nuclei are also made in the laboratory, including those with more nucleons than the heaviest ones found naturally on earth (see e.g., [84]). A list of known elements together with their chemical names and abbreviations is given in Table 1-2.

To a first-order approximation, stable nuclei have $N = Z$, with neutron number the same as proton number. The best example is perhaps the $A = 2$ system. Here, we find that the only stable nucleus is the deuteron, made of one proton and one neutron. Di-proton and di-neutron are both known to be unstable. From this observation we can conclude that the force between a neutron and a proton is attractive on the whole, but not necessarily that between a pair of neutrons or a pair of protons.

As we go to heavier nuclei, the number of protons increases. Since Coulomb force has a long range, its (negative) contribution to the binding energy increases quadratically with charge. In contrast, nuclear force is effective only between a few neighboring nucleons. As a result, the attractive contribution increases only linearly with A. To partially offset the Coulomb effect, stable nuclei are found with an excess of neutrons over protons. The neutron excess $(N - Z)$ increases slowly with nucleon number A. For example, the most stable nucleus for $Z = 40$ is ^{90}Zr with $N = 50$. The neutron excess in this case is 10. For $Z = 82$, we find ^{206}Pb as the most stable isotope with $N = 126$, a neutron excess of 44. For $Z > 82$, all the known nuclei are unstable. If we view the (negative of) nuclear binding energy as a function of N and Z, the stable and long-lived nuclei are found in a valley in such a two-dimensional plot, as shown in Fig. 1-1. This is sometimes referred to as the "valley of stability," At low values of N and Z, the bottom of the valley lies along the line with $N = Z$. As we go to heavier nuclei, the valley shifts gradually to $N > Z$.

Table 1-2: Known elements.

Z	Symbol	Name	Z	Symbol	Name	Z	Symbol	Name
1	H	Hydrogen	39	Y	Yttrium	77	Ir	Iridium
2	He	Helium	40	Zr	Zirconium	78	Pt	Platinum
3	Li	Lithium	41	Nb	Niobium	79	Au	Gold
4	Be	Beryllium	42	Mo	Molybdenum	80	Hg	Mercury
5	B	Boron	43	Tc	Technetium	81	Tl	Thallium
6	C	Carbon	44	Ru	Ruthenium	82	Pb	Lead
7	N	Nitrogen	45	Rh	Rhodium	83	Bi	Bismuth
8	O	Oxygen	46	Pd	Palladium	84	Po	Polonium
9	F	Fluorine	47	Ag	Silver	85	At	Astatine
10	Ne	Neon	48	Cd	Cadmium	86	Rn	Radon
11	Na	Sodium	49	In	Indium	87	Fr	Francium
12	Mg	Magnesium	50	Sn	Tin	88	Ra	Radium
13	Al	Aluminum	51	Sb	Antimony	89	Ac	Actinium
14	Si	Silicon	52	Te	Tellurium	90	Th	Thorium
15	P	Phosphorus	53	I	Iodine	91	Pa	Protactinium
16	S	Sulfur	54	Xe	Xenon	92	U	Uranium
17	Cl	Chlorine	55	Cs	Cesium	93	Np	Neptunium
18	Ar	Argon	56	Ba	Barium	94	Pu	Plutonium
19	K	Potassium	57	La	Lanthanum	95	Am	Americium
20	Ca	Calcium	58	Ce	Cerium	96	Cm	Curium
21	Sc	Scandium	59	Pr	Praseodymium	97	Bk	Berkelium
22	Ti	Titanium	60	Nd	Neodymium	98	Cf	Californium
23	V	Vanadium	61	Pm	Promethium	99	Es	Einsteinium
24	Cr	Chromium	62	Sm	Samarium	100	Fm	Fermium
25	Mn	Manganese	63	Eu	Europium	101	Md	Mendelevium
26	Fe	Iron	64	Gd	Gadolinium	102	No	Nobelium
27	Co	Cobalt	65	Tb	Terbium	103	Lr	Lawrencium
28	Ni	Nickel	66	Dy	Dysprosium	104	Rf	Rutherfordium
29	Cu	Copper	67	Ho	Holmium	105	Db	Dubnium
30	Zn	Zinc	68	Er	Erbium	106	Sg	Seaborgium
31	Ga	Gallium	69	Tm	Thulium	107	Bh	Bohrium
32	Ge	Germanium	70	Yb	Ytterbium	108	Hs	Hassium
33	As	Arsenic	71	Lu	Lutetium	109	Mt	Meitnerium
34	Se	Selenium	72	Hf	Hafnium	110	–	$^{271}110$
35	Br	Bromine	73	Ta	Tantalum	111	–	$^{272}111$
36	Kr	Krypton	74	W	Tungsten	112	–	$^{277}112$
37	Rb	Rubidium	75	Re	Rhenium			
38	Sr	Strontium	76	Os	Osmium			

The newly identified elements of $Z = 110$ to $Z = 112$ have not yet been assigned official names.

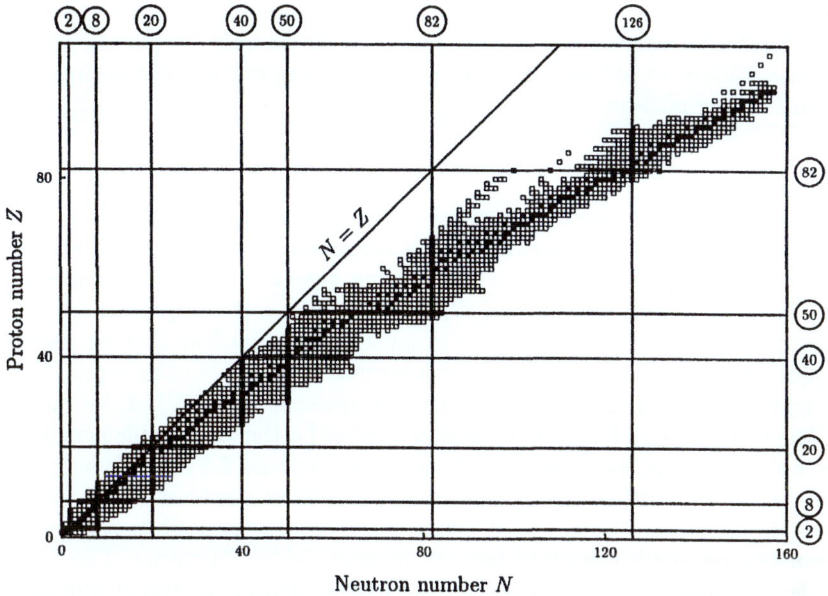

Figure 1-1: Distribution of stable and long-lived nuclei as a function of neutron and proton numbers. Stable nuclei are shown as filled squares and they exist between long-lived ones (empty squares) that are unstable against β-decay, nucleon emission, and α-particle decay.

In most cases, the number of stable nuclei for a given N, Z, or A is fairly small, and the lifetimes of unstable ones on both sides of the stable ones decrease rapidly as we move away from the central region. For nuclei with a few more neutrons than those in the valley of stability, β^--decay by electron emission is energetically favored. Similarly, for nuclei with a few "extra" protons, the rates of β^+-decay by positron emission determines their lifetimes. As the number of neutrons or protons becomes too large compared with those for stable nuclei in the same region, particle emission takes over as the dominant mode of decay and the lifetimes decrease dramatically as strong interaction becomes involved. By the time we get to the upper end (large N and Z) of the valley of stability, nuclei become unstable toward α-decay and fission as well.

The local variations in the "width" of the valley of stability, that is, the number of stable nuclei for a given Z, N, or A, reflect finer details in the nature of nuclear force. For example, there are more even-even (even N and even Z) stable nuclei than odd-mass and odd-odd nuclei, a result of pairing interaction, to be discussed in more detail in Chapter 7. There, we shall also see the reason why the largest numbers of stable nuclei are found near the "magic numbers."

Binding energy. A more detailed examination of the binding energies of stable nuclei shows some additional interesting features. For simplicity, let us consider only the most stable nucleus for a given nucleon number. The binding energy, $E_B(Z,N)$, is the amount it takes to remove all Z protons and N neutrons from the nucleus and is given by the mass difference between the nucleus and the sum of those of the (free) nucleons that make up the nucleus,

$$E_B(Z,N) = \left\{ZM_{\rm H} + NM_n - M(Z,N)\right\}c^2 \tag{1-1}$$

Here $M(Z,N)$ is the mass of the neutral atom, $M_{\rm H}$ is the mass of a hydrogen atom, and M_n is the mass of a free neutron. It is conventional to use neutral atoms as the basis for tabulating nuclear masses and binding energies, as mass measurements are usually carried out with most, if not all, of the atomic electrons present.

Because of the short-range nature of nuclear force, nuclear binding energy, to a first approximation, increases linearly with nucleon number. For this reason, it is more meaningful to consider the binding energy per nucleon, $E_B(Z,N)/A$, for our purpose here. The variation as a function of nucleon number for the most stable member of each isobar is shown in Fig. 1-2. The maximum value is around 8.5 MeV, found at $A \approx 56$. For heavier nuclei, binding energy per nucleon decreases slowly with increasing A due to rising Coulomb repulsion. As a result, energy is released when a heavy nucleus undergoes fission and is converted into two or more lighter fragments. This is the basic principle behind nuclear fission reactors. For light nuclei, the reverse is true and energy is released by fusing two together to form a heavier one. This is the main source of energy radiated from stars and the cause behind nucleosynthesis of elements up to $A \approx 56$.

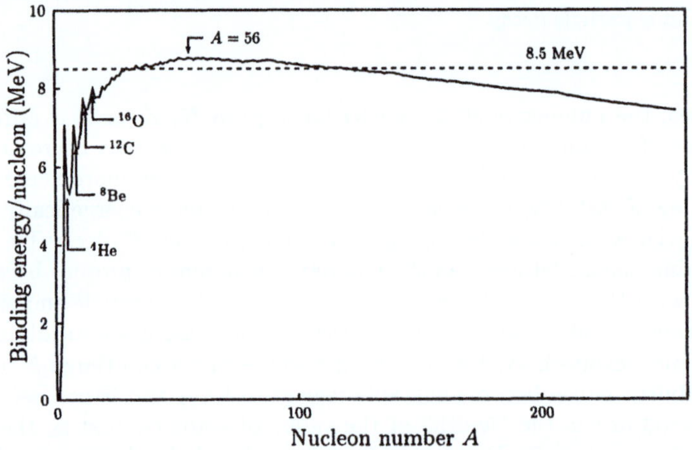

Figure 1-2: Average binding energy per nucleon as a function of nucleon number A for the most stable nucleus of each nucleon number.

The sharp rise in the binding energy per nucleon for light nuclei ($A \leq 20$) comes from increasing number of nucleon pairs. A closer examination shows that the trend is not a smooth one and the values are larger for the $4n$ nuclei, those with $A = 4 \times n$ for $n = 0, 1, 2, \ldots$. Since $N = Z$ for these light, stable nuclei, the $4n$ nuclei may be viewed as if they are made of α-particles. The fact that their average binding energies per nucleon are larger than their neighbors implies that nucleons like to form α-particle clusters in nuclei. This can be seen quantitatively by looking at the values for nuclei, with $2 \leq A \leq 25$, shown in Table 1-3. For the $4n$ nuclei, the difference between the total binding energy and the sum of those for n α-particles is also given:

$$\Delta E \equiv E_B(N, Z) - nE_B(^4\text{He})$$

For $n = 2$, we find that the value is negative, showing that ^{8}Be is unstable with respect to α-particle emission. For the others in the list, the value increases with n. In fact, if we divided ΔE by the number of α-particle pairs, given by $n(n-1)/2$, the result is roughly constant, with a value around 2 MeV. This gives us a picture that, at least for light nuclei, a large part of the binding energy lies in forming α-particle clusters, around 7 MeV per nucleon, as can be seen from the binding energy of ^{4}He. The much smaller reminder, around 1 MeV per nucleon or 2 MeV between a pair of α-clusters, goes to the binding between clusters. This phenomenon is usually referred to as the "saturation of nuclear force." That is, nuclear force is strongest between the members of a group of two protons and two neutrons, and as a result, nucleons prefer to form α-particle clusters in nuclei. It is a reflection of a fundamental symmetry of nuclear force, known as SU_4 or Wigner supermultiplet symmetry. As the number of nucleons increases, the "excess" in binding energy per nucleon of $4n$ nuclei is no longer visible. Beyond ^{16}O, the increase in the binding among four nucleons in forming a cluster is averaged over a larger number of nucleons in the simple way we are examining the question here. However, the saturation effect persists to heavy nuclei. This may be seen by the local increase in the energy required to take away a nucleon, shown later in Fig. 7-2.

Table 1-3: Binding energies (MeV) for some stable light nuclei.

Symbol	E_B	E_B/A	ΔE	Symbol	E_B	E_B/A	Symbol	E_B	E_B/A
^{2}H	2.22	1.11	—	^{3}H	8.48	2.83	^{3}He	7.72	2.57
^{4}He	28.30	7.07	—	^{5}He	27.41	5.48	^{5}Li	26.33	5.27
^{6}Li	32.00	5.33	—	^{7}Li	39.25	5.61	^{7}Be	37.60	5.37
^{8}Be	56.50	7.06	−0.09	^{9}Be	58.17	6.46	^{9}B	56.31	6.26
^{10}B	64.75	6.48	—	^{11}B	76.21	6.93	^{11}C	73.44	6.68
^{12}C	92.16	7.68	7.27	^{13}C	97.11	7.47	^{13}N	94.11	7.24
^{14}N	104.66	7.48	—	^{15}N	115.49	7.70	^{15}O	111.96	7.46
^{16}O	127.62	7.98	14.44	^{17}O	131.76	7.75	^{17}F	128.22	7.54
^{18}F	137.37	7.63	—	^{19}F	147.80	7.78	^{19}Ne	143.78	7.57
^{20}Ne	160.65	8.03	19.17	^{21}Ne	167.41	7.97	^{21}Na	163.08	7.77
^{22}Na	174.15	7.92	—	^{23}Na	186.57	8.11	^{23}Mg	181.73	7.90
^{24}Mg	198.26	8.26	28.48	^{25}Mg	205.59	8.22	^{25}Al	200.53	8.02

Nuclear radius and nuclear density. In addition to binding energy, the general trend of nuclear size shows also a simple dependence on nucleon number. For the most part, the nuclear radius is given by

$$R = r_0 A^{1/3} \tag{1-2}$$

with $r_0 = 1.2$ fm (1 fm, or femtometer, equals 10^{-15} m). This means that the volume is linearly proportional to A and that the nucleons are not compressed in size in spite of the large forces acting between them. In fact, one has to go to some extreme situations, such as a black hole or during the collapse of a large star prior to a supernova explosion, before nucleons can be compressed much beyond what is known as the nuclear matter density $\rho_0 \sim 0.16 \pm 0.02$ nucleons/fm^3, a value that is 3×10^{14} times the density of water. We can also arrive at the same order of magnitude from the fact that the mass of a neutron star is typically around 1 solar mass ($\sim 10^{30}$ kg) and the radius roughly 10 km.

In finite nuclei, the average density is somewhat smaller than ρ_0. Using Eq. (1-2), we arrive at $\rho \sim 0.12$ nucleons/fm^3. This is attributed to a large diffused surface region where the density drops off to zero more or less exponentially. For many purposes, the radial distribution of nuclear density may be represented by a Woods-Saxon form,

$$\rho(r) = \frac{\rho_0}{1 + \exp\{(r - c)/z\}} \tag{1-3}$$

Here z is a parameter that measures the "diffuseness" of the nuclear surface, with typical values around 0.5 fm, and c is the distance from the center to the point where the density drops to a half value. Some of the typical values found in nuclei are listed later in Table 4-1

Nuclear shape. For stable nuclei, the nuclear shape is essentially spherical. As we shall see later in §4-9, this is an effort to minimize the surface energy, in analogy to a drop of fluid. However, small departures from spheres are observed, for example, in the region $150 < A < 190$. One way to quantify these "deformations" is to use the ratio

$$\delta = \frac{\Delta R}{R}$$

where R is the average nuclear radius given by Eq. (1-2) and, for the case of an ellipsoidal shape nucleus, ΔR is the difference between semi-major and semi-minor axes. For a sphere, $\Delta R = 0$. In nuclei, the typical value of δ does not exceed 0.1 for low-lying states. However, large deformations can be created in the laboratory by fusing two nuclei together. In this way, values of δ around 0.6 (that is, semi-major axis twice the semi-minor axis) have been observed. This is the case of superdeformation, to which we shall return in §9-2.

One of the reasons for nuclear deformation is the competition between Coulomb and nuclear forces. Since the strength of the Coulomb force is inversely proportional to the square of the distance, a nucleus can decrease its total energy (and increase its binding energy) by putting protons as far away from each other as possible. For the same volume, a deformed shape is preferred as a result. Nuclear force, on the other hand, tries to keep the shape spherical so that the short-range attraction can be more effective. Since nuclear forces are stronger, light nuclei on the whole are spherical. However, once we go to intermediate values of A and beyond, the saturation property cuts off further increase in the binding energy per nucleon with increasing A due to nuclear force. As a result, slight deformation can actually increase the binding energy by decreasing the Coulomb contribution.

Density of excited states. The binding energy defined in Eq. (1-1) is only that for the ground state of a nucleus. In general, a nucleus has a number of excited states as well. For these states, it is customary to use a slightly different scale and measure the energies relative to the ground state as the zero point. If we examine the spectra for different nuclei, we find that each one is sufficiently unique that it can be used as a signature to identify the nucleus, similar to the case of atomic spectra. In spite of the individual characteristics, there are certain general features in the distribution of excited states that are worth noting.

Nuclei are made of nucleons. Being fermions, Pauli exclusion principle demands that each nucleon must occupy a different single-particle state. In the limit that interactions can be ignored, the ground state of a nucleus is one with nucleons filling up all the single-particle states in order of their energies, starting from the lowest one. This is similar to a Fermi gas, one with all the molecules made of identical, noninteracting fermions. At zero temperature, the fermions settle in the lowest possible single-particle states and the energy of the highest filled one is known as the Fermi level. The only way to put excitation energy into such a system is to promote some of the particles below the Fermi surface to the unoccupied ones above. At low excitations, there is only enough energy to put a few such particles from states just below the Fermi surface to those just above. As there are not too many different independent ways to carry out this operation, the density of states, the number of excited states per unit energy, is small. As we increase the excitation energy, more particles can be promoted and the number of different ways to form many-body states increases, resulting in higher level density. Based on such a simple picture, Bethe [30] in 1937 obtained the following formula for the density of states at excitation energy E:

$$\rho_A(E) = \frac{1}{12a^{1/4}E^{5/4}}\, e^{2\sqrt{aE}} \tag{1-4}$$

generally known as the Fermi gas model formula. The quantity a is the *level-density parameter*. A derivation of Eq. (1-4) can be found, for example, in Ref. [152].

Interaction between nucleons modifies the energy spectrum from such a simple, smooth form. The location of each excited state is now a complicated function of the nuclear interaction and the nucleons. Nevertheless, the general form given by the Fermi gas model remains to be essentially correct. The main effect of interaction may be separated into two parts. The first is a change in the relative positions of individual levels. From a certain point of view, we can say that the interaction introduces a "fluctuation" in the spectrum over the smooth form given by the Fermi gas model. Depending on one's interest, the fluctuations can be all that is important in a study if one's focus is on the position of a particular level or a group of levels. On the other hand, if the concern is with general features, such as the amount of energy that can be stored in an excited nucleus under certain conditions, only the smooth part of the spectrum is of primary importance.

A second consequence because of interaction is a shift in the energy scale by some amount Δ. Since excitation energy is measured from the ground state, any change to the latter produces a constant shift of the whole spectrum. In general, interaction tends to lower the ground state energy from the value given by a noninteracting model. Because of such a change in the energy scale, the level-density formula in many applications takes on the form

$$\rho_A(E) = \frac{1}{12a^{1/4}(E-\Delta)^{5/4}} e^{2\sqrt{a(E-\Delta)}} \tag{1-5}$$

This is known as the *back-shifted* Fermi gas model formula. Here, both a and Δ are considered as adjustable parameters to be determined by fitting to known data [53]. An example for the nucleus ^{56}Fe is shown in Fig. 1-3.

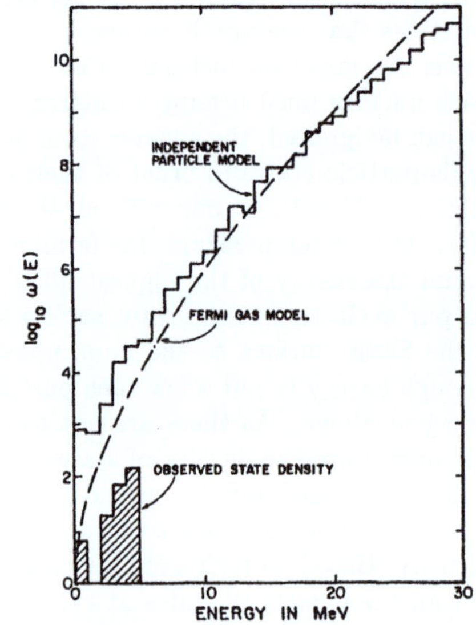

Figure 1-3: State density of ^{56}Fe obtained using Eq. (1-4) (smooth curve) and an independent particle model (staircase). The observed values (shaded histogram) are lower than the calculated ones, as the ground state energy is depressed by two-body correlations. This effect may be accounted for by the back-shifted Fermi gas formula given in Eq. (1-5).

Scattering cross section. In studying atomic nuclei, we often resort to scattering of one particle off another. This comes from the necessity that we are examining objects of dimension on the order of femtometers (10^{-15} m). The wavelength of visible light, on the other hand, is much longer, on the order of 10^{-7} m. To go down to length scales of interest to subatomic physics, far shorter wavelengths than visible light and, hence, much higher energies are needed, and this can be achieved most readily by scattering. A feeling of the energies required in a scattering experiment to reach a given length scale may be obtained by examining the corresponding de Broglie wavelength:

$$\lambda = \frac{h}{p} \xrightarrow[v \to c]{} \frac{hc}{E}$$

Table 1-4 lists the values for photons, electrons, and protons at typical energies used in nuclear experiments.

Table 1-4: De Broglie wavelengths of γ-ray, electron, and proton.

Energy (MeV)	Wavelength (fm)		
	Photon	Electron	Proton
0.1	1.2×10^4	3701	90
0.5	2.5×10^3	1421	40
1	1.2×10^3	872	29
10	1.2×10^2	118	9.0
100	1.2×10	12	2.8
1,000	1.2	1.2	0.73
10,000	1.2×10^{-1}	1.2×10^{-1}	1.1×10^{-1}

The probability for a projectile scattering off a target particle is usually expressed in terms of a quantity called "cross section." The total cross section σ in a reaction is defined in the following way: Consider a single incident particle moving outside the range of any interaction along a straight line toward the target. If the velocity is v, the particle sweeps in time t a cylinder of volume $vt\mathcal{A}$, where $\mathcal{A}$ is the area covered by the particle. The scattering probability P is given by the ratio of the area block by the target particles and $\mathcal{A}$. If the number of target particles per unit volume is n and the target thickness is T, the number of target particles "seen" by the beam particle is $n\mathcal{A}T$. The scattering probability is then

$$P = \frac{n\mathcal{A}T\sigma}{\mathcal{A}} = \sigma nT$$

Since n and T have, respectively, dimensions inverse length cubed and length, the total cross section σ must have the dimension of length squared, or area, as P is dimensionless.

The total cross section is often not the quantity measured directly in an experiment, as it requires all the scattered particle to be detected (hence, the name *total* cross section). The angular distribution of the scattered particles is actually a more useful quantity, as it provides us with more information. In the same way as above, we can define the differential scattering cross section $d\sigma/d\Omega$ in terms of the probability $P(\theta,\phi)$ for a scattered particle to arrive at a detector that is located at angles (θ,ϕ) and subtends a solid angle $\Delta\Omega$ at the center of the target by the relation

$$P(\theta,\phi) = \frac{d\sigma}{d\Omega} nT$$

The connection between differential and total cross section is given by integrating over all solid angles:

$$\sigma = \int_0^{4\pi} \frac{d\sigma}{d\Omega} d\Omega = \int_0^{2\pi}\int_0^{\pi} \frac{d\sigma}{d\Omega} \sin\theta \, d\theta \, d\phi$$

In §B-2, we shall redefine the same differential cross section in term of the wave functions involved in a reaction.

Reaction types. The usual type of final state we wish to deal with in a reaction is two body. In other words, before the reaction, we have a projectile particle a incident on a target particle A. After the reaction, a particle b is scattered away, leaving behind a residual particle B. The reaction may be represented in either one of the following two ways:

$$A(a,b)B \qquad \text{or} \qquad a + A \longrightarrow b + B$$

For example, if a proton incidents on a ^{48}Ca target and a neutron is observed to emerge from the reaction, the residual nucleus is ^{48}Sc. The reaction may be written as

$$^{48}\text{Ca}(p,n)^{48}\text{Sc} \qquad \text{or} \qquad p + {}^{48}\text{Ca} \rightarrow n + {}^{48}\text{Sc}$$

Other reactions may also take place in bombarding a ^{48}Ca target by a beam of protons. For example, a proton may emerge, leaving the ^{48}Ca nucleus in an excited state. The reaction may be expressed as

$$^{48}\text{Ca}(p,p')^{48}\text{Ca}^* \qquad \text{or} \qquad p + {}^{48}\text{Ca} \rightarrow p' + {}^{48}\text{Ca}^*$$

Here the asterisk indicates that, after the reaction, ^{48}Ca is in an excited state and the prime on the proton says that the energy is different from the incident amount.

Each one of these combinations is a different *exit channel* for proton-^{48}Ca scattering, and the possible, or "open," exit channels are governed by conservation laws and selection rules operating in the scattering. In general, the number of open channels increases very fast with increasing energy available in the reaction.

The allowed exit channel is not restricted to final states consisting of two particles. For example, an experiment may be carried out using a deuteron as the incident particle instead of the proton in the above example. A possible exit channel may involve a breakup of the deuteron into a proton and a neutron. The reaction is represented as

$$^{48}\text{Ca}(d,pn)^{48}\text{Ca} \qquad \text{or} \qquad d + {}^{48}\text{Ca} \rightarrow p + n + {}^{48}\text{Ca}$$

To simplify the discussion, we shall for the most part ignore reactions involving three or more particles in the final state. Furthermore, the distinction between projectile and target nuclei and that between the scattered particle and the residual nucleus is useful only in fixed-target experiments in which the target is stationary in the laboratory. For colliding beam experiments, in which the two particles in the incident channel are moving toward each other, the separation is not meaningful. For most of our discussions, we shall be working in the center of mass of the two-body system, and the distinction reduces to a simple question of semantics.

In an *elastic* scattering, both the incident and target particles remain in their original states, usually their respective ground states. Elastic scattering is, in general, the simplest from a reaction point of view. For example, elastic scattering of electrons is used to map the charge density distribution of a nucleus. Since the interaction is mainly electromagnetic, it is possible to infer from the results how nuclear charge distribution differs from that for a point particle.

Inelastic scattering is the process where a part of the incident kinetic energy is used to excite the nuclei involved or to create new particles. The most obvious example is

Coulomb scattering where the target nucleus is raised to an excited state by electromagnetic interaction, the inverse of electromagnetic decay. As another example, the reaction

$$\nu_e + {}^{37}\mathrm{Cl} \to e^- + {}^{37}\mathrm{Ar}$$

is the inverse of β^--decay of ^{37}Ar and is used in detecting solar neutrinos.

When two nuclei interact, it is possible to transfer one or more nucleons between them. For example, if a deuteron is incident on a ^{16}O target, the loosely bound neutron in the projectile may be attracted by the target nucleus and becomes attached to it as a result. The scattered particle is now a proton and the residual nucleus becomes ^{17}O. Such a reaction, $^{16}\mathrm{O}(d,p)^{17}\mathrm{O}$, is called a stripping reaction, as a neutron is stripped from the projectile. The inverse is a pickup reaction, for example, $^{17}\mathrm{O}(^3\mathrm{He},^4\mathrm{He})^{16}\mathrm{O}$, whereby a neutron in the target ^{17}O is picked up by the ^{3}He projectile. The scattered particle is now ^{4}He, and ^{16}O becomes the residual nucleus. More complicated nucleon transfer reactions may be induced using heavy ions.

Nuclear fusion may be considered as the extreme of nucleon transfer reactions. In this case, two heavy ions are brought into close proximity to each other so that nuclear force can act between the nucleons in the two ions, forming a compound nucleus as the intermediate state. Under favorable circumstances, some of the excess energy in the system may be discarded by emitting γ-rays and nucleons, resulting in a final state that may be considered as a nucleus. For example, the yet-to-be named superheavy element ^{277}112 is obtained in this way from the irradiation of $^{208}_{82}$Pb by $^{70}_{30}$Zn [84].

Alternatively, the final state may be an unusual one in a known nucleus. Since the collision of two heavy ions often involves large quantities of angular momentum, the final state is very likely to retain a significant fraction and ends up in a state of high spin. For example, the reaction $^{155}_{64}\mathrm{Gd}(^{16}_{8}\mathrm{O},4\mathrm{n})^{167}_{72}\mathrm{Hf}$ produces ^{167}Hf nuclei by "fusing" ^{16}O with ^{155}Gd. Ignoring angular momentum carried away by the four neutrons (and several γ-rays), we can make an estimate of the amount available in the final system. If the center-of-mass energy of the ^{16}O beam is $E_{\mathrm{cm}} = 75$ MeV and the impact parameter $b = 10$ fm, we have the result

$$\ell = mv_{\mathrm{cm}}b = b\sqrt{2mE_{\mathrm{cm}}} \sim 80\hbar$$

by starting from the classical definition $\boldsymbol{\ell} = \boldsymbol{r} \times \boldsymbol{p}$ with $\boldsymbol{p} = m\boldsymbol{v}$. This is sufficient to create states of very high spin values, such as $\frac{73}{2}\hbar$, observed in ^{167}Hf formed in this way. For comparison, the ground state spin of ^{167}Hf is only $\frac{5}{2}\hbar$. The only way for such large spins to exist in a nucleus with only 167 nucleons is for a significant fraction of the nucleons to act coherently as a single unit. This is an example of collective behavior in a nucleus that takes the nuclear shape far from the nearly spherical ones normally observed for ground states.

The usual consideration for creating such exotic states is that the energy involved must be sufficiently high to overcome the Coulomb barrier between the two ions. This is necessary for the two groups of nucleons to come into contact with each other for fusion to take place. At the same time, one does not want to inject any more energy into the system than necessary, as any excess has to be discarded in order for the final system to live long enough to be detected. The value of $E_{\mathrm{cm}} = 75$ MeV is roughly what is used in practice for "light" ions such as ^{16}O. The value of $b = 10$ fm is also a reasonable choice, as it is essentially half the distance between the centers of the two ions when they are just in contact with each other.

1-4 Commonly Used Units and Constants

In atomic nuclei, we are dealing with length scales that are extremely small and time scales that are extremely short, compared with standard measures in daily life. Instead of the meter, a more suitable unit of length, as we have seen earlier, is the femtometer, abbreviated as fm (1 fm = 10^{-15} m). For example, the typical size of a nucleus is of the order of 1 fm. The same is also true for the range of nuclear force. For nuclear reaction cross sections, a derived unit, the barn, equal to 10^{-28} m^2, is often used. Typical values are often given in millibarns, 10^{-3} b or 10^{-1} fm^2.

A wide range of time scales enters into nuclear physics. In Table 1-1 we have seen that the typical reaction time for strong interaction is 10^{-23} second, or 10^{-23} s using the standard abbreviation for seconds. At the other end of the scale, we find naturally occurring radioactive elements that were made prior to the formation of the solar system. The lifetimes of these radioactive nuclei must be of the order of 10^9 years or longer, as anything with much shorter lives would have almost completely decayed away.

For states that live on the order of 10^{-15} to 10^{-23} s, the width of its energy distribution Γ is sometimes used to characterize the lifetimes. Because of the uncertainty principle, $\Delta E \Delta t = \hbar$, a state that lives only for a time Δt can have its energy measured only up to an uncertainty no better than $\Delta E \sim \hbar/\Delta t$. This gives a width $\Gamma = \hbar/\overline{T}$ in the probability distribution of the observed energy of the state. Here, $\overline{T}$ is the lifetime, or mean life, of the state. Since $\hbar = 6.58 \times 10^{22}$ MeV-s, lifetimes of the order of 10^{-23} s correspond to Γ of the order of 100 MeV, and a time scale on the order of 10^{-15} s corresponds to a width on the order of 1 eV.

The mass of a nucleon is 1.67×10^{-27} kg, with neutrons more massive than protons by about 0.14%. A convenient unit for mass is the atomic mass unit, commonly abbreviated as u, or amu, and 1 u is $1.6605402 \times 10^{-27}$ kg. It is defined using the neutral ^{12}C atom as the standard,

$$\mathrm{u} = \frac{\text{mass of } ^{12}\text{C atom}}{12} = \frac{1\ \text{kg}}{N_A} = 1.6605402(10) \times 10^{-27}\ \text{kg} = 931.49432(28)\ \text{MeV}/c^2$$

where $N_A = 6.0221367(36) \times 10^{26}$ (kg mol)$^{-1}$ is Avogadro's number and the values inside the parentheses indicate the uncertainties in the last digits. In terms of atomic mass unit, the masses of a free proton and a free neutron are, respectively,

$$M_p = 1.007276470(12)\ \mathrm{u} \qquad M_n = 1.008664898(12)\ \mathrm{u}$$

By definition, the mass of ^{12}C is exactly 12 u.

Since binding energy is a small fraction of the rest mass energy of a nucleus, atomic masses in atomic mass units are usually not very different numerically from the number of nucleons $A = N + Z$. It is sometimes convenient to express nuclear masses in terms of the *mass excess*, $\Delta(Z, N)$ (also referred to on occasions as mass defect), defined in the following manner:

$$\Delta(Z, N) \equiv \{M(Z, N) \text{ in u} - A\} \times 931.49432\ \text{MeV}$$

where multiplication by 931.49432 converts the quantity from atomic mass units to energy units in MeV. For a hydrogen atom, the mass excess is

$$\Delta(H) = (1.007276470 - 1) \times 931.49432 + 0.51110 = 7.2891 \text{ MeV}$$

and for a free neutron,

$$\Delta(n) = (1.008664904 - 1) \times 931.49432 = 8.0713 \text{ MeV}$$

Given the mass excess of a nucleus, the binding energy in Eq. (1-1) may be expressed as

$$E_B(Z,N) = Z\Delta(H) + N\Delta(n) - \Delta(Z,N)$$

In some tables of binding energy, the values are given in terms of mass excesses.

Instead of mass, it is sometimes preferable to work in terms of the equivalent rest mass energy. The commonly used unit of energy in nuclear physics, as we have already seen, is MeV, or million electron-volts, and 1 MeV is $1.60217733 \times 10^{-13}$ J. For example, the rest mass energy of a neutron is 939.56563 MeV. For some of the higher energy events, it is more suitable to use instead GeV (10^9 eV), which is 1000 times larger than MeV. For example, the order of magnitude for a nucleon mass is 1 GeV. A few other derived units are also in use to measure other nuclear properties, such as nuclear magneton μ_N for magnetic dipole moment. We shall define each one of them as they appear in the discussion.

Universal constants, such as Planck's constant h, speed of light c, and electric charge e, enter quite often into calculations involving nuclei. For electric charge, we shall use e, the charge carried by a proton as the unit. For Planck's constant, $\hbar = h/2\pi$ turns out to be more convenient on most occasions. In fact, the combination $\hbar c = 197.3$ MeV-fm enters naturally in a variety of calculations. For example, in our earlier discussion on de Broglie wavelength, the calculation can be carried much easier in terms of $\hbar c$ in the following way:

$$\lambda = \frac{h}{p} = \frac{2\pi\hbar c}{pc} \xrightarrow[v \to c]{} \frac{2\pi\hbar c}{E}$$

Here p is the momentum and E the energy of the particle. Similarly, in our estimate of the angular momentum ℓ carried by two colliding heavy ions at impact parameter b, the value in units of $\hbar$ may be evaluated as

$$\frac{\ell}{\hbar} = mv_0 b = b\frac{\sqrt{2mE}}{\hbar} = b\frac{\sqrt{2mc^2E}}{\hbar c}$$

We see that, in the final expression, the mass is converted into rest mass energy mc^2 and the denominator becomes $\hbar c$.

Formulas involving electromagnetism are complicated by the fact that both centimeter-gram-second (cgs) and Système International (SI) units are in common usage. We shall write them with an "extra" factor in square brackets that "converts" the equation from cgs units to SI units. That is, the equation is in cgs units if the factor is not there and in SI units if included. Thus, electrostatic potential $V_C(R)$ between two point particles, one with charge ze and the other with Ze, separated by distance R, is given by

$$V_c(R) = \left[\frac{1}{4\pi\epsilon_0}\right]\frac{(ze)(Ze)}{R} \tag{1-6}$$

The formula is in cgs units if the factor in square brackets is ignored and in SI units when the factor is included. To avoid any dependence on the system of electromagnetic units adopted, we can make use of the fine structure constant

$$\alpha = \left[\frac{1}{4\pi\epsilon_0}\right]\frac{e^2}{\hbar c} \approx \frac{1}{137}$$

to replace factors in Eq. (1-6) that depend on the system adopted. Thus, we can write

$$V_c(R) = \alpha\hbar c\frac{zZ}{R} \approx 1.44\frac{zZ}{R(\text{fm})}\ \text{MeV}$$

In the final form, R is given in terms of femtometers so that we can use the numerical value $\alpha\hbar c \approx 1.44$ MeV-fm. In a similar way, the Coulomb energy for a spherical nucleus with Z protons and radius R is given by

$$E_C = \frac{3}{5}\left[\frac{1}{4\pi\epsilon_0}\right]\frac{(Ze)^2}{R} = \frac{3}{5}\alpha\hbar c\frac{Z^2}{R} \longrightarrow \frac{3}{5}\alpha\hbar c\frac{Z(Z-1)}{R} \tag{1-7}$$

Here, we assume that the charge is distributed evenly throughout the spherical volume. The final form of the expression excludes the Coulomb energy associated with each one of the Z individual protons.

Problems

1-1. Given that the radius of a nucleon inside a nucleus is $R = 1.2$ fm, calculate the density of nuclear matter. From this, evaluate the radius of the sun (mass = 2×10^{30} kg) if it collapses into a neutron star without losing any of its present mass.

1-2. From the uncertainty relation, find the minimum kinetic energy of a nucleon in ^{208}Pb. Use $R = 1.2A^{1/3}$ fm for nuclear radius.

1-3. If the cross section for neutrino interaction with iron is $\sigma \approx 10^{-48}$ m^2, find the mean free path of a neutrino in solid iron.

1-4. Use conservation of energy and momentum to calculate the maximum kinetic energies for electrons released in the decay of a free neutron,

$$n \rightarrow p + e^- + \overline{\nu}_e$$

and in the decay of a free muon,

$$\mu^- \rightarrow e^- + \nu_\mu + \overline{\nu}_e$$

Consider the particles are initially at rest in the laboratory.

1-5. If the density distribution of ^{184}W is given by the form shown Eq. (1-3), find the average density of the nucleus using the values of c and z given in Table 4-1.

1-6. For ^{56}Fe, the level-density parameter is found to be $a = 7.2$ MeV^{-1}. Evaluate the level density of ^{56}Fe at excitation energy $E = 20$ MeV.

Chapter 2

Nucleon Structure

All nuclei are made of neutrons and protons, the two lightest members of the baryon family. Nucleons are, however, not elementary particles. Partly for this reason, a significant fraction of the present-day interest in nuclear physics is related in one way or another to the underlying quark's degree of freedom. Such a study is, in turn, a part of the larger subject of quantum chromodynamics (QCD), the study of quarks and the interaction between them. It is still too early at this stage of the development of QCD to demand a complete description of nuclear physics starting from first principles; nevertheless, an understanding of the nucleus cannot be achieved without some awareness of quarks and their interactions. We shall attempt here only an introduction to certain aspects of strong interaction essential to nuclear physics.

There are also good practical reasons to examine the relationship between quarks before those between nucleons. One of the dominant considerations in subatomic physics is the role of symmetries. In this respect, there are many similarities and connections between quarks and nucleons, as expected. In some cases it is easier to study these symmetry principles using quarks rather than nucleons, in part because the number of quarks inside a hadron is much more restricted than the possible number of nucleons inside a nucleus. For this reason as well, we shall devote a large part of this chapter to the symmetry relations between strongly interacting fermions using quarks as an example.

2-1 Quarks and Leptons

The search for the fundamental building blocks of all matter in the universe has always been a central issue in physics. As our understanding of physical laws improves, our view changes on what constitutes the elementary particles, particles that cannot be made as composites of others. These days, the accepted view is that all matter is made of quarks and leptons. The only additions to the list are photons, $W^{\pm}$ and Z^0 bosons, gluons, and gravitons, particles mediating electromagnetic, weak, strong, and gravitational interactions, respectively.

Quarks. Quarks are the basic building blocks of hadrons, particles interacting with each other through strong interaction. In nuclear physics, we are mostly concerned with the lightest members of the hadron family: nucleons, which make up all the nuclei, and pions, which constitute the main carriers of nuclear force. There are six different kinds, or *flavors*, of quarks: u (up), d (down), c (charm), s (strange), t (top), and b (beauty, or bottom). These six particles may be arranged according to their masses into three pairs, with one member of each pair having a charge $\frac{2}{3}e$ and the other $-\frac{1}{3}e$, as shown in Table 2-1.

Table 2-1: Quarks and leptons.

Quarks			
$Q/e=\frac{2}{3}$	u	c	t
$Q/e=-\frac{1}{3}$	d	s	b
Leptons			
$Q/e=-1$	e	μ	τ
$Q/e=0$	ν_e	ν_μ	ν_τ

Since quarks have not been observed in isolation—they appear either as bound quark-antiquark pairs in the form of mesons or bound groups of three quarks in the form of baryons—the names assigned to them, up, down, strange, etc., are only mnemonic symbols to identify the different species. The word "flavor" is used, for convenience, to distinguish between different types of quark, not because it has anything to do with taste. Besides flavor, quarks also come in three different *colors,* for example, red, green, and blue. Color and flavor are quantum-mechanical labels, or quantum numbers, very similar to spin and parity, required to differentiate between the different states in which a quark finds itself. Since there are no classical analogues to flavor and color degrees of freedom, there are no observables that can be directly associated with them. In this respect, they are similar to the parity label of a state which must be "observed" through indirect evidence. For quarks, observation of any of their properties is made even harder by the fact that they appear only in groups of two or more. However, there is by now a large amount of evidence for the presence of flavor, color, and other degrees of freedom associated with quarks, and we shall examine some of these properties in this chapter.

Leptons. Although quarks make up the bulk of observed mass in the universe, they are not the only elementary building block of particles with finite rest masses. *Leptons*, or light particles, are not made of quarks. They participate in electromagnetic and weak interactions but not in strong interaction. The number of different types of known leptons is also six and can also be arranged into three pairs, as shown in Table 2-1. The electron (e), the muon (μ), and the tau lepton (τ) carry a charge $-e$ each, but the electron neutrino (ν_e), the muon neutrino (ν_μ), and the tau neutrino (ν_τ) are neutral. The masses of leptons are much less than those of quarks, with $m_ec^2 = 0.511$ MeV, $m_\mu c^2 = 106$ MeV, and $m_\tau c^2 = 1784$ MeV. The neutrinos are known to be much lighter and their rest masses may even be zero. A large amount of effort has been devoted in recent years to measuring the mass of ν_e. The best estimate at the moment is that it is a few electron-volts or less, although much larger values for the upper limit have also been reported. For the other two types of neutrinos, only the upper limits of their masses are known: $m_{\nu_\mu} < 0.25$ MeV and $m_{\nu_\tau} < 70$ MeV.

In nuclear physics, leptons make their presence felt through nuclear β-decay and other weak transitions. In general, only electrons and electron neutrinos are involved; occasionally muons may enter, such as in the case of a muonic atom where a muon replaces one of the electrons in the atom. Because of its larger mass and its more recent discovery, the τ-lepton has yet to enter nuclear physics studies.

Lepton number conservation. The number of leptons is conserved in a reaction. For example, a free neutron decays with a lifetime of 886.7 ± 1.9 s through the reaction

$$n \rightarrow p + e^- + \overline{\nu}_e \tag{2-1}$$

The bar over ν_e indicates that it is an electron antineutrino, the antiparticle of an electron neutrino. On the left-hand side of the equation, only a neutron is present. Since there is no lepton, we can assign $L = 0$ as its *lepton number.* On the right-hand side of the equation, we have one electron, which carries a lepton number $L = 1$. An antiparticle is given a particle number of the same magnitude as the particle with which it is associated but with the opposite sign. This is necessary since an antiparticle can annihilate a particle to form a state with no particle. Hence, the lepton number of $\overline{\nu}_e$ is -1. The total lepton number on the right-hand side of Eq. (2-1) is then $L = 1+(-1) = 0$. With these assignments, we find that the lepton number is conserved in the reaction.

Conservation of lepton numbers in Eq. (2-1) depends on the recognition that the neutral lepton produced in the reaction is observed to be an antineutrino rather than a neutrino. This is not merely a gimmick to balance the lepton number of the two sides of the equation. The two types of neutrinos, ν_e and $\overline{\nu}_e$, are two different particles, related to each other by a transformation between a particle and its antiparticle, or *charge conjugation.* Electron neutrinos, ν_e, can be obtained, for example, from the reaction

$$p_{\text{bound}} \rightarrow n + e^+ + \nu_e \tag{2-2}$$

Such a process is not energetically possible for a free proton, the nucleus of a hydrogen atom, as a free neutron is more massive ($M_n c^2 = 939.566$ MeV) than a free proton ($M_p c^2 = 938.272$ MeV). However, a proton bound within a nucleus, p_{bound}, can undergo the reaction of Eq. (2-2). The necessary energy conservation is now between the parent nucleus, having the bound proton as one of its nucleons, and the daughter nucleus containing the neutron. As long as there is enough energy difference between the parent and daughter nuclei to create the two leptons, a positron e^+ and an electron neutrino ν_e, the reaction is possible (see §5-5 for detail). Since a positron is the antiparticle of an electron, its lepton number $L = -1$. To conserve charge, the charged lepton on the right-hand side of Eq. (2-2) must be a positron, and to conserve lepton number, the reaction must be accompanied by an electron neutrino in the final state.

If ν_e and $\overline{\nu}_e$ were the same particle, we could make use of the electron neutrino obtained from the reaction given by Eq. (2-2) to induce the inverse of that of Eq. (2-1),

$$\nu_e + p \rightarrow e^+ + n \tag{2-3}$$

To conserve charge, the charged lepton on the right-hand side must be a positron with $L = -1$. However, the reaction is not observed to take place. On the other hand, the reaction

$$\overline{\nu}_e + p \rightarrow e^+ + n \tag{2-4}$$

is observed. This establishes that ν_e and $\overline{\nu}_e$ are two different particles as well as confirms that the lepton number is conserved (see §5-6 for more detail). In fact, the conservation of leptons (i.e., that leptons cannot be created or annihilated except in pairs each one consisting of a lepton and an antilepton) is a fundamental conservation law, not too different from the conservation of energy and momentum. Our convention of assigning lepton numbers starts by giving $L = +1$ to an electron. Once this is fixed, all the other lepton numbers are determined by conservation requirements.

Particles that are distinct from their antiparticles are called Dirac particles. This is to distinguish them from Majorana particles, which are the same as their antiparticles. As we shall see in §5-6, one of the interests in double β-decay, nuclear decay through the emission of two electrons or positrons, is to find out whether neutrinos can be Majorana particles. So far all the evidence seems to suggest that they are strictly Dirac particles.

The conservation of lepton numbers applies separately to each one of the three groups of leptons, e and ν_e, μ and ν_μ, and τ and ν_τ. That is, the number of leptons in the electron family L_e, the number of leptons in the muon family L_μ, and the number of leptons in the tau family L_τ are conserved separately in a reaction. For example, muons decay with a mean life of 2.2 μs through the reaction

$$\mu^- \rightarrow e^- + \overline{\nu}_e + \nu_\mu \tag{2-5}$$

Since only a muon appears on the left-hand side of this reaction, we have $L_e = 0$ and $L_\mu = 1$ (as well as $L_\tau = 0$). On the right-hand side, the muon number is conserved by the presence of ν_μ. The electron number must also be zero on the right-hand side to conserve L_e, and this requires $\overline{\nu}_e$ to appear with e^-. The fact that the reaction produces two neutrinos, a muon neutrino and an electron antineutrino, rather than, for example, ν_e and $\overline{\nu}_e$ or two γ-rays, is good evidence for the conservation of L_e and L_μ separately. For most interests in nuclear physics we are concerned primarily with the leptons in the electron family.

Baryon number conservation. The number of quarks of each type, u, d, s, etc., is also conserved in strong interaction processes. That is, one type of quark cannot be changed into another. The exception happens in weak interaction processes. This is equivalent to saying that flavor is a good quantum number only in the limit that weak force can be ignored. Unless we are dealing with the quark contents of hadrons, it is more convenient to examine instead the baryon number, which is known to be conserved under the influence of weak interaction as well. The only exception is the possible decay of protons through reactions such as

$$p \rightarrow e^+ + \pi^0 \tag{2-6}$$

allowed under theories for grand unification of all forces. At present, the observed limit on the lifetime of a proton is longer than 10^{25} years. As a result, we shall not be concerned with this possibility and we shall take the baryon number to be conserved in all the reactions of interest to us.

There is no conservation law for the number of mesons. If there is enough energy available, they can decay into other mesons, baryon and antibaryon pairs, lepton and antilepton pairs, or γ-rays. The lightest members of the meson family are the pions with rest mass around 140 MeV/c^2. It is stable on the time scale of strong interaction, as it cannot decay into another hadron. However, through weak interactions, charged pions decay predominantly to muons,

$$\pi^+ \longrightarrow \mu^+ + \nu_u \qquad\qquad \pi^- \longrightarrow \mu^- + \overline{\nu}_u \tag{2-7}$$

with a mean life of 2.6×10^{-8} s, and a neutral pion decays 99% of the time to two γ-rays,

$$\pi^0 \longrightarrow \gamma + \gamma \tag{2-8}$$

with a mean life of 8.4×10^{-17} s. Both lifetimes are much longer, by something around 6 to 14 orders of magnitude, than the typical time scale for strong interactions. Note also that in all three modes of decay the lepton numbers are conserved and, as we shall see later, the total number of quarks is also conserved.

2-2 Quarks, the Basic Building Block of Hadrons

Quark masses. Among the six quarks listed in Table 2-2, the least massive members are the u- and d-quarks. These two are believed to have essentially the same mass, in the range of a few MeV/c^2. The lightest baryons, nucleons and Δ-particles, and the lightest mesons, pions, must be made exclusively of these two quarks and their antiquarks. The s-quark is more massive. The unique feature of the s-quark is that it carries a quantum number called *strangeness* and is therefore a necessary constituent of particles with nonzero strangeness, such the K-mesons, or kaons, and the baryon Λ. The c-quark is even more massive. It was first found through the discovery of the J/ψ-meson in 1974 as a narrow resonance in the annihilation of a positron with an electron at 3.1 GeV center-of-mass energy. Since a meson is made of a quark and an

Table 2-2: Quantum numbers of quarks.

Flavor	A	t	t_0	$\mathcal{S}$	$\mathcal{C}$	$\mathcal{B}$	$\mathcal{T}$	$Q(e)$	Mc^2(GeV)
u (up)	$\frac{1}{3}$	$\frac{1}{2}$	$\frac{1}{2}$	0	0	0	0	$+\frac{2}{3}$	0.002 − 0.008
d (down)	$\frac{1}{3}$	$\frac{1}{2}$	$-\frac{1}{2}$	0	0	0	0	$-\frac{1}{3}$	0.005 − 0.015
s (strange)	$\frac{1}{3}$	0	0	−1	0	0	0	$-\frac{1}{3}$	0.1 − 0.3
c (charm)	$\frac{1}{3}$	0	0	0	1	0	0	$+\frac{2}{3}$	1.0 − 1.6
b (beauty)	$\frac{1}{3}$	0	0	0	0	−1	0	$-\frac{1}{3}$	4.1 − 4.5
t (top)	$\frac{1}{3}$	0	0	0	0	0	1	$+\frac{2}{3}$	180 ± 12

A: baryon number t: isospin $\mathcal{S}$: strangeness
$\mathcal{C}$: charm $\mathcal{B}$: beauty $\mathcal{T}$: top

antiquark, a new quark, heavier than u, d, and s, the three known at the time, must be postulated in order to understand this new meson. This is the c-, or charm, quark, having a mass far greater than those of u, d, and s. The existence of the c-quark was subsequently confirmed by other experiments, including the discovery of excited states of J/ψ. The birth of the c-quark prompted the search for even heavier quarks. In this way, the presence of a b-quark was found in the Υ-meson at 10 GeV. The mass of the t-quark has only been measured recently [97]. At the same time, one may also wonder whether there is a fourth generation of quarks beyond the three known ones.

Associated with each quark there is an antiquark. All the known hadrons are made of these six quarks and their antiquarks. The properties of quarks are deduced from measurements made on mesons and baryons, as observations on isolated quarks cannot be carried out. The masses, magnetic moments, and other properties of quarks are inferred from what we know of the properties of mesons and baryons (see, e.g., [22]). Currently, our ability to make such deductions relies on our incomplete understanding of QCD. It is especially inadequate at low energies, where the majority of the experimental observations are made. For example, to obtain the masses of quarks from the known hadron masses, we need to know the strength of the interaction between quarks that binds them inside the hadron. Since this is poorly known, the quark masses listed in Table 2-2 represent only the best estimates and may or may not be closely related to their true masses. Furthermore, different ways of making the estimate result in different values.

Fermions and bosons. Hadrons are subdivided into two classes, baryons and mesons. Besides nucleons, we have Δ, Λ, and a large number of heavier particles in the baryon family. Among mesons, we have already encountered pions, kaons, J/ψ, and Υ, and there are many others.

Baryons are distinguished by the fact that they are fermions, particles that obey Fermi-Dirac statistics. Because of this property, two identical baryons cannot occupy the same quantum-mechanical state. The fact that baryons are fermions implies that quarks must also be fermions, as it is impossible to construct fermions except from odd numbers of fermions. Furthermore, if we accept that a quark cannot exist as a free particle, the lightest fermion in the hadron family must be made of three quarks.

As fermions, baryons must have half-integer intrinsic spins. For example, the intrinsic spin of a nucleon is $\frac{1}{2}$ and that of a Δ-particle is $\frac{3}{2}$. This implies that quarks must also have half-integer intrinsic spins. In addition, the quarks are residing in states with definite orbital angular momenta, just like electrons in an atom. The energy of three quarks in a baryon depends on the interaction between them and this, in turn, depends on the total spin and angular momentum. We shall come back for a brief look of this question later in the quark model of hadrons (§2-7).

Among the baryons, we are mostly concerned with the lightest pair, the neutron and the proton. From charge considerations alone, we can deduce that a proton, carrying a charge $+e$, must be made of two u-quarks, each having a charge of $\frac{2}{3}e$, and one d-quark, $-\frac{1}{3}e$. The quark wave function of a proton may be represented as

$$|p\rangle = |uud\rangle \tag{2-9}$$

Similarly, the quark wave function of a neutron is

$$|n\rangle = |udd\rangle \tag{2-10}$$

so that the total charge of a neutron in units of e is $\frac{2}{3} - \frac{1}{3} - \frac{1}{3} = 0$. Nuclear physics is usually not concerned with any of the heavier baryons, except perhaps for Δ- and Λ-particles. This comes because we are normally dealing with very low-energy phenomena, a few giga-electron-volts per nucleon or less. As a result, there is usually inadequate energy to excite nucleons to become heavier baryons.

Bosons, particles obeying Bose-Einstein statistics, may be made from even number of fermions. This means that mesons are constructed of an even number of quarks. Since, on the one hand, bosons can be created or annihilated under suitable conditions and, on the other hand, the number of quarks is conserved in strong interaction processes, a meson must be made of an equal number of quarks (q) and antiquarks ($\overline{q}$). The simplest meson is, therefore, made of a quark-antiquark pair ($q\overline{q}$). For example, pions, the lightest members among the mesons, are made of a quark, either u or d, and an antiquark, either $\overline{u}$ or $\overline{d}$.

Quark charge. Many hadrons are observed to carry electric charge. This leads to the conclusion that quarks must also carry charge. In nature all observed charges are in multiples of $e = 1.60217733 \times 10^{-19}$ C, with the charge on an electron being $-e$ and on a proton $+e$. The most convenient assignment of charge to the quarks is for u-, c-, and t-quarks to have $+\frac{2}{3}e$ and d-, s-, and b-quarks to have $-\frac{1}{3}e$. The assignment of multiples of $\frac{1}{3}e$ to quarks seems, on the surface, to violate the notion that e is a fundamental or indivisible unit of charge. However, there is no reason to assume that $\frac{1}{3}e$ cannot be the more fundamental unit instead of e. Furthermore, there is no problem, as quarks do not exist freely and all the observed charges are in integer multiples of e.

2-3 Isospin

The nucleon. A proton and a neutron may be considered as two different aspects of the same particle, the nucleon. Both of them have spin $\frac{1}{2}$ and their masses, 939.566 MeV/c^2 for a neutron and 938.272 MeV/c^2 for a proton, differ only by about 0.1%. The main distinction between these two particles is in their electromagnetic properties: namely, charge and magnetic dipole moment (see §2-8). If we are dealing only with strong interactions, such differences are immaterial. That is, in the absence of electromagnetic interaction, a proton cannot be distinguished from a neutron. This is similar to the case of particles with different values of m_s, projections of the intrinsic spin $\boldsymbol{s}$ on the quantization axis. Consider a spin-$\frac{1}{2}$ particle. In the absence of a magnetic field $\boldsymbol{B}$, particles with the two possible values, $\pm\frac{1}{2}$, of m_s are degenerate in energy and, consequently, are indistinguishable from each other. On the other hand, once a magnetic field is introduced, the degeneracy is removed and particles are observed to have different energies depending on whether their intrinsic spins $\boldsymbol{s}$ are aligned parallel or antiparallel to $\boldsymbol{B}$. The difference between a proton and a neutron is analogous to the difference between particles with $m_s = \pm\frac{1}{2}$ if we substitute the Coulomb field with a magnetic field.

If protons and neutrons are considered as identical particles, we need a new label to distinguish between them. For this purpose, the concept of *isospin* is introduced. Since there are only two possible states for a nucleon, the proton state and the neutron state, we can assign isospin $t = \frac{1}{2}$ to a nucleon, based on the analogy that a spin-$\frac{1}{2}$ system can have two different substates. The two nucleons are distinguished by $t_0 = \pm\frac{1}{2}$, the expectation value of the third component of isospin operator $\boldsymbol{t}$. It is a matter of convention whether we consider the $|t{=}\frac{1}{2}, t_0{=}+\frac{1}{2}\rangle$ state to be a proton state and the $|t{=}\frac{1}{2}, t_0{=}-\frac{1}{2}\rangle$ state to be a neutron state, or the other way around. Both conventions are in use and we shall adopt the more popular one with

$$|p\rangle \equiv |t{=}\tfrac{1}{2}, t_0{=}+\tfrac{1}{2}\rangle \qquad\qquad |n\rangle \equiv |t{=}\tfrac{1}{2}, t_0{=}-\tfrac{1}{2}\rangle \tag{2-11}$$

where $|p\rangle$ and $|n\rangle$ represent, respectively, the wave functions of a proton and a neutron. For a nucleus consisting of several nucleons, the total isospin is given by the vector sum of that for each individual nucleon,

$$\boldsymbol{T} = \sum_{i=1}^{A} \boldsymbol{t}(i) \tag{2-12}$$

where A is the number of nucleons. This is identical to the rule for angular momentum addition.

In the absence of electromagnetic interaction, we expect isospin to be a constant of motion. That is, the eigenstates of the Hamiltonian can also be the eigenstates of $\boldsymbol{t}^2$, the square of the isospin operator, as well as the third component $\boldsymbol{t}_0$. As a result, each eigenstate may also be labeled by t and t_0, with $t(t+1)$ as the expectation value of $\boldsymbol{t}^2$ and t_0, that of $\boldsymbol{t}_0$ for the eigenstate. In dealing with nuclei, the main source of isospin symmetry breaking comes from Coulomb interaction that acts only between protons. A less severe but nevertheless noticeable source is the difference between the masses of the neutral and charged mesons exchanged between two nucleons (see §3-6). The possibility of more fundamental isospin-breaking terms in the nuclear force, for instance, due to a possible small difference between the masses of u- and d-quarks, is not yet well established but has not been completely ruled out either.

From a purely mathematical point of view, spin and isospin are similar in structure. Let us concentrate on isospin-$\frac{1}{2}$ systems for the moment and study them by analogy with spin-$\frac{1}{2}$ systems. A particle with $s = \frac{1}{2}$ and projection along the quantization axis $m = +\frac{1}{2}$ may be represented by a two-component column matrix in the following way:

$$|s{=}\tfrac{1}{2}, m{=}+\tfrac{1}{2}\rangle = \begin{pmatrix} 1 \\ 0 \end{pmatrix} \tag{2-13}$$

Similarly, the corresponding $m = -\frac{1}{2}$ state may be represented as

$$|s{=}\tfrac{1}{2}, m{=}-\tfrac{1}{2}\rangle = \begin{pmatrix} 0 \\ 1 \end{pmatrix} \tag{2-14}$$

The isospin wave functions of nucleons may be written in an analogous way,

$$|p\rangle = |t{=}\tfrac{1}{2}, t_0{=}+\tfrac{1}{2}\rangle = \begin{pmatrix} 1 \\ 0 \end{pmatrix}_t \tag{2-15}$$

$$|n\rangle = |t{=}\tfrac{1}{2}, t_0{=}-\tfrac{1}{2}\rangle = \begin{pmatrix} 0 \\ 1 \end{pmatrix}_t \tag{2-16}$$

where the subscript t, which we shall omit in the future unless required for reasons of clarity, identifies that the column matrices are for isospin. Using the convention that a proton has $t_0 = +\frac{1}{2}$ and a neutron has $t_0 = -\frac{1}{2}$, we can relate the *charge number* Q, electric charge in units of e, for a nucleon to t_0,

$$Q = t_0 + \tfrac{1}{2}$$

When we extend the concept of isospin to antiparticles and to systems of several nucleons, the relation between Q and t_0 depends also on A, the number of baryons in the system,

$$Q = t_0 + \tfrac{1}{2}A \tag{2-17}$$

A more general relation involving strangeness and other quantum numbers is given later in Eq. (2-37).

Isospin operators for $t = \frac{1}{2}$ systems can be constructed from Pauli matrices $\boldsymbol{\sigma}$ in the same way as angular momentum operators for a spin-$\frac{1}{2}$ system. For example, we can write

$$\tau_1 = \begin{pmatrix} 0 & 1 \\ 1 & 0 \end{pmatrix} \qquad \tau_2 = \begin{pmatrix} 0 & -i \\ i & 0 \end{pmatrix} \qquad \tau_3 = \begin{pmatrix} 1 & 0 \\ 0 & -1 \end{pmatrix} \tag{2-18}$$

for the x-, y-, and z-components of isospin operator $\boldsymbol{\tau}$. The matrices obey the relation

$$\tau_i \tau_j = \delta_{ij} \boldsymbol{I} + i\epsilon_{ijk}\tau_k \tag{2-19}$$

Here $\boldsymbol{I}$ is the 2×2 unit matrix and ϵ_{ijk} is the three-dimensional Levi-Civita symbol, with $\epsilon_{ijk} = 1$ if the order of i, j, and k is an even permutation of 1, 2, and 3; -1 if the order is an odd permutation; and zero if two or more of the three indices are the same.

For a nucleon it is easily seen that the wave functions given by Eqs. (2-15) and (2-16) are the eigenfunctions of the τ_3 operator, or τ_0 operator in spherical representation,

$$\tau_0 \begin{pmatrix} 1 \\ 0 \end{pmatrix} = \tau_3 \begin{pmatrix} 1 \\ 0 \end{pmatrix} = \begin{pmatrix} 1 & 0 \\ 0 & -1 \end{pmatrix} \begin{pmatrix} 1 \\ 0 \end{pmatrix} = + \begin{pmatrix} 1 \\ 0 \end{pmatrix}$$

$$\tau_0 \begin{pmatrix} 0 \\ 1 \end{pmatrix} = \tau_3 \begin{pmatrix} 0 \\ 1 \end{pmatrix} = \begin{pmatrix} 1 & 0 \\ 0 & -1 \end{pmatrix} \begin{pmatrix} 0 \\ 1 \end{pmatrix} = - \begin{pmatrix} 0 \\ 1 \end{pmatrix}$$

The value of the third component of isospin, t_0, is equal to half of the expectation value of τ_0, the same relation as that between m_s and σ_0 for spin-$\frac{1}{2}$ particles. By the same token, the expectation value of $\boldsymbol{\tau}^2$ is 3, four times the value of $t(t+1)$ for a nucleon.

From the form of $\boldsymbol{\tau}$ given in Eq. (2-18), we can construct isospin-raising (τ_+) and isospin-lowering (τ_-) operators that transform, respectively, a neutron to a proton and a proton to a neutron,

$$\tau_+ = \tfrac{1}{2}(\tau_1 + i\tau_2) = \begin{pmatrix} 0 & 1 \\ 0 & 0 \end{pmatrix} \qquad \tau_- = \tfrac{1}{2}(\tau_1 - i\tau_2) = \begin{pmatrix} 0 & 0 \\ 1 & 0 \end{pmatrix} \tag{2-20}$$

In the same way as angular momentum raising and lowering operators, $\tau_\pm$ changes the value of t_0 without affecting isospin t or any other parts of the wave function,

$$\tau_\pm | t, t_0 \rangle = \sqrt{t(t+1) - t_0(t_0 \pm 1)}\; | t, t_0 \pm 1 \rangle \tag{2-21}$$

The definition used here for τ_+ and τ_- is the more general form used in textbooks and differs slightly from the convention for spherical tensor operators given in §A-2.

For nuclei made of several nucleons, isospin operators may be constructed out of the single-nucleon operators $\boldsymbol{I}$, τ_1, τ_2, and τ_0 $(= \tau_3)$. For example,

$$T_0 = \frac{1}{2}\sum_{i=1}^{A} \tau_0(i) \tag{2-22}$$

where $\tau_0(i)$ acts only on the isospin wave function of the ith nucleon.

The usefulness of isospin is not restricted to the economy gained in treating formally a proton and a neutron as two different states of the same particle. Since isospin is a constant of motion in strong interaction processes, it is a fundamental symmetry, essentially on the same footing as flavor, parity, etc. Isospin is useful in classifying hadrons in general. For example, as we shall see in §2-5, pions come in three different charge states, π^+, π^0, and π^-. They may be treated as the three projections, $t_0 = +1$, 0, -1, of an isospin $t = 1$ system. Since pions are not baryons, the baryon number $A = 0$. We see that the relation between charge number Q and the third component of isospin given in Eq. (2-17) holds here as well. In §2-7 we shall see the case of a quartet of baryons, the Δ-particles, that appear in four different charge states, Δ^{++}, Δ^+, Δ^0, and Δ^-, with charge number $Q = 2, 1, 0, -1$. It is therefore a $t = \frac{3}{2}$ system of baryons. We shall return later for a discussion of the isospin wave function of hadrons and nuclei.

2-4 Isospin of Antiparticles

Particles and antiparticles. An antiparticle may be characterized by the property that it can annihilate the particle with which it is associated. Energy and momentum conservation are maintained in the process, for example, by the emission of two γ-rays or the creation of a different particle-antiparticle pair. Since the final state of an annihilation process is electrically neutral, a particle and its antiparticle must have opposite charges to conserve electric charge. For example, an electron has charge $-e$, and its antiparticle, the positron, has charge $+e$. Similarly, the conservation of other scalar quantum numbers, such as lepton number and baryon number, requires the corresponding labels for particles and antiparticles to be equal in magnitude but opposite in sign, as we have seen in earlier examples. By the same token, the intrinsic parity of an antiparticle is shown in §A-1 to be opposite to that of its particle. For vector quantities, such as intrinsic spin and isospin, the rules of angular momentum addition require that the magnitudes be the same for a particle and its antiparticle so that they can be coupled together to form scalars.

Let us take the case of proton-antiproton annihilation at rest with the emission of two photons as an example:

$$p + \overline{p} \rightarrow \gamma + \gamma$$

Since a photon is an isospin zero, or *isoscalar,* particle, the total isospin on the right-hand side of the reaction is zero. Conservation of isospin requires that the proton and the antiproton are coupled to a $T = 0$ state on the left-hand side. Since $t = \frac{1}{2}$ for a proton, the antiproton $\bar{p}$ must also have $t = \frac{1}{2}$. The third component of isospin for a proton is $t_0 = +\frac{1}{2}$ by the convention we have adopted. For an isoscalar system, the sum of the third components of isospin for all the quantities involved must also be zero. From this, we conclude that $t_0 = -\frac{1}{2}$ for an antiproton, the opposite in sign from that for its particle.

We see that the relation between charge number Q and the third component of isospin given by Eq. (2-17) applies to antiparticles as well. For an antiproton, the baryon number $A = -1$, and we obtain the correct result of $Q = -1$ from Eq. (2-17) using t_0 value of $-\frac{1}{2}$ deduced above.

Charge conjugation. The operation that transforms the wave function of a particle to that of an antiparticle is called *charge conjugation*. It changes the sign of the charge of a particle without affecting any of the properties unrelated to charge. In relativistic quantum mechanics, this implies a transformation between a particle and its antiparticle, hence the name particle-antiparticle transformation. Let $|p\rangle$ and $|n\rangle$ represent, respectively, the wave functions of a proton and a neutron. In terms of second-quantized creation operators $a^\dagger_{tt_0}$ for a particle, we may express these wave functions as

$$|p\rangle = a^\dagger_{1/2,+1/2}|0\rangle \qquad |n\rangle = a^\dagger_{1/2,-1/2}|0\rangle \tag{2-23}$$

where $|0\rangle$ is the wave function for vacuum. In the expression, we have displayed only the isospin ranks and suppressed all other labels for simplicity. The wave functions of an antiproton $|\bar{p}\rangle$ and an antineutron $|\bar{n}\rangle$ may be constructed in a similar way, using the creation operator $b^\dagger_{tt_0}$ for an antiparticle,

$$|\bar{n}\rangle = b^\dagger_{1/2,+1/2}|0\rangle \qquad |\bar{p}\rangle = b^\dagger_{1/2,-1/2}|0\rangle \tag{2-24}$$

Here we have made use of the fact that, on transforming a particle to an antiparticle (and vice versa), the charge, and hence the projection of isospin on the quantization axis, changes sign. If particles and antiparticles are unrelated to each other, $a^\dagger_{tt_0}$ and $b^\dagger_{tt_0}$ are completely different operators defined, respectively, by Eqs. (2-23) and (2-24). However, particles and antiparticles can transform into each other through charge conjugation, C, and as a result, operators $a^\dagger$ and $b^\dagger$ are not independent of each other.

In addition to isospin (and spin), the wave functions of a particle and an antiparticle can also differ by a phase factor. There are several ways to obtain this factor. If we take $a^\dagger_{tt_0}$ and $b^\dagger_{tt_0}$ as operators with a definite irreducible spherical tensor rank t, the phase factor is fixed by their transformation properties under a rotation in the isospin space. For second-quantized operators, we have the relation

$$b^\dagger_{tt_0} = (-1)^{t-t_0} a_{t,-t_0} \tag{2-25}$$

The phase factor arises from the fact that operators $a^\dagger_{tt_0}$ and $a_{t,-t_0}$ are not Hermitian conjugates of each other without the factor $(-1)^{t-t_0}$. (For a more detailed discussion, see, e.g., Refs. [34, 50].)

With the relation between the second-quantized operators given by Eq. (2-25), we find that, under charge conjugation,

$$\begin{aligned} |p\rangle &\xrightarrow[C]{} (-1)^{1/2+1/2}|\bar{p}\rangle = -|\bar{p}\rangle \\ |n\rangle &\xrightarrow[C]{} (-1)^{1/2-1/2}|\bar{n}\rangle = +|\bar{n}\rangle \end{aligned} \tag{2-26}$$

The same considerations apply to other particles as well. For example, since a u-quark has isospin ranks $(t, t_0) = (\frac{1}{2}, +\frac{1}{2})$ and a d-quark has ranks $(\frac{1}{2}, -\frac{1}{2})$, the transformations to their antiparticles under charge conjugation are

$$\begin{aligned} |u\rangle &\xrightarrow[C]{} (-1)^{1/2+1/2}|\bar{u}\rangle = -|\bar{u}\rangle \\ |d\rangle &\xrightarrow[C]{} (-1)^{1/2-1/2}|\bar{d}\rangle = +|\bar{d}\rangle \end{aligned} \tag{2-27}$$

These phase factors are used in writing the quark wave functions for pions in the next section.

2-5 Isospin of Quarks

One of the consequences of treating a proton and a neutron as two different isospin states of a nucleon is that we can change a proton into a neutron, and vice versa, using the isospin-lowering and isospin-raising operators given in Eq. (2-20),

$$\begin{aligned} \tau_+|n\rangle &= \begin{pmatrix} 0 & 1 \\ 0 & 0 \end{pmatrix}\begin{pmatrix} 0 \\ 1 \end{pmatrix} = \begin{pmatrix} 1 \\ 0 \end{pmatrix} = |p\rangle \\ \tau_-|p\rangle &= \begin{pmatrix} 0 & 0 \\ 1 & 0 \end{pmatrix}\begin{pmatrix} 1 \\ 0 \end{pmatrix} = \begin{pmatrix} 0 \\ 1 \end{pmatrix} = |n\rangle \end{aligned} \tag{2-28}$$

In terms of quarks, we have already seen in Eqs. (2-9) and (2-10) that

$$|p\rangle = |uud\rangle \qquad\qquad |n\rangle = |udd\rangle$$

When we substitute these results into Eq. (2-28), we obtain

$$\tau_+|udd\rangle = |uud\rangle \qquad\qquad \tau_-|uud\rangle = |udd\rangle \tag{2-29}$$

Since a proton and a neutron are considered here to be identical to each other except for the third component of their isospin, the other parts of the wave functions are not changed by isospin operations. In terms of quarks, the only difference between a proton and a neutron is the replacement of one of the two u-quarks by a d-quark. The relations expressed by Eq. (2-29), therefore, imply that, when an isospin-raising operator acts on the quarks, it transforms a d-quark to a u-quark, and the other way around for an isospin-lowering operator. Since no other quarks are involved here, we conclude that d- and u-quarks also form an isospin doublet, analogous to the proton-neutron pair. Furthermore, since the third component of isospin is a scalar quantity, the sum of t_0 of two u-quarks and one d-quark in a proton must be $+\frac{1}{2}$ and that of one u-quark and two d-quarks in a neutron must be $-\frac{1}{2}$. To satisfy both requirements, we must assign

$t_0 = +\frac{1}{2}$ to a u-quark and $t_0 = -\frac{1}{2}$ to a d-quark. The relation between charge number Q and t_0 is still given by Eq. (2-17), when we note that a quark has baryon number $A = \frac{1}{3}$, from the fact that it takes three quarks to make a baryon.

More formally, we can write the $\tau_\pm$ operator for a nucleon as the sum of isospin-raising or isospin-lowering operators acting on each one of the three quarks,

$$\tau_\pm(\text{nucleon}) \longrightarrow \sum_{i=1}^{3} \tau_\pm(q_i)$$

where $\tau_\pm(q_i)$ acts on the isospin of the ith quark only. Ignoring for the moment any antisymmetrization requirement between the three quarks in a nucleon, we can write the first relation of Eq. (2-29) in the following way:

$$\tau_+|n\rangle = \left\{\tau_+(q_1) + \tau_+(q_2) + \tau_+(q_3)\right\} |u(1)d(2)d(3)\rangle$$

where we have assumed that the first quark in the neutron is a u-quark and the remaining two are d-quarks. Since $\tau_+|u\rangle = 0$ (a u-quark has $t_0 = +\frac{1}{2}$), the first term vanishes. The second and third terms give the results

$$\begin{aligned}\tau_+(q_2)|u(1)d(2)d(3)\rangle &= |u(1)u(2)d(3)\rangle \\ \tau_+(q_3)|u(1)d(2)d(3)\rangle &= |u(1)d(2)u(3)\rangle\end{aligned}$$

Upon antisymmetrization these two terms produce identical results which we shall represent generically as $|uud\rangle$.

Quark wave functions of pions. We can check the isospin assignment to the u- and d-quarks by examining the structure of mesons formed of these two quarks and their antiquarks. It is simplest to start from π^- with $t = 1$ and $t_0 = -1$. Since we cannot use any quarks other than u, d, $\overline{u}$, and $\overline{d}$, the only way to form a $t_0 = -1$ system is to take the $\overline{u}d$ combination. We can easily deduce that this pair of quarks must form a $t = 1$ system by elimination. Two isospin-$\frac{1}{2}$ particles can only be coupled to total isospin 0 and 1. The $\overline{u}d$ system cannot be $t = 0$ as it has $t_0 = -1$. As a result, we can make the identification

$$|\pi^-\rangle = |\overline{u}d\rangle \tag{2-30}$$

as there is no other way to form a $t = 1$, $t_0 = -1$ state with u, d, $\overline{u}$, and $\overline{d}$.

In general, it is possible to find several different linearly independent components corresponding to the same t and t_0. The appropriate combination for a given situation is guided by isospin-coupling rules. Furthermore, the wave function must be antisymmetric among the quarks and is an eigenstate of the Hamiltonian. For our interest in this section, we shall only be concerned with isospin coupling.

From the wave function of π^-, we can construct that for π^0 using an isospin-raising operator,

$$|\pi^0\rangle = \frac{1}{N}\tau_+|\pi^-\rangle = \frac{1}{N}\sum_{i=1,2} \tau_+(q_i)|\overline{u}d\rangle \tag{2-31}$$

where N is the normalization factor to be determined later. The operator τ_+ acts on the wave function of each quark. We have already seen that

$$\tau_+|d\rangle = |u\rangle \tag{2-32}$$

However, for the antiquarks,

$$\tau_+|\overline{u}\rangle = -|\overline{d}\rangle \tag{2-33}$$

The additional negative sign comes from the symmetry requirement under charge conjugation, as discussed in the previous section.

The normalization factor N in Eq. (2-31) may be determined using Eq. (2-21). Since $|\pi^-\rangle$ is a $t = 1$, $t_0 = -1$ system,

$$\tau_+|1,-1\rangle = \sqrt{2}\,|1,0\rangle$$

we obtain a value $N = \sqrt{2}$. The final result for the wave function of π^0 is then

$$|\pi^0\rangle = \frac{1}{\sqrt{2}}\Big\{|u\overline{u}\rangle - |d\overline{d}\,\rangle\Big\} \tag{2-34}$$

The same result can also be obtained by coupling the isospin of the two particles using the Clebsch-Gordan coefficients described in §A-3. Since the values of both coefficients to couple two isospin-$\frac{1}{2}$ particles to total isospin 1 are

$$\langle\tfrac{1}{2},\pm\tfrac{1}{2};\tfrac{1}{2},\mp\tfrac{1}{2}|10\rangle = \frac{1}{\sqrt{2}}$$

we obtain the same result as given in Eq. (2-34) after inserting an "extra" minus sign arising from the transformation from $|d\rangle$ to $|\overline{d}\,\rangle$ under charge conjugation.

The wave function for π^+ in terms of quarks is

$$|\pi^+\rangle = -|u\overline{d}\,\rangle \tag{2-35}$$

This result may be arrived at either by applying an isospin-raising operator on the quark wave function of π^0 obtained above or by constructing a $(t, t_0) = (1, +1)$ system in the same manner as we have just done for the π^--system. Again, the overall minus sign comes from charge conjugation between d and $\overline{d}$.

One question still remains concerning the $t_0 = 0$ wave function for a quark-antiquark pair. There are two different ways, $u\overline{u}$ and $d\overline{d}$, to form a $t_0 = 0$ state from the two quarks and two antiquarks provided. Besides the one given in Eq. (2-34), we can also take the linear combination

$$|\eta_0\rangle = \frac{1}{\sqrt{2}}\Big\{|u\overline{u}\rangle + |d\overline{d}\,\rangle\Big\} \tag{2-36}$$

It is orthogonal to $|\pi^0\rangle$ and must therefore describe a meson other than π^0. Both $u\overline{u}$ and $d\overline{d}$ have $t_0 = 0$ but a mixture of $(t, t_0) = (1, 0)$ and $(t, t_0) = (0, 0)$. A particular linear combination was taken in Eq. (2-34) so as to have the correct isospin of $t = 1$ for the π^0-meson. The linear combination given in Eq. (2-36) is a different one and must correspond to an isospin zero system as a result, a fact that can also be seen from the explicit values of the Clebsch-Gordan coefficients required to construct a $t = 0$ system. Such an isospin-singlet meson may be identified with the η-meson, which has a rest mass of ~550 MeV/c^2.

The four particles, π^+, π^0, π^-, and η, exhaust all the observed mesons in the form of a quark-antiquark pair that can be constructed out of u, d, $\overline{u}$, and $\overline{d}$ in their lowest possible energy states. To obtain other mesons, we must either introduce excitations in the quark-antiquark system or invoke s- and other more massive quarks. We shall return to this point later.

Other quarks. Let us examine briefly the isospin of the other quarks, c, s, t, and b and their antiparticles. It is perhaps tempting to assume that each one of the remaining two pairs forms also an isospin doublet. This, however, is not the case. As can be seen from Table 2-2, these four quarks are isoscalar particles. Two questions are raised here: How are the assignments of $t = 0$ made to these quarks? What is the relation between their values of Q and t_0? Since quarks are not particles observed in isolation, assignment of isospin (as well as other quantum numbers) must be carried out through the hadrons they make up. This is what we have done for the u- and d-quarks and we shall see how it can be carried out for the s-quarks as an example.

After u- and d-quarks, the next one in order of increasing mass is the s-quark. They are found in hadrons with nonzero strangeness $\mathcal{S}$. For our purpose here we can regard $\mathcal{S}$ as a label to identify the number of strange antiquarks in the hadron. The lightest strange mesons are the kaons, or K-mesons. They come as two isospin doublets ($t = \frac{1}{2}$ systems), one consisting of $K^+(u\overline{s})$ and $K^0(d\overline{s})$, and the other of $K^-(\overline{u}s)$ and $\overline{K}^0(\overline{d}s)$. Since u- and d-quarks have isospin $t = \frac{1}{2}$, the s-quark must have integer isospin 0 or 1 in order to form kaons with $t = \frac{1}{2}$. The assignment of $t = 1$ may be ruled out on the grounds that, if this were true, we should be able to form $t = \frac{3}{2}$ strange mesons, for example, made of an s-quark and an antiquark, either $\overline{u}$ or $\overline{d}$. The fact that such mesons have not been observed implies that the isospin of the s-quark is zero. The assignment of isospin to the other quarks may be carried out in a similar way and we shall not go into the steps here.

With the assignment of $t = 0$ to the heavy quarks, we need now to modify the relation between the charge number Q and t_0. Equation (2-17) was derived for u- and d-quarks and must be changed now, as the other quarks have different relations between Q and t_0. The more general form of Eq. (2-17) is given by

$$Q = t_0 + \tfrac{1}{2}(A + \mathcal{S} + \mathcal{C} + \mathcal{B} + \mathcal{T}) \tag{2-37}$$

where assignments of baryon number A, strangeness $\mathcal{S}$, charm quantum number $\mathcal{C}$, beauty quantum number $\mathcal{B}$, and top quantum number $\mathcal{T}$ for each of the six flavors of quarks are given in Table 2-2.

2-6 Strangeness and Other Quantum Numbers

In strong interaction processes, the total number of each type of quark, u, d, s, c, b, and t, is conserved. However, through weak interactions, quarks can be transformed from one flavor to another, such as the example shown in Fig. 2-1. In terms of observed particles, the flavor degree of freedom in quarks shows its presence by separating hadrons into different groups, with transitions between groups allowed only through weak interactions. As a result, transition rates between members of the same group, characterized by the fact that they have the same quark content, are fast and typical of strong interaction processes. On the other hand, transitions between members of different groups

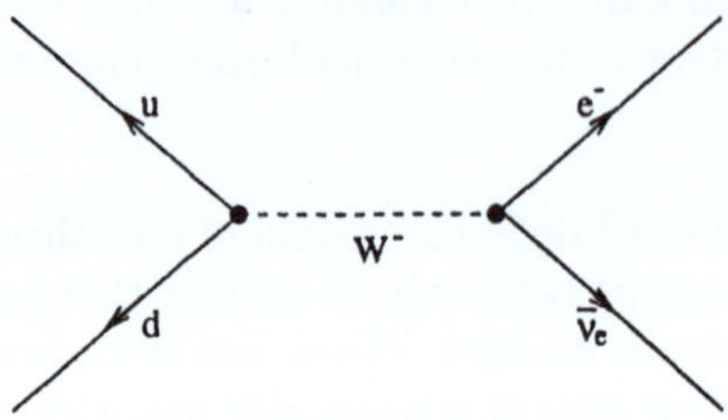

Figure 2-1: Transformation of a d-quark into a u-quark through weak interaction. The virtual W^--boson emitted decays into a pair of leptons, e^- and $\overline{\nu}_e$.

involve the transformation of one type of quark to another and are much slower, with lifetimes more typical of weak interaction processes.

Each group of hadrons is characterized by the number of quarks of each flavor, and transitions from one group to another involve a change in one or more of these numbers. For example, the K^+-meson, with mass 494 MeV/c^2, is made of the quark-antiquark pair $u\overline{s}$. The dominant mode of decay, 63.5% of the time, is into leptons, $\mu^+ + \overline{\nu}_\mu$. A less prominent mode, 21.2% of the time, is into a pair of pions $\pi^+ + \pi^0$. In either decay mode, the total number of quarks is conserved. However, there is no strange antiquark among the end products of the decay. One way to "remove" the strange antiquark $\overline{s}$ without changing the net number of quarks involved is to let it decay to a $\overline{u}$-quark, which then annihilates with the u-quark in K^+ to produce a pair of leptons. Alternatively, the $\overline{s}$-quark may β-decay to a $\overline{d}$-quark instead and form a part of the pions in the end product. The mean life of the K^+-meson, 1.2×10^{-8} s, is typical for weak decays.

Strangeness, charm, and beauty. Among the baryons, Λ is a particle with quark content (uds) and mass 1116 MeV. It is produced in reactions such as

$$\pi^- + p \rightarrow \Lambda + K^0$$

where the meson K^0, with quark structure $d\overline{s}$, is the isospin partner of K^+. On the left-hand side of the reaction, there is no strange quark, as both π^- and p are made exclusively of u's and d's. On the right-hand side, we see that the presence of an s-quark in the Λ-particle is accompanied by an $\overline{s}$ in the K^0 meson. In terms of the observed hadrons, we find that the production of a Λ and other hadrons containing an s-quark is always accompanied by another hadron containing an $\overline{s}$. This type of association may be accounted for by assigning a *strangeness* quantum number $\mathcal{S}$ to count the number of strange quarks. In strong interaction processes, we say that strangeness is a conserved quantity to indicate the fact that the numbers of s and $\overline{s}$ produced must be the same. For historical reasons, an s-quark is assigned $\mathcal{S} = -1$ and $\overline{s}$ is assigned $\mathcal{S} = +1$ (and $\mathcal{S} = 0$ for all other quarks).

Similar to strangeness, we can assign a *charm* quantum number $\mathcal{C}$ to account for the number of c-quarks, with $\mathcal{C} = 1$ for a c-quark and $\mathcal{C} = -1$ for a $\overline{c}$-quark (and $\mathcal{C} = 0$ for all other quarks). To account for the number of b-quarks, a *beauty* quantum number

$\mathcal{B}$ is used with $\mathcal{B} = -1$ for a b-quark and $\mathcal{B} = +1$ for a $\bar{b}$-quark (and $\mathcal{B} = 0$ for all other quarks). Thus, for example, a hadron with n c-quarks has $\mathcal{C} = n$ and a hadron with m b-quarks has $\mathcal{B} = -m$. Whether a quark or an antiquark of a given flavor should take on the positive sign for the quantum number representing that flavor is somewhat arbitrary. The convention described here is the one commonly used and satisfies the relation between Q and t_0 given in Eq. (2-37).

Earlier, we saw the long lifetime of kaons as an example of strangeness conservation. Similarly, long lifetimes of the order 10^{-13} s are observed for the analogous situation of charm and beauty conservation in the decay of D- and B-mesons, the lightest mesons containing, respectively, a c- and b-quark or their antiquarks. The rest masses of D (1869 MeV for $D^{\pm}$) and B (5278 MeV for $B^{\pm}$) are, however, much larger than those for K-mesons, reflecting the heavier masses of c- and b-quarks. Relatively long lifetimes are also observed for mesons made of heavy quark-antiquark pairs as, for example, those shown in Table 2-3. Since these particles are not stable, they are observed as resonances when their production cross sections are plotted as functions of the bombarding energy. For this reason, it is more common to characterize the stability of such "particles" by the widths Γ of their resonance curves, related to their mean life $\overline{T}$ through the uncertainty relation

$$\Gamma = \frac{\hbar}{\overline{T}}$$

where $\hbar = 6.58 \times 10^{-22}$ MeV-s is the Planck constant.

Table 2-3: Lifetimes of ϕ-, J/ψ-, and Υ-mesons.

Meson	Rest mass energy MeV	Width Γ (MeV)	Mean life $\overline{T}$(s)	quark content
ϕ	1019.41±0.01	4.43 ±0.05	1.49×10^{-22}	$s\bar{s}$
J/ψ	3096.88±0.04	0.087±0.005	0.75×10^{-20}	$c\bar{c}$
Υ	9460.37±0.21	0.053±0.002	1.24×10^{-20}	$b\bar{b}$

It is worthwhile noting that lifetimes for J/ψ- and Υ-mesons are much longer and the widths Γ narrower than expected. This is caused by the special circumstance that there is not enough energy available for a J/ψ-particle ($c\bar{c}$), with rest mass 3097 MeV/c^2, to decay into a $D^+(c\bar{d})$ and a $D^-(\bar{c}d)$ particle, the lightest members of quarks containing a charm quark, as their combined rest mass energy is 2×1869 MeV. Similarly, a Υ-particle (rest mass 9460 MeV/c^2) cannot decay to a pair of mesons containing b-quarks, as the lightest pair, a $B^+(\bar{b}u)$ and a $B^-(b\bar{u})$, has a combined rest mass energy of 2×5278 MeV. As a result, J/ψ and Υ must decay through much slower processes involving three or more lighter hadrons, as shown, for example, in Fig. 2-2(a), and into lepton pairs by weak interaction. In contrast, the analogous ϕ-meson (rest mass 1019 MeV/c^2), made of $s\bar{s}$, can decay to a K^+ and a K^- with a combined rest mass energy of 2×493.6 MeV. The narrow widths of J/ψ- and Υ-particles are quite astonishing in view of the high energies involved. As a result, they are useful as energy calibrations and

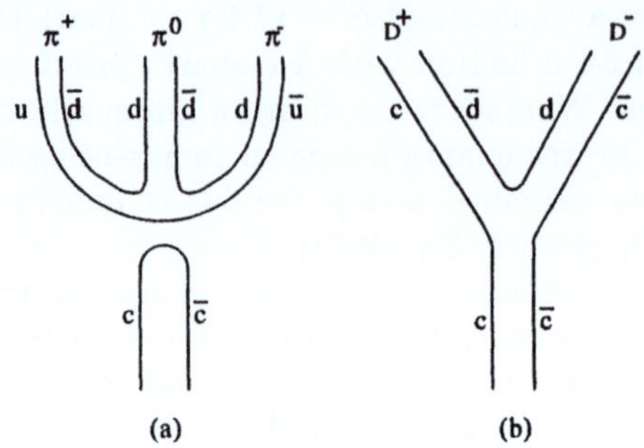

Figure 2-2: Example of decay into hadrons for J/ψ-meson, made of $c\bar{c}$. The three-pion process shown in (a) is allowed whereas transition to D^+D^- shown in (b) is forbidden, as the total mass of the final product is greater than that of J/ψ.

as signatures of special events in high-energy nuclear physics as, for example, those discussed in Chapter 9.

Color. Besides flavor, each quark has another important degree of freedom, known as *color.* The need of this additional quantum-mechanical label can be seen most readily by examining the quark wave function of a Δ-particle. As we have seen earlier, Δ is an isospin $t = \frac{3}{2}$ particle with four different charge states, Δ^{++}, Δ^{+}, Δ^{0}, and Δ^{-}. Since it is a nonstrange baryon ($S = 0$), it must be made of u- and d-quarks alone. For Δ^{++}, the member with the highest charge state, there is only one possible combination of quarks, (uuu), to make a baryon with $Q = 2$. The intrinsic parity of Δ is known to be positive. This, together with other evidence, requires the spatial part of the wave function for the three u-quarks in Δ^{++} to be symmetric if we permute any two of them. The intrinsic spin of Δ is $\frac{3}{2}$ and, hence, the intrinsic spin part of the wave function for the three u-quarks is also symmetric. Similarly, the isospin part of the wave must also be symmetric in order to have $t = \frac{3}{2}$. The net result is that the product of space, intrinsic spin, and isospin parts of the wave function for Δ^{++} is symmetric under a permutation among the three u-quarks. On the other hand, quarks are fermions and the Pauli exclusion principle requires that the total wave function of the three identical quarks be antisymmetric with respect to a permutation of any two of the three quarks. The wave function we have obtained so far for Δ^{++} is in contradiction to this fundamental principle of quantum mechanics. There are two possible ways to get out of this dilemma: Either the Pauli principle is wrong, which is very unlikely, or else we have missed one of the degrees of freedom for quarks.

This new degree of freedom is given the name "color" and hence the name *quantum chromodynamics* for the theory dealing with strong interactions involving "colored" quarks. To account for this new degree of freedom, a color is assigned to each quark, for example, R (red), G (green), and B (blue). From the example of the Δ^{++}-particle we can deduce that the quarks in hadrons must be antisymmetric in the color degree of freedom; that is, the net color in a hadron must vanish. We can also reach the same conclusion from another point of view. Since color has not been an observed property, all hadrons must be colorless objects: The color degree of freedom of the constituents inside a hadron must somehow neutralize each other. For mesons, this is easy to achieve, as an antiquark must have the opposite color for a quark. For baryons made of three identical quarks, such as Δ^{++}, the different colors cancel by being antisymmetric with respect to each other.

In nuclear physics, we are usually not involved explicitly with the color degree of freedom. However, the Δ-particle is important. It was discovered by Fermi and Anderson in 1949 as a resonance in π^+-scattering off protons at pion kinetic energy $T_\pi = 195$ MeV, as shown in Fig. 2-3. It corresponds to a mass of the pion-proton system of 1232 MeV. Since this takes place in the $\ell = 1$ reaction channel with both spin and isospin $\frac{3}{2}$, it is also known as the P_{33}-resonance. Because it is a very strong resonance at relatively low energy, nucleons inside a nucleus may be excited fairly easily to become a Δ-particle, and as we shall see later, such excitations may have a strong influence in processes involving energies comparable to those required to change a nucleon into a Δ-particle.

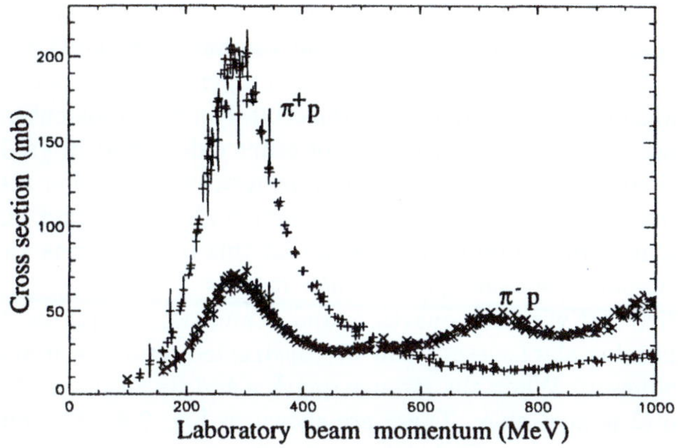

Figure 2-3: Total cross section of charged pions scattering off protons. The strong P_{33}-resonance in $\pi^+ + p$ reaction occurs in the $(J^\pi, T) = (\frac{3}{2}^+, \frac{3}{2})$ channel with $\ell = 1$. The $\pi^- + p$ cross section at the same energy is much smaller, as the system is a mixture of $T = \frac{3}{2}$ and $T = \frac{1}{2}$. The data are taken from Ref. [22].

2-7 Static Quark Model of Hadrons

A quark model of the hadrons should, in principle, involve all six different flavors. This can be rather complicated, as a large number of particles can be constructed from six different quarks and six different antiquarks. Fortunately c-, b-, and t-quarks are so much more massive than u-, d-, and s-quarks that they are important primarily in heavy hadrons. For most particles of interest at low energies, only the three light quarks, u, d, and s, and their antiquarks, are involved. For this reason, it is quite adequate for us to consider a model consisting of only these three quarks and their antiquarks.

Mesons and pseudoscalar mesons. Let us start with the simpler case of mesons. Although mesons can be made with any number of quark-antiquark ($q\overline{q}$)-pairs, most of the observed ones may be understood by considering only a single ($q\overline{q}$)-pair. A simple quark model of mesons, therefore, involves a quark and an antiquark moving with respect to each other with orbital angular momentum $\boldsymbol{\ell}$. The total angular momentum, or spin, of the system is $\boldsymbol{J} = \boldsymbol{\ell} + \boldsymbol{S}$, where $\boldsymbol{S} = \boldsymbol{s}_q + \boldsymbol{s}_{\overline{q}}$ is the sum of the intrinsic spins of the quark and the antiquark. Since $s_q = s_{\overline{q}} = \frac{1}{2}$, the possible value of S for a $q\overline{q}$-system is either 0 (singlet state) or 1 (triplet state). As for the spatial part of the wave function, it has been found that mesons with relative orbital angular momentum $\ell = 0$ are lower in energy, the same as in the case of atomic levels. We shall restrict ourselves to these low-lying ones as they include essentially all those of interest to nuclear physics.

We have already seen that pions are the least massive particles among mesons. Since both orbital angular momentum ℓ and total intrinsic spin S are known to be zero, the spin J of a pion is also zero. They are therefore "scalar" particles, as their wave functions are invariant under a rotation of the spatial coordinate system. However, unlike ordinary scalars, their wave functions change sign under a parity transformation. This may be seen in the following way. The parity of the pion is given by the product of the intrinsic parities of the quark (+1) and the antiquark (−1) and the parity of the spatial wave function of the ($q\overline{q}$)-pair. The property of the spatial wave function under a parity transformation is related to the orbital angular momentum ℓ and is given by $(-1)^{\ell}$, the same as spherical harmonics of order ℓ discussed in §A-1. Since $\ell = 0$, the parity of the complete pion wave function is negative. The pion therefore behaves like a pseudoscalar quantity, one that is invariant under a rotation but changes sign under an inversion of the coordinate system. For this reason, pions and other $J = 0$, negative-parity mesons are called *pseudoscalar* mesons.

We saw earlier that with two quarks, u and d, and two antiquarks, $\overline{u}$ and $\overline{d}$, a total of $2 \times 2 = 4$ (pseudoscalar) mesons can be constructed with $\ell = 0$ and $S = 0$: three pions and one η-meson. When the strange quark s and its antiquark $\overline{s}$ are included in addition, the total is now nine. These are shown in Fig. 2-4. The nine mesons may be separated into two groups. Eight of the nine form an octet, the members of which transform into each other under a rotation in the flavor space. That is, when we make an interchange among u, d, and s, the wave functions of the eight mesons transform

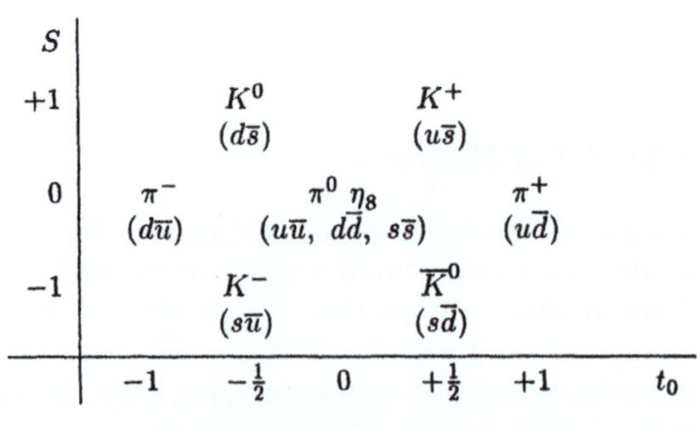

Figure 2-4: Pseudoscalar mesons.

as an irreducible representation of the SU_3 group, special unitary group of dimension 3. Mathematically the transformation is very similar to, for instance, a rotation of the spatial coordinate axes by some Euler angles. The various components of a spherical tensor of a given rank, e.g., spherical harmonics $Y_{\ell m}(\theta, \phi)$, differing only in the values of m, are modified because of the rotation. However, the relation between different components of $Y_{\ell m}(\theta, \phi)$ is such that, in the rotated system, the new $Y_{\ell m}(\theta, \phi)$ can always be expressed in terms of the spherical harmonics of the same order ℓ in the old system, as shown in §A-2. In group theoretical language, the $2\ell+1$ spherical harmonics of the same ℓ but different m form an irreducible representation. Members of the meson octet also form such a group representation except that the rotation is in the (three-dimensional) flavor space consisting of u-, d-, and s-quarks, and the transformation is from quarks of one flavor to another.

The remaining meson, η_0, is invariant under any such interchanges among the three quarks and forms an irreducible representation by itself. In this way, the nine mesons in the model flavor space of u-, d-, and s-quarks, and their antiquarks, may be classified into an octet and a singlet according to their SU_3 symmetry in flavor transformation. We shall soon see that, although this symmetry in the SU_3(flavor) is not exactly preserved in strong interactions, it is nevertheless useful as a classification scheme for both mesons and baryons.

It is a simple matter to write down the wave functions of the nine mesons in terms of $(q\overline{q})$-pairs. The pion wave functions have already been given in §2-5. There is no ambiguity in constructing the kaon wave functions, as each one must involve either an s or an $\overline{s}$. The flavor of the other quark for the $S = 1$ kaons, or an antiquark for the $S = -1$ kaons, must be either u or d, or $\overline{u}$ or $\overline{d}$, and the choice is completely determined by the charge carried by each kaon. The results are shown in Fig. 2-4.

The wave functions of the two isoscalar mesons, η_8 and η_0, are slightly more complicated and must be deduced using, for example, symmetry arguments. Since η_0 is invariant under a transformation among the three flavors, its wave function must be a linear combination of $(u\overline{u})$, $(d\overline{d})$, and $(s\overline{s})$, with equal weight:

$$|\eta_0\rangle = \frac{1}{\sqrt{3}}\left\{|u\overline{u}\rangle + |d\overline{d}\,\rangle + |s\overline{s}\rangle\right\} \tag{2-38}$$

where the factor $1/\sqrt{3}$ comes from the normalization requirement. The η_0-meson is, then, an "extension" of the η_0-meson constructed out of u- and d-quarks (and their antiquarks) given in Eq. (2-36). Similar to the two-flavor case, the wave function of η_8, the isoscalar meson in the octet, may be obtained by requiring it to be an isoscalar and orthogonal to both $|\,\pi^0\,\rangle$ and $|\,\eta_0\,\rangle$. The result is

$$|\eta_8\rangle = \frac{1}{\sqrt{6}}\left\{|u\overline{u}\rangle + |d\overline{d}\,\rangle - 2|s\overline{s}\rangle\right\} \tag{2-39}$$

The derivation is left as an exercise (see Problem 2-5).

Two isospin $t = 0$ pseudoscalar mesons are known at low energies, the η-meson with mass 548.8 MeV/c^2 and the η'-meson with mass 957.5 MeV/c^2. Since the SU_3(flavor) symmetry is not an exact one, the observed mesons are mixtures of η_0 and η_8 given above. The mixing coefficient is usually expressed in terms of an angle θ, known as the

Cabibbo angle,

$$\eta = \eta_8 \cos\theta + \eta_0 \sin\theta \qquad\qquad \eta' = -\eta_8 \sin\theta + \eta_0 \cos\theta$$

For the pseudoscalar mesons, the value is $\theta \simeq 10°$.

Vector mesons. Instead of $S = 0$, the total intrinsic spin of the quark-antiquark pair in a meson may be coupled to $S = 1$. For $\ell = 0$, the spin of the pions produced is now $J = 1$. The parity, however, remains negative. Similar to the pseudoscalar mesons, we now have a set of nine vector mesons whose wave functions behave like an ordinary vector under a transformation of the spatial coordinate system.

The structure of the set of vector mesons, as far as their symmetry under a rotation in the flavor space is concerned, is the same as that of the pseudoscalar mesons, as can be seen by comparing Fig. 2-5 with Fig. 2-4. Corresponding to the pions, we have an isospin triplet of ρ-mesons, and instead of the strange pseudoscalar mesons K^0, K^+, K^-, and $\overline{K}^0$, we now have the strange vector mesons K^{*0}, K^{*+}, K^{*-}, and $\overline{K}^{*0}$. The two isoscalar vector mesons with definite SU_3 symmetry are ϕ_0 and ϕ_8. The observed isoscalar vector mesons, ϕ and ω, have much larger SU_3(flavor) mixing, with Cabibbo angle $\theta \sim 40°$, compared to $\theta \sim 10°$ for the pseudoscalar mesons.

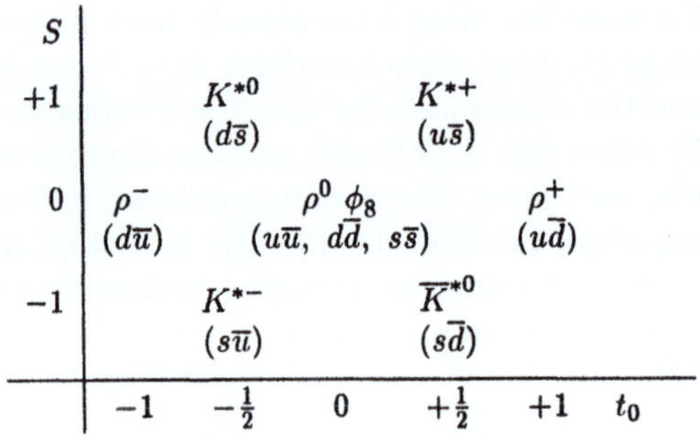

Figure 2-5: $J^\pi = 1^-$ vector mesons.

The vector mesons are more massive than their pseudoscalar counterparts. For example, the ρ-meson has a rest mass energy of 767 MeV and the ω-meson has 782 MeV. In contrast, the pion rest mass energies are 140 MeV for $\pi^\pm$ and 135 MeV for π^0. As far as their wave functions are concerned, the vector and pseudoscalar mesons differ only in their total intrinsic spin, with $S = 1$ for the former and $S = 0$ for the latter. The large difference in their masses must come from the differences in the interaction between a quark and an antiquark in the $S = 0$ and $S = 1$ states. We see here an example of the important role of the interaction between quarks which we have ignored for the sake of simplicity in most of our discussions.

Because of their larger masses, the ρ- and ω-mesons can decay via strong interactions to pions with lifetimes at least six orders of magnitude shorter than those of pions. The ρ-meson transforms to two pions with a mean life of 4×10^{-24} s (or width $\Gamma = 153$ MeV) and the ω-meson goes 90% of the time to three pions with a lifetime of

8×10^{-23} s ($\Gamma = 8.5$ MeV). As we shall see later, both ρ- and ω-mesons play special roles in the interaction between nucleons.

Baryons. With three flavors, we can construct a total of $3 \times 3 \times 3 = 27$ baryons for a given set of (ℓ, S)-values. They can be classified according to their SU_3(flavor) symmetries into four groups consisting of 10, 8, 8, and 1 members. The group of 10 baryons (decuplet) is completely symmetric under a transformation in flavor, and the group of one baryon (singlet) is completely antisymmetric. The other two groups, consisting of eight members each (octets), have mixed SU_3(flavor) symmetry, neither completely symmetric nor completely antisymmetric. Similar to the case of mesons, we can make use of the SU_3(flavor) symmetry to construct the quark wave functions for these baryons.

The baryon wave functions are slightly more complicated to derive than those for mesons for the simple reason that we are now dealing with products of three objects instead of two. It is convenient to treat all the quarks as identical particles distinguished only by their flavor and color labels. Since hadrons are color neutral objects, their quark wave functions must be totally antisymmetric in color. As a result, the rest of the wave function, formed of a product of flavor, spin, and spatial parts, must be symmetric under a permutation of any two of the quarks.

Consider first the decuplet. The 10 members of the group, together with their quark contents, are listed in Fig. 2-6. Because they are completely symmetric in flavor, it is relatively simple to construct the quark wave functions. We have already encountered one of the members in this group, Δ^{++}, in introducing the color degree of freedom. As mentioned previously, both the intrinsic spin and isotopic parts of the Δ-particle wave function must be completely symmetric in order to couple to a state with maximum total spin $S = \frac{3}{2}$ and isospin $t = \frac{3}{2}$. Furthermore, for a symmetric product of spin, isospin, and spatial parts, the spatial part of the wave function of three u-quarks must also be symmetric.

With Δ^{++} given by $|uuu\rangle$ from isospin considerations, the wave functions of the

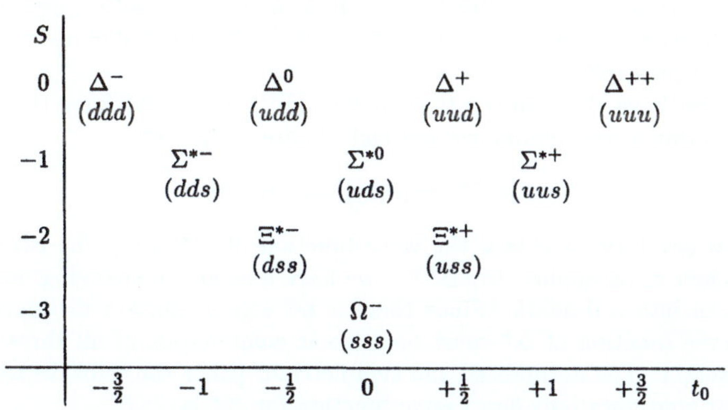

Figure 2-6: $J^\pi = \frac{3}{2}^+$ baryon decuplet.

other three members of the isospin quartet, Δ^+, Δ^0, and Δ^-, may be obtained using the isospin-lowering operator τ_- on $|\Delta^{++}\rangle$, in the same way as we have done earlier in obtaining the pion wave functions. With each application of the τ_--operator, one of the u-quarks is changed into a d-quark. This gives us the correct isospin structure of all four members of the Δ. However, unlike the pion case where two nonidentical fermions are involved, a quark and an antiquark, we are dealing here with three identical particles. In addition to isospin coupling, we must also ensure proper symmetry between the quarks under a permutation between any two of them.

When two identical particles are said to be in a symmetrical state under an interchange, we mean that the wave function remains the same when we permute the particle labels 1 and 2. Consider as an example the following symmetrical wave function of two fermions:

$$\Psi_S(1,2) = \frac{1}{\sqrt{2}}\{\xi(1)\zeta(2) + \zeta(1)\xi(2)\} \tag{2-40}$$

where ξ and ζ are single-particle wave functions and $1/\sqrt{2}$ is the normalization factor for the case where the two single-particle wave functions are different from each other. Under a permutation between 1 and 2, the two particles exchange the single-particle states they occupy. Let us denote this operation by operator P_{12}. It is obvious that, for the wave function $\Psi_S(1,2)$ defined above, we have

$$P_{12}\Psi_S(1,2) = \Psi_S(2,1) = \frac{1}{\sqrt{2}}\{\xi(2)\zeta(1) + \zeta(2)\xi(1)\} = \Psi_S(1,2)$$

Similarly an antisymmetric two-particle wave function $\Psi_A(1,2)$ may be written as

$$\Psi_A(1,2) = \frac{1}{\sqrt{2}}\{\xi(1)\zeta(2) - \zeta(1)\xi(2)\} \tag{2-41}$$

By inspection, we see that

$$P_{12}\Psi_A(1,2) = \Psi_A(2,1) = \frac{1}{\sqrt{2}}\{\xi(2)\zeta(1) - \zeta(2)\xi(1)\} = -\Psi_A(1,2)$$

For $\zeta = \xi$, the symmetric wave function reduces to $\xi(1)\xi(2)$ (with appropriate change in the normalization) and the antisymmetric wave function vanishes, as required by the Pauli exclusion principle.

For Δ^{++} we have the simple situation that all three quarks have the same flavor. The wave function is the symmetric product of three u-quarks,

$$|\Delta^{++}\rangle = |u(1)\rangle|u(2)\rangle|u(3)\rangle \tag{2-42}$$

We shall now see how to obtain the wave function of Δ^+ using the isospin-lowering operator. When τ_- is applied to $|\Delta^{++}\rangle$, we have a choice of changing any one of the three u-quarks into a d-quark. Since there is no way to make a distinction between them, the wave function of Δ^+ must be a linear combination of all three possibilities with equal weight. The normalized and symmetrized (since the color degree of freedom is outside our considerations here) wave function for Δ^+ is

$$|\Delta^+\rangle = \frac{1}{\sqrt{3}}\Big\{|d(1)\rangle|u(2)\rangle|u(3)\rangle + |u(1)\rangle|d(2)\rangle|u(3)\rangle + |u(1)\rangle|u(2)\rangle|d(3)\rangle\Big\}$$

To simplify the notation, we shall write the same equation in a shorthand form,

$$|\Delta^+\rangle = \frac{1}{\sqrt{3}}\Big\{|duu\rangle + |udu\rangle + |uud\rangle\Big\}$$

where it is implied that the first symbol in each term is for quark number 1, the second one for number 2, and the third one for number 3. In cases where we wish only to indicate the quark content of a hadron without displaying the permutation symmetry explicitly, the notation can be shortened further to (uud), for example, as done in Fig. 2-6.

Following this rule, the wave functions of Δ^0 and Δ^- may be written in the following manner:

$$\begin{aligned} |\Delta^0\rangle &= \frac{1}{\sqrt{3}}\Big\{|ddu\rangle + |dud\rangle + |udd\rangle\Big\} \\ |\Delta^-\rangle &= |ddd\rangle \end{aligned}$$

They are obtained by applying the $\boldsymbol{\tau}_-$-operator to the wave function of Δ^+, once for $|\,\Delta^0\,\rangle$ and twice for $|\,\Delta^-\,\rangle$. Alternatively, we can start from the only possibility to construct the wave function for $|\,\Delta^-\,\rangle$, the $t_0 = -\frac{3}{2}$ member of the isospin quartet, as we have done earlier for Δ^{++}, and apply the isospin-raising operator to produce $|\,\Delta^0\,\rangle$.

The wave functions of the three strangeness $\mathcal{S} = -1$ baryons in the decuplet may be obtained by starting with $|\,\Sigma^{*+}\,\rangle$, the $t_0 = 1$ member of the isospin triplet. We can use $|\,\Delta^{++}\,\rangle$ given in Eq. (2-42) as the starting point and replace one of the three u-quarks with an s-quark. This is similar to the way we obtained $|\,\Delta^+\,\rangle$ from $|\,\Delta^{++}\,\rangle$ above by replacing a u-quark with a d-quark. Here, instead of isospin, we are lowering the strangeness by replacing a d-quark with an s-quark. Again, from the symmetry requirement, the normalized wave function is

$$|\Sigma^{*+}\rangle = \frac{1}{\sqrt{3}}\Big\{|\,suu\,\rangle + |\,usu\,\rangle + |\,uus\,\rangle\Big\} \qquad (2\text{-}43)$$

Next, we apply $\boldsymbol{\tau}_-$ operator on $|\,\Sigma^{*+}\,\rangle$ to obtain $|\,\Sigma^{*0}\,\rangle$ and, thence, $|\,\Sigma^{*-}\,\rangle$. Since the s-quark is an isospin zero particle, it vanishes when acted upon by the $\boldsymbol{\tau}_-$- (or $\boldsymbol{\tau}_+$-) operator. The only effect of the isospin-lowering operation is to change one of the u-quarks to a d-quark. As a result, we obtain

$$\begin{aligned} |\Sigma^{*0}\rangle &= \frac{1}{\sqrt{6}}\Big\{|dus\rangle + |uds\rangle + |dsu\rangle + |usd\rangle + |sdu\rangle + |sud\rangle\Big\} \\ |\Sigma^{*-}\rangle &= \frac{1}{\sqrt{3}}\Big\{|dds\rangle + |dsd\rangle + |sdd\rangle\Big\} \end{aligned}$$

This completes the wave functions for the three $\mathcal{S} = -1$ members.

It is trivial to obtain the wave functions for the strangeness $\mathcal{S} = -2$ members of the decuplet, since now only one of the three quarks carries a nonzero isospin and, as a result, only an isospin doublet can be constructed. Their wave functions are given by the following expressions:

$$\begin{aligned} |\Xi^{*+}\rangle &= \frac{1}{\sqrt{3}}\Big\{|uss\rangle + |sus\rangle + |ssu\rangle\Big\} \\ |\Xi^{*-}\rangle &= \frac{1}{\sqrt{3}}\Big\{|dss\rangle + |sds\rangle + |ssd\rangle\Big\} \end{aligned}$$

For $\mathcal{S} = -3$ there is only one possibility,

$$|\Omega^-\rangle = |\, sss\,\rangle$$

Although it is an isoscalar particle, and consequently $t_0 = 0$, it is a charged particle. This can be seen from Eq. (2-37). Since the baryon number is $A = 1$ and strangeness $\mathcal{S} = -3$, the charge number of the particle is -1. The same result can also be obtained from the fact that each s-quark carries a charge $-\frac{1}{3}e$. The particle therefore carries one unit of negative charge and, hence, the negative sign in the superscript.

Baryon singlet. A state of three quarks completely antisymmetric in flavor is also simple to construct. The quark content in this case must be uds, one of each flavor. However, there are $3! = 6$ possible choices: three choices in arranging which one of the three quarks has flavor label u and two choices in arranging which one of the remaining two quarks has flavor label d. (The last one takes on label s.) There is no reason to favor any one of the six possibilities and a linear combination of all six is required. The particular choice, however, must be antisymmetric with respect to a permutation between any two quarks in order to satisfy the requirement of being a singlet state. We can arrive at the correct linear combination by starting from any of the six terms, for instance, $u(1)d(2)s(3)$. To this, we add terms generated from it by applying all the possible linearly independent permutations among the three indices 1, 2, and 3. For the three odd permutations P_{12}, P_{23}, and P_{31} producing the arrangements (dus), (usd), and (sdu), we must take them with the negative sign in order to satisfy the requirement of being symmetric. Similarly, the two even permutations $P_{12}P_{23}$ and $P_{31}P_{23}$ that generate the arrangements (dsu) and (sud) must be taken with the positive sign. The normalized singlet quark wave function is then

$$|\Lambda_1\rangle = \frac{1}{\sqrt{6}}\Big\{|uds\rangle + |dsu\rangle + |sud\rangle - |dus\rangle - |usd\rangle - |sdu\rangle\Big\}$$

Except for an overall sign, this is the only unique way to construct the required antisymmetric linear combination. Since there is no other way to construct a wave function with the same symmetry, $|\,\Lambda_1\,\rangle$ forms an irreducible SU_3(flavor) representation by itself.

Let us examine the symmetry of the isospin part of the wave function. Since the s-quark is an isoscalar quantity, the isospin of the wave function is determined by u- and d-quarks. To illustrate this point, we can rewrite the wave function in the following form by putting the s-quark always at the end:

$$\begin{aligned}|\Lambda_1\rangle \;=\; \frac{1}{\sqrt{6}}\Big\{&\big(|\,u(1)\,\rangle|\,d(2)\,\rangle - |\,d(1)\,\rangle|\,u(2)\,\rangle\big)|\,s(3)\,\rangle \\ &+\big(|\,u(2)\,\rangle|\,d(3)\,\rangle - |\,d(2)\,\rangle|\,u(3)\,\rangle\big)|\,s(1)\,\rangle \\ &+\big(|\,u(3)\,\rangle|\,d(1)\,\rangle - |\,d(3)\,\rangle|\,u(1)\,\rangle\big)|\,s(2)\,\rangle\Big\}\end{aligned}$$

We can recognize that each one of the linear combinations in u and d has $t = 0$, as it is antisymmetric in the isospin-carrying parts. The singlet SU_3(flavor) representation therefore describes an isoscalar particle. Since the isospin is completely determined by the symmetry in flavor of the quark wave function, it does not constitute an independent degree of freedom in fixing the wave function.

So far we have not explicitly put in the intrinsic spin part of the wave function. An example will be given later for the nucleon wave function. Here, we shall simply state the result that $J^\pi = \frac{1}{2}^+$ for the Λ_1-baryon.

Since the SU_3(flavor) symmetry is not an exact one, the observed strangeness $\mathcal{S} = -1$, isoscalar, $J^\pi = \frac{1}{2}^+$ particle Λ is a mixture of $|\Lambda_1\rangle$ and $|\Lambda_8\rangle$, the latter being a member of the $J^\pi = \frac{1}{2}^+$ baryon octet. This is similar to the admixture in pseudoscalar and in vector meson wave functions we have seen earlier.

Baryon octet. The remaining 16 members of the 27 possible baryons constructed from u-, d-, and s-quarks have mixed symmetry in flavor. They may be classified as two octets distinguished by their symmetries under a simultaneous interchange of both flavor and spin. We shall be interested only in the lower energy octet, as it contains protons and neutrons as members. The waves function for each member in this group is symmetric under the combined exchange of flavor and intrinsic spin, as the three quarks must be antisymmetric in color. The members of the octet together with their quark contents are shown in Fig. 2-7.

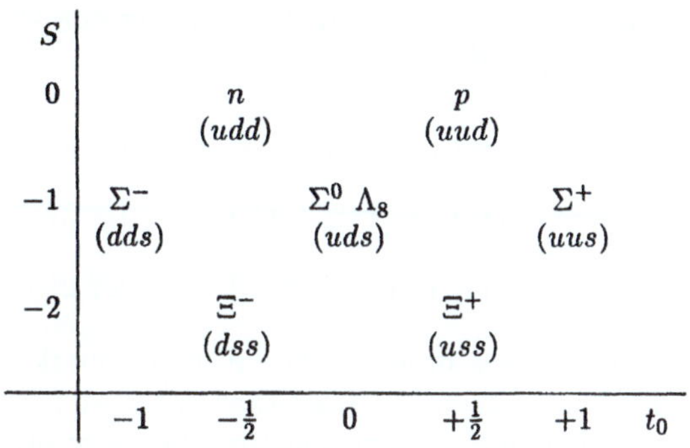

Figure 2-7: $J^\pi = \frac{1}{2}^+$ baryon octet.

Let us construct the proton wave function as an example. Since the intrinsic spin and parity of a proton are $\frac{1}{2}^+$, and the orbital angular momentum is 0, we can start by coupling the intrinsic spins of the three quarks to the value $\frac{1}{2}$. There are several ways to achieve this, and we shall take the simplest one by coupling the first two quarks to spin 0 and then couple the third one with spin up to form a system with $(S, S_0) = (\frac{1}{2}, +\frac{1}{2})$,

$$|\tfrac{1}{2}, +\tfrac{1}{2}\rangle = \frac{1}{\sqrt{2}}\Big(|q(1)\uparrow\rangle|q(2)\downarrow\rangle - |q(1)\downarrow\rangle|q(2)\uparrow\rangle\Big)|q(3)\uparrow\rangle \tag{2-44}$$

Here, the up-arrow symbol represents a quark with intrinsic spin up $(+\frac{1}{2})$ and the down-arrow symbol a quark with spin down $(-\frac{1}{2})$. The assignment of flavor to each of the quarks will be made later. A second possibility to form an $(S, S_0) = (\frac{1}{2}, +\frac{1}{2})$ system is to couple the first two quarks to spin 1 instead of 0, as we have done above. This choice is complicated by the fact that the final system is a mixture of total spin $\frac{3}{2}$ and $\frac{1}{2}$. To project out the desired $S = \frac{1}{2}$ part, a linear combination must be taken of the two possibilities, $\{(q_1q_2)_{1,1}(q_3)_{1/2,-1/2}\}$ and $\{(q_1q_2)_{1,0}(q_3)_{1/2,+1/2}\}$, where the first of the two subscripts indicates S and the second the value of its third component.

The combined symmetry of the spin and flavor parts of the wave function may be determined after assigning a flavor to each one of the quarks involved, subject to the condition that, for a proton, each term must consist of two u-quarks and one d-quark. Let us start by giving the first two quarks different flavors. Equation (2-44) becomes

$$|\tfrac{1}{2},+\tfrac{1}{2}\rangle = \frac{1}{\sqrt{2}}\Big(|u(1)\uparrow\rangle|d(2)\downarrow\rangle - |u(1)\downarrow\rangle|d(2)\uparrow\rangle\Big)|u(3)\uparrow\rangle$$

The combination of spin and flavor may be symmetrized in two stages. First we shall carry out the process only for the first two quarks and obtain

$$\begin{aligned}|\tfrac{1}{2},+\tfrac{1}{2}\rangle &= \tfrac{1}{2}\Big(|u(1)\uparrow\rangle|d(2)\downarrow\rangle - |u(1)\downarrow\rangle|d(2)\uparrow\rangle \\ &\quad +|d(1)\downarrow\rangle|u(2)\uparrow\rangle - |d(1)\uparrow\rangle|u(2)\downarrow\rangle\Big)|u(3)\uparrow\rangle\end{aligned} \tag{2-45}$$

Next, we shall generate the others by applying permutations P_{31} and P_{32} on each of the four terms in Eq. (2-45). This gives us a total of 12 terms. On grouping identical terms together, we obtain the quark wave function for a proton with spin orientations of all the quarks indicated explicitly,

$$\begin{aligned}|p\rangle &= \frac{1}{\sqrt{18}}\Big\{2\Big(|u\uparrow u\uparrow d\downarrow\rangle + |u\uparrow d\downarrow u\uparrow\rangle + |d\downarrow u\uparrow u\uparrow\rangle\Big) \\ &\quad -\Big(|u\uparrow u\downarrow d\uparrow\rangle + |u\uparrow d\uparrow u\downarrow\rangle + |d\uparrow u\uparrow u\downarrow\rangle \\ &\quad +|u\downarrow u\uparrow d\uparrow\rangle + |u\downarrow d\uparrow u\uparrow\rangle + |d\uparrow u\downarrow u\uparrow\rangle\Big)\Big\}\end{aligned} \tag{2-46}$$

To simplify the notation, we have dropped the labels for quark number and rely on the order each quark appears instead. The fact that this wave function is symmetrical under a simultaneous interchange of flavor and spin between any two quarks can be established by inspection.

The neutron wave function can be written down from that for a proton by simply substituting all the u-quarks by d-quarks and vice versa. Similarly, the wave functions of the strangeness $\mathcal{S} < 0$ members of the octet can be built from that of the proton, in the same way as we have done for members of the decuplet by starting from $|\Delta^{++}\rangle$. These are left as exercises.

2-8 Magnetic Dipole Moment of the Baryon Octet

The hadron wave functions obtained in the previous section are based solely on symmetry considerations. Since they are not the eigenfunctions of a realistic Hamiltonian involving interaction between quarks, we cannot expect them to be able to describe any of the dynamic properties with good accuracy. Nevertheless, calculations for some simple quantities can be carried out, and the results will show whether they are useful as zero-order approximations to the wave functions. Besides quantities used already in obtaining the wave functions, such as charge number, spin, isospin, and strangeness, the magnetic dipole moment is one we can use to test our model. We shall only deal with members of the baryon octet given in Fig. 2-7, as more are known about them.

The magnetic dipole moment of a baryon comes from two sources, the intrinsic dipole moments of the constituent quarks and the orbital motion of the quarks. For the baryon octet of interest here, all the members have $J^\pi = \frac{1}{2}^+$. In the simple model adopted in the previous section for our discussion, the three quarks are symmetric in the spatial parts of their wave functions, with relative motion $\boldsymbol{\ell} = 0$. As a result, contributions from quark orbital motion may be ignored.

Quark magnetic dipole moments. Associated with the intrinsic spin $\boldsymbol{s}$ of a particle, there is an intrinsic magnetic dipole $\boldsymbol{\mu}$. The ratio between the two quantities is a constant g, known as the gyromagnetic ratio. In terms of operators, we have the relation

$$\boldsymbol{\mu} = g\boldsymbol{s}\mu_{\rm D} \tag{2-47}$$

For quarks, it is convenient to measure the dipole moment in terms of

$$\mu_{\rm D} = \frac{q\hbar[c]}{2m_q c}$$

This is identical in form to the definition for nuclear magneton μ_N except, here, q is the quark charge $e/3$ and m_q, the quark mass, is used instead of nucleon mass (and the factor $[c]$ in the numerator converts the result from cgs to SI units).

For Dirac particles, i.e., particles devoid of internal structure, we have $g = 2$ for those with intrinsic spin $s = \frac{1}{2}$. In practice, no particle is observed to be completely without some "structure" associated with it. For example, electrons and muons emit and absorb virtual photons. The contributions from these virtual processes give rise to an "anomalous" magnetic dipole moment such that the observed value of g is $2 \times 1.001159652193(10)$ for an electron and $2 \times 1.001165923(8)$ for a muon. The small corrections to the simple Dirac particle values for the charged leptons are well understood and can be calculated to very high precision in quantum electrodynamics.

Since a quark is an elementary particle, we can take it as a simple Dirac particle to start with. The relation between intrinsic magnetic moment and spin is given by Eq. (2-47) with $g = 2$. However, we do not know the quark masses; it is therefore not possible to deduce the values of $\boldsymbol{\mu}$ in any simple way. (For this reason there is no point to consider corrections to g due to anomalous magnetic dipole moment either.) However, if we assume that the masses of u- and d-quarks are equal, the ratio between their magnetic dipole moments is given by the ratio of their charges. This gives us the result

$$\mu_u = -2\mu_d \tag{2-48}$$

As we shall soon see, this is useful in getting an idea for the values of the intrinsic magnetic dipole moments of the u- and d-quarks.

Nucleons. The contribution to the magnetic dipole moment of a baryon from quark intrinsic dipole moments depends also on the orientation of the spin of each quark. Since the quark orbital motion does not enter here, the magnetic dipole moment is given by the number of quarks of each flavor in each one of the two possible spin orientations. For a proton, we can count the numbers of $u\uparrow$, $u\downarrow$, $d\uparrow$, and $d\downarrow$ explicitly using the wave function given earlier in Eq. (2-46). The results are simply the sums of the squares of the coefficients in the wave function for each one of the four possible combinations of flavor and spin orientations: $\frac{5}{3}$ for the number of u-quarks with spin up, $\frac{1}{3}$ for the number of u-quarks with spin down, $\frac{1}{3}$ for the number of d-quarks with spin up, and $\frac{2}{3}$ for the number of d-quarks with spin down. The net contribution from u-quarks to the proton magnetic dipole moment is then $\frac{5}{3} - \frac{1}{3} = \frac{4}{3}$, and that from d-quarks is $\frac{1}{3} - \frac{2}{3} = -\frac{1}{3}$. In this simple model, the final result for a proton is then

$$\mu_p = \tfrac{4}{3}\mu_u - \tfrac{1}{3}\mu_d$$

For a neutron, we can again interchange the roles of u- and d-quarks in the expression above and obtain the result,

$$\mu_n = \tfrac{4}{3}\mu_d - \tfrac{1}{3}\mu_u$$

If we now make the assumption that the masses of the u- and d-quarks involved are equal and their ratio of magnetic dipole moments is given by Eq. (2-48), we obtain the ratio between those of a neutron and a proton,

$$\frac{\mu_n}{\mu_p} = \frac{\frac{4}{3}\mu_d - \frac{1}{3}\mu_u}{\frac{4}{3}\mu_u - \frac{1}{3}\mu_d} = \frac{2\mu_d}{-3\mu_d} = -\frac{2}{3}$$

This is in good agreement with the observed value of $-1.913/2.793 = -0.685$.

Baryons with $S>0$. For the other six members of the octet, there is at least one s-quark involved. As a result, we need to include the contributions from the intrinsic magnetic dipole moment of strange quarks. Since the s-quark is known to be more massive than the u- and d-quarks, we cannot easily relate its intrinsic magnetic dipole moment to those of the u- or d-quarks in the same way as we have done in Eq. (2-48) between u- and d-quarks. On the other hand, eight magnetic dipole moments are known for the members of the octet and all of them are given in terms of the intrinsic magnetic dipole moments of the three quarks in this simple model. As a result, a least-squares-fitting procedure may be used to deduce the three unknown quark values from these eight pieces of data. To carry out this procedure, we must first express the baryon magnetic dipole moments in terms of those for the three quarks, as we have done above for the nucleons.

Although we do not have the quark wave functions written out in detail for the $S < 0$ members of the octet as we have done in Eq. (2-46) for the proton, we can nevertheless count the number of quarks of each flavor with spin up and that with spin down, starting with the quark content of each baryon given in Fig. 2-7. This is particularly simple for those strange baryons involving only two different flavors, i.e., those made of s- and u-quarks only or s- and d-quarks only. For example, the Σ^+-baryon is made of two u-quarks and one s-quark. Compared with the proton, (uud), we see that the only difference between the quark structure of these two baryons is that, in the place of a d-quark in proton, we have an s-quark in Σ^+. Since all members of the octet have the same combined symmetry for spin and flavor, the proton and Σ^+

must have very similar quark wave functions, except for the replacement of d with s. Hence, the magnetic dipole moment of the Σ^+-baryon is given by

$$\mu_{\Sigma^+} = \tfrac{4}{3}\mu_u - \tfrac{1}{3}\mu_s$$

Similarly, the quark content of Σ^- is (dds). Comparing it with a neutron, we find that the expression for the magnetic dipole moment of a Σ^--baryon is the same except that, in the place of u-quarks in a neutron, we have the contributions from s-quarks. This gives us the result

$$\mu_{\Sigma^-} = \tfrac{4}{3}\mu_d - \tfrac{1}{3}\mu_s$$

Using similar methods, the expressions for the magnetic dipole moments of the two $S = -2$ members of the octet, $\Xi^-(dss)$ and $\Xi^+(uss)$, can be obtained and the results are given in Table 2-4.

Table 2-4: Magnetic dipole moment of baryon octet.

Octet member	Quark content			Best fit	Observed
	u	d	s	μ_N	μ_N
p	$\frac{4}{3}$	$-\frac{1}{3}$	0	2.793	2.792847386(63)
n	$-\frac{1}{3}$	$\frac{4}{3}$	0	−1.913	−1.91304275(45)
Λ	0	0	1	−0.613	−0.613(4)
Σ^+	$\frac{4}{3}$	0	$-\frac{1}{3}$	2.674	2.458(10)
Σ^-	0	$\frac{4}{3}$	$-\frac{1}{3}$	−1.092	−1.160(25)
Ξ^0	$-\frac{1}{3}$	0	$\frac{4}{3}$	−1.435	−1.250(14)
Ξ^-	0	$-\frac{1}{3}$	$\frac{4}{3}$	−0.493	−0.6507(25)
$\Sigma^0 \to \Lambda$	$-\sqrt{\frac{1}{3}}$	$\sqrt{\frac{1}{3}}$	0	−1.630	−1.61(8)
Ω^-			3	−1.839	−2.02(5)
u	1			1.852	
d		1		−0.972	
s			1	−0.613	

For the two remaining members of the octet, Σ^0 and Λ_8, the quark contents are (uds) for both. Since three different flavors are involved, a slightly different approach is required. For both hadrons, we can make use of their isospin difference to derive their wave functions. For this purpose we can ignore the s-quark for the moment, as it is an isoscalar particle not involved in any isospin considerations.

Let us start with Λ_8. Since it is an isospin singlet, we have $t = 0$. In the discussions given earlier for the quark wave functions of π^0- and η_0-mesons, we have seen that a $t_0 = 0$ system, consisting of a u- and a d-quark, is a mixture of isospin 1 and 0. To project out the isospin $t = 0$ part, we need an antisymmetric linear combination of the two possible arrangements of u and d,

$$|(t, t_0) = (0, 0)\rangle = \frac{1}{\sqrt{2}}\{|u(1)d(2)\rangle - |d(1)u(2)\rangle\}$$

The spins of these two quarks cannot be both up, as such an arrangement will be antisymmetric under the simultaneous exchange of spin and flavor. The only possibility is therefore

$$|(t,t_0)=(0,0);(s,m_s)=(0,0)\rangle = \tfrac{1}{2}\Big\{\big(|u(1)\!\uparrow d(2)\!\downarrow\rangle - |d(1)\!\uparrow u(2)\!\downarrow\rangle\big) + \big(|d(1)\!\downarrow u(2)\!\uparrow\rangle - |u(1)\!\downarrow d(2)\!\uparrow\rangle\big)\Big\}$$

Note that the total spin of the two quarks is also zero as a result of the symmetry requirement.

We can now couple the s-quark to the product and form a spin-$\frac{1}{2}$ system of three quarks,

$$|(t,t_0)=(0,0);(s,m_s)=(\tfrac{1}{2},+\tfrac{1}{2})\rangle = \tfrac{1}{2}\Big\{\big(|u(1)\!\uparrow d(2)\!\downarrow\rangle - |d(1)\!\uparrow u(2)\!\downarrow\rangle\big) + \big(|d(1)\!\downarrow u(2)\!\uparrow\rangle - |u(1)\!\downarrow d(2)\!\uparrow\rangle\big)\Big\}|s(3)\!\uparrow\rangle \tag{2-49}$$

The wave function is not properly antisymmetrized with respect to the third quark. However, for the purpose of calculating the magnetic dipole moment, this is not necessary; all we need to do is to count the number of quarks of each flavor with spin up and the corresponding number with spin down, and this is independent of the symmetrization among the three quarks beyond those given in Eq. (2-49) above. Furthermore, it is also evident from the structure of the wave function that the net contributions from both u- and d-quarks are zero, as there are equal numbers of each with spin pointing up as there are with spin pointing down. As a result, we obtain

$$\mu_\Lambda = \mu_s$$

for the magnetic dipole moment of Λ_8. Because of the crudeness of the model used here, there is no point in considering any SU_3(flavor) symmetry-breaking effects and the resulting difference between Λ_8 and the observed Λ-baryon.

The isospin of Σ^0 is unity, as it is a member of a triplet, Σ^+, Σ^0, and Σ^-. The quark wave function is somewhat more complicated than what we have obtained for Λ_8, as the u- and d-quarks must now be coupled to a spin-1 state. We shall leave the calculation of μ_{Σ^0} in terms of quark magnetic dipole moments as an exercise, since the value of μ_{Σ^0} is not known and we cannot make use of it in our calculation. On the other hand, the decay of Σ^0 through the reaction

$$\Sigma^0 \to \Lambda + \gamma$$

is similar to a magnetic dipole transition (see §5-3) and, as a result, the transition probability is proportional to $|\mu_{\Sigma^0\to\Lambda}|^2$. The matrix element of the magnetic dipole transition operator $\boldsymbol{O}(M1)$ has the value

$$\mu_{\Sigma^0\to\Lambda} = \langle\Sigma^0|\boldsymbol{O}(M1)|\Lambda\rangle = -\frac{1}{\sqrt{3}}(\mu_u - \mu_d)$$

Since the experiment measures the square of the transition matrix element, only the absolute value is determined [116]. The sign is known from independent sources to be negative. As a result the value of $\mu_{\Sigma^0\to\Lambda}$ may be used as a piece of information for our calculation.

Table 2-4 summarizes the contribution from each one of the three quarks to the magnetic dipole moments for the members of the baryon octet together with $\mu_{\Sigma^0\to\Lambda}$. The observed values are listed in the last column in units of nuclear magnetons, $\mu_N = e\hbar/(2M_pc)$ in cgs units or $e\hbar/(2M_p)$ in SI units. The values of μ_u, μ_d, and μ_s are deduced by fitting them to the eight measured dipole moments. Since the accuracies that can be achieved for the measured values of the various baryons differ by a large margin, the eight pieces of data that went into the calculation as input have been weighed inversely according to their experimental uncertainties, given in parentheses in the table. The results of the calculation are shown under the column labeled "Best fit." The calculated values agree quite well with observation, especially in view of the crude model used. Except for Ξ^-, the discrepancies are less than $0.2\mu_N$. This close agreement has two implications. The first is that the model used to deduce the moments in terms of those of the three quarks is a reasonable one, otherwise much larger differences would have resulted. The second is that the values deduced for the quark magnetic dipole moments are physically meaningful.

We expect several corrections to our simple analysis. One of the assumptions we have made is that the wave functions have only $\ell = 0$ components. This is true if orbital angular momentum is conserved by the interaction between quarks. As we shall see in the next chapter, in an analogous discussion on the deuteron ground state, the orbital angular momentum is not necessarily a constant of motion. Consequently, it is unreasonable for us to expect that the ground states of the members of the baryon octet be purely $\ell = 0$. In general, some configuration mixing from $\ell > 0$ terms is present, and this may be the most important correction to our simple model. A more detailed discussion can be found in a status report by Brekke and Rosner [36].

The values of the magnetic dipole moments of the three lighter quarks obtained from the least-squares-fitting procedure are given at the bottom of the table. Although there are no observed values to compare with, we can, nevertheless, get a rough idea whether the results are reasonable. For example, the ratio $\mu_u/\mu_d = -1.91$ is fairly close to the value of -2 obtained earlier by assuming that the masses of u- and d-quarks are identical to each other and both of them have the same gyromagnetic ratios.

2-9 Hadron Mass and Quark-Quark Interaction

Another striking feature in hadron spectroscopy is the systematics in their masses. In Table 2-5, the observed values for some of the low-lying members are given, together with their uncertainties in the last digits in parentheses. First of all, we notice that the masses for the members of the $J^\pi = \frac{1}{2}^+$ baryon octet are well correlated in value with their strangeness quantum numbers. That is, the mass differences between members with the same strangeness are much smaller than those between members of different strangeness. For example, the difference between a proton and a neutron is less than 2 MeV/c^2, whereas the difference between a Λ-baryon ($\mathcal{S} = -1$) and a neutron is around 176 MeV/c^2, and that between a Ξ^0-baryon ($\mathcal{S} = -2$) and a Λ-baryon is around 200 MeV/c^2. The obvious conclusion one can draw from such comparisons is that the rest mass energies of the underlying u- and d-quarks are the same within a few mega-electron-volts and that the value of the s-quark is larger than those of the u- and d-quarks by 100 to 200 MeV. Further support for s-quarks being more massive can be found in the mass differences between members of the baryon decuplet, between members of the pseudoscalar mesons, and between members of the vector mesons. As for the charm and beauty quarks, we have already seen evidence from the masses of J/ψ- and Υ-mesons in §2-6 that they are even more massive. From a comparison of the masses of hadrons made of different quarks, the masses of the various quarks can be deduced. The generally accepted values were given earlier in Table 2-2.

Table 2-5: Low-lying hadron masses in MeV/c^2.

Baryons		Mesons	
$S=0$		Pseudoscalar mesons	
p	938.27231(28)	$\pi^{\pm}$	139.56995(35)
n	939.56563(28)	π^0	134.9764(6)
$S=-1$		$K^{\pm}$	493.677(13)
Λ	1115.684(6)	$K^0, \overline{K}^0$	497.672(31)
Σ^+	1189.37(7)	η	547.45(19)
Σ^0	1192.55(8)	η'	957.77(14)
Σ^-	1197.436(33)	Vector mesons	
$S=-2$		ρ	768.5(6)
Ξ^0	1314.9(6)	K^*	891.59(24)
Ξ^-	1321.32(13)	ω	781.94(12)
		ϕ	1019.413(8)

The small mass differences between hadrons having the same strangeness can come from either electromagnetic effects or a small difference between the masses of u- and d-quarks. However, our present knowledge of the strong interaction is not able to elucidate on this question. In spite of our ignorance, the small difference between the masses of proton and neutron and between $\pi^{\pm}$ and π^0 are important in understanding certain nuclear phenomena, such as isospin symmetry breaking in the nuclear force.

It is worthwhile to emphasize here again that, since quarks have not been observed in isolation outside hadrons, the values deduced from hadron spectra are not their true masses. The observed hadron masses depend on the intrinsic masses of the quarks as well as the binding energy between the quarks. If the binding energies are known, it is a trivial matter to obtain the quark masses from those of hadrons. As we shall see later in the analogous situation of nuclear masses, binding energy calculations require a knowledge of the interaction between the constituents. Even in the nuclear case, it is not easy to obtain high precisions, partly because of our incomplete understanding of the interaction between nucleons and partly because of the inherent difficulties of the many-body problem.

For quarks, the situation is further complicated by several factors. First, the quark-quark interaction is known to be very strong at energies of concern to us here. We have seen an example of this from the mass differences between the pseudoscalar mesons (sum of quark intrinsic spins $S = 0$) and the vector mesons ($S = 1$). For example, the quark contents of the π- and ρ-mesons are the same, but their masses are quite different, with $m_\pi c^2 \approx 140$ MeV and $m_\rho c^2 \approx 767$ MeV, respectively. The large difference must be attributed mainly to the dependence of the interaction on the total intrinsic spin of the quark-antiquark pair. This is quite different from the usual situation in quantum systems where the interesting physics often arises from small parts of the complete interaction. For example, in atomic physics the main contribution to the binding energies of electrons comes from the electrostatic attraction between the nucleus and each one of the electrons. Most of the other properties of an atom, on the other hand, are sensitive mainly to small *perturbations* caused by the interaction between electrons. As a result, a number of perturbative methods have been developed over the years and they are found to be quite successful in handling such problems.

For quarks, the interaction is very strong at low energies where nuclear physics operates and where most of the experimental observations are made. Because of what is generally known as *asymptotic freedom*, the quark-quark interaction is weak only at extremely high energies. As a result, perturbative techniques apply to QCD only at such extremes, far beyond the realm of nuclear physics and low-lying hadron spectroscopy. For the low-energy regions, methods other than perturbative approaches must be applied before we can properly link QCD calculations to observations. We shall see one such example in the form of lattice QCD calculation in §9-3.

Second, there is the question of *confinement*. Again, since quarks are not observed in isolation, their mutual interaction must have a component that grows stronger as the distance of separation between them increases. This is opposite to our experience in the macroscopic world, where interactions, such as gravitational and electromagnetic, grow weaker as the distance of separation between the interacting objects is increased (and the relation is given by the inverse square law). As a result, we need to devise new methods to handle the problem. One way is to impose confinement "artificially" as a boundary condition. In other words, the quarks are considered to be inside a "bag" that prevents them from escaping to the outside. Such a *bag model*, together with its many variants, has made impressive contributions in improving our understanding of the structure of hadrons and in linking the quark-quark interaction with the interaction between nucleons. We shall see a simple application of such a phenomenological model when we come to the question of phase transition from hadronic matter to quark-gluon plasma in §9-3.

Problems

2-1. Show that conservation of energy and momentum requires at least two γ-rays to be emitted in the annihilation of an electron by a positron.

2-2. Show that the nucleon isospin wave functions given in Eqs. (2-15) and (2-16) are the eigenfunctions of the operator

$$\boldsymbol{\tau}^2 = \tau_1^2 + \tau_2^2 + \tau_3^2$$

with eigenvalues 3. Express $\boldsymbol{\tau}^2$ in terms of τ_+, τ_-, and τ_0 and calculate again, in terms of these operators, the expectation values of $\boldsymbol{\tau}^2$ for the isospin wave function of a proton and a neutron.

2-3. Antiprotons are created when a beam of high-energy protons strikes a hydrogen target. In the laboratory system what is the minimum proton kinetic energy required for the reaction to take place?

2-4. Construct the quark wave function of π^- by applying an isospin-lowering operator to the wave function of π^0 given in Eq. (2-34). Use the same technique to construct the quark wave function of Σ^{*0} by applying an isospin-lowering operator to that of Σ^{*+} given in Eq. (2-43).

2-5. The meson η_8 is a neutral, isospin-singlet particle made of a linear combination of quark-antiquark pairs taken from u-, d-, and s-quarks and their antiquarks. Construct the quark wave function of η_8 by requiring it to be normalized and orthogonal to those of π^0 and η_0 given in Eqs. (2-34) and (2-38).

2-6. An electron is moving in a circular orbit. Show that the magnetic dipole moment generated by the orbital motion is given by the relation

$$\boldsymbol{\mu} = -\frac{e\hbar[c]}{2m_e c}\boldsymbol{\ell}$$

where $\boldsymbol{\ell}$ is the angular momentum in units of $\hbar$ and the factor $[c]$ converts the formula from cgs to SI units. Assume that the charge and mass of the electron are distributed uniformly along the orbit and ignore the contributions from the intrinsic magnetic dipole moment.

2-7. The Σ^0-particle is a baryon made of a u-quark, a d-quark, and an s-quark coupled to total intrinsic spin $S = \frac{1}{2}$ and isospin $t = 1$. Assume that the orbital angular momentum $\ell = 0$; show that the magnetic dipole moment of Σ^0 is given in the quark model by

$$\mu_{\Sigma^0} = \tfrac{2}{3}(\mu_u + \mu_d) - \tfrac{1}{3}\mu_s$$

where μ_u, μ_d, and μ_s are, respectively, the intrinsic magnetic dipole moments of the u-, d-, and s-quark.

2-8. Use the quark model to show that the magnetic dipole moments of vector mesons ρ^+ and ρ^- are equal in magnitude but opposite in sign.

2-9. If the magnetic dipole moment of a u-quark is $1.852\mu_N$, of a d-quark, $-0.972\mu_N$, and of an s-quark, $-0.613\mu_N$, what are the values of magnetic dipole moments of their antiquarks? The ρ^+-meson is a vector meson with $J^\pi = 1^-$ and isospin $T = 1$. Calculate the magnetic dipole moment using the values of quarks assuming the orbital angular momentum to be the lowest value possible.

Chapter 3

Nuclear Force and Two-Nucleon Systems

The interaction between two nucleons is one of the central questions in physics and its importance goes beyond the properties of nuclei. In a 1953 *Scientific American* article, Bethe (page 59 of Ref. [31]) estimated that "in the past quarter century physicists have devoted a huge amount of experimentation and mental labor to this problem—probably more man-hours than have been given to any other scientific question in the history of mankind." In the intervening years after Bethe wrote these words, even more effort has been expended on the topic than before and much progress has been made. We now know that nucleons are not elementary particles and their interactions derive from the force acting between quarks that make them up. While quantum chromodynamics gives a fairly good description of the structure of hadrons in terms of quarks, it is far less certain how the interaction between nucleons is quantitatively related to the fundamental quark-quark interaction.

In this chapter, we shall examine the problem from a mostly phenomenological point of view. We shall concentrate on two-nucleon systems and make use of their simplicity to illustrate some of the challenges we face in nuclear studies. First we shall examine the deuteron, the only bound system formed of two nucleons. Far more information is provided by the scattering of one nucleon off another, and we shall see what we can learn about the nucleon-nucleon interaction from such studies.

3-1 The Deuteron

Binding energy. The deuteron is a very unique nucleus in many respects. It is only loosely bound, having a binding energy much less than the average value between a pair of nucleons in all the other stable nuclei. We have seen in Eq. (1-1) that the binding energy E_B of a nucleus is given by the mass difference between the neutral atom and the sum of the masses of free neutrons and protons in the form of hydrogen atoms. For a deuteron, the mass M_d is 1876.1244 MeV/c^2. The binding energy is then the difference between M_d and the sum of those for a neutron, M_n, and a hydrogen atom, M_H:

$$
\begin{array}{rl}
M_n c^2 = & 939.5656 \text{ MeV} \\
+\ M_{\rm H} c^2 = & 938.7833 \text{ MeV} \\
\hline
= & 1878.3489 \text{ MeV} \\
-\ M_d c^2 = & 1876.1244 \text{ MeV} \\
\hline
E_B = & 2.2245 \text{ MeV}
\end{array}
\tag{3-1}
$$

A more precise value, E_B =2.22457312(22) MeV, is obtained from radiative capture of a neutron by hydrogen. In this reaction, represented as $p(n,\gamma)d$, a slow neutron is captured by a hydrogen atom followed by the emission of a γ-ray (see Ref. [78] for details). If the energy of the incident neutron is negligible, the energy of the γ-ray emitted gives the deuteron binding energy. Since it is usually far easier to determine γ-ray energies accurately than measurements of atomic masses, binding energies are often better known than absolute masses.

Partly because of the small binding energy, the deuteron has no excited state; all observations on the deuteron are made on the ground state. The results of the more important measured quantities are listed in Table 3-1. In spite of the small number of independent pieces of data available, we stand to learn a great deal about the two-nucleon system from the deuteron. Furthermore, because of their fundamental importance, many careful and sophisticated measurements have been carried out and the available values represent some of the best that can be obtained for the type of measurement. In this section we shall make use only of spin, parity, and isospin, leaving the study of the magnetic dipole moment and electric quadrupole moment to the next two sections.

Table 3-1: Ground state properties of deuteron.

Ground State Property	Value
Binding energy, E_B	2.22457312(22) MeV
Spin and parity, J^π	1^+
Isospin, T	0
Magnetic dipole moment, μ_d	0.857438230(24) μ_N
Electric quadrupole moment, Q_d	0.28590(30) $e$fm^2
Radius, r_d	1.963(4) fm

Note: Uncertainties in last digits of the measured values are given in parentheses.

Spin and parity. The parity of a state describes the behavior of its wave function under a reflection of the coordinate system through the origin, as shown in §A-1. For the deuteron, it is known that the parity is positive. Let us see what we can learn from this piece of experimental information. For this purpose, it is useful to separate the wave function into a product of three parts: the intrinsic wave function of the proton, the intrinsic wave function of the neutron, and the orbital wave function for the relative motion between the proton and the neutron. Since a proton and a neutron are just two different states of a nucleon, their intrinsic wave functions have the same parity. As a result, the product of their intrinsic wave functions has positive parity, regardless of the parity of the nucleon. This leaves the parity of the deuteron to be determined solely by the relative motion between the two nucleons.

For states with a definite orbital angular momentum $\boldsymbol{L}$, the angular dependence in the wave function is given by spherical harmonics $Y_{LM}(\theta\phi)$. Under an inversion of the coordinate system, spherical harmonics transform according to the relation

$$Y_{LM}(\theta,\phi) \xrightarrow[P]{} Y_{LM}(\pi-\theta,\pi+\phi) = (-1)^L Y_{LM}(\theta,\phi)$$

The parity of $Y_{LM}(\theta,\phi)$ is therefore $(-1)^L$. The fact that the deuteron parity is positive implies that the orbital angular momentum must be even.

The spin of the ground state of deuteron is $J=1$, where $\boldsymbol{J}=\boldsymbol{L}+\boldsymbol{S}$. The possible values of $\boldsymbol{S}$, the sum of the intrinsic spins of the two nucleons, are 0 and 1. We can eliminate $S=0$, as it is impossible to couple it with even values of L to form a $J=1$ state. Furthermore, we can also rule out any L values greater than 2 by the same argument. From the fact that the spin and parity of a deuteron are $J^\pi=1^+$, we find that the only possible values of (L,S) are $(0,1)$ and $(2,1)$. We shall see later that the dominant part of the ground state wave function is the $L=0$ component. However, the small $L=2$ admixture is important to understand certain properties of deuteron and nuclear force.

Isospin. Through symmetry arguments, we can also deduce the isospin T for the deuteron. Since the projection of isospin on the quantization axis is $t_0=+\frac{1}{2}$ for a proton and $-\frac{1}{2}$ for a neutron, the deuteron is a state with the sum of the isospin projections $T_0=0$. The isospin of such a system of two nucleons can be coupled together to either $T=0$ or $T=1$, as we saw earlier in §2-7. For a light nucleus such as the deuteron, the isospin is expected to be a good quantum number and the ground state of the deuteron can take on only one of these two values.

If, again, we regard a proton and a neutron as two different isospin states of a nucleon, a deuteron may be treated as made up of two identical particles. The total wave function of such a system is required to be antisymmetric under a permutation of the indices of the two (Fermi-Dirac) particles,

$$P_{12}\Psi(1,2) = \Psi(2,1) = -\Psi(1,2) \tag{3-2}$$

The wave function $\Psi(1,2)$ may be decomposed into a product of spatial, spin, and isospin parts. For the spatial part, a permutation of the indices means that

$$\boldsymbol{r} \equiv \boldsymbol{r}_1 - \boldsymbol{r}_2 \xrightarrow[P_{12}]{} -\boldsymbol{r}$$

In a spherical polar coordinate system, this corresponds to the transformation

$$(r,\theta,\phi) \xrightarrow[P_{12}]{} (r,\pi-\theta,\phi+\pi)$$

Since the radial coordinate r is unchanged by the transformation, the symmetry of the spatial wave function is given by the angular dependence and, consequently, that of the spherical harmonics. The transformation is, then, mathematically the same as that under a parity change. For $L = 0, 2$, the spatial wave function is symmetric under permutation as a result.

It is also easy to see that the intrinsic spin part of the deuteron wave function is even in the $S = 1$ state. Consider the state with $M_S = 1$ among the triplet of possible states of $M_S = 0, \pm 1$ for $S = 1$. The intrinsic spin wave function may be written as the product of those for the two nucleons,

$$|S{=}1, M_S{=}1\rangle = |s{=}\tfrac{1}{2}, m_s{=}\tfrac{1}{2}\rangle_1 |s{=}\tfrac{1}{2}, m_s{=}\tfrac{1}{2}\rangle_2 \tag{3-3}$$

The function on the right-hand side of the equation is obviously even under a permutation of the indices of the two nucleons indicated by the subscripts. Since there is no other way to construct an $(S, M_S) = (1, 1)$ state, the function given by Eq. (3-3) must be the intrinsic spin wave function for the state. The wave functions of the other two $S = 1$ states, with $M_S = 0, -1$, may be generated from the $M_S = 1$ state using an angular momentum lowering operator. Since the operator is symmetric with respect to the two nucleons, the resulting wave functions retain the symmetry of the $(S, M_S) = (1, 1)$ state we started with. Consequently, they are also symmetric under a permutation of the two nucleons. From this, we establish that the intrinsic spin part of the deuteron wave function is even under a permutation of the two nucleons.

With both spatial and spin parts of the wave function symmetric, the isospin part must be antisymmetric in order to maintain the product of all three to be antisymmetric under a permutation of the two nucleons, as required by the Pauli exclusion principle. The algebras of intrinsic spin and isospin are the same. From the discussion above on the intrinsic spin wave function, we can conclude that the $T = 1$ state of two nucleons is symmetric under permutation. On the other hand, the antisymmetric linear combination

$$|T{=}0, T_0{=}0\rangle = \frac{1}{\sqrt{2}}\Big\{|t{=}\tfrac{1}{2}, t_0{=}+\tfrac{1}{2}\rangle_1 |t{=}\tfrac{1}{2}, t_0{=}-\tfrac{1}{2}\rangle_2 - |t{=}\tfrac{1}{2}, t_0{=}+\tfrac{1}{2}\rangle_2 |t{=}\tfrac{1}{2}, t_0{=}-\tfrac{1}{2}\rangle_1\Big\} \tag{3-4}$$

describes a $T = 0$ state. This can be seen either by examining the explicit values of the Clebsch-Gordan coefficients involved or by the fact that the right-hand side of Eq. (3-4) vanishes when either an isospin-raising or an isospin-lowering operator is applied to it. The requirement that the isospin part of the two-nucleon system is antisymmetric then implies that the deuteron ground state is in a $T = 0$ state.

We can also arrive at the same conclusion by a different set of arguments. If the ground state of the deuteron were $T = 1$, we expect to find similar bound states in the other two $T = 1$ two-nucleon systems, the two-proton system ($T_0 = 1$) and the two-neutron system ($T_0 = -1$). However, no such bound states have been observed. We can perhaps eliminate the possibility of a two-proton bound state on the grounds that Coulomb repulsion between two protons is of the order of 1 MeV at the distance of the deuteron radius. Since this value is a large fraction of the deuteron binding energy, it is difficult to expect that a bound state can be formed of two protons. This limitation, however, does not apply to a system of two neutrons. Since no bound state is observed for two neutrons either, we come to the conclusion that it is not possible to have a $T = 1$ bound state for two nucleons in general. The neutron-proton system can be either an isoscalar ($T = 0$) or an isovector ($T = 1$) state. Since there does not seem to be a bound state for the $T = 1$ system, the deuteron ground state must have isospin $T = 0$. We may also conclude from the same argument that there is an isospin dependence in the force between a pair of nucleons that is attractive only in the $T = 0$ state.

In summary, we have established using symmetry considerations that the deuteron ground state has $S = 1$ and $T = 0$. There remain, however, two possibilities, $L = 0$ and $L = 2$, for the spatial part of the wave function. In spectroscopic notation, the $L = 0$, $S = 1$ state is represented as $^3S_1$ (triplet-S state) and the $L = 2$, $S = 1$ as $^3D_1$ (triplet-D state). If L and S are good quantum numbers, i.e., if the nuclear Hamiltonian H commutes with both $\boldsymbol{L}^2$ and $\boldsymbol{S}^2$, the deuteron ground state would have to be in either one of these two states. There is, however, no fundamental reason to expect that this has to be true. In fact, we shall soon see that there is clear evidence that both the $^3S_1$- and the $^3D_1$-components must be present in deuteron. This, in turn, leads to the conclusion that the nuclear force mixes different L-components in an eigenstate.

3-2 Deuteron Magnetic Dipole Moment

Magnetic dipole operator. The magnetic dipole moment of a nucleus arises from a combination of two different sources. First, each nucleon has an intrinsic magnetic dipole moment coming from the intrinsic spin and the orbital motion of quarks (see §2-8). Second, since each proton carries a net positive charge, its orbital motion constitutes an electric current loop. If, for simplicity, we assume that the proton charge is distributed evenly along its orbit, we can use classical electromagnetic theory to obtain its contribution to the magnetic dipole moment of a nucleus,

$$\boldsymbol{\mu}_i^{(\text{orbital})} = \frac{e\hbar[c]}{2M_p c}\boldsymbol{\ell}_i \tag{3-5}$$

where $\boldsymbol{\ell}_i$ is the orbital angular momentum of the ith proton in units of $\hbar$ and M_p is its mass (see Problem 2-6). As usual, Eq. (3-5) is in cgs units if the factor inside the square brackets is ignored and in SI units if included.

It is more convenient to express the contributions to the nuclear magnetic dipole moment from individual nucleons in terms of gyromagnetic ratios $g(i)$. For orbital motion, we can define $g_\ell(i)$ by the relation

$$\boldsymbol{\mu}_i^{(\text{orbital})} = g_\ell(i)\,\boldsymbol{\ell}_i \tag{3-6}$$

with

$$g_\ell(i) = \begin{cases} 1\mu_N & \text{for a proton} \\ 0 & \text{for a neutron} \end{cases}$$

to reflect the fact that only protons carry net charge and, consequently, can contribute to the nuclear magnetic dipole moment. Use of the nuclear magneton μ_N as the unit avoids any explicit dependence in the appearance of the equation on the system of electromagnetic units used. Similarly, contributions from the intrinsic spin of each nucleon may be expressed as

$$\boldsymbol{\mu}_i^{(\text{spin})} = g_s(i)\,\boldsymbol{s}_i \tag{3-7}$$

Since $s = \frac{1}{2}$, the gyromagnetic ratio for a free nucleon is

$$g_s(i) = \begin{cases} g_p = 2\mu_p = 5.585695\mu_N & \text{for a proton} \\ g_n = 2\mu_n = -3.826085\mu_N & \text{for a neutron} \end{cases} \tag{3-8}$$

Here, we have assumed that the structure of a bound nucleon inside a nucleus is the same as in its free state. As a result, we may use g_p and g_n, the "bare" nucleon values, as those for $g_s(i)$ in nuclei as well.

In terms of gyromagnetic ratios, the magnetic dipole operator may be written as a function of the orbital angular momentum operator $\boldsymbol{\ell}_i$ and the intrinsic spin operator $\boldsymbol{s}_i$ of each nucleon. For a deuteron, only two nucleons are involved, and the operator takes on a particular simple form,

$$\boldsymbol{\mu}_d = g_p\boldsymbol{s}_p + g_n\boldsymbol{s}_n + \boldsymbol{\ell}_p$$

where $\boldsymbol{\ell}_p$ is the angular momentum operator for the proton and $\boldsymbol{s}_p$ and $\boldsymbol{s}_n$ are, respectively, the intrinsic spin operators acting on the proton and the neutron wave functions. To simplify the expression, we have made use of the fact that g_ℓ is 1 for a proton and 0 for a neutron. Since the masses of a proton and a neutron are roughly equal to each other, we may assume that each one of the two nucleons carries one-half the orbital angular momentum associated with their relative motion, i.e., $\boldsymbol{\ell}_p = \frac{1}{2}\boldsymbol{L}$. The final result is then

$$\boldsymbol{\mu}_d = g_p\boldsymbol{s}_p + g_n\boldsymbol{s}_n + \tfrac{1}{2}\boldsymbol{L} \tag{3-9}$$

where $\boldsymbol{L}$ is the deuteron orbital angular momentum operator.

Contribution from the $^3S_1$-state. For the deuteron $^3S_1$-state, $L = 0$, and the expectation value of the magnetic dipole operator reduces to a sum of the intrinsic dipole moments of a proton and a neutron,

$$\mu_d(^3S_1) = \mu_p + \mu_n = 0.879805\mu_N \tag{3-10}$$

The details of this calculation are given later in Eq. (3-14). The final result of Eq. (3-10) is almost the same as the observed value of $0.857438\mu_N$. The small difference,

$$\mu_d - \mu_d(^3S_1) = 0.857438 - 0.879805 = -0.022367\mu_N$$

is, however, worth more careful consideration.

We can think of at least three possible causes for the small departure of the measured value from the expectation one in the $^3S_1$-state:

(1) The internal structures of the proton and the neutron are modified by the fact that the two nucleons are in a bound state. As a result, the gyromagnetic ratios for intrinsic spin may be different from g_p and g_n given in Eq. (3-8) for free nucleons.

(2) There are contributions from charged (virtual) mesons exchanged between the proton and the neutron, and these have not been included in Eq. (3-10).

(3) There is a small admixture of the $^3D_1$-state in the deuteron ground state.

Item 1 is extremely unlikely, as the deuteron is only a loosely bound system. The binding energy of 2.22 MeV can hardly be expected to affect the motion of quarks inside a nucleon that are bound by energies of the order of hundreds of mega-electron-volts. The effect of *mesonic current* suggested in item 2 is possible. In fact, it has been shown that mesonic currents are important in understanding magnetic dipole moments of odd-mass nuclei (see §4-7). However, we shall not discuss the topic here, partly for the reason that item 3 is more likely to be the major cause for the discrepancy between μ_d and $\mu_d(^3S_1)$. Furthermore, it is not easy to distinguish between items 2 and 3 from a more fundamental, field-theoretical point of view. For simplicity, we shall consider only item 3 and treat the $^3D_1$-admixture as the source for the small discrepancy between the observed and the calculated $^3S_1$-state value.

Expectation value of the magnetic dipole operator. Let us calculate next the expectation value of $\boldsymbol{\mu}_d$ in the deuteron $^3D_1$-state using the form given in Eq. (3-9). The matrix element depends also on M, the projection of spin $\boldsymbol{J}$ of the state on the z-axis. By convention, the magnetic moment, similar to other static electromagnetic moments, is defined as the expectation value of the z-component, or the $q = 0$ component in spherical tensor notation of §A-2, of the operator in the substate of maximum M, i.e., $M = J$. For the magnetic deuteron dipole moment, we have

$$\mu_d = \langle J, M = J|\boldsymbol{\mu}_{d,q=0}|J, M = J\rangle \tag{3-11}$$

Since both $\boldsymbol{\mu}_{d,q=0}$ and $\boldsymbol{J}_0$ ($q = 0$ component of $\boldsymbol{J}$) are similar operators, as far as their angular momentum properties are concerned, their expectation values must be proportional to each other, as shown in §A-6.

The constant of proportionality is given by the Landé formula, Eq. (A-20), in terms of the expectation value of the projection of $\boldsymbol{\mu}$ along $\boldsymbol{J}$,

$$\begin{aligned}\langle J, M|\mu_0|J, M\rangle &= \frac{1}{J(J+1)}\langle J, M|(\boldsymbol{\mu}\cdot\boldsymbol{J})\,J_0|J, M\rangle \\ &= \frac{M}{J(J+1)}\langle J, M|(\boldsymbol{\mu}\cdot\boldsymbol{J})|J, M\rangle\end{aligned} \tag{3-12}$$

To calculate the expectation value of the scalar product between $\boldsymbol{\mu}$ and $\boldsymbol{J}$, we shall first rewrite Eq. (3-9) in terms of $\boldsymbol{S} = \boldsymbol{s}_p + \boldsymbol{s}_n$ and $\boldsymbol{J} = \boldsymbol{L} + \boldsymbol{S}$,

$$\boldsymbol{\mu}_d = \tfrac{1}{2}\big\{(g_p + g_n)\boldsymbol{S} + (g_p - g_n)(\boldsymbol{s}_p - \boldsymbol{s}_n) + \boldsymbol{L}\big\} \tag{3-13}$$

Since the operator $\boldsymbol{s}_p - \boldsymbol{s}_n$ acts on proton and neutron spins with opposite signs, it can only connect between two states, one with $S = 1$ and the other with $S = 0$, and as a result, cannot contribute to the expectation value of interest to us here. This reduces $\boldsymbol{\mu}_d$ to a function of $\boldsymbol{L}$ and $\boldsymbol{S}$ only.

On substituting the simplified form of μ_d of Eq. (3-13) into (3-12), we obtain the result

$$\langle J, M|\mu_0|J, M\rangle = \frac{M}{J(J+1)}\langle J, M|\tfrac{1}{2}\{(g_p + g_n)(\boldsymbol{S}\cdot\boldsymbol{J}) + (\boldsymbol{L}\cdot\boldsymbol{J})\}|J, M\rangle$$

The scalar products on the right-hand side may be written in terms of $\boldsymbol{J}^2$, $\boldsymbol{L}^2$ and $\boldsymbol{S}^2$,

$$\boldsymbol{S}\cdot\boldsymbol{J} = \boldsymbol{S}\cdot(\boldsymbol{L}+\boldsymbol{S}) = \boldsymbol{S}^2 + \tfrac{1}{2}(\boldsymbol{J}^2 - \boldsymbol{L}^2 - \boldsymbol{S}^2) = \tfrac{1}{2}(\boldsymbol{J}^2 - \boldsymbol{L}^2 + \boldsymbol{S}^2)$$
$$\boldsymbol{L}\cdot\boldsymbol{J} = \boldsymbol{L}\cdot(\boldsymbol{L}+\boldsymbol{S}) = \boldsymbol{L}^2 + \tfrac{1}{2}(\boldsymbol{J}^2 - \boldsymbol{L}^2 - \boldsymbol{S}^2) = \tfrac{1}{2}(\boldsymbol{J}^2 + \boldsymbol{L}^2 - \boldsymbol{S}^2)$$

and the value of the magnetic dipole moment in a state of given J, L, S and $M = J$ reduces to

$$\begin{aligned}\mu_d &= \frac{1}{4(J+1)}\Big\{(g_p+g_n)(J(J+1) - L(L+1) + S(S+1)) \\ &\qquad + (J(J+1) + L(L+1) - S(S+1))\Big\} \end{aligned} \tag{3-14}$$

For the $^3S_1$-state, $L = 0$ and $J = S = 1$, we recover the result $\mu_d(^3S_1) = \mu_p + \mu_n$ given earlier in Eq. (3-10). For the $^3D_1$-state, we have $L = 2$ and this gives us the result

$$\mu_d(^3D_1) = \tfrac{1}{8}\Big\{(g_p+g_n)(-2) + 6\Big\} = 0.310\mu_N \tag{3-15}$$

The difference from the measured value of $0.857\mu_N$ is even larger than that for the $^3S_1$-state, making it unlikely for the deuteron ground state to be a purely $^3D_1$-state. This leads to an admixture of $^3S_1$- and $^3D_1$-states as the most likely candidate for the deuteron wave function, as far as the magnetic dipole moment is concerned.

Admixture of $^3D_1$-state. We can make a simple estimate of the amount of $^3D_1$-component in the deuteron ground state using the measured value of μ_d and the calculated values of $\mu_d(^3S_1)$ and $\mu_d(^3D_1)$ obtained above. For a linear combination of $^3S_1$- and $^3D_1$-components, the deuteron wave function may be written as

$$|\psi_d\rangle = a\,|^3S_1\rangle + b\,|^3D_1\rangle \tag{3-16}$$

with the normalization condition

$$a^2 + b^2 = 1 \tag{3-17}$$

Since there is no off-diagonal matrix element of $\boldsymbol{\mu}$ between $^3S_1$- and $^3D_1$-states, the deuteron magnetic dipole moment is given by

$$\mu_d = a^2\mu_d(^3S_1) + b^2\mu_d(^3D_1) = 0.857\mu_N \tag{3-18}$$

The values of a and b may be obtained by solving Eqs. (3-17) and (3-18) together. The value for b^2 turns out to be around 0.04, suggesting that there is a 4% admixture of the $^3D_1$-component in the deuteron ground state. As we shall see later, this is consistent, though somewhat on the low side, with the range of values obtained from other measured properties of the deuteron.

3-3 Deuteron Electric Quadrupole Moment

In electrostatics, the potential due to an arbitrary charge distribution at points far away from the source may be characterized by the moments of a multipole expansion of the source. For a microscopic object like a nucleus, it is not possible to observe the distribution directly, and the measured multipole moments give us some information on the source. For the charge distribution in a nucleus, the lowest nonvanishing multipole moment is the quadrupole, as the expectation value of the electric dipole operator, as well as all the other odd multipoles, vanishes due to the fact that the operators change sign under space inversion (see §4-6). For the deuteron, the measured quadrupole moment is $Q_d = 0.28590\ e\text{fm}^2$, as shown in Table 3-1.

Quadrupole operator. For a spherical nucleus, the expectation values of the squares of the distance from the center to the surface along x-, y-, and z-directions are equal to each other,

$$\langle x^2\rangle = \langle y^2\rangle = \langle z^2\rangle$$

As a result, the expectation value of $r^2 = x^2 + y^2 + z^2$ is

$$\langle r^2\rangle = \langle x^2 + y^2 + z^2\rangle \xrightarrow[\text{spherical}]{} 3\langle z^2\rangle$$

The electric quadrupole operator, which measures the lowest order departure from a spherical charge distribution in a nucleus, is defined in terms of the difference between $3z^2$ and r^2,

$$\boldsymbol{Q}_0 = e(3z^2 - r^2) \tag{3-19}$$

For a spherical nucleus, we have $\langle \boldsymbol{Q}_0\rangle = 0$ as a result. If a nucleus bugles out along the equatorial direction and flattens in the polar region, $\langle z^2\rangle$ is smaller than the average expectation value of the squares of the distance along the other two axes and the quadrupole moment is negative. A positive quadrupole moment of $0.29\ e\text{fm}^2$ indicates that the deuteron is slightly elongated along the z-axis, like an olive.

The operator $\boldsymbol{Q}_0$ is a spherical tensor of rank 2 (see §A-2), carrying two units of angular momentum. In terms of spherical harmonics,

$$\boldsymbol{Q}_0 = e(3z^2 - r^2) = er^2(3\cos^2\theta - 1) = \sqrt{\frac{16\pi}{5}}\, er^2 Y_{20}(\theta,\phi) \tag{3-20}$$

The electric quadrupole moment of a nuclear state is defined as the expectation value of $\boldsymbol{Q}_0$ in the substate of maximum M,

$$Q_A = \langle J\, M{=}J|\boldsymbol{Q}_0|J\, M{=}J\rangle \tag{3-21}$$

similar to the definition of magnetic dipole moment seen earlier.

Based on angular momentum considerations, any nuclear state with $J < 1$ cannot have a quadrupole moment different from zero. The expectation value $\langle J, M|\boldsymbol{Q}_0|J, M\rangle$ vanishes if the three angular momenta involved, J, 2, and J, cannot be coupled together to form a closed triangle. At the same time, since $\boldsymbol{Q}_0$ operates only in the coordinate space, it is independent of the total intrinsic spin S. This means that the orbital angular momentum L of the state must also be greater or equal to 1. For this reason, the expectation value of Q_0 vanishes in the $^3S_1$-state. The existence of a nonvanishing quadrupole moment is therefore a direct evidence of the presence of the $^3D_1$-component in the deuteron ground state.

Expectation value of the quadrupole operator. Let us work out the connection between the spatial part of the wave function and the quadrupole moment, assuming for the time being that the deuteron is in a state of definite orbital angular momentum L. Such a wave function $|LS;JM\rangle$ may be represented by the product of a spatial part $|LM_L\rangle$ and an intrinsic spin part $|SM_S\rangle$, coupled together to total angular momentum (J, M),

$$|LS;JM\rangle = \sum_{M_L M_S} \langle LM_L SM_S|JM\rangle |LM_L\rangle |SM_S\rangle \tag{3-22}$$

where $\langle LM_L SM_S|JM\rangle$ is the Clebsch-Gordan coefficient. For the expectation value of Q_0, we have

$$\begin{aligned} Q_d(L) &= \langle LS;JM|Q_0|LS;JM\rangle \\ &= \sum_{M_L M_S} \sum_{M'_L M'_S} \langle LM_L SM_S|JM\rangle \langle LM'_L SM'_S|JM\rangle \langle LM_L SM_S|Q_0|LM'_L SM'_S\rangle \end{aligned}$$

Since the operator does not act on intrinsic spin, we can remove this part of the wave function from the matrix element by making use of the orthogonal relation between the intrinsic spin wave functions,

$$\langle SM_S|SM'_S\rangle = \delta_{M_S M'_S}$$

As a result, the expectation value of Q_0 reduces to a matrix element involving only the spatial part of the wave function,

$$Q_d(L) = \sum_{M_L} \langle LM_L\, S\,(M-M_L)|JM\rangle^2 \langle LM_L|Q_0|LM_L\rangle \tag{3-23}$$

where we have made use of the property that the Clebsch-Gordan coefficients vanish if $M_S \neq (M - M_L)$.

We may simplify the expression on the right-hand side of Eq. (3-23) further by writing the spatial wave function as a product of radial and angular parts,

$$|LM_L\rangle = R_L(r)\, Y_{LM_L}(\theta\phi)$$

where the angular part is given by spherical harmonics $Y_{LM_L}(\theta\phi)$ and the radial part satisfies the normalization condition,

$$\int_0^\infty R_L^*(r) R_L(r)\, r^2\, dr = 1$$

Using the explicit form of Q_0 given in Eq. (3-19), we obtain the result

$$\begin{aligned} \langle LM_L|Q_0|LM_L\rangle &= e\sqrt{\frac{16\pi}{5}} \int_0^\infty R_L^*(r) r^2 R_L(r) r^2\, dr \\ &\quad \times \int_0^{2\pi} \int_0^\pi Y_{LM_L}^*(\theta\phi) Y_{20}(\theta\phi) Y_{LM_L}(\theta\phi) \sin\theta\, d\theta\, d\phi \end{aligned} \tag{3-24}$$

The radial integral here is the expectation value of r^2 and therefore must be a positive quantity. However, we cannot evaluate its value without making some assumptions on the radial wave function.

The angular integral over three spherical harmonics may be expressed in terms of $3j$-symbols,

$$\int_0^{2\pi}\int_0^{\pi} Y^*_{LM_L}(\theta\phi)Y_{20}(\theta\phi)\,Y_{LM_L}(\theta\phi)\sin\theta\,d\theta\,d\phi$$
$$= (-1)^{M_L}\begin{pmatrix} L & 2 & L \\ -M_L & 0 & M_L\end{pmatrix}(2L+1)\sqrt{\frac{5}{4\pi}}\begin{pmatrix} L & 2 & L \\ 0 & 0 & 0\end{pmatrix} \tag{3-25}$$

For $L = 2$, the numerical values for this integral are $-1/7\sqrt{5/\pi}$, $+1/14\sqrt{5/\pi}$, and $+1/7\sqrt{5/\pi}$, respectively, for $M_L = 2, 1, 0$. Before we can insert these values into Eq. (3-24) and obtain a value for $Q_d(L = 2)$, we also need the square of the three Clebsch-Gordan coefficients $\langle LM_L\,S\,(M-M_L)|JM\rangle^2$ for $S = 1$, $L = 2$, $M = J = 1$, and $M_L = 2, 1, 0$. These can be found using Table A-1, and the values are $\frac{6}{10}$, $\frac{3}{10}$, and $\frac{1}{10}$, respectively. With these, we obtain the result

$$\begin{aligned} Q_d(^3D_1) &= \langle ^3D_1 M = 1|Q_0|^3D_1 M = 1\rangle \\ &= \tfrac{6}{10}\langle L = 2, M_L = 2|Q_0|L = 2, M_L = 2\rangle \\ &\quad +\tfrac{3}{10}\langle L = 2, M_L = 1|Q_0|L = 2, M_L = 1\rangle \\ &\quad +\tfrac{1}{10}\langle L = 2, M_L = 0|Q_0|L = 2, M_L = 0\rangle \\ &= -\tfrac{1}{5}e\langle r^2\rangle_D \end{aligned}$$

where

$$\langle r^2\rangle_D = \int_0^\infty R^*_D(r)r^2R_D(r)r^2\,dr$$

If, as an estimate, we take the value of $\langle r^2\rangle_D$ as the square of the deuteron radius, we obtain $Q_d(^3D_1) = -0.77$ efm^2. Since even the sign disagrees with the measured value, it is unlikely that the deuteron wave function is made up entirely of the $^3D_1$-state.

For a more realistic model, we shall take a linear combination of $^3S_1$- and $^3D_1$-components, as we have done earlier in Eq. (3-16) for the magnetic dipole moment. The deuteron electric quadrupole moment now has the form

$$\begin{aligned} Q_d &= a^2\langle ^3S_1 M = 1|Q_0|^3S_1 M = 1\rangle + b^2\langle ^3D_1 M = 1|Q_0|^3D_1 M = 1\rangle \\ &\quad +2ab\langle ^3S_1 M = 1|Q_0|^3D_1 M = 1\rangle \end{aligned}$$

The first term vanishes, since $L = 0$. The main contribution is likely to come from the last term, as the $^3D_1$-component is only a few percent of the total and $|a| > |b|$ as a result. This term involves an off-diagonal radial integral,

$$\langle ^3S_1 M = 1|Q_0|^3D_1 M = 1\rangle \propto \int_0^\infty R^*_S(r)r^2R_D(r)\,r^2\,dr$$

The value is sensitive to the shapes of the radial wave functions and, as a result, it is difficult to put a firm value on the amount of $^3D_1$-component in the deuteron ground state from electric quadrupole moment calculations. Most of the estimates put $|b|^2$ to be in the range of 4% to 7%.

3-4 Tensor Force and the Deuteron D-state

We have seen that both magnetic dipole and electric quadrupole moments support the idea that the deuteron ground state is a linear combination of $^3S_1$- and $^3D_1$-states. The orbital angular momentum is, therefore, not a good quantum number and the nucleon-nucleon interaction potential does not commute with the operator $\boldsymbol{L}^2$. In this way, presence of the $^3D_1$-component in the deuteron ground state provides us with one clear piece of information concerning the property of nuclear force.

Deuteron Hamiltonian. The Hamiltonian for the deuteron problem may be written in the form

$$H = -\frac{\hbar^2}{2\mu}\nabla^2 + V \tag{3-26}$$

where the first term is the kinetic energy in the center of mass of the two-nucleon system and μ is the reduced mass. The second term expresses the interaction between the two nucleons in terms of a potential V. The ground state wave function, ψ_d, is the eigenfunction of the Schrödinger equation

$$H|\psi_d\rangle = E_d|\psi_d\rangle$$

with eigenvalue E_d. In §3-1 we deduced from symmetry arguments that only $L = 0$, 2 can contribute to the wave function, and the eigenfunction has the form given in Eq. (3-16),

$$|\psi_d\rangle = a\,|^3S_1\rangle + b\,|^3D_1\rangle$$

where a and b are coefficients to be determined by solving the Schrödinger equation.

It is convenient for us to think in terms of a matrix approach to the eigenvalue problem (see also §7-1). Since we are only interested in finding the amount of mixing between $^3S_1$- and $^3D_1$-states, we may use these two states alone as the basis to construct the Hamiltonian matrix,

$$\{H\} = \begin{pmatrix} H_{11} & H_{12} \\ H_{21} & H_{22} \end{pmatrix} \tag{3-27}$$

where the matrix elements are given by the following definitions:

$$H_{11} = \langle^3S_1|H|^3S_1\rangle \qquad H_{22} = \langle^3D_1|H|^3D_1\rangle \qquad H_{12} = H_{21} = \langle^3D_1|H|^3S_1\rangle$$

On diagonalizing this real, symmetric matrix, we obtain the energy E_d and the coefficients a and b. However, this is not our interest here; we are more concerned with the type of Hamiltonian that can cause a mixing between $^3S_1$- and $^3D_1$-states.

If the off-diagonal matrix elements H_{12} and H_{21} vanish, the Hamiltonian matrix is diagonal. The two eigenstates are then $|\,^3S_1\,\rangle$ and $|\,^3D_1\,\rangle$ without any mixing between them. The fact that the deuteron ground state is a linear combination of these two basis states implies that the off-diagonal matrix elements are not zero:

$$\langle^3D_1|H|^3S_1\rangle \neq 0 \tag{3-28}$$

Since the kinetic energy term of the Hamiltonian contributes only to the diagonal matrix elements in the two-dimensional Hilbert space we are working in here, Eq. (3-28) must be the result of the interaction potential V. This leads to the conclusion

$$\langle^3D_1|V|^3S_1\rangle \neq 0 \tag{3-29}$$

That is, the nuclear potential is not diagonal in a basis span by states with definite orbital angular momentum and can, therefore, mix $^3S_1$- and $^3D_1$-states.

In order to have a nonvanishing off-diagonal matrix element, the potential V must have a spatial part that is a spherical tensor of rank 2 so as to be able to connect an S- to a D-state, as required by Eq. (3-29). Again, let us express the deuteron wave function as a product of spatial, intrinsic spin, and isospin parts. Similarly, the matrix element of V above may also be written as a product of three matrix elements: one in coordinate space, one in intrinsic spin space, and one in isospin space. For simplicity, we shall ignore any possible dependence on the isospin in the matrix element of V in the following discussion.

Since the nuclear Hamiltonian conserves the total angular momentum of the system, the potential V must be a scalar in spin J. However, if the spherical rank of the spatial part of the operator is 2, we must find an operator of the same rank for the intrinsic spin part so that a scalar product of these two rank 2 operators can be constructed. For this purpose, let us examine first the possible spherical tensor operators that we have in the intrinsic spin space.

Spin operator. For a spin-$\frac{1}{2}$ system, an arbitrary operator may be expressed in terms of Pauli matrices,

$$\boldsymbol{\sigma}_x = \begin{pmatrix} 0 & 1 \\ 1 & 0 \end{pmatrix} \qquad \boldsymbol{\sigma}_y = \begin{pmatrix} 0 & -i \\ i & 0 \end{pmatrix} \qquad \boldsymbol{\sigma}_z = \begin{pmatrix} 1 & 0 \\ 0 & -1 \end{pmatrix} \tag{3-30}$$

together with the two-dimensional unit matrix. In terms of spherical components, the Pauli matrices may be written as

$$\begin{aligned} \boldsymbol{\sigma}_{+1} &= -\frac{1}{\sqrt{2}}(\boldsymbol{\sigma}_x + i\boldsymbol{\sigma}_y) = \sqrt{2}\begin{pmatrix} 0 & -1 \\ 0 & 0 \end{pmatrix} \\ \boldsymbol{\sigma}_{-1} &= +\frac{1}{\sqrt{2}}(\boldsymbol{\sigma}_x - i\boldsymbol{\sigma}_y) = \sqrt{2}\begin{pmatrix} 0 & 0 \\ 1 & 0 \end{pmatrix} \\ \boldsymbol{\sigma}_0 &= \boldsymbol{\sigma}_z = \begin{pmatrix} 1 & 0 \\ 0 & -1 \end{pmatrix} \end{aligned} \tag{3-31}$$

These form the three components of an operator acting only on the intrinsic spin part of the wave function of a nucleon and carrying one unit of angular momentum, similar to other vector operators such as $\boldsymbol{L}$.

A two-body operator in the nucleon intrinsic spin space may be constructed from a product of $\boldsymbol{\sigma}(1)$ and $\boldsymbol{\sigma}(2)$, respectively, the intrinsic spin operator for particles 1 and 2. Since each one is a vector, the spherical tensor rank of the product is the vector sum of $\boldsymbol{\sigma}(1)$ and $\boldsymbol{\sigma}(2)$ and may therefore carry zero, one, or two units of angular momentum, as shown in §A-3. The first two possibilities are, respectively, analogous to the usual scalar and vector products in multiplying ordinary vectors. The last possibility is a new one and is often loosely referred to as the "tensor product" of two vector operators. The name should not be confused with the more general product of two spherical tensors.

The scalar product of two vectors is a familiar quantity. In a Cartesian coordinate system, a scalar two-body operator in the intrinsic spin space may be expressed as

$$\boldsymbol{\sigma}(1) \cdot \boldsymbol{\sigma}(2) = \boldsymbol{\sigma}_x(1)\boldsymbol{\sigma}_x(2) + \boldsymbol{\sigma}_y(1)\boldsymbol{\sigma}_y(2) + \boldsymbol{\sigma}_z(1)\boldsymbol{\sigma}_z(2) \tag{3-32}$$

In terms of the spherical components given in Eq. (3-31), the same product may be written in the following way:

$$\boldsymbol{\sigma}(1)\cdot\boldsymbol{\sigma}(2)=\sigma_0(1)\sigma_0(2)-\sigma_{+1}(1)\sigma_{-1}(2)-\sigma_{-1}(1)\sigma_{+1}(2) \tag{3-33}$$

as done in Eq. (A-19). We can also make use of the explicit values of the following Clebsch-Gordan coefficients given in §A-3,

$$\langle 1{+}1\,1{-}1|00\rangle=\langle 1{-}1\,1{+}1|00\rangle=+\frac{1}{\sqrt{3}} \qquad \langle 1010|00\rangle=-\frac{1}{\sqrt{3}}$$

and write the result in spherical tensor notation,

$$\boldsymbol{\sigma}(1)\cdot\boldsymbol{\sigma}(2)=-\sqrt{3}\sum_q\langle 1q\,1\,-q|00\rangle\sigma_q(1)\sigma_{-q}(2)=-\sqrt{3}(\boldsymbol{\sigma}(1)\times\boldsymbol{\sigma}(2))_{00} \tag{3-34}$$

where q, the index of summation, takes on values -1, 0, and $+1$. The last form expresses the two-body operator in intrinsic spin space as a product of two Pauli spin operators with angular momentum coupled together to zero.

In general, a product of two operators with tensorial ranks r and s, coupled together to form a tensor of rank t, may be written as a coupled product,

$$(T_r\times U_s)_{tm}\equiv\sum_{pq}\langle rpsq|tm\rangle\, T_{rp}U_{sq} \tag{3-35}$$

as done in Eq. (A-10). Thus, the vector product of $\boldsymbol{\sigma}(1)$ and $\boldsymbol{\sigma}(2)$ is given by

$$(\boldsymbol{\sigma}(1)\times\boldsymbol{\sigma}(2))_{1m}=\sum_{pq}\langle 1p1q|1m\rangle\sigma_p(1)\sigma_q(2) \tag{3-36}$$

It is left as an exercise to show that this is equivalent to the vector product of $\boldsymbol{\sigma}(1)$ with $\boldsymbol{\sigma}(2)$ in Cartesian coordinates (see Problem 3-5).

By the same token, we can also use Eq. (3-35) to write the components of the second-rank tensor product of $\boldsymbol{\sigma}(1)$ and $\boldsymbol{\sigma}(2)$,

$$(\boldsymbol{\sigma}(1)\times\boldsymbol{\sigma}(2))_{2m}=\sum_{pq}\langle 1p1q|2m\rangle\sigma_p(1)\sigma_q(2)$$

Each component may be expressed explicitly as

$$\begin{aligned}
(\boldsymbol{\sigma}(1)\times\boldsymbol{\sigma}(2))_{20} &= \frac{1}{\sqrt{6}}\{\sigma_1(1)\sigma_{-1}(2)+\sigma_{-1}(1)\sigma_1(2)+2\sigma_0(1)\sigma_0(2)\}\\
(\boldsymbol{\sigma}(1)\times\boldsymbol{\sigma}(2))_{2\pm1} &= \frac{1}{\sqrt{2}}\{\sigma_{\pm1}(1)\sigma_0(2)+\sigma_0(1)\sigma_{\pm1}(2)\}\\
(\boldsymbol{\sigma}(1)\times\boldsymbol{\sigma}(2))_{2\pm2} &= \sigma_{\pm1}(1)\sigma_{\pm1}(2)
\end{aligned} \tag{3-37}$$

using the values of the Clebsch-Gordan coefficients given in Table 3-2.

We can now return to the operator for the intrinsic spin part of the nuclear potential V. Since the product of $\boldsymbol{\sigma}(1)$ and $\boldsymbol{\sigma}(2)$ can only be coupled together to form a scalar, a vector, or an operator of spherical tensor rank 2, the maximum rank of V in intrinsic spin space is 2. Furthermore, since the intrinsic spin part and the orbital angular

Table 3-2: Values of Clebsch-Gordan coefficients $\langle 1p1q|2m\rangle$.

m	p	q	$\langle 1p1q\|2m\rangle$	m	p	q	$\langle 1p1q\|2m\rangle$
0	1	−1	$\sqrt{\frac{1}{6}}$	1	1	0	$\sqrt{\frac{1}{2}}$
0	−1	1	$\sqrt{\frac{1}{6}}$	1	0	1	$\sqrt{\frac{1}{2}}$
0	0	0	$\sqrt{\frac{2}{3}}$	−2	−1	−1	1
−1	−1	0	$\sqrt{\frac{1}{2}}$	2	1	1	1
−1	0	−1	$\sqrt{\frac{1}{2}}$				

momentum part must have the same rank to form a scalar product in J, the maximum rank of the orbital angular momentum part must also be 2. As we have seen earlier, this is adequate for our purposes since, from the admixture of the $^3D_1$-component in the deuteron ground state, we have concluded in Eq. (3-29) that there must be a component in V with spherical tensor rank 2 in L.

Tensor operator. An operator formed by the scalar product of a second-rank operator in intrinsic spin space and a similar one in coordinate space is often referred to as a *tensor operator*. It is generally written in the form

$$S_{12} = \frac{3}{r^2}(\boldsymbol{\sigma}_1 \cdot \boldsymbol{r})(\boldsymbol{\sigma}_2 \cdot \boldsymbol{r}) - \boldsymbol{\sigma}_1 \cdot \boldsymbol{\sigma}_2 \tag{3-38}$$

where we have used subscripts to indicate on which of the two nucleons each Pauli spin operator acts. We shall follow this practice for single-particle operators in general wherever there is no need to indicate the spherical tensor component of the operator. The context will always make it clear whether the subscript on an operator is for spherical tensor rank or an index for particle number. The form of the tensor operator given in Eq. (3-38) is only the L- and S-dependent parts: The strength of the force as well as its radial and isospin dependence must be put in separately.

The fact that there is a small admixture of $^3D_1$-component in the predominantly $^3S_1$ deuteron ground state implies that there must be a tensor component in the nucleon-nucleon potential. Although we cannot say much more about this component of the nuclear force from the deuteron properties alone, the clear evidence for such a term is an indication of the richness in the deuteron problem. In the next section, we shall see that, besides the tensor force, the nuclear potential contains also terms that have tensorial ranks 0 as well as 1 in intrinsic spin and spatial coordinates as well as other operators of rank 2.

3-5 Symmetry and Nuclear Force

Nucleons interact with each other through two-body interactions. That is, the force between nucleons acts only between a pair of them at a time. The absence of one-body terms in the potential can be seen by contrasting with atomic electrons. In an atom, the electrons are bound to a central electrostatic potential provided by the protons in the nucleus. As a result, there is a force acting on an electron even in cases where it is the only one present, such as the hydrogen atom. This is not true for nuclei, as there is no external source to provide a force on the individual nucleon. The only one-body operator in a nuclear Hamiltonian is the kinetic energy arising from the motion of each nucleon. We shall see later, for example in §7-3, that one may on occasion introduce an "effective" one-body term in the nuclear potential to approximate the average effect from all the other nucleons in the nucleus. The source of such an effective one-body potential is, however, the two-body interaction between nucleons.

On the other hand, it is not possible to rule out completely three-body and higher particle-rank terms in the nuclear interaction. A three-body force is one which is felt only when there are at least three particles present, such as that represented later by Fig. 3-8(g). For example, in a three-nucleon system such as a triton, the nucleus of tritium, or a ^{3}He made of two protons and one neutron, a two-body force acts between nucleons 1 and 2, between nucleons 2 and 3, and between nucleons 3 and 1. If, after taking away the sum of the interactions between these three pairs, there is still a residual force left in the system, we can then say that there is a three-body force between nucleons. All the available evidence indicates that such a term, if present, must be very much weaker than the two-body force. With the possible exception of three-nucleon systems, it is unlikely that our present experimental equipment and theoretical knowledge can detect the presence of any three-body forces in nuclei. For this reason we shall ignore any possible three-body forces from now on. The same applies to other many-body terms in the nuclear potential.

One way to study nuclear two-body interactions is to make use of two-nucleon systems such as the deuteron. However, as we have already seen, the deuteron is a very limited system having only one bound state. For a more comprehensive investigation, we must resort to scattering of one nucleon off another. Before going into the details of nucleon-nucleon scattering, it is advantageous to examine first some of the restrictions imposed on the nuclear interaction by the symmetry requirement on a two-nucleon system.

Charge independence. We shall assume that nuclear force is *charge independent;* that is, the only difference in the interaction between a pair of protons and a pair of neutrons is the Coulomb interaction between protons. This is, again, an assumption based on experimental evidence. There is no fundamental reason to rule out a charge-symmetry-breaking term in the nuclear force itself. As we shall see in the next section, the difference in mass between charged and neutral pions alone implies the possibility of a small but significant difference between proton-neutron interaction and the interaction between a pair of protons or a pair of neutrons. On the other hand, from a practical point of view, there is perhaps no need to be concerned with any possible violation of charge symmetry in nuclear force. At the moment, all the evidence puts the term to be smaller than the accuracies we can achieve in handling the much stronger charge-independence-breaking effect due to electromagnetic interaction. For simplicity, we shall ignore, from now on, any charge dependence that may be present in nuclear force. Furthermore, for our discussion of nuclear force here, we shall also ignore any charge symmetry breaking coming from electromagnetic origin.

Since charge is related to the third component of the isospin operator $\boldsymbol{T}$, the charge independence of nuclear force implies the commutation relation,

$$[H, \boldsymbol{T}_0] = 0 \tag{3-39}$$

where H is the nuclear Hamiltonian. This, in turn, means that the eigenvectors $|\,\psi\,\rangle$ of a nuclear Hamiltonian can also be the eigenvectors of $\boldsymbol{T}_0$ at the same time,

$$\boldsymbol{T}_0|\psi\rangle = \tfrac{1}{2}(Z - N)|\psi\rangle \tag{3-40}$$

where Z is the proton number and N the neutron number of the nucleus.

Isospin invariance. In addition to $\boldsymbol{T}_0$, the nuclear Hamiltonian commutes also with the square of the isospin operator,

$$[H, \boldsymbol{T}^2] = 0 \tag{3-41}$$

In other words, the eigenfunctions of the nuclear Hamiltonian are also eigenfunctions of the operator $\boldsymbol{T}^2$,

$$\boldsymbol{T}^2|\psi\rangle = T(T+1)|\psi\rangle \tag{3-42}$$

Physically, it means that the wave function of a state with a definite isospin T is unchanged if we replace some of the protons by neutrons, and vice versa. Such a transformation between the two states of a nucleon takes us from one member of the isobar, a nucleus of the same nucleon number, to another. Since nothing else is changed, these two states must have essentially the same properties except for a difference in the proton and neutron numbers. Mathematically, the wave functions of two such states are related to each other through an isospin rotation that can be realized using raising or lowering operators.

A group of states related by a rotation in the isospin space are known as *isobaric analogue states* (IAS) of each other. Many examples are known in light nuclei. As illustration, the energy level spectra for two $A = 11$ nuclei and two $A = 21$ nuclei are shown in Fig. 3-1. In both examples, the members of each pair are, furthermore, *mirror nuclei* of each other; that is, the number of protons in one is equal to the number of neutrons in the other, and vice versa. In a sense, they are the image of each other in a mirror that turns protons into neutrons and neutrons into protons. In addition to energy level positions, many other properties of states in mirror nuclei are found to be very similar to each other. Most of the small differences found may be attributed to the Coulomb interaction, which we choose to ignore in our discussion here.

In heavy nuclei, the Coulomb effect increases because of the larger numbers of protons. Since electromagnetic interactions do not conserve isospin, we find that nuclear states are no longer eigenstates of $\boldsymbol{T}^2$. In contrast to Eq. (3-42), we have

$$\boldsymbol{T}^2|\psi_i\rangle = \sum_k a_k|\psi_k\rangle$$

That is, when $\boldsymbol{T}^2$ acts on an eigenstate of the nuclear Hamiltonian, the result is a linear combination of eigenstates of different isospin. The situation may also be described by saying that the strength of the IAS is split among several states. When this happens, it can be difficult to find any direct evidence for the presence of IAS and the concept

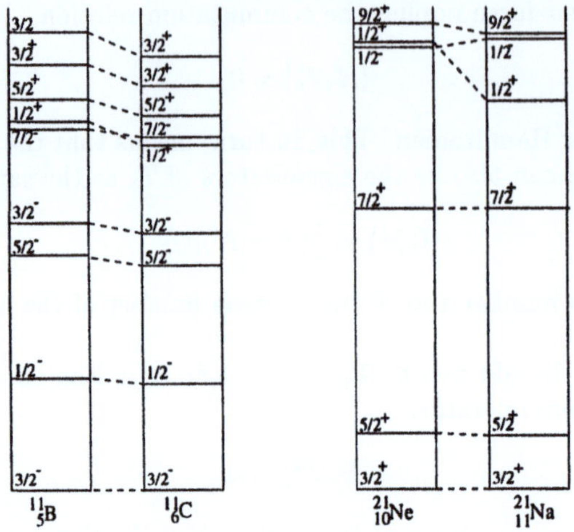

Figure 3-1: A comparison of the low-lying spectra of members of the $A = 11$ and $A = 21$ isobars, showing the similarity in their level structure. (Plotted using data from Ref. [95].)

of isospin ceases to be useful. In spite of the complications caused by electromagnetic interactions, the evidence for isospin invariance of the nuclear force itself is quite strong. As we shall soon see, we may be able to make use of isospin symmetry to limit the possible forms the nuclear potential can take.

Isospin operators. Isospin invariance, however, does not imply isospin independence of nuclear force. We have already seen evidence that the nuclear force is different depending on the isospin of the two-nucleon system. For example, a bound state is found for $T = 0$, the deuteron, but not for $T = 1$. Let us examine the possible forms of isospin operators that satisfy these conditions and can be used in a nucleon-nucleon potential.

For a single nucleon, the isospin operator $\boldsymbol{\tau}$ may be written in terms of Pauli matrices,

$$\tau_1 = \begin{pmatrix} 0 & 1 \\ 1 & 0 \end{pmatrix} \qquad \tau_2 = \begin{pmatrix} 0 & -i \\ i & 0 \end{pmatrix} \qquad \tau_3 = \begin{pmatrix} 1 & 0 \\ 0 & -1 \end{pmatrix} \tag{3-43}$$

as we have done in Eq. (2-18). Alternatively, they can be expressed in terms of spherical components analogous to Eq. (3-31), except that the operators here act on the isospin part of the wave function. Since there are only two isospin states for a nucleon, the only possible (single-nucleon) operators for the isospin part of the wave function are $\boldsymbol{\tau}$ and the identity operator,

$$\mathbf{1} = \begin{pmatrix} 1 & 0 \\ 0 & 1 \end{pmatrix} \tag{3-44}$$

All other single-nucleon isospin operators can be expressed in terms of $\mathbf{1}$ and $\boldsymbol{\tau}$. For example, the eigenvalue of $\boldsymbol{\tau}^2$ in the space of a nucleon is always 3, as can be seen by an explicit calculation using the nucleon isospin wave functions given in Eqs. (2-15) and

(2-16). As a result, the operator $\boldsymbol{\tau}^2$ may be replaced by the form 3 times the identity operator **1**.

For a system consisting of A nucleons, the isospin operator is the sum over those acting on individual nucleons,

$$\boldsymbol{T} = \frac{1}{2}\sum_{i=1}^{A} \boldsymbol{\tau}_i \tag{3-45}$$

For the two-nucleon system we are mainly interested in here, we can write

$$\boldsymbol{T} = \tfrac{1}{2}(\boldsymbol{\tau}_1 + \boldsymbol{\tau}_2) \tag{3-46}$$

Since nuclear force is two body in nature and a scalar in isospin space, we must construct two-body operators using operators **1** and $\boldsymbol{\tau}$ acting on each one of the two nucleons. The operator $\boldsymbol{T}$ is unsuitable for our purpose here, as it is one body and acts on one nucleon at a time. Furthermore, it is a vector in isospin space. One way to construct a two-body, isoscalar operator is to take a scalar product of $\boldsymbol{T}$ with itself. From Eq. (3-46), we have

$$\boldsymbol{T}^2 \equiv \boldsymbol{T}\cdot\boldsymbol{T} = \tfrac{1}{4}(\boldsymbol{\tau}_1{}^2 + \boldsymbol{\tau}_2{}^2 + 2\boldsymbol{\tau}_1 \cdot \boldsymbol{\tau}_2) \tag{3-47}$$

The first two terms on the right-hand side are one-body operators, seen by the fact that they do not vanish even when there is only one nucleon present. Only the third term, $\boldsymbol{\tau}_1 \cdot \boldsymbol{\tau}_2$, is a two-body operator, as it vanishes unless it is acting on a state with both nucleons 1 and 2 present. The operator $\boldsymbol{T}^2$ therefore has mixed particle rank of 1 and 2. The only purely two-body operators we are left with in the isospin space are the unity operator and $\boldsymbol{\tau}_1 \cdot \boldsymbol{\tau}_2$. All other two-body isoscalar operators may be expressed as functions of these two.

From Eq. (3-47), we have the relation,

$$\boldsymbol{\tau}_1 \cdot \boldsymbol{\tau}_2 = 2\boldsymbol{T}^2 - \tfrac{1}{2}\boldsymbol{\tau}_1{}^2 - \tfrac{1}{2}\boldsymbol{\tau}_2{}^2$$

For a single nucleon (isospin $\frac{1}{2}$), the expectation of $\boldsymbol{\tau}^2$ is 3, as we have seen earlier. The expectation value of $\boldsymbol{\tau}_1 \cdot \boldsymbol{\tau}_2$ in the space of two nucleons with total isospin T is then

$$\langle T|\boldsymbol{\tau}_1 \cdot \boldsymbol{\tau}_2|T\rangle = \begin{cases} -3 & \text{for } T = 0 \\ 1 & \text{for } T = 1 \end{cases} \tag{3-48}$$

Since the expectation value is different, the operator $\boldsymbol{\tau}_1 \cdot \boldsymbol{\tau}_2$ is able to distinguish a two-nucleon state with isospin $T = 0$ from one with $T = 1$. In contrast, the identity operator has the same expectation value, unity, in both $T = 0$ and $T = 1$ states. We shall see later that the difference between these two operators is adequate to give a proper isospin dependence for nucleon-nucleon potentials.

Other symmetries and general form of nuclear potential. The force between two nucleons must be invariant under a translation in space of the two-nucleon system as a whole. In other words, the interaction can only depend on the relative position of the two nucleons and not on their absolute positions with respect to some arbitrary coordinate system. This requirement is generally referred to as translational invariance, and it implies that only the relative coordinate between the two nucleons

$$\boldsymbol{r} = \boldsymbol{r}_1 - \boldsymbol{r}_2 \tag{3-49}$$

can enter as one of the arguments.

The nuclear potential V may depend on the momenta $\boldsymbol{p}_1$ and $\boldsymbol{p}_2$ of the two particles. On the other hand, since the sum $\boldsymbol{P} = \boldsymbol{p}_1 + \boldsymbol{p}_2$ corresponds to the center-of-mass momentum of the two-nucleon system as a whole, it cannot appear as an argument of the interaction between the two particles. The only possible momentum dependence V can have is on the relative value between two nucleons, defined by

$$\boldsymbol{p} = \tfrac{1}{2}(\boldsymbol{p}_1 - \boldsymbol{p}_2) \tag{3-50}$$

This is known as Galilean invariance of the two-nucleon system.

In addition to isospin, translational, and Galilean invariances, a nuclear potential must also remain unchanged under a rotation of the coordinate system, time reversal, space reflection (parity), and a permutation between the two nucleons. In terms of independent operators, the potential can only be a function of $\boldsymbol{\sigma}_1$, $\boldsymbol{\sigma}_2$, $\boldsymbol{\tau}_1$, and $\boldsymbol{\tau}_2$ in addition to $\boldsymbol{r}$ and $\boldsymbol{p}$. As we have demonstrated with isospin operators, only a very limited number of linearly independent two-body operators, satisfying the symmetry requirements for a nuclear potential, can be constructed using a given set of single-nucleon operators. For example, the single-particle orbital angular momentum operator $\boldsymbol{\ell}$ is not an independent operator from those in the set given above, as it is the vector product of $\boldsymbol{r}$ and $\boldsymbol{p}$. Okubo and Marshak [114] have shown that the most general two-body potential under these conditions must take on the form

$$\begin{aligned} V(r;\boldsymbol{\sigma}_1,\boldsymbol{\sigma}_2,\boldsymbol{\tau}_1,\boldsymbol{\tau}_2) \;=\;& V_0(r) + V_\sigma(r)\boldsymbol{\sigma}_1\cdot\boldsymbol{\sigma}_2 + V_\tau(r)\boldsymbol{\tau}_1\cdot\boldsymbol{\tau}_2 + V_{\sigma\tau}(r)(\boldsymbol{\sigma}_1\cdot\boldsymbol{\sigma}_2)(\boldsymbol{\tau}_1\cdot\boldsymbol{\tau}_2) \\ &+V_{LS}(r)\boldsymbol{L}\cdot\boldsymbol{S} + V_{LS\tau}(r)(\boldsymbol{L}\cdot\boldsymbol{S})(\boldsymbol{\tau}_1\cdot\boldsymbol{\tau}_2) \\ &+V_T(r)S_{12} + V_{T\tau}(r)S_{12}\,\boldsymbol{\tau}_1\cdot\boldsymbol{\tau}_2 \\ &+V_Q(r)Q_{12} + V_{Q\tau}(r)Q_{12}\,\boldsymbol{\tau}_1\cdot\boldsymbol{\tau}_2 \\ &+V_{PP}(r)(\boldsymbol{\sigma}_1\cdot\boldsymbol{p})(\boldsymbol{\sigma}_2\cdot\boldsymbol{p}) + V_{PP\tau}(r)(\boldsymbol{\sigma}_1\cdot\boldsymbol{p})(\boldsymbol{\sigma}_2\cdot\boldsymbol{p})(\boldsymbol{\tau}_1\cdot\boldsymbol{\tau}_2) \end{aligned} \tag{3-51}$$

with 12 terms. In addition to the tensor operator S_{12} given earlier in Eq. (3-38), we have two other operators that are constructed from elementary single-nucleon operators: the two-body spin-orbit operator,

$$\boldsymbol{L}\cdot\boldsymbol{S} = \tfrac{1}{2}(\boldsymbol{\ell}_1 + \boldsymbol{\ell}_2)\cdot(\boldsymbol{\sigma}_1 + \boldsymbol{\sigma}_2) \tag{3-52}$$

and the quadratic spin-orbit operator,

$$Q_{12} = \tfrac{1}{2}\big\{(\boldsymbol{\sigma}_1\cdot\boldsymbol{L})(\boldsymbol{\sigma}_2\cdot\boldsymbol{L}) + (\boldsymbol{\sigma}_2\cdot\boldsymbol{L})(\boldsymbol{\sigma}_1\cdot\boldsymbol{L})\big\} \tag{3-53}$$

The radial dependence and strength of each one of the 12 terms are given by the 12 functions $V_0(r)$, $V_\sigma(r)$, … . To determine the forms of these functions, we will need information in addition to those generated from symmetry arguments above. For example, we can make use of our knowledge of the basic nature of the nuclear force, such as the meson exchange picture of Yukawa to be discussed in the next section, or we can use a semi-empirical procedure and fit some assumed forms of the radial dependence to experimental data. When our understanding of QCD is fully developed in the future, it should also be possible to determine these functions from first principles.

The 12 terms in Eq. (3-51) may be divided into five groups. The first four,

$$V_{\text{central}} = V_0(r) + V_\sigma(r)\boldsymbol{\sigma}_1 \cdot \boldsymbol{\sigma}_2 + V_\tau(r)\boldsymbol{\tau}_1 \cdot \boldsymbol{\tau}_2 + V_{\sigma\tau}(r)(\boldsymbol{\sigma}_1 \cdot \boldsymbol{\sigma}_2)(\boldsymbol{\tau}_1 \cdot \boldsymbol{\tau}_2) \tag{3-54}$$

are the *central force* terms, as the tensorial ranks of the spatial parts of all four operators are zero. Similar to isospin, there are only two two-body operators for intrinsic spin: unity and $\boldsymbol{\sigma}_1 \cdot \boldsymbol{\sigma}_2$. Analogous to Eq. (3-48), two-nucleon states with total intrinsic spin S are distinguished by

$$\langle S|\boldsymbol{\sigma}_1 \cdot \boldsymbol{\sigma}_2|S\rangle = \begin{cases} -3 & \text{for } S = 0 \\ 1 & \text{for } S = 1 \end{cases} \tag{3-55}$$

The product of two independent two-body intrinsic spin operators and two similar ones for isospin gives us the four central force terms. As we can see from Eq. (3-54), the first term $V_0(r)$ depends only on the radial distance r. The second term has, in addition, a dependence on the intrinsic spin but not isospin. The third has isospin dependence but no intrinsic spin dependence. Only the fourth term has both intrinsic spin and isospin dependence. However, all four terms are scalars in intrinsic spin and, hence, in orbital angular momentum as well. A central force, therefore, commutes with $\boldsymbol{S}^2$, $\boldsymbol{L}^2$, and $\boldsymbol{J}^2$.

The other terms in Eq. (3-51) do not necessarily preserve the total intrinsic spin and the total orbital angular momentum of a two-nucleon system. In the presence of these terms, the two-nucleon system is invariant only in the combined space of L and S labeled by J. The dependence of the nuclear force on the two-body spin-orbit operator is expressed by fifth and sixth terms in Eq. (3-51),

$$V_{\text{spin-orbit}} = V_{LS}(r)\boldsymbol{L} \cdot \boldsymbol{S} + V_{LS\tau}(r)(\boldsymbol{L} \cdot \boldsymbol{S})(\boldsymbol{\tau}_1 \cdot \boldsymbol{\tau}_2) \tag{3-56}$$

The reason that two separate components are needed here (as well as the other terms to be discussed below) comes from the possibility that the radial dependence of the isospin-dependent and the isospin-independent parts may be different from each other, as for example the result of different mesons being exchanged. The spatial part of the two-body spin-orbit operator involves $\boldsymbol{L}$. Since it does not change sign under an inversion of the spatial coordinate system, it is an axial vector. In order to maintain parity invariance as well as rotational invariance for V, only a scalar product with another axial vector may enter here. It is easy to see that the operator $\boldsymbol{L}^2$ is not suitable for this purpose, as it conserves both L and S and is, therefore, a part of the central force. The only other possibility is the product $\boldsymbol{L} \cdot \boldsymbol{S}$. (See Problem 3-8 for other possible forms.)

The two-body spin-orbit operator, however, cannot connect two states with different orbital angular momenta. In other words,

$$\langle LS|\boldsymbol{L} \cdot \boldsymbol{S}|L'S'\rangle = 0 \quad \text{for} \quad L' \neq L$$

This comes from a combination of two reasons. From angular momentum coupling requirements, the matrix element $\langle LS|\boldsymbol{L} \cdot \boldsymbol{S}|L'S'\rangle$ vanishes if $|L' - L| > 1$, as the operator $\boldsymbol{L}$ carries only one unit of orbital angular momentum. On the other hand, the parity of the orbital part of the wave function of a state with angular momentum L is $(-1)^L$. Under a space reflection, operators $\boldsymbol{L}$ and $\boldsymbol{S}$ do not change sign. The matrix element $\langle LS|\boldsymbol{L}\cdot\boldsymbol{S}|L'S'\rangle$, however, changes sign if $L' = L\pm1$ and must therefore vanish. The net result is that the spin-orbit term is nonzero only between states of the same orbital angular momentum. However, the same constraint does not apply to the total intrinsic spin. As a result, $V_{\text{spin-orbit}}$ is not a central potential.

The next pair (seventh and eighth) of terms in Eq. (3-51) involve tensor force which we have already encountered in §3-4. The quadratic spin-orbit terms $V_Q(r)Q_{12}$ and $V_{Q\tau}(r)Q_{12}\boldsymbol{\tau}_1\cdot\boldsymbol{\tau}_2$ enter only when there is momentum dependence in the potential. The last two terms, $V_{PP}(r)(\boldsymbol{\sigma}_1\cdot\boldsymbol{p})(\boldsymbol{\sigma}_2\cdot\boldsymbol{p})$ and $V_{PP\tau}(r)(\boldsymbol{\sigma}_1\cdot\boldsymbol{p})(\boldsymbol{\sigma}_2\cdot\boldsymbol{p})(\boldsymbol{\tau}_1\cdot\boldsymbol{\tau}_2)$, are often dropped since, for elastic scattering, they can be expressed as a linear combination of other terms. Their contributions, therefore, cannot be determined using elastic scattering, from which we obtain most of the information on nucleon-nucleon interaction.

Returning now to the deuteron system, we see that if only the central force terms, given by strengths V_0, V_σ, V_τ, and $V_{\sigma\tau}$, are present in the nuclear potential, both L and S are good quantum numbers. The same is also true for the spin-orbit terms for reasons mentioned earlier. Among the remaining terms, the simplest one that can admix the ${}^3S_1$- and ${}^3D_1$-states is the tensor force. The presence of the ${}^3D_1$-component in the ground state wave function provides the clearest indication of the presence of such a term in the nuclear force.

3-6 Yukawa Theory of Nuclear Interaction

The meson exchange idea introduced by Yukawa in 1934 is a good starting point to examine nucleon-nucleon interaction beyond what we can learn through symmetry arguments in the previous section. In the Yukawa picture, the interaction between two nucleons is mediated by the exchange of mesons. Although it is not straightforward to draw a quantitative connection with the underlying quark structure of the hadrons, the theory does make it possible to relate nuclear interaction with various other hadronic processes, such as the strength of meson-nucleon interaction. On a more empirical level, the Yukawa idea provides us with a reasonable form for the radial dependence of nuclear potentials. Such expressions may be used, for example, as the starting point for semi-empirical approaches. Our focus in this section will be mainly on the origin of the meson exchange idea itself. We shall leave any applications to the last section after we have first taken a look at the experimental information derived from nucleon-nucleon scattering.

A proper derivation of a potential based on boson exchange requires a relativistic quantum field theory treatment that is beyond our present scope. However, the essence may be obtained by drawing an analogy with classical electrodynamics. The electrostatic potential $\phi(\boldsymbol{r})$ in a source-free region is a solution of Laplace's equation,

$$\nabla^2\phi(\boldsymbol{r}) = 0 \tag{3-57}$$

In the presence of a point source with charge q, located at the origin, the equation takes on the form

$$\nabla^2\phi(\boldsymbol{r}) = -\left[\frac{1}{4\pi\epsilon_0}\right]4\pi q\delta(\boldsymbol{r}) \tag{3-58}$$

The solution is the familiar Coulomb potential,

$$\phi(\boldsymbol{r}) = \left[\frac{1}{4\pi\epsilon_0}\right]\frac{q}{r} \tag{3-59}$$

When the electromagnetic field is quantized, photons emerge as the field quanta and the charge becomes the source of the field.

Nuclear force differs from its electromagnetic counterpart in several respects. The most important one is perhaps the short range, and we shall see evidence in support of this in the next chapter. For now, we are concerned mainly with the question of finding an equation similar to Eq. (3-58), and its analogue in quantum field theory, for a short-range nuclear potential. The equation must also be invariant under a Lorentz transformation so that it is correct in the relativistic limit as well. This rules out the Schrödinger equation, which applies only in the nonrelativistic limit. The field quantum exchanged between the nucleons must be a boson, as only bosons can be created and annihilated singly. A fermion, on the other hand, must be created and annihilated together with its antiparticle. The Dirac equation is therefore unsuitable, as it is an equation for spin-$\frac{1}{2}$ particles (i.e., fermions). This leaves the Klein-Gordon equation as the prime candidate.

The relativistic energy-momentum relation is given by the equation

$$E^2 = p^2c^2 + m^2c^4$$

We can quantize this equation in the same way as in nonrelativistic quantum mechanics by replacing energy E with operator $i\hbar(\partial/\partial t)$ and momentum $\boldsymbol{p}$ with operator $-i\hbar\nabla$,

$$-\hbar^2\frac{\partial^2}{\partial t^2}\phi(\boldsymbol{r}) = \left(-\hbar^2c^2\nabla^2 + m^2c^4\right)\phi(\boldsymbol{r}) \tag{3-60}$$

Here, m is now the mass of the field quantum. After dividing both sides of the equation by $(\hbar c)^2$ and rearranging terms, we obtain the familiar Klein-Gordon equation,

$$\left(\nabla^2 - \frac{1}{c^2}\frac{\partial^2}{\partial t^2}\right)\phi(\boldsymbol{r}) = \frac{m^2c^2}{\hbar^2}\phi(\boldsymbol{r}) \tag{3-61}$$

This is only the analogue of Eq. (3-57), as it does not yet contain a source term for field quanta. This point may be further demonstrated by letting the mass of the field quantum m go to zero and ignoring the time dependence. The result is the same as Eq. (3-57).

To include a source, we must find the equivalent of Poisson's equation (3-58) by adding a source term to Eq. (3-61). For simplicity, we shall consider only the static limit and ignore terms involving time derivatives. For a point source with strength g located at the origin, this is given by

$$\nabla^2\phi(\boldsymbol{r}) = \frac{m^2c^2}{\hbar^2}\phi(\boldsymbol{r}) - g\delta(\boldsymbol{r}) \tag{3-62}$$

The solution for this equation,

$$\phi(\boldsymbol{r}) = \frac{g}{4\pi r}e^{-mcr/\hbar} \tag{3-63}$$

has the well-known form of a Yukawa potential and reduces to Eq. (3-59) on letting $m = 0$ and $g = [(4\pi\epsilon_0)^{-1}]4\pi q$. On the other hand, if the field quantum has a finite mass, we find that the strength of the potential drops by $\sim 1/e$ at a distance $r_0 = \hbar/mc$. The quantity r_0 may be taken as a measure of the range of the force mediated by a boson with mass m. For pions ($m \sim 140$ MeV/c^2), the value of r_0 is around 1.4 fm. We shall see later that one-pion exchange gives a good representation of the long-range part of the nuclear potential.

3-7 Nucleon-Nucleon Scattering Phase Shifts

The form of the nucleon-nucleon interaction potential given in Eq. (3-51) was obtained using properties of the deuteron and symmetries in the two-nucleon system. To make further progress, we need additional experimental information, and this is provided by the scattering of one nucleon off another at different energies.

Nucleon-nucleon scattering. In principle there are four types of scattering measurements involving two nucleons that can be carried out. The scattering of an incident proton off a proton (*pp*-scattering) is the simplest one of the four from an experimental point of view, as it is relatively easy to accelerate protons and to construct targets containing hydrogen (proton). For neutron scattering, there are two major sources for incident beam. At low energies, neutrons from nuclear reactors may be used. At higher energies, one can make use of neutrons produced by a beam of protons, for instance, through a (p,n) reaction on a ^{7}Li target. However, both the intensity and the energy resolution of neutron beams obtained in these ways are much more limited than those for proton beams. As a result, neutron scattering is, in general, a more difficult experiment than those with protons. The scattering of neutrons off proton targets (*np*-scattering) and the corresponding *pn*-scattering are however important in that the reaction takes place in the $T = 0$ channel as well. In contrast, *pp*- (and *nn*-) scattering can only provide information on the $T = 1$ state of two nucleons.

In addition to *pp*- and *np*-measurements, one can, in principle, carry out *pn*- and *nn*-scattering experiments as well. Here, instead of using protons as the target, a "neutron target" is used. Free neutrons are unstable, with a half-life on the order of 10 min. It is therefore impossible to construct a "fixed" neutron target, in contrast to protons where material consisting of hydrogen may be used. There are, in principle, two methods of getting around this limitation. One way is to carry out a "colliding beam" experiment. In place of a target fixed in the laboratory, a second neutron beam is used and, instead of having an incident beam scattering from a fixed target, two beams of particles are directed toward each other. Scattering takes place when the particles in the two beams collide. To be practical, such an experiment requires high intensities in both beams, and currently, highly intense beams of neutrons are not easily available.

The other way is to "simulate" a fixed neutron target using deuterium. Since the deuteron is a loosely bound system of a neutron and a proton, the desired *pn*- or *nn*-scattering results can be obtained by carrying out the corresponding *pd*- or *nd*-scattering experiments. The contribution due to protons in the deuterium target may be removed by subtracting from the measured values the corresponding results obtained in *pp*- or *np*-scattering. This procedure is correct provided that:

(1) The subtraction procedure can be carried out with sufficient precision. This requires that the corresponding scattering data on a proton target are available with comparable or better accuracy and that the effect of deuteron binding energy can be corrected in a satisfactory manner. In general, both points are relatively easy to achieve.

(2) Three-body effects are negligible. When a nucleon interacts with a deuteron, the entire system now consists of three nucleons. If there are fundamental three-body forces, their contributions will be present in *pd*- and *nd*-scattering but not in the scattering of one nucleon off another one. Hence, proton-deuteron scattering, for example, may not be the simple sum of *pp*- and *pn*-contributions alone. As we saw earlier, this may not be a problem, as three-body forces, if they exist, are expected to be weak.

The information obtained from *pn*- and *nn*-scattering may not be any different from that in *np*- and *pp*-scattering. For example, the only difference between *pn*- and *np*-scattering is whether the neutron or the proton is the target. Under time-reversal invariance, these two arrangements are expected to give identical results.

As we have seen earlier, both *pp* and *nn* are $T = 1$ systems. If nuclear force is charge independent, the results of *pp*- and *nn*-scattering can only be different by the contribution made by Coulomb interaction. Since the latter is well known, a comparison of *pp*- and *nn*-scattering results can, in principle, test the charge independence of nuclear interaction. However, the accuracy that can be achieved with *nn*-scattering is still inadequate for such a task. In the next section we shall see that there is a possible test at low energies where high precision is relatively easier to attain.

To simplify the notation, we shall use the symbol *NN* from now on to represent a system of two nucleons when there is no need to differentiate between neutrons and protons and the symbol *np* to represent both *np* and *pn* unless further distinction is required by the occasion. Furthermore, we shall assume that Coulomb contribution, where present, has already been taken out and we can therefore ignore it in the discussion.

Our primary interest here is to relate scattering data to the *NN*-interaction potential. A large collection of measured values at a variety of bombarding energies and scattering angles have been accumulated over the years. Instead of relating the potential V directly to the scattering results, it is more common to reduce the experimental information to phase shifts δ_ℓ for different ℓ-partial waves. The merit of a particular potential is often judged by comparing the calculated phase shifts with those extracted from experimental data, such as the example shown later in Fig. 3-3. For this reason we shall briefly review first the subject of partial-wave analysis for *NN*-scattering. A more detailed discussion is given in §B-2.

Scattering cross section. The quantity measured in a scattering experiment is the number of counts registered by a detector at angle (θ, ϕ). The counting rate depends on the solid angle subtended by the detector at the scattering center, the intensity of the incident beam, the number of target nuclei involved, and the differential cross section $d\sigma/d\Omega$. Our primary interest is in $d\sigma/d\Omega$, a function of the bombarding energy as well as the scattering angle.

In the nonrelativistic limit, the scattering of one particle off another is described by the Schrödinger equation. In the center of mass, the wave function is the solution of the equation

$$-\frac{\hbar^2}{2\mu}\nabla^2\psi + (V - E)\psi = 0 \tag{3-64}$$

where μ is the reduced mass in the two-nucleon system. We shall be mainly concerned with short-range nuclear forces here (see §B-5 for Coulomb scattering), and consequently we can assume that $V = 0$ except in a very small region where scattering takes place.

In the asymptotic region, far away from the small volume where V is different from zero, the wave function has the form

$$\psi(r, \theta, \phi) \xrightarrow[r\to\infty]{} e^{ikz} + f(\theta, \phi)\frac{e^{ikr}}{r} \tag{3-65}$$

where the term $\exp\{ikz\}$ represents the incident plane wave and the part of the incident beam unaffected by the reaction. The scattered wave is given by a spherical function, $r^{-1}e^{ikr}$, radiating outward from the scattering center. The probability of scattering to direction (θ, ϕ) is specified by the scattering amplitude $f(\theta, \phi)$. For simplicity, we shall consider first only elastic scattering, and as a result, the wave number k in the center of mass of the two particles has the same magnitude before and after the scattering. The differential scattering cross section at angles (θ, ϕ) is given by Eq. (B-7) as the square of the scattering amplitude,

$$\frac{d\sigma}{d\Omega}(\theta, \phi) = |f(\theta, \phi)|^2 \tag{3-66}$$

As shown in Fig. 3-2, the geometry of a scattering arrangement is such that it is convenient to place the origin of the coordinate system at the center of the scattering region and take the direction of the incident beam as the positive direction along the z-axis. The incident wave vector $\boldsymbol{k}$ and the scattered wave vector $\boldsymbol{k}'$ define a plane, the scattering plane.

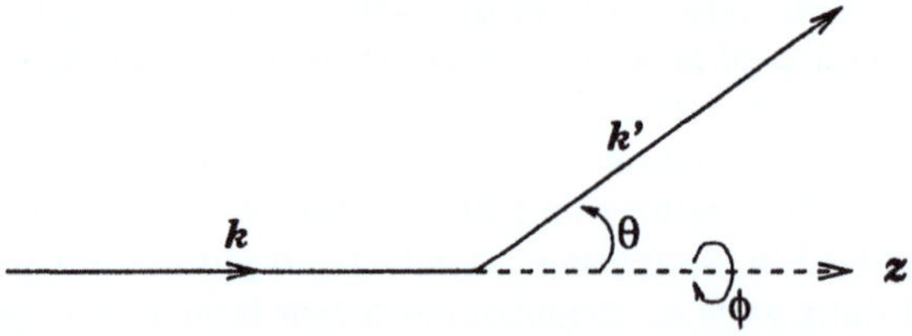

Figure 3-2: Schematic diagram of a scattering arrangement. The scattering angle θ is between wave vectors $\boldsymbol{k}$, along the direction of the projectile, and $\boldsymbol{k}'$, that of the scattered particle. The result is independent of the azimuthal angle ϕ unless the orientation of the spin of one of the particles involved is known.

If nucleons in the incident beam and in the target are not polarized—that is, there is no preferred direction in space with which the intrinsic spins are aligned—the scattering is invariant with respect to a rotation around the z-axis. In such cases, the cross section is independent of the azimuthal angle ϕ and the differential cross section is a function of the scattering angle θ alone. We shall return later to the more general case where the orientation of the intrinsic spin is also detected for one or more nucleons involved in the scattering.

Partial-wave analysis. For a central potential, the relative angular momentum ℓ between the two scattering nucleons is a conserved quantity. Under such conditions, it is useful to expand the wave function as a sum over the contributions from different partial waves, each with a definite ℓ-value,

$$\psi(r,\theta) = \sum_{\ell=0}^{\infty} a_\ell Y_{\ell 0}(\theta) R_\ell(k,r) \tag{3-67}$$

Here a_ℓ are the expansion coefficients. Only spherical harmonics $Y_{\ell m}(\theta,\phi)$ with $m=0$ appears in the expansion since, in the absence of polarization, the wave function is independent of the azimuthal angle ϕ. We have explicitly included the wave number k here in the argument of the radial wave function $R_\ell(k,r)$ so as to emphasize the dependence on energy.

For a free particle, $V=0$, and the radial wave function reduces to

$$R_\ell(k,r) \xrightarrow{\text{free}} j_\ell(kr) \xrightarrow[r\to\infty]{\text{free}} \frac{1}{kr}\sin(kr - \tfrac{1}{2}\ell\pi) \tag{3-68}$$

where $k=\sqrt{2\mu E}/\hbar$ and $j_\ell(\rho)$ is the spherical Bessel function of order ℓ. If only elastic scattering is allowed by the potential, the probability current density in each partial-wave channel is conserved. The only effect the potential can have on the wave function is a change in the phase angle. In other words,

$$R_\ell(k,r) \xrightarrow[r\to\infty]{\text{scatt.}} \frac{1}{kr}\sin(kr - \tfrac{1}{2}\ell\pi + \delta_\ell) \tag{3-69}$$

where δ_ℓ is the *phase shift* in the ℓth partial-wave channel. (For more details, see §B-2.)

Using Eq. (B-16), the scattering amplitude may be expressed in terms of δ_ℓ as

$$f(\theta) = \frac{\sqrt{4\pi}}{k}\sum_{\ell=o}^{\infty}\sqrt{2\ell+1}\, e^{\imath\delta_\ell}\sin\delta_\ell Y_{\ell 0}(\theta) \tag{3-70}$$

Using Eq. (3-66), the differential scattering cross section may be written in terms of the phase shifts,

$$\frac{d\sigma}{d\Omega} = \frac{4\pi}{k^2}\left|\sum_{\ell=0}^{\infty}\sqrt{2\ell+1}\, e^{\imath\delta_\ell}\sin\delta_\ell Y_{\ell 0}(\theta)\right|^2$$

as given in Eq. (B-17). The scattering cross section, the integral of $d\sigma/d\Omega$ over all solid angles, becomes

$$\sigma = \int \frac{d\sigma}{d\Omega}d\Omega = \frac{4\pi}{k^2}\sum_{\ell=0}^{\infty}(2\ell+1)\sin^2\delta_\ell(k) \tag{3-71}$$

Decomposition into partial waves is a useful way to analyze the scattering results for a given bombarding energy. In particular, only a few of the low-order partial waves can contribute to the scattering at low energies, as shown in §B-2. For realistic nuclear potentials, the orbital angular momentum is not conserved. A partial-wave expansion remains to be useful as only a limited number of ℓ-values can be admixed by the noncentral forces. We shall see how to handle such cases later.

Nucleon-nucleon scattering phase shifts. Realistic nucleon-nucleon scattering differs in several important respects from the simple, central potential processes discussed above. First, we found out earlier in §3-5 that nuclear potential is a function of also the total intrinsic spin of the two nucleons. As a result, the total angular momentum $\boldsymbol{J} = \boldsymbol{\ell} + \boldsymbol{S}$, rather than the orbital angular momentum $\boldsymbol{\ell}$, is conserved in the scattering. For two nucleons, the value of total intrinsic spin S can be either 0 or 1. To determine the value of S, we need to detect the orientations, or *polarizations*, of the spins of the nucleons involved. In fact, the information on NN-scattering is incomplete unless polarizations are also observed. Second, with sufficient energy, scattering can excite the internal degrees of freedom in nucleons, for example, by changing one of them to a Δ-particle through such reactions as

$$p + p \rightarrow \Delta^{++} + n$$

or producing secondary particles, such as pions,

$$p + p \rightarrow p + n + \pi^+$$

and baryon-antibaryon pairs,

$$p + p \rightarrow p + p + p + \overline{p}$$

These are inelastic scattering events, as part of the incident kinetic energy is converted into excitation energies or mass of the particles created.

Since we are dealing with identical fermions, the scattering of two nucleons can take place only in a state that is totally antisymmetric with respect to a permutation of the two particles, in the same way as discussed earlier for the deuteron. For pp-scattering, we have $T = 1$ and the two nucleons are symmetric, as far as their total isospin wave function is concerned. If the intrinsic spins of the two protons are coupled together to $S = 0$ (antisymmetric state), the relative orbital wave function must be in a symmetric state and, as a result, only even ℓ-values are allowed. For $S = 0$, we have $J = \ell$, and the partial waves for the lowest two orders of pp-scattering are ${}^1S_0$ ($\ell = 0$) and ${}^1D_2$ ($\ell = 2$). The phase shifts extracted from measured pp-scattering data for these two partial waves at bombarding energy less than 300 MeV in the laboratory are shown in Fig. 3-3(a) as illustrative examples. Only the real part of the phase shifts are given. At laboratory energy less than 300 MeV, contributions from inelastic scattering are still relatively unimportant and the imaginary parts of the phase shifts extracted from measured scattering cross sections are small.

By the same token, partial waves for triplet ($S = 1$) pp-scattering have odd ℓ-values. The lowest order in this case is a p-wave ($\ell = 1$). When $\ell = 1$ is coupled with $S = 1$, three states, with $J = 0$, 1, 2, are produced. The phase shifts for two of the triplet of states, ${}^3P_0$ and ${}^3P_1$, are also shown in Fig. 3-3(a). There is no admixture between the

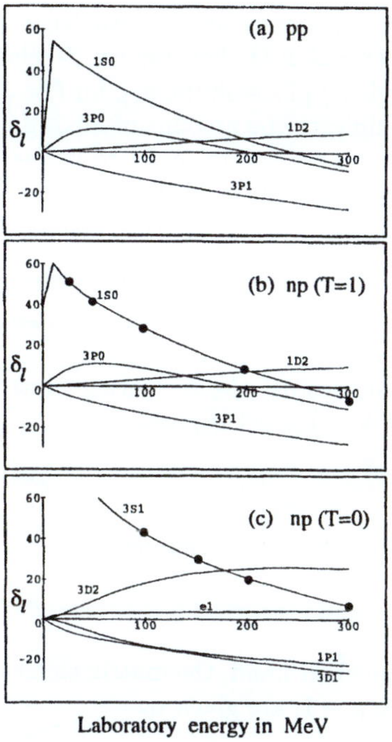

Figure 3-3: Real part of nucleon-nucleon scattering phase shifts in degrees for low-order partial waves [10]: (*a*) proton-proton scattering with contributions from the Coulomb potential removed, (*b*) isovector neutron-proton scattering, and (*c*) isoscalar neutron-proton scattering. Filled circles in the $^1S_0$ and $^3S_1$ phase shifts of *np*-scattering are the calculated results using a Paris potential [138].

two $J = 0$ states, $^3P_0$ and $^1S_0$, as they are of different parity. As a result, we find that both ℓ and S are good quantum numbers here by default.

The *np*-system may be coupled together to either isospin $T = 0$ or $T = 1$. For $T = 0$, the two nucleons are antisymmetric in isospin. In this case, the $S = 0$ states must have odd ℓ-values in order to be antisymmetric in the total wave function. The lowest order partial wave here is $\ell = 1$ and the phase shifts for $^1P_1$-scattering extracted from experimental data are shown in Fig. 3-3(c). In order for p-wave *np*-scattering to be in the $S = 1$ state, it is necessary for the total isospin to be $T = 1$. The phase shifts in this case are expected to be identical to those found in *pp*-scattering, if nuclear force is charge independent and Coulomb effects are removed. An examination of the two sets of empirical p-wave phase shifts, $^3P_0$ and $^3P_1$, given in Fig. 3-3(b), shows that they are only slightly different from the corresponding values given in Fig. 3-3(a) for *pp*-scattering. It is not clear whether the small differences come from the way the phase shifts are extracted from experimental scattering cross sections or they are indications of a weak charge dependence in the nuclear force. We shall return to this point in the next section in a discussion on the difference between the scattering lengths for *pp*- and *np*-scattering.

The other $T = 0$ phase shifts in the *np*-system, shown in Fig. 3-3(c), are for triplet ($S = 1$), even ℓ-scattering. This is the first time we encounter a mixing of different ℓ-partial waves. Up to now, each phase shift has been characterized by a definite ℓ-value (as well as J- and S-values) even though the orbital angular momentum is not fundamentally a good quantum number. Mixing of different ℓ-partial waves has not taken place because of parity and other invariance conditions. As in the deuteron case, the tensor force can mix two triplet states of the same J but different in ℓ by two units ($\ell = J \pm 1$). For a given J-value, the scattering is now specified by two (energy-dependent) phase shifts, $\delta_{J>}$ for $\ell = J - 1$ and $\delta_{J<}$ for $\ell = J + 1$, as well as a parameter ϵ_J to indicate the amount of mixing between the two at a given energy.

There are several ways to define the parameter ϵ_J. The usual convention used in the literature today is that of Stapp, Ypsilantis, and Metropolis [132]. In this system of definitions, the scattering matrix (see §B-6) for a given J is written in the form

$$\{S\} = \begin{pmatrix} e^{2i\delta_{J>}} \cos 2\epsilon_J & ie^{i(\delta_{J>}+\delta_{J<})} \sin 2\epsilon_J \\ ie^{i(\delta_{J>}+\delta_{J<})} \sin 2\epsilon_J & e^{2i\delta_{J<}} \cos 2\epsilon_J \end{pmatrix} \tag{3-72}$$

In other words, the scattering matrix element from $\ell = J + 1$ channel to the same $\ell = J + 1$ channel is given by

$$e^{2i\delta_J} = e^{2i\delta_{J>}} \cos 2\epsilon_J$$

and from $\ell = J - 1$ channel to the same $\ell = J - 1$ channel,

$$e^{2i\delta_J} = e^{2i\delta_{J<}} \cos 2\epsilon_J$$

On the other hand, the matrix elements from $\ell = J - 1$ to $\ell = J + 1$, and from $\ell = J + 1$ to $\ell = J - 1$, are given by

$$e^{2i\delta_J} = e^{i(\delta_{J>}+\delta_{J<})} \sin 2\epsilon_J$$

These are generally referred to as the *nuclear bar* phase shifts. For the triplet $J = 1$ state, the values of ϵ_1 deduced from experimental data are shown as a part of Fig. 3-3(c) for illustration.

Spin polarization in nucleon-nucleon scattering. Because of spin dependence in the nuclear potential, the scattering cross sections between nucleons are different depending on whether the sum of their intrinsic spins is coupled to $S = 0, 1$. To observe S, it is necessary to detect the orientation of nucleon intrinsic spins. Since each nucleon is a spin-$\frac{1}{2}$ particle, its projection on the quantization axis can either be $+\frac{1}{2}$ or $-\frac{1}{2}$. If the spins of all the nucleons in the incident beam are aligned in a particular direction, the beam is said to be a *polarized* one. Similarly, if the spins of the target nucleons are oriented along a given direction, the target is a polarized one. When the orientation of spins is taken into account, there are four possible combinations for the two nucleons in the initial state, as well as the final state, in a nucleon-nucleon scattering experiment. These four may be represented as $|+\frac{1}{2},+\frac{1}{2}\rangle$, $|-\frac{1}{2},+\frac{1}{2}\rangle$, $|+\frac{1}{2},-\frac{1}{2}\rangle$ and $|-\frac{1}{2},-\frac{1}{2}\rangle$. Since spin orientations can be changed by spin dependence in the interaction, there are 16 possible different polarization measurements that can be carried out, corresponding to starting from any one of the four incident spin combinations to any one of the four final ones.

Mathematically, we may write the scattering amplitude as a 4×4 matrix, with each of the elements representing the probability for one of the 16 possible arrangements for the scattering. These 16 quantities are not independent of each other. Because of time reversal and other symmetries inherent in the scattering, only five matrix elements are unique for each type of nucleon pair, i.e., *np*- or *pp*-scattering. This can be seen by writing the five independent scattering amplitudes in the following manner:

$$\begin{aligned} f_1 &= f_{++,++} = f_{--,--} \qquad & f_2 &= f_{++,--} = f_{--,++} \\ f_3 &= f_{+-,+-} = f_{-+,-+} \qquad & f_4 &= f_{+-,-+} = f_{-+,+-} \\ f_5 &= f_{++,+-} = f_{-+,--} = f_{--,+-} = f_{-+,++} \\ &= f_{--,-+} = f_{+-,++} = f_{++,-+} = f_{+-,--} \end{aligned} \tag{3-73}$$

where we have used + and − in the subscript to stand, respectively, for $+\frac{1}{2}$ and $-\frac{1}{2}$ projections of the spins of two nucleons in the initial and in the final states.

Instead of scattering amplitudes f, it is more common to express *NN*-scattering as the matrix element of the t-matrix operator defined in Eq. (B-64). An element of the t-matrix is related to the scattering amplitude in the following way:

$$f_{\boldsymbol{k}'\boldsymbol{k}} = -\frac{\mu}{2\pi\hbar^2}\langle \boldsymbol{k}'|t|\boldsymbol{k}\rangle \tag{3-74}$$

where $|\,\boldsymbol{k}\,\rangle$ and $|\,\boldsymbol{k}'\,\rangle$ are, respectively, the initial and final states of the two nucleons. In the place of f_1 to f_5, the t-matrix for nucleon-nucleon interaction is often written as a function of five coefficients, A, B, C, E, and F, in the form of an operator:

$$\begin{aligned} t_{\boldsymbol{k}'\boldsymbol{k}}(1,2) &= A + B\sigma_n(1)\sigma_n(2) + C\{\sigma_n(1) + \sigma_n(2)\} \\ &\quad + E\sigma_q(1)\sigma_q(2) + F\sigma_p(1)\sigma_p(2) \end{aligned} \tag{3-75}$$

The three directional vectors $\boldsymbol{n}$, $\boldsymbol{p}$, and $\boldsymbol{q}$, along which the nucleon spin components are taken, are defined in terms $\boldsymbol{k}$ and $\boldsymbol{k}'$,

$$\hat{\boldsymbol{n}} = \frac{\boldsymbol{k}\times\boldsymbol{k}'}{|\boldsymbol{k}\times\boldsymbol{k}'|} \qquad \hat{\boldsymbol{q}} = \frac{\boldsymbol{k}'-\boldsymbol{k}}{|\boldsymbol{k}'-\boldsymbol{k}|} \qquad \hat{\boldsymbol{p}} = \hat{\boldsymbol{q}}\times\hat{\boldsymbol{n}} \tag{3-76}$$

The relation between the five coefficients A to F and the five scattering amplitudes f_1 to f_5, as well as other common ways of writing the *NN*-scattering t-matrix, can be found in standard references such as Bystricky, Lehar, and Winternitz [41]. Instead of *np*- and *pp*-pairs, decomposition of the *NN*-scattering amplitude into five independent quantities may also be carried out in terms of $T = 0$ and $T = 1$ states of the two nucleons.

The amount of independent information obtained from scattering is greatly increased with polarization measurements. The experiments are, unfortunately, far more difficult to carry out than ordinary scattering measurements. Polarized beams are fairly common these days, and it is relatively easy to carry out analyzing power (A_y) measurements. Here, only beam is polarized. Because of spin dependence, the differential cross section at a fixed scattering angle θ may be different depending on whether the spin of the incident nucleon is polarized along the unit vector $\hat{\boldsymbol{n}}$ defined in Eq. (3-76) or antiparallel to it. Such a difference is characterized by the analyzing power. This supplies one of the five independent quantities in the scattering. The sum of the same two differential cross sections supplies the other.

For additional information, the polarization of the scattered particle must be measured. The only efficient way to detect the polarization of a nucleon is to carry out a second scattering off a target of known analyzing power. The asymmetry in the differential cross section of the second scattering provides us with information on the polarization direction of the nucleon after the first scattering. Since the cross section is low in general for nuclear processes, a second scattering greatly complicates the experimental setup and reduces the rate of data collection. In spite of such difficulties, more and more high-precision data involving the polarization of the scattered particle are becoming available. Valuable information can also be obtained using polarized targets. However, this requires sophisticated low-temperature techniques to "freeze" the spin orientations of the nucleons with respect to some given spatial direction, such as that provided by an external magnetic field. Data involving such targets are still quite rare as a result.

Inelastic scattering. With sufficient kinetic energy available in the center of mass of the two-nucleon system, inelastic reactions become possible. Since the mass of the lightest meson, $\pi^{\pm}$, is around 140 MeV/c^2, we expect pion production to take place once the bombarding energy is above the threshold (see Fig. 3-4). As the energy increases, excitation of the internal degrees of freedom of the nucleon as well as the production of secondary particles become increasingly more likely. Inelastic scattering represents a loss of flux from the incident channel, and as far as the incident channel is concerned, the probability amplitude is no longer conserved. Such a situation may be described by a complex scattering potential.

Both the scattering amplitude and the phase shifts produced by a complex potential are also complex in general. Let us define a scattering amplitude f_ℓ for the ℓth partial

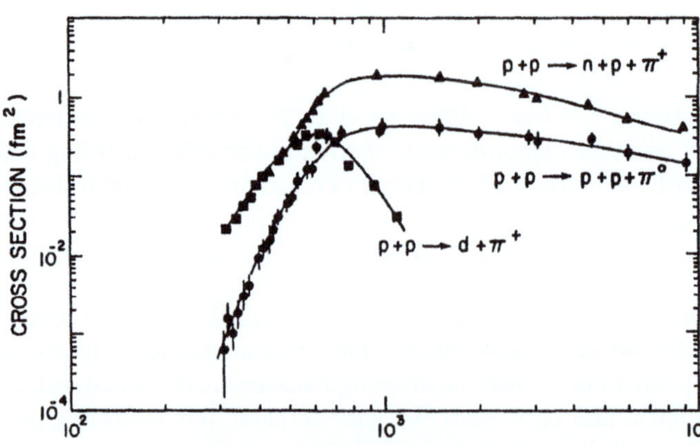

Figure 3-4: Energy dependence of the total cross section for pion production in pp-scattering leading to final states $d + \pi^+$, $p + n + \pi^+$, and $p + p + \pi^0$. (Adapted from Ref. [90].)

wave by the expression

$$f(\theta) = \sum_{\ell=o}^{\infty} \sqrt{4\pi(2\ell+1)}\, f_\ell Y_{\ell 0}(\theta) \tag{3-77}$$

For purely elastic scattering,

$$f_\ell = \frac{1}{k} e^{i\delta_\ell} \sin\delta_\ell = \frac{1}{2ik}(e^{2i\delta_\ell} - 1) \tag{3-78}$$

as we have seen earlier. If inelastic scattering is also taking place, the scattering amplitude becomes complex and may be expressed in terms of two real numbers: η_ℓ (the inelasticity parameter) and ζ_ℓ, defined by

$$f_\ell = \frac{1}{2ik}(\eta_\ell e^{2i\zeta_\ell} - 1) \tag{3-79}$$

The energy dependence of complex scattering amplitudes is often displayed in the form of Argand diagrams in terms of the locus of the point

$$z = kf_\ell = \frac{1}{2i}(\eta_\ell e^{2i\zeta_\ell} - 1) \tag{3-80}$$

in the complex plane [22, 152].

An examination of the values deduced from experimental nucleon-nucleon scattering data shows that the phase shifts are essentially purely real until the energy is above 300 MeV in the laboratory (~150 MeV in the center of mass). At much higher energies, the real and imaginary parts become comparable to each other, as more and more inelastic channels are open. Complete lists of phase shifts up to 1 GeV in laboratory scattering energy are available, for example, from Arndt, Hyslop, and Roper [10].

3-8 Low-Energy Scattering Parameters

Effective range analysis. If we make a partial-wave expansion of the scattering wave function, as given in Eq. (3-67) for example, and substitute the results into the Schrödinger equation (3-64), we obtain an equation for the radial wave function for orbital angular momentum ℓ:

$$-\frac{\hbar^2}{2\mu}\left\{\frac{1}{r^2}\frac{d}{dr}r^2\frac{d}{dr} - \frac{\ell(\ell+1)}{r^2}\right\}R_\ell(k,r) + V(r)R_\ell(k,r) = ER_\ell(k,r) \tag{3-81}$$

The term $\ell(\ell+1)/r^2$ comes from the angular part of the kinetic energy and is sometimes referred to as a centrifugal barrier, as it is repulsive to an incoming particle. The "effective potential" experienced by the scattering particle is then

$$\widetilde{V}(r) = V(r) + \frac{\hbar^2}{2\mu}\frac{\ell(\ell+1)}{r^2}$$

as given in Eq. (B-24). Because of the barrier, scattering at low energies is dominated by partial waves for small ℓ-values. In particular, for $E < 10$ MeV, nucleon-nucleon scattering is essentially given by s-waves ($\ell = 0$) alone, as can be seen from Fig. 3-3 from the fact that only δ_0, the $\ell = 0$ phase shifts, are significantly different from zero.

When the kinetic energy $E \to 0$, the total cross section remains finite for nucleon-nucleon scattering. The limiting value is often characterized by a length parameter a defined by the relation

$$\lim_{E\to 0} \sigma = 4\pi a^2 \tag{3-82}$$

The quantity a is known as the *scattering length*, and it is often convenient to discuss extremely low energy scattering in terms of it instead of the s-wave phase shift. The two quantities are related in the following way, as shown in Eq. (B-28):

$$a = \lim_{k\to 0} \Re\Big\{-\frac{1}{k}e^{i\delta_0}\sin\delta_0\Big\}$$

where $k^2 = 2\mu E/\hbar^2$ and $\Re$ indicates the real part. The energy dependence of δ_0 at low energies is given by the *effective range* parameter r_e, defined by the relation

$$k\cot\delta_0 = -\frac{1}{a} + \frac{1}{2}r_e k^2 \tag{3-83}$$

A more detailed discussion of these parameters and their relation to the nucleon-nucleon interaction potential is given in §B-3.

Scattering length and effective range provide a useful way to parametrize information on low-energy nucleon-nucleon scattering. Furthermore, these parameters may be related to observations other than *NN*-scattering, such as deuteron binding energy. In addition, very accurate results can be obtained for the *np*-system by scattering slow neutrons off protons in hydrogen atoms bound in H_2 molecules. For these reasons, a great deal of attention is devoted to the measurement and understanding of these parameters.

Neutron scattering off hydrogen molecules. The hydrogen molecule, H_2, is a *homonuclear* molecule, a diatomic molecule made of two identical nuclei. Since the distance between the two atoms is large (7.8×10^{-11} m), compared with the range of nuclear force, we do not need to consider any nuclear interaction between the two protons in a H_2 molecule. On the other hand, being identical particles, they must obey the Pauli exclusion principle. Like other two-nucleon systems, the allowed states for two protons in a hydrogen molecule must be antisymmetric in the product of their orbital and spin wave functions. For this reason, the spin orientations of the two protons are correlated with their relative orbital angular momentum, and such a correlation may be exploited for neutron-proton scattering length measurements.

There are two low-lying states for a hydrogen molecule. The lower one in energy is the *para*-hydrogen state, in which the two protons are symmetric relative to each other in their spatial wave function. The higher energy state is the *ortho*-hydrogen state, in which the two protons are antisymmetric in spatial wave function. For an ortho-hydrogen, it is necessary that the intrinsic spins of the two protons be coupled together to $S_H = 1$ to satisfy the Pauli principle. For this arrangement, M_S, the projection of $\boldsymbol{S}_H$ on the quantization axis, can take on any one of three values, -1, 0, and $+1$, and, consequently, there are three possible states associated with each ortho-hydrogen molecule. The total intrinsic spin of the two protons in a para-hydrogen, in contrast, is $S_H = 0$ and there is only one possible state. At room temperature, the thermal energy is much higher than the difference between the para- and ortho-hydrogen, and we expect ortho-hydrogen to have three times the statistical weight of para-hydrogen in a sample at thermal equilibrium. On the other hand, at low temperatures, hydrogen molecules tend to go into the lowest possible energy state and, as a result, are almost completely in the lower energy para-hydrogen state. Thus the relative amount of para- and ortho-hydrogen in a sample may be controlled by varying the temperature of the sample.

Measurements of low-energy neutron scattering from hydrogen molecules can be carried out with high precision partly because of the intense neutron flux available from, for example, reactors. By lowering the energy, the wavelength of incident neutrons can be made sufficiently long so that scattering off the two protons in a hydrogen molecule is a coherent one. Low-energy neutrons are also useful in that very little energy is transferred to the hydrogen target. Energy received by a molecule may cause transitions from para- to ortho-hydrogen states, and this reduces the accuracy that can be achieved in a measurement. For these reasons, the neutron energy is kept low, around 10 meV (1 meV $= 10^{-3}$ eV), corresponding to the average thermal energy at temperature 100 K. At such low energies, contributions from the effective range term in Eq. (3-83) may be ignored, and the scattering is characterized by the two *np*-scattering lengths.

For $\ell = 0$, a neutron-proton system is either in its singlet state with $S = 0$ and $T = 1$ or in its triplet state with $S = 1$ and $T = 0$. The scattering length in the form of an operator may be expressed in the following way:

$$\boldsymbol{a} = \tfrac{1}{4}(3a_t + a_s) + (a_t - a_s)\boldsymbol{s}_n \cdot \boldsymbol{s}_p \tag{3-84}$$

where a_t is the scattering length in the triplet state and a_s that for the singlet state. The operators $\boldsymbol{s}_n = \frac{1}{2}\boldsymbol{\sigma}_n$ and $\boldsymbol{s}_p = \frac{1}{2}\boldsymbol{\sigma}_p$ act, respectively, on the intrinsic spins of the neutron and the proton. Similar to $\boldsymbol{\sigma}_1 \cdot \boldsymbol{\sigma}_2$ given in Eq. (3-55), the scalar product $\boldsymbol{s}_n \cdot \boldsymbol{s}_p$ is sensitive to the sum of intrinsic spins of the neutron-proton system. It is easy to check that the expectation value of $\boldsymbol{a}$ in Eq. (3-84) is a_t in a triplet state and a_s in a singlet state for a neutron-proton system.

Returning now to the hydrogen molecule, we may write the operator for the sum of the intrinsic spins of the two protons as

$$\boldsymbol{S}_H = \boldsymbol{s}_{p_1} + \boldsymbol{s}_{p_2} \tag{3-85}$$

In terms of $\boldsymbol{S}_H$, the scattering length for a slow neutron from two protons in a hydrogen molecule may be written in a form similar to Eq. (3-84) above,

$$\boldsymbol{a}_H = \tfrac{1}{2}(3a_t + a_s) + (a_t - a_s)\boldsymbol{s}_n \cdot \boldsymbol{S}_H \tag{3-86}$$

The scattering cross section is given by the expectation value of $\boldsymbol{a}_H$ squared,

$$\begin{aligned} \sigma_H &= 4\pi\langle a_H^2\rangle \\ &= 4\pi\Big\{\tfrac{1}{4}(3a_t + a_s)^2 + (3a_t + a_s)(a_t - a_s)\langle \boldsymbol{s}_n \cdot \boldsymbol{S}_H\rangle + (a_t - a_s)^2\langle(\boldsymbol{s}_n \cdot \boldsymbol{S}_H)^2\rangle\Big\} \end{aligned} \tag{3-87}$$

For unpolarized incident neutrons, the second term in the final form vanishes on averaging over all possible orientations of the intrinsic spin of an incident neutron. The third term may be simplified by applying the same argument in the following way. First, we expand the operator in terms of the Cartesian components of the intrinsic spins,

$$\begin{aligned}(\boldsymbol{s}_n \cdot \boldsymbol{S}_H)^2 &= s_{nx}^2 S_{Hx}^2 + s_{ny}^2 S_{Hy}^2 + s_{nz}^2 S_{Hz}^2 \\ &\quad + 2s_{nx}s_{ny}S_{Hx}S_{Hy} + 2s_{ny}s_{nz}S_{Hy}S_{Hz} + 2s_{nz}s_{nx}S_{Hz}S_{Hx}\end{aligned}$$

The expectation values of the last three terms in the expression are again zero for an unpolarized neutron beam. For the first three terms we note that, since $\boldsymbol{s}^2 = s_x^2 + s_y^2 + s_z^2$ and $\langle \boldsymbol{s}^2 \rangle = \frac{3}{4}$, we have

$$\langle s_{nx}^2 \rangle = \langle s_{ny}^2 \rangle = \langle s_{nz}^2 \rangle = \tfrac{1}{4}$$

As a result,

$$\langle s_{nx}^2 S_{Hx}^2 + s_{ny}^2 S_{Hy}^2 + s_{nz}^2 S_{Hz}^2 \rangle = \tfrac{1}{4}\langle S_{Hx}^2 + S_{Hy}^2 + S_{Hy}^2 \rangle = \tfrac{1}{4}\langle \boldsymbol{S}_H^2 \rangle = \tfrac{1}{4} S_H(S_H+1)$$

and

$$\sigma_H = \pi\{(3a_t + a_s)^2 + (a_t - a_s)^2 S_H(S_H+1)\} \tag{3-88}$$

For para-hydrogen, we have $S_H = 0$, and the cross section is

$$\sigma_{\text{para}} = \pi(3a_t + a_s)^2 \tag{3-89}$$

For ortho-hydrogen, we have $S_H = 1$, and the result is

$$\sigma_{\text{ortho}} = \pi(3a_t + a_s)^2 + 2\pi(a_t - a_s)^2 \tag{3-90}$$

From the values of σ_{para} and σ_{ortho} measured with slow neutron scattering off hydrogen molecules, the values of scattering lengths a_s and a_t may be deduced, and the results are listed in Table 3-3. Data of similar quality can also be obtained from coherent scattering of slow neutrons from protons bound in crystals, from crystal diffraction, and from reflection of slow neutrons by liquid hydrocarbon mirrors.

Table 3-3: Nucleon-nucleon scattering length (a) and effective range (r_e).

		$S=0,\ T=1$ (fm)	$S=1,\ T=0$ (fm)
pp	a	-17.1 ± 0.2	—
	r_e	2.794 ± 0.015	—
nn	a	-16.6 ± 0.6	—
	r_e	2.84 ± 0.03	—
np	a	-23.715 ± 0.015	5.423 ± 0.005
	r_e	2.73 ± 0.03	1.73 ± 0.02

Neutron-proton scattering length. Let us examine first the singlet scattering length for the np-system. Since this is a system with isospin $T = 1$, we can compare its value with a_{pp} and a_{nn}, the scattering lengths for pp- and nn-scattering. The signs in all three cases are negative. Using the definition for the sign given in §B-3, this means that there is no bound state for two nucleons in $T = 1$, a fact we encountered earlier in the discussion on deuterons.

The *pp*-scattering length a_{pp} is easily measured from low-energy proton scattering off a hydrogen target. However, since the cross section for Coulomb scattering, given later in Eq. (4-7), is inversely proportional to the square of the energy, the observed *pp*-scattering at low energies is dominated by electromagnetic effects. In principle, one can subtract the contributions of the Coulomb term from the measured values. However, the accuracy one can achieve in practice is rather limited, as the cross section for nuclear scattering is only a very small part of the total measured value. For example, the scattering length corresponding to the measured cross section is -7.823 ± 0.01 fm, and after correction for Coulomb effects, the *pp*-scattering length is -17.1 ± 0.2 fm (see, e.g., Ref. [112]).

Measurements of the *nn*-scattering length are complicated by the absence of fixed neutron targets. Several different types of experiment have been carried out to deduce the value of a_{nn} using either deuterons or tritons in reactions, such as

$$n + d \to p + n + n \qquad\qquad n + t \to d + n + n$$
$$d + d \to p + p + n + n \qquad\qquad t + d \to {}^3\text{He} + n + n$$
$$t + t \to \alpha + n + n$$

With the availability of good quality pion beams in recent years, the reaction

$$\pi^- + d \to \gamma + n + n$$

has also been used to reduce the measured uncertainty of a_{nn}. Here, instead of relying on scattering of neutrons off neutrons, the value of a_{nn} is obtained from "final state interaction," that is, changes in the observed reaction cross section due to interaction between the two emerging neutrons. The observed values of a_{nn} are displayed in Fig. 3-5

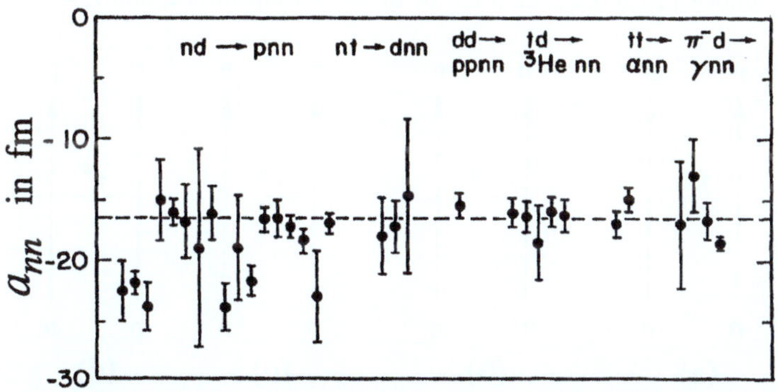

Figure 3-5: Distribution of the measured values of *nn*-scattering length a_{nn} in chronological order from left to right. The average value of -16.6 ± 0.6 fm is indicated by the dashed line. (Taken from Ref. [73].)

in chronological order, and the average of these values, shown as the dashed line in the figure, is listed in Table 3-3.

A comparison of the three values of $T = 1$ scattering length is of interest. First we find that the value $a_{np} = -23.715 \pm 0.015$ fm is noticeably larger than those for a_{pp} and a_{nn}. It is tempting to treat the difference as a possible indication of charge dependence in nuclear force. However, most of the difference may be explained by the following argument. At low scattering energies, the two nucleons are never very close to each other and, as a result, only the long-range part of the nuclear force is operating. The nuclear interaction here is dominated by the exchange of a single pion. For a pair of protons and a pair of neutrons, only a neutral pion can be exchanged. On the other hand, a charged pion can also be exchanged in the interaction between a neutron and a proton, as shown in Fig. 3-6. This can take place by the proton emitting a π^+ and changing itself into a neutron while the original neutron becomes a proton on absorbing the positive pion. Alternatively, the neutron may emit a π^- and change into a proton while the original proton converts itself into a neutron on absorbing the pion. These are "exchange" processes, as the "identities" of the neutron and proton are interchanged. Since in quantum mechanics it is not possible to follow the trajectory of a particle as it interacts with another indistinguishable particle, there is no way to associate either one of the two nucleons in the final state with a particular one in the initial state. As a result, we cannot distinguish an exchange process from a direct one in which a neutral pion is exchanged and the contributions of both processes must be included in an *np*-scattering. Because of the small mass difference between charged and neutral pions,

$$m_{\pi^\pm} - m_{\pi^0} = 4.6 \text{ MeV}$$

we expect a small difference in the interaction between a proton and a neutron from that for two identical nucleons. In this way, most of the difference between the *np*-scattering length in the triplet state and the scattering lengths for *pp*- and *nn*-systems can be accounted for (for more details, see, e.g., Ref. [149]). The observed difference of a_{np} from a_{pp} and a_{nn}, therefore, cannot be taken as an indication of a fundamental charge dependence in the nuclear force or in the strong interaction itself.

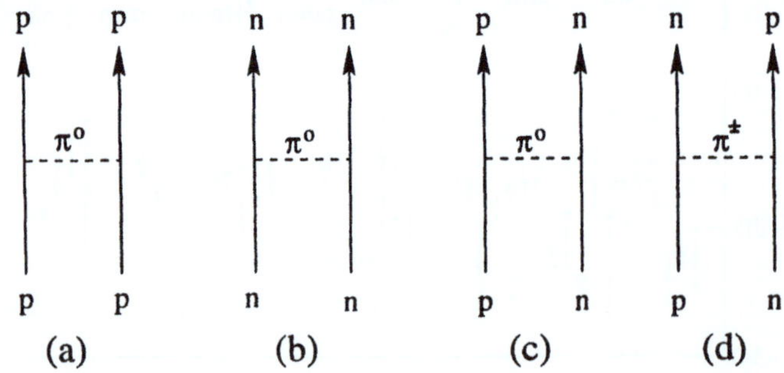

Figure 3-6: One-pion exchange diagrams: (*a*) *pp*-interaction, (*b*) *nn*-interaction, (*c*) direct term, and (*d*) exchange term for *np*-interaction.

The difference between $a_{pp} = -17.1 \pm 0.2$ fm and $a_{nn} = -16.6 \pm 0.6$ fm is not significant at this time because of the large uncertainty associated with the measured value of a_{nn}. Within experimental error, *pp*- and *nn*-scattering lengths are equal to each other and the results support charge independence of nuclear force. However, it is worth noting that the measurement of *nn*-scattering length by Gabioud et al. [70] gave a value $a_{nn} = -18.6 \pm 0.5$ fm from $\pi^- + d \rightarrow \gamma + n + n$ reactions. Using improved techniques, the experimental uncertainty in this measurement is reduced compared with previous results. Since the value obtained differs from a_{pp} by more than one standard deviation of experimental uncertainty, the charge independence of nuclear force may again be in question. A new type of experiment involving direct scattering of neutrons from neutrons using two intense colliding beams of neutrons has been planned at the Los Alamos National Laboratory [73].

For $T = 0$, the scattering length can only be measured on the triplet *np*-system. The large number of significant figures in the value 5.423 ± 0.05 fm is a reflection of the accuracy that can be achieved in slow neutron scattering. The positive sign indicates that there is a bound state, which we have already seen as the deuteron ground state. The fact that the value is significantly different from that for $T = 1$ is a clear indication of the isospin dependence of nuclear force.

The values for the effective range may be obtained from low-energy nucleon-nucleon scattering as well as, for example, photodisintegration of deuterons or slow neutron capture by protons. The best known values are given in Table 3-3 for comparison. Again, we find evidence for isospin dependence but there is no indication of any contradiction to the assumption of charge independence of nuclear force. The accuracies of the measured values are, however, somewhat poorer than the corresponding scattering length measurements, in particular, for the *np*-system. This is not surprising, as we are no longer in the extremely low-energy region where high precision is possible, as we have seen with scattering length measurements.

3-9 The Nuclear Potential

One-pion exchange potential. When Yukawa's idea of a simple one-pion exchange potential (OPEP) was applied to nuclear force, it was found that it could fit experimental data only for internucleon distances greater than 2 fm. In retrospect this is not surprising. As we have seen earlier in §3-6, the pion mass is around 140 MeV/c^2, corresponding to a range of approximately 1.4 fm. At shorter distances, contributions from sources other than single-pion exchange enter into the picture. This can also be seen, for example, from the values of *s*-wave phase shifts shown in Fig. 3-3. At low energies, the values are large and positive, indicating that the force is an attractive one. As the energy is increased to around 250 MeV in the laboratory for the $^1S_0$-channel and to just above 300 MeV for the $^3S_1$-channel, the phase shifts become negative, showing that the force is now repulsive. This is generally interpreted as evidence of a *hard core* in nucleon-nucleon interaction when the two nucleons are within a distance of the order of a femtometer between their centers.

From a quark picture, such a strongly repulsive, short-range term in the interaction between nucleons is to be expected. Being fermions, each one of the three quarks inside a nucleon must occupy one of the three lowest available states. When two nucleons are close together, a large fraction of their volumes overlap each other. As a result, the six quarks in a two-nucleon system can no longer be considered as two separate groups consisting of three quarks each. The Pauli exclusion principle between the quarks demands that three of the six quarks must go to states above the lowest ones already occupied by the other three. A large amount of energy is required to make this transition. From the point of view of nucleon-nucleon scattering, this additional energy shows up as a great resistance for the two nucleons to come very close to each other, almost as if there is some sort of impenetrable barrier between them. Unfortunately, it is not easy to obtain a quantitative predication. To start with, 300 MeV is a very low energy for QCD to carry out any sort of reliable calculations. As a result, we have no precise way yet of evaluating the range or the strength of the repulsive core from fundamental considerations.

At the hadron level, it is also difficult to generalize the one-pion-exchange picture, designed to understand the long-range part of the force. In addition, the OPEP also has difficulties in relating the strength of nucleon-nucleon interactions to the observed magnitude of pion-nucleon interactions. If two nucleons interact with each other through the exchange of a virtual pion, the strength of the interaction must be related to the probability of a nucleon emitting and absorbing real pions. Such probabilities are, in turn, connected to the *coupling constant*, or interaction strength, between a pion and a nucleon. In the language of field theory, the strength is characterized by the pion-nucleon coupling constant $g_{\pi N}$, analogous to the factor g in Eq. (3-62). Models of nuclear force built solely upon the one-pion-exchange picture have found that, in general, the value of $g_{\pi N}$ obtained, for example, from pion-nucleon scattering cannot be used directly to calculate the coupling constant for nucleon-nucleon scattering. As a result, the value for two nucleons is often treated as a parameter, adjusted to fit the nucleon-nucleon scattering data.

One-boson exchange potential. Our present view is that nuclear force may be divided into three parts, as illustrated schematically in Fig. 3-7. The long-range part ($r > 2$ fm) is dominated by one-pion exchange. If exchanges of a single pion are important, there is no reason to exclude similar processes involving two or more pions and mesons heavier than pions. The range of interaction associated with these more massive bosons is shorter, and for this reason, the intermediate-range part of the nuclear force (1 fm$< r <$ 2 fm) comes mainly from exchanges of single heavier mesons and two pions. The hard core in the interaction ($r \lesssim 1$ fm) is made of heavy meson exchanges, multipion exchanges, as well as QCD effects.

It is helpful to use pictures based on Feynman diagrams in field theory to represent the various boson exchange terms. The exchange of one pion between two nucleons may be represented by the diagram given in Fig. 3-8(a). An implicit assumption in the diagram is that the time axis is in the vertical direction. Two nucleons with momenta $\boldsymbol{p}_1$ and $\boldsymbol{p}_2$, represented by the two solid lines, are moving freely until time t_1 when a (virtual) pion is emitted by nucleon 1. The pion emission, represented by the dashed line, changes the momentum of nucleon 1 from $\boldsymbol{p}_1$ to $\boldsymbol{p}'_1$. At time t_2, the pion is absorbed by nucleon 2 and the momentum of the nucleon is changed from $\boldsymbol{p}_2$ to $\boldsymbol{p}'_2$ as a result. For simplicity, the diagram is often abbreviated in the form shown in Fig. 3-8(b).

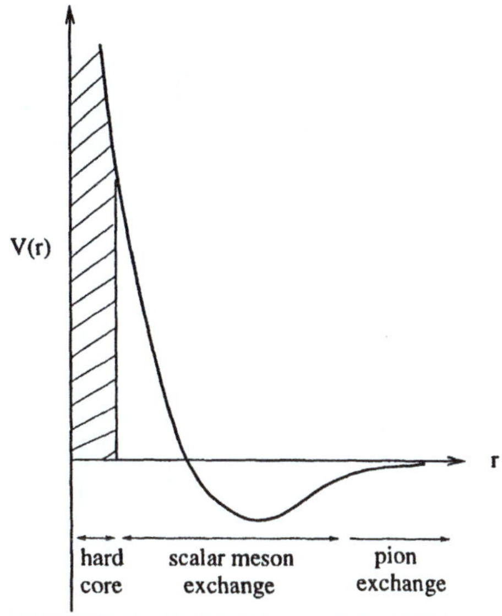

Figure 3-7: Schematic diagram showing the different parts of a nucleon-nucleon potential as a function of distance r between two nucleons. The hard core radius is around 0.4 fm and it takes energy >1 GeV to bring two nucleons closer than (twice) this distance. The main part of the attraction lies at intermediate ranges, at radius ~1 fm, and is believed to be dominated by the exchange of scalar mesons. The long-range part, starting at around 2 fm, is due to the single-pion exchange.

Following the same rules, a two-pion exchange process may be represented by that shown in Fig. 3-8(c). On possible form of two-pion exchange term is given by Fig. 3-8(d), in which the intermediate state of one of the nucleons becomes a Δ-particle, shown as a double line, as a result of absorbing the pion. Since a ρ-meson decays into two pions with a mean life of only 4×10^{-24} s, the exchange of a ρ-meson, shown in Fig. 3-8(e), may be considered as a special type of two-pion exchange term. Similarly, the exchange of an ω-meson is a type of three-pion exchange (not shown) as ω decays to three pions with a mean life of 8×10^{-23} s. Figure 3-8(f) is another type of two-pion exchange term where both pions are emitted before either one is absorbed. In contrast, the two pions in Fig. 3-8(c) are emitted and absorbed one after another. As a side interest, a three-body force may arise, for example, as the result of a nucleon emitting a ρ-meson. The two pions from the decay are absorbed by two different nucleons, as shown in Fig. 3-8(g). As a result, there are three nucleons involved in the process.

Nucleon-nucleon potentials. There are two general approaches to construct a potential that has the correct form for long-, intermediate-, and short-range parts. The first is a phenomenological one which generalizes the one-pion exchange idea to a one-boson exchange (OBE) picture. To keep the form simple, only the exchange of a single boson is allowed. In addition to pions, heavier mesons are introduced to account for the intermediate range. To compensate for multimeson exchanges, the strength for each type of meson exchange is left as a parameter to be determined by fitting, for example, NN-scattering data. The hard core is put in explicitly "by hand" without any reference to its source. The strength of such an approach lies in its simplicity. There are, however, several problems. The lifetimes of many of the mesons involved are sufficiently short that the validity of a model involving the exchange of these particles without considering their decay is not very sound. Furthermore, in order to fit experimental data with a minimum number of terms, the range of each OBE term and

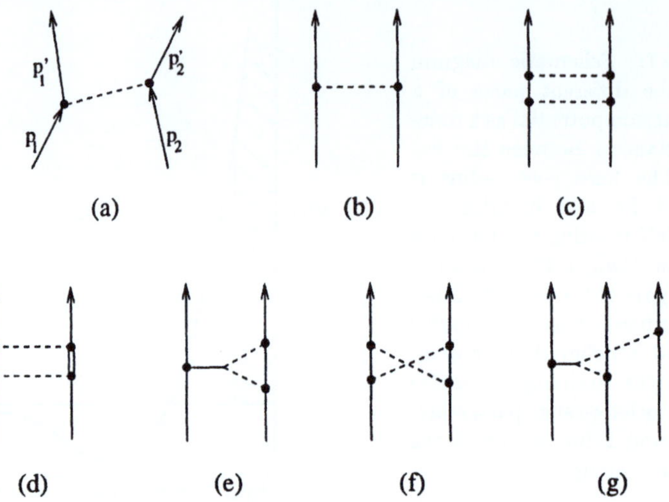

Figure 3-8: Diagrammatic representation of meson exchange between two nucleons: (a, b) one-pion exchange, (c) two-pion exchange, (d) two-pion exchange with intermediate state involving a Δ-particle, (e) ρ-meson exchange, and (f) another type of two-pion exchange term with both pions emitted before either one reabsorbed. An example of three-body force is shown by (g).

consequently the masses of the mesons exchanged often become adjustable parameters as well, with little or no relation to real mesons. These "fundamental" objections to such phenomenological potentials, however, should not detract us from their successes in a variety of applications.

A second approach in constructing a nucleon-nucleon potential is to make use of our knowledge of hadrons as much as possible and treat phenomenologically only those aspects, mainly short-range interactions, of which we have incomplete knowledge. Such a program was carried out, for example, by the Paris group [138] and the Bonn group [101] with great success. The one- and two-pion exchange parts of the potential are well known and both groups used essentially the same approach. For the less well known short-range parts, different techniques were employed.

It is perhaps of interest to examine three important differences in the two potentials, in part to see the possible future direction in the development of nuclear force studies. The first is the treatment of three- and four-pion exchanges that form a part of the short-range interaction. Here the Paris potential used a phenomenological approach and determined some of the parameters involved by fitting them to known data. The Bonn potential made an estimate of the effect instead.

A second difference in the two potentials is in the treatment of the Δ-particle. As we have seen earlier in Fig. 2-3, a strong resonance in the scattering of π^+ off protons is found at laboratory pion energy 195 MeV. Such a dominating feature in pion-nucleon reaction must have a profound influence on the nucleon-nucleon interaction.

For example, a nucleon may be excited to become a Δ-particle in the intermediate state, as shown in Fig. 3-8(d). Since it is distinguishable from a nucleon, the Δ-particle is not affected by the Pauli exclusion principle with respect to nucleons in the nucleus. For applications involving nucleon energies above 300 MeV in the laboratory, the formation of a Δ-particle is expected to play a significant role and must be included as a part of the potential. On the other hand, it is not easy to incorporate such a strong inelastic channel in a potential except by putting in the resonance explicitly, an approach adopted by the Bonn group.

The interactions between two antinucleons, and between a nucleon and an antinucleon, are also integral parts of a nucleon-nucleon potential. This is especially true if we take a fully relativistic approach where both nucleons and antinucleons appear together in the same wave function. Furthermore, experimental data are available for scattering of antinucleons off nucleons and nuclei (see, e.g., Fig. 8-9). Studies of such scattering using nucleon-nucleon potential, with antinucleons incorporated as a part, can tell us more about the two-nucleon system than considering nucleons and antinucleons as totally separate entities. There are several different ways to carry out the extensions to include antinucleons and, in this respect, the Paris and Bonn potentials differ also from each other.

In spite of these differences, it is important to realize that, at low energies where most of the experimental data are taken, calculations using both potentials have produced very similar results. For instance, the values of $^1S_0$ and $^3S_1$ phase shifts obtained with both potentials are essentially indistinguishable from each other and only a representative example is shown in Fig. 3-3 as illustration. The close agreement between the two sets of calculated results and with values extracted from NN-scattering data is a demonstration of the degree of understanding already achieved in nucleon-nucleon interaction.

Nucleon-nucleon interaction for bound nucleons. One of the reasons for having a nucleon-nucleon potential is to make use of it in nuclear structure and nuclear reaction studies. For this purpose it is not essential, in principle, to have a potential. Most of the applications require only many-body matrix elements of the nuclear interaction. A two-body force acts between two nucleons at a time. Many-body matrix elements of such an interaction can always be expressed in terms of two-body matrix elements, similar in form to the nucleon-nucleon t-matrix given in Eq. (3-75). These matrix elements are, however, different from those for free nucleons in two important aspects. In the first place, there may be a difference in the interaction between a pair of nucleons inside a nucleus from that between a pair of free ones. In this chapter we have dealt mainly with the latter category. As we shall see in §7-5, interaction between bound nucleons is modified by the presence of other nucleons in the same nucleus and may be different from that operating between free or *bare* nucleons discussed here.

A second problem is the question of whether a nuclear potential can be specified completely within the two-nucleon space. When two free nucleons interact, energy and momentum are conserved within the two-particle system. Let us consider only the nonrelativistic limit for simplicity. The momenta of the two nucleons, p_1 and p_2, are restricted by the relation

$$\frac{p_1^2}{2\mu_1} + \frac{p_2^2}{2\mu_2} = E \tag{3-91}$$

where μ_1 and μ_2 are the reduced masses of the two particles in their center of mass. In other words, the sum of the momenta of the two nucleons is confined to lie on a spherical "shell" in momentum space with the square of the radius, $p_x^2+p_y^2+p_z^2 = 2\mu E$, determined by the total available kinetic energy E in the center of mass. Under such circumstances, the two-body interaction t-matrix elements are said to be "on the energy shell" and are called *on-shell* matrix elements. Once nucleons are bound to a nucleus, energy-momentum conservation applies to the nucleus as a whole and the momenta of a pair of nucleons inside a nucleus are no longer restricted by Eq. (3-91). Interaction between two nucleons is "off the energy shell," or *off-shell* for short, if the condition given by Eq. (3-91) is not satisfied, i.e., the sum of the momenta squared is not constrained by the kinetic energy of their relative motion. Such off-shell interactions are usually built into a nucleon-nucleon interaction potential. On the other hand, since off-shell conditions do not exist for two free nucleons, we have no way of determining these parts of the potential using NN-scattering. In this sense, the nuclear potential cannot be completely specified by studies made on systems of two free nucleons alone. By the same token, purely phenomenological potentials with parameters fitted to data on two free nucleons have no way of knowing *a priori* whether they are correct for off-shell effects.

Again there are two possible ways to solve the problem of the off-shell behavior of a nuclear potential. The first is to have a theory connecting off-shell effects to those on-shell, a relation that is implicit in all the models of nuclear potential. If we have the correct association between these two types, the off-shell behavior of a potential is completely determined once the on-shell matrix elements are given. However, we have not yet arrived at this level of understanding of nucleon-nucleon interaction. In the absence of such a theory, an alternative is to take a semi-empirical approach and determine the off-shell matrix elements by comparing them with data sensitive to such effects. Unfortunately, such investigations must be carried out on systems with more than two nucleons; however, not too many quantities have been found that are useful for this purpose. An unambiguous determination of the off-shell behavior of nuclear force is still to be developed.

Relation with quark-quark interaction. Although it is generally accepted that the force between nucleons is a facet of the strong interaction between quarks, a quantitative connection between nuclear force and quark-quark interaction is still lacking. The root of the problem is the difficulty of carrying out QCD calculations at the low energies where nuclear physics operates.

To study nuclear interaction in a quark model, we need a system of at least six quarks. The force between the quarks must be such that it satisfies the condition of confinement; that is, unless the two nucleons are very close together, the six quarks are clustered tightly into two separate groups, or "bags," of three quarks each. At large enough distances compared with the average value between quarks inside a nucleon, the force between these two bags of quarks must have a form consistent with that given successfully by meson exchange. At intermediate distances, the force must be attractive and not too different from what is given by models involving the exchange of several pions and heavier mesons. With our present level of knowledge of low-energy strong interaction, it is not difficult to demonstrate such a relation must exist between quark-quark and nucleon-nucleon interactions; however, a proper derivation of nuclear force from QCD is still being developed.

Qualitatively we can see how nuclear force may arise from a quark-quark interaction by making an analogy with the force between chemical molecules. The fundamental force here is electromagnetic. Charge distributions in many molecules are spherically symmetric so that there should not be any net electrostatic force left to act between two such molecules. However, since we know that such molecules do condense into liquids and solids, there must be a residual force between them, generally known as van der Waals force. It is useful to see how such a force arises between molecules so that we may gain some insight into the question of how force between nucleons comes from the interaction between quarks.

Suppose a neutral molecule acquires an electric dipole moment $\boldsymbol{p}$, for instance, as a result of fluctuation in its shape and, consequently, in its charge distribution. The electrostatic potential at a point $\boldsymbol{r}$ from the center of a dipole is given by the expression

$$\phi(\boldsymbol{r}) = \left[\frac{1}{4\pi\epsilon_0}\right]\frac{\hat{\boldsymbol{r}}\cdot\boldsymbol{p}}{r^2} = \left[\frac{1}{4\pi\epsilon_0}\right]\frac{p\cos\theta}{r^2} \tag{3-92}$$

where the angle θ is between vectors $\boldsymbol{p}$ and $\hat{\boldsymbol{r}}$ ($\hat{\boldsymbol{r}} = \boldsymbol{r}/r$ and $r = |\boldsymbol{r}|$) and the factors within square brackets are needed if we wish to work with SI units. The electric field from such a dipole,

$$\boldsymbol{E}(r,\theta) = -\nabla\phi = -\left[\frac{1}{4\pi\epsilon_0}\right]\left\{\frac{\boldsymbol{p}}{r^3} - 3\Big(\frac{p\cos\theta}{r^4}\Big)\boldsymbol{r}\right\} \tag{3-93}$$

induces a dipole moment $\boldsymbol{p}'$ in another molecule with a magnitude proportional to the polarizability χ,

$$\boldsymbol{p}' = \chi\boldsymbol{E}$$

As a result, a dipole-dipole interaction arises between these two molecules with a strength

$$V(\boldsymbol{r}) = -\boldsymbol{p}'\cdot\boldsymbol{E} = -\Big[\Big(\frac{1}{4\pi\epsilon_0}\Big)^2\Big]\chi(1+3\cos^2\theta)\frac{p^2}{r^6} \tag{3-94}$$

Note that the interaction energy is always negative regardless of the orientation of the first dipole assumed at the start. Consequently we have a force that is always attractive and varies as r^{-7}.

For a spherically symmetric molecule, the dipole moment is zero on the average, $\langle\boldsymbol{p}\rangle = 0$. However, because of fluctuation, the instantaneous value of $\boldsymbol{p}$ may be different from zero (i.e., $\langle\boldsymbol{p}^2\rangle \neq 0$), resulting in an attractive van der Waals force. In the same way, an attractive force between two nucleons can also arise because of fluctuation. Instead of electrostatic force, we are dealing with the "color" force between quarks. Although strong interactions confine quarks within nucleons, a *color* van der Waals force can, in principle, occur between nucleons, just as a dipole-dipole force appears between a pair of molecules. In this way, we can see how a nuclear force arises from the residual interaction between quarks in two nucleons. Although the idea of a color van der Waals force is pleasing, the actual form it produces has a range much longer than what is observed. Currently, the color van der Waals force does not seem to be a correct model for nuclear interaction without modifications.

Problems

3-1. Find the possible range of values for the depth of a one-dimensional square well, 3 fm wide, that has only one bound state for a nucleon.

3-2. If the surface of a deformed nucleus is given by the equation

$$x^2 + y^2 + 1.2z^2 = R^2$$

where $R = 1.2A^{1/3}$ femtometers, calculate classically the electric quadrupole moment, assuming $A = 200$, $Z = 80$, and the nuclear density is uniform inside the surface and zero outside.

3-3. Carry out the angular part of the integration

$$\int_0^{2\pi}\int_0^{\pi}(Y_{LM}(\theta\phi))^* Y_{20}(\theta\phi)\, Y_{LM}(\theta\phi)\sin\theta\, d\theta\, d\phi$$

in Eq. (3-25) for the expectation value of the quadrupole operator in the $L = 2$ and $M = 2$ state using the explicit forms of the spherical harmonics.

3-4. For an infinite three-dimensional harmonic oscillator potential well, with oscillator frequency ω, the radial wave functions for the lowest s-state and the lowest d-state are, respectively,

$$R_{1s}(r) = 2\nu^{3/4}\pi^{-1/4}e^{-\nu r^2/2} \qquad R_{1d}(r) = \frac{4}{\sqrt{15}}\nu^{7/4}\pi^{-1/4}r^2 e^{-\nu r^2/2}$$

where the oscillator length parameter $\nu = M\omega/\hbar$, with M as the mass of a nucleon. Find the root-mean-square radii in each of these states taking $\hbar\omega =$ 15 MeV. Compare the values obtained with the measured deuteron radius. For the radial wave function given above, what is the value of the off-diagonal matrix element $\langle R_{1s}|r^2|R_{1d}\rangle$? Use this model to calculate the deuteron quadrupole moment, assuming that the wave function is predominantly made of the ${}^3S_1$-state with a 4% admixture of the ${}^3D_1$-state.

3-5. Rewrite the right-hand side of following the rank 1 spherical tensor product

$$(\boldsymbol{\sigma}(1)\times\boldsymbol{\sigma}(2))_{1m} = \sum_{pq}\langle 1p1q|1m\rangle\sigma_p(1)\sigma_q(2)$$

given in Eq. (3-36) in terms of the Cartesian components of Pauli matrices $\boldsymbol{\sigma}(1)$ and $\boldsymbol{\sigma}(2)$ for nucleons 1 and 2, respectively. Show that, in Cartesian coordinates, it has the same form as an ordinary vector product of the two vectors.

3-6. Show that the spherical tensor rank of the operator

$$S_{12} = \frac{3}{r^2}(\boldsymbol{\sigma}_1 \cdot \boldsymbol{r})(\boldsymbol{\sigma}_2 \cdot \boldsymbol{r}) - \boldsymbol{\sigma}_1 \cdot \boldsymbol{\sigma}_2$$

is 2 in intrinsic spin space. That is, it may be written as an operator of the form $\sum_{qq'} \langle j_1 q j_2 q' | \lambda M \rangle \sigma_{1_q} \sigma_{2_{q'}}$ with $\lambda = 2$ and $j_1 = j_2 = 1$. Here $\langle j_1 q j_2 q' | \lambda M \rangle$ is the Clebsch-Gordan coefficient.

3-7. In classical electrodynamics, the scalar field $\phi(\boldsymbol{r})$ produced by an electron located at the origin is given by the Poisson equation

$$\nabla^2 \phi(\boldsymbol{r}) = -4\pi e \delta(\boldsymbol{r})$$

Show that the radial dependence of the field is given by

$$\phi(r) = \frac{e}{r}$$

For a nucleon, the scalar field satisfies the Klein-Gordon equation

$$\left(\nabla^2 - \frac{1}{r_0^2}\right)\phi(\boldsymbol{r}) = 4\pi g \delta(\boldsymbol{r})$$

Show that the radial dependence of the field is given by

$$\phi(r) = -g\frac{e^{-r/r_0}}{r}$$

Derive that the range r_0 is given by the relation $r_0 = \hbar/mc$ using the fact that the boson, with mass m, is a virtual particle and can therefore exist only for a time Δt given by the Heisenberg uncertainty relation.

3-8. For a velocity-independent two-body potential, the only two-body scalars that can be formed using operators $\boldsymbol{r} = \boldsymbol{r}_1 - \boldsymbol{r}_2$, $\boldsymbol{S} = \boldsymbol{\sigma}_1 + \boldsymbol{\sigma}_2$, and $\boldsymbol{T} = \boldsymbol{\tau}_1 + \boldsymbol{\tau}_2$ are r, $\boldsymbol{\sigma}_1 \cdot \boldsymbol{\sigma}_2$, $\boldsymbol{\tau}_1 \cdot \boldsymbol{\tau}_2$, $\boldsymbol{\sigma}_1 \cdot \boldsymbol{\sigma}_2 \boldsymbol{\tau}_1 \cdot \boldsymbol{\tau}_2$ and S_{12}, where $S_{12} \equiv 3(\boldsymbol{r} \cdot \boldsymbol{\sigma}_1)(\boldsymbol{r} \cdot \boldsymbol{\sigma}_2)/r^2 - (\boldsymbol{\sigma}_1 \cdot \boldsymbol{\sigma}_2)$. Show that the operators

(a) $\boldsymbol{S} \cdot \boldsymbol{S}$ (b) $(\boldsymbol{r} \cdot \boldsymbol{S})^2$
(c) $(\boldsymbol{r} \times \boldsymbol{S}) \cdot (\boldsymbol{r} \times \boldsymbol{S})$ (d) $(\boldsymbol{r} \times (\boldsymbol{\sigma}_1 - \boldsymbol{\sigma}_2)) \cdot (\boldsymbol{r} \times (\boldsymbol{\sigma}_1 - \boldsymbol{\sigma}_2))$

can be reduced to functions of these scalars. Give the symmetry argument of why scalar products $\boldsymbol{r} \cdot \boldsymbol{S}$ and $\boldsymbol{r} \cdot \boldsymbol{T}$ are not allowed for a nuclear potential.

With velocity or momentum dependence, the only additional operator required is $\boldsymbol{L} \cdot \boldsymbol{S}$, where $\boldsymbol{L} = \boldsymbol{r} \times \boldsymbol{p}$ and $\boldsymbol{p} = \frac{1}{2}(\boldsymbol{p}_1 - \boldsymbol{p}_2)$. Show that the following terms do not form independent scalars either:

(e) $\boldsymbol{p} \cdot \boldsymbol{p}$, $\boldsymbol{L} \cdot \boldsymbol{L}$, $(\boldsymbol{L} \cdot \boldsymbol{S})^2$ (f) $\boldsymbol{r} \times \boldsymbol{L} \cdot \boldsymbol{p}$ (g) $(\boldsymbol{L} \cdot \boldsymbol{S})(\boldsymbol{L} \cdot \boldsymbol{L})$
(h) $(\boldsymbol{r} \cdot \boldsymbol{p})(\boldsymbol{r} \cdot \boldsymbol{S})$ (i) $(\boldsymbol{r} \cdot \boldsymbol{p})(\boldsymbol{L} \cdot \boldsymbol{S})$

3-9. Calculate the s-wave phase shift of a neutron scattered by an attractive square-well potential of depth V_0 and width W. Obtain the scattering length a and effective range r_0 in terms of V_0 and W.

3-10. Show that the angular distribution of an s-wave scattering is isotropic. If the only nonzero phase shifts in a hypothetical scattering of a particle off another are s- and p-waves, find the angular distribution of the scattering cross section assuming that the particle is a neutron, the s-wave phase shift is $\delta_0 = 45°$, the p-wave phase shift $\delta_1 = 30°$, and the scattering takes place with laboratory energy 5 MeV. Plot the results for scattering angle between 0° and 180°.

3-11. If, instead of the observed value of $J^\pi = 1^+$, the deuteron ground state were $J^\pi = 0^-$, what are now the possible values of orbital angular momentum L, sum of intrinsic spin S, and isospin T in this hypothetical state? What are the implications for nuclear force if this were true?

3-12. Show that, for any function $f(r)$ of r,

$$\nabla(\boldsymbol{\sigma}\cdot\nabla f) = \hat{r}(\boldsymbol{\sigma}\cdot\hat{r})\left[\frac{\partial^2 f}{\partial r^2} - \frac{1}{r}\frac{\partial f}{\partial r}\right] + \boldsymbol{\sigma}\frac{1}{r}\frac{\partial f}{\partial r}$$

where $\hat{r}$ is an unit vector and $\nabla_2 = -\nabla_1 = \nabla$.

3-13. At distances sufficiently large that overlap between their densities may be ignored, the interaction between two nucleons may be shown to be similar to that between two point dipoles,

$$V(r) \sim (\boldsymbol{\sigma}_1\cdot\nabla_1)(\boldsymbol{\sigma}_2\cdot\nabla_2)f(r)$$

Under the assumption of one-pion exchange, we may take the radial dependence to have the form

$$f(r) = \frac{e^{-r/r_0}}{r}$$

where

$$r_0 = \frac{\hbar c}{m_\pi c^2}$$

is the range. The strength of the potential may be related to the pion-nucleon coupling constant g ($g^2/\hbar c \simeq 0.081 \pm 0.002$). Except for isospin dependence, which we shall ignore here for simplicity, the potential may be written as

$$V(r) = -g^2 r_0^2(\boldsymbol{\sigma}_1\cdot\nabla_1)(\boldsymbol{\sigma}_2\cdot\nabla_2)\frac{e^{-r/r_0}}{r}$$

Use the result of Problem 3-12 above to show that $V(r)$ can be expressed in terms of the tensor operator S_{12} given in Eq. (3-38),

$$V(r) = \frac{g^2}{3}\left\{\left[\left(1 + \frac{3r_0}{r} + \frac{3r_0^2}{r^2}\right)S_{12} + \boldsymbol{\sigma}_1\cdot\boldsymbol{\sigma}_2\right]\frac{e^{-r/r_0}}{r} - 4\pi r_0^2\delta(r)\boldsymbol{\sigma}_1\cdot\boldsymbol{\sigma}_2\right\}$$

where $r = |\boldsymbol{r}_1 - \boldsymbol{r}_2|$.

Chapter 4

Bulk Properties of Nuclei

In general, observations made on atomic nuclei can be separated into four categories: energies, static moments, transition probabilities, and reaction rates. In this chapter, we shall be concerned primarily with the first two. In particular, we shall examine the energy, spin, isospin, and static moments of nuclei in their ground states. From their variations across the periodic table, we can gain some useful insight into the bulk properties of nuclei.

4-1 Electron Scattering Form Factor

The best tool to study the density distribution of nuclei is electron scattering. Besides being a point particle, the electron also can be accelerated easily. It interacts with nuclei predominantly through electromagnetic interaction. This is an advantage, as the interaction is well known and the results are relatively easy to interpret. On the other hand, the scattering is only sensitive to charge and can only probe the distribution of protons in the nucleus.

Form factor. The density distribution of a nucleus is given by the square of the wave function $\Psi(\boldsymbol{r})$, usually that for the ground state,

$$\rho(\boldsymbol{r}) = |\Psi(\boldsymbol{r})|^2$$

For the charge in a nucleus with proton number Z, it is more convenient to define the charge density distribution as

$$\rho_{\text{ch}}(\boldsymbol{r}) = Z|\Psi(\boldsymbol{r})|^2 \tag{4-1}$$

In this way the dependence on the total charge is made explicit and the wave function $\Psi(\boldsymbol{r})$ can be taken to be normalized to unity.

The actual quantity obtained in an electron scattering experiment is the Fourier transform of $\rho_{\text{ch}}(\boldsymbol{r})$, given by the integral

$$F(\boldsymbol{q}) = \int \rho_{\text{ch}}(\boldsymbol{r}) e^{i\boldsymbol{q}\cdot\boldsymbol{r}}\, dV \tag{4-2}$$

This is known as the charge, or *longitudinal,* form factor, to distinguish from the transverse form factor to be discussed later.

Usually we are only interested in the radial dependence of the density. As a result, we can average over the angles and consider only $\rho_{ch}(r)$. The angular part of the integration in Eq. (4-2) can be carried out explicitly in this case and the (radial) charge form factor reduces to the expression

$$F(q^2) = \frac{4\pi}{q}\int \rho_{ch}(r)\sin(qr)r\,dr \tag{4-3}$$

It is a function of q^2, the square of the momentum transfer to be defined later in Eq. (4-13), as q^2 (rather than q) is a proper Lorentz scalar. We shall also see in Eq. (4-19) that $F(q^2)$ is an even function of q.

In terms of $F(q^2)$, the cross section for elastic scattering of electrons off a spin $J = 0$ nucleus may be expressed as

$$\frac{d\sigma}{d\Omega} = \Big(\frac{d\sigma}{d\Omega}\Big)_{\text{point}}|F(q^2)|^2 \tag{4-4}$$

where $(d\sigma/d\Omega)_{\text{point}}$ is the differential cross section for scattering off a point particle carrying the same amount of charge. This gives a physical meaning to the (square of the) form factor as the ratio of the observed scattering cross section to the expected value for a point nucleus. The density distribution is obtained from $F(q^2)$ by applying an inverse of the transformation given in Eq. (4-2),

$$\rho_{ch}(r) = \frac{1}{2\pi^2 r}\int_0^\infty F(q^2)\sin(qr)q\,dq \tag{4-5}$$

Examples of charge density distribution obtained this way are shown in Fig. 4-1.

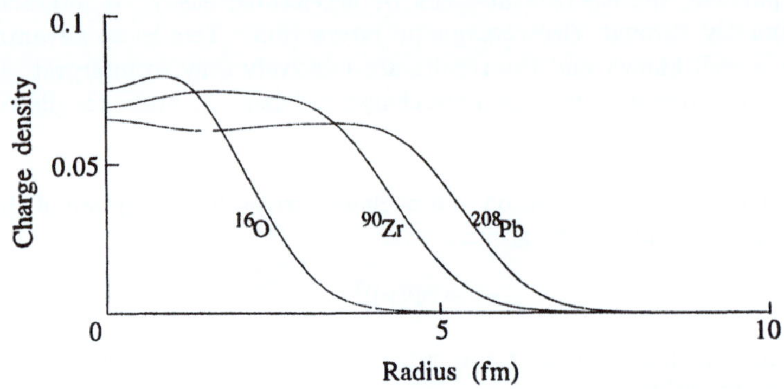

Figure 4-1: Charge densities of ^{16}O, ^{40}Ca, and ^{206}Pb obtained using Eq. (4-21). The Fourier-Bessel coefficients come from fits to electron scattering data [51].

For elastic scattering off $J = 0$ states, only the "electric" part of the interaction can contribute. On the other hand, states with $J > 0$ have usually nonzero magnetic moments that can interact with the intrinsic magnetic dipole moment of the electron. As a result, we have an additional term. In the place of Eq. (4-4), the cross section is now given by

$$\frac{d\sigma}{d\Omega} = \Big(\frac{d\sigma}{d\Omega}\Big)_{\text{point}}\Big\{|F(q^2)|^2 + \Big(\frac{1}{2} + \tan^2\frac{\theta}{2}\Big)|F_T(q^2)|^2\Big\} \tag{4-6}$$

where θ is the scattering angle. The additional contribution is characterized by the *transverse* form factor $F_T(q^2)$. In terms of operators, longitudinal and transverse form factors are related to electromagnetic moments given in §4-6, and transition probabilities between states are given in §5-3.

Scattering off point particles. A convenient starting point for discussing charged particle scattering is the Rutherford formula for differential cross sections at scattering angle θ,

$$\left(\frac{d\sigma}{d\Omega}\right)_{\text{Rutherford}} = \left\{\left[\frac{1}{4\pi\epsilon_0}\right]\frac{zZe^2}{4T\sin^2\frac{\theta}{2}}\right\}^2 = \left\{\frac{\alpha\hbar czZ}{4T}\frac{1}{\sin^2\frac{\theta}{2}}\right\}^2 \tag{4-7}$$

It is intended for a point projectile carrying a charge ze scattering off a point particle target with charge Ze. The original application was to analyze α-particle scattering off nuclei. Since the projectile was obtained from the decay of long-lived, naturally occurring radioactive nuclei, the kinetic energy T in the center of mass was low. Furthermore, the ground state spin of an α-particle is 0, and as a result, we can ignore any influence due to spin. Departures in the observed scattering can then be attributed to finite nuclear size.

We cannot simply use the Rutherford formula as the quantity $(d\sigma/d\Omega)_{\text{point}}$ in Eq. (4-6) without some modifications. First, we are primarily interested in electrons with de Broglie wavelength on the order of nuclear dimension or shorter. The kinetic energy required is much higher than the rest mass energy of ~0.5 MeV. As a result, a relativistic form of the Coulomb scattering formula must be used. Second, electrons are Dirac particles, each one with intrinsic spin $\boldsymbol{s} = \frac{1}{2}$. The magnetic moment associated with it makes a contribution to the scattering in addition to the purely electrostatic one given by Eq. (4-7). The replacement is the Mott formula

$$\begin{aligned}\left(\frac{d\sigma}{d\Omega}\right)_{\text{Mott}} &= \left\{\left[\frac{1}{4\pi\epsilon_0}\right]\frac{e^2ZE}{2p^2c^2\sin^2\frac{\theta}{2}}\right\}^2\left\{1-\frac{p^2c^2}{E^2}\sin^2\frac{\theta}{2}\right\}\\ &= \left\{\frac{\alpha\hbar cZE}{2p^2c^2\sin^2\frac{\theta}{2}}\right\}^2\left\{1-\beta^2\sin^2\frac{\theta}{2}\right\}\end{aligned} \tag{4-8}$$

It gives the differential cross section for scattering relativistic electrons off point-charge particles. Here p is the magnitude of the momentum, $E = \sqrt{(pc)^2 + (m_ec^2)^2}$ is the total relativistic energy of the incident electron, and $\beta = v/c$. In the nonrelativistic limit, we have $E \approx m_ec^2$ and the kinetic energy of the incident electron becomes $T = p^2/2m_e$. This gives us

$$\frac{E}{2p^2c^2} \xrightarrow[v\ll c]{} \frac{1}{4T} \qquad\qquad \beta \xrightarrow[v\ll c]{} 0$$

and the Mott formula reduces to that of Rutherford.

Momentum transfer. It is often useful to express the scattering result as a function of the *momentum transfer* $\hbar\boldsymbol{q}$ from the electron to the nucleus. In the nonrelativistic limit, we can define a three-momentum transfer

$$\boldsymbol{q} = \boldsymbol{k}_a - \boldsymbol{k}_b \tag{4-9}$$

where $\hbar\boldsymbol{k}_a$ is the incident electron momentum and $\hbar\boldsymbol{k}_b$ is the final electron momentum. If the electron energy is high, it is more appropriate to use, instead, the Mandelstam variable t. Let us represent the four-momenta of incident electron a and scattered electron b in the center of mass by

$$\varrho_a = \Big(\frac{iE_a}{c}, k_{ax}, k_{ay}, k_{az}\Big) \qquad \varrho_b = \Big(\frac{iE_b}{c}, k_{bx}, k_{by}, k_{bz}\Big) \tag{4-10}$$

where E_a and E_b are, respectively, the center-of-mass energies of a and b. The four-momentum transfer is given by the difference between ϱ_a and ϱ_b. The square of this quantity is a Lorentz scalar given by

$$\begin{aligned} t &= -(\varrho_a - \varrho_b)^2 \\ &= \frac{(E_a - E_b)^2}{\hbar^2 c^2} - (\boldsymbol{k}_a - \boldsymbol{k}_b)^2 \\ &= m_a^2c^2 + m_b^2c^2 - 2\frac{E_aE_b}{c^2} + 2k_a\, k_b \cos\theta \\ &\xrightarrow[m_a=m_b]{} -2k_ak_b + 2k_a\, k_b\cos\theta = -4k_ak_b\sin^2\frac{\theta}{2} \end{aligned} \tag{4-11}$$

For elastic scattering, $m_a = m_b$, $k_a = k_b \equiv k$, and $E_a = E_b$. In this case Eq. (4-11) may be rewritten as

$$t = -2k^2(1 - \cos\theta) = -4k^2\sin^2\frac{\theta}{2} \tag{4-12}$$

The square of momentum transfer is reduced to

$$q^2 = -t = 4k^2\sin^2\frac{\theta}{2} \simeq \Big(\frac{2E}{\hbar c}\Big)^2\sin^2\frac{\theta}{2} \tag{4-13}$$

with the magnitude of q simply related to the scattering angle.

It is also possible to express the differential cross section in terms of q^2 instead of the solid angle. From Eq. (4-13), we have the relation

$$d\Omega = 2\pi d(\cos\theta) = \frac{\pi}{k^2}dq^2 \tag{4-14}$$

In the limit $E \to pc = \hbar kc$, the Rutherford cross section may be approximated as

$$\Big(\frac{d\sigma}{dq^2}\Big)_{\text{Rutherford}} \approx \frac{4\pi Z^2\alpha^2}{q^4} \tag{4-15}$$

This result demonstrates that the differential cross section is mainly a function of the momentum transfer without any explicit dependence on the energy.

Dirac formula. Since a nucleus is much heavier than an electron, it is often more convenient to express the scattering cross section in the laboratory frame of reference in which the target is at rest. The difference between the energies E_a and E_b of, respectively, the incident and scattered electron is the energy taken away by the recoil of the target particle as a result of the scattering,

$$\frac{E_b}{E_a} = \frac{1}{1 + (2E_a/Mc^2)\sin^2(\theta/2)} \tag{4-16}$$

where M is the mass of the target particle.

In the limit that the electron rest mass may be ignored, the cross section in the laboratory for elastic scattering of unpolarized electrons off spinless ($J = 0$), point-charge particles is given by

$$\left(\frac{d\sigma}{d\Omega}\right)_{\text{point}} = \left(\frac{Z\alpha\hbar c}{2E_a \sin^2(\theta/2)}\right)^2 \frac{E_b}{E_a} \cos^2\frac{\theta}{2} \tag{4-17}$$

For targets with a finite spin, there is an additional contribution coming from "magnetic" scattering. Instead of the above expression, we have

$$\left(\frac{d\sigma}{d\Omega}\right)_{\text{Dirac}} = \left(\frac{Z\alpha\hbar c}{2E_a \sin^2(\theta/2)}\right)^2 \frac{E_b}{E_a}\left\{\cos^2\frac{\theta}{2} + \frac{(\hbar q)^2}{2(Mc)^2}\sin^2\frac{\theta}{2}\right\} \tag{4-18}$$

This is known as the Dirac formula and is used as the point particle scattering cross section in, for example, extracting form factors. To differentiate between the two terms in the expression, the first one is called the *electric* term and the second one the *magnetic* term. The relative contributions of these two terms may be found in the following way. From Eq. (4-13), we see that

$$\frac{(\hbar q)^2}{2(Mc)^2} \simeq 2\left(\frac{E}{Mc^2}\right)^2 \sin^2\frac{\theta}{2}$$

is much less than unity if the electron energy is much smaller than the rest mass energy of the target particle ($E \ll Mc^2$). As a result, the magnetic scattering term in Eq. (4-18) may be ignored in elastic scattering. The exceptions are found at high energies and backward angles. In the latter case, the $\cos^2(\theta/2)$ factor reduces contributions from the electric term compared with the $\sin^2(\theta/2)$ dependence for the magnetic term. For inelastic scattering, the magnetic term dominates in cases where the electric term is forbidden by selection rules.

4-2 Charge Radius and Charge Density

Charge radius. From the charge form factor $F(q^2)$ deduced from electron scattering experiments, we can obtain the root-mean-square (rms) radius of a nucleus. At low momentum transfers, $F(q^2)$ may be expanded as an infinite series in q^2. To achieve this, we shall start with Eq. (4-3) and write the sine function in the integrand on the right-hand side as a power series in terms of its argument (qr). Upon integrating term by term, we obtain the result

$$\begin{aligned} F(q^2) &= \frac{4\pi}{q}\int \rho_{\text{ch}}(r)\left\{qr - \frac{1}{3!}(qr)^3 + \cdots\right\} r\,dr \\ &= \int \rho_{\text{ch}}(r) 4\pi r^2 dr - \frac{1}{6}q^2 \int r^2 \rho_{\text{ch}}(r) 4\pi r^2\,dr + \cdots \\ &= Z\left\{1 - \frac{1}{6}q^2 < r^2 > + \cdots\right\} \end{aligned} \tag{4-19}$$

where the overall factor Z comes from the normalization chosen for charge densities given in Eq. (4-1). Since the sine function involves only odd powers of q, the final result contains only even powers.

The radial integral in the second term of Eq. (4-19) is the expectation value of the radius squared. At low q^2, the behavior of the form factor is dominated by this term, and the slope of $F(q^2)$ in this region gives us the expectation value for the square of the radius. A plot of $\langle r^2\rangle^{1/2}$ deduced from electron scattering and other measurements is given as a function of nucleon number A in Fig. 4-2. To see the dependence of $\langle r^2\rangle^{1/2}$ on $A^{1/3}$, the figure is made in terms of the ratio $\langle r^2\rangle^{1/2}/A^{1/3}$. Except for small A, we see that $\langle r^2\rangle^{1/2}/A^{1/3}$ is roughly constant, with a value of 0.97 ± 0.04 fm. This is a direct evidence of the notion that the nucleus is made of an "incompressible fluid," with the volume increasing linearly with nucleon number A and radius $A^{1/3}$.

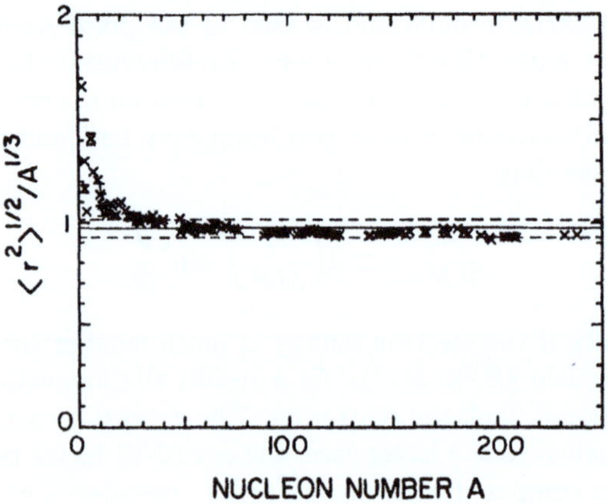

Figure 4-2: Distribution of $\langle r^2\rangle^{1/2}/A^{1/3}$ as a function of nucleon number A using values of $\langle r^2\rangle^{1/2}$ deduced from electron scattering data [51]. The horizontal lines are for $\langle r^2\rangle^{1/2} = (0.97 \pm 0.04) \times A^{1/3}$ femtometers.

Note that the value of $\langle r^2\rangle^{1/2}$ is not the nuclear radius R. This may be illustrated by the example of a constant-density sphere of radius R with

$$\rho(r) = \begin{cases} \rho_0 & \text{for } r \le R \\ 0 & r > R \end{cases}$$

The volume of the sphere is then

$$V = 4\pi \int_0^\infty \rho(r)\, r^2\, dr = \tfrac{4}{3}\pi R^3 \rho_0$$

and the expectation value of r^2 is given as

$$\langle r^2\rangle = \frac{3}{R^3\rho_0}\int_0^\infty \rho(r)\, r^4\, dr = \tfrac{3}{5}R^2 \tag{4-20}$$

This gives $R = \{(5/3)\langle r^2\rangle\}^{1/2} = 1.29\,\langle r^2\rangle^{1/2}$. For more realistic radial density distributions, such as the ones given below, the ratio $R^2/\langle r^2\rangle$ turns out to be slightly smaller than $\frac{5}{3}$. For the value of $\langle r^2\rangle^{1/2} \approx 0.97A^{1/3}$ obtained from Fig. 4-2, we come to the result $R = r_0A^{1/3}$ with $r_0 = 1.2$ fm, as given in Eq. (1-2).

Fourier-Bessel coefficients. Nuclear charge densities are often tabulated in terms of Fourier-Bessel coefficients. The density up to some cutoff radius R_c is expressed in terms of $j_0(\xi)$, with spherical Bessel function of order zero,

$$\rho_{\rm ch}(r) = \begin{cases} \sum_k a_k j_0(k\pi r/R_c) & \text{for } r \le R_c \\ 0 & \text{for } r > R_c \end{cases} \tag{4-21}$$

The parameters a_k are known as Fourier-Bessel coefficients. Only a spherical Bessel function of order zero enters here, as the charge density operator is a scalar, carrying no angular momentum. For inelastic transitions involving multipole excitation of order λ, spherical Bessel function $j_\lambda(\xi)$ takes the place of $j_0(\xi)$.

The Fourier-Bessel coefficients may be expressed in terms of the density in the following way. Since

$$j_0(\xi) = \frac{\sin\xi}{\xi}$$

and

$$\int_0^1 \sin(m\pi x)\sin(n\pi x)\,dx = \tfrac{1}{2}\delta_{mn} \qquad \text{for integer } m \text{ and } n$$

the Fourier-Bessel coefficients are related to the charge density through the integral

$$a_m = \frac{2m^2\pi^2}{R_c^3}\int_0^{R_c} \rho(r) j_0(m\pi\frac{r}{R_c})\, r^2\,dr$$

In practice, the form factor $F(q^2)$ can only be measured up to some maximum momentum transfer. For this reason, the density $\rho(r)$ can be determined only up to a certain precision. This, in turn, implies that there is only a finite number of Fourier-Bessel coefficients that can be found from a given measurement. The accuracy achieved in using a finite number of Fourier-Bessel coefficients to represent a charge density depends somewhat on the choice of the cutoff radius R_c as well. Usually R_c is taken to be just slightly beyond where the density essentially drops off to zero. For light nuclei, a value around 8 fm is often used and for heavy nuclei, 12 fm.

Other forms of charge density. For many practical applications, density distributions in terms of Fourier-Bessel coefficients are still too complicated. Furthermore, the density is essentially constant except in the surface ($r \approx R$) region, as can be seen from examples shown in Fig. 4-1. This is particularly true for heavy nuclei where the nuclear size is large enough for the central region to be significant.

Because of the short-range nature of nuclear force, nucleons near the surface of the nucleus are less tightly bound than those inside, for the simple reason that there are fewer nucleons in the vicinity with which to interact. As a result, nuclear density drops off more or less exponentially in the surface region. A density distribution with a constant central region and a diffused edge may be represented by expressions involving fewer parameters than the number of Fourier-Bessel coefficients in Eq. (4-20). Several such forms are commonly used in both nuclear reaction and nuclear structure studies.

The radial dependence of a density distribution with a diffused edge may be written as

$$\rho_{2pF}(r) = \frac{\rho_0}{1 + \exp\{(r - c)/z\}} \tag{4-22}$$

The two parameters c and z are determined, for instance, by fitting to densities derived from measured form factors, and the factor ρ_0 is given by normalization. Equation (4-22) is generally known as a two-parameter Fermi or Woods-Saxon form. The meaning of c may be interpreted as the radius of the distribution to a point where the density drops to half of its central value, and z is the *diffuseness*, related to the thickness of the surface region (see Problem 4-5). Examples of the values extracted from observed charge densities are listed in Table 4-1 for illustration.

Table 4-1: Sample values of $\langle r^2\rangle^{1/2}$ and the parameters for two- and three-parameter Fermi forms of charge density distribution.

Nucleus	$\langle r^2\rangle^{1/2}$	c	z	w
^{16}O	2.730±0.025	2.608	0.513	−0.051
^{28}Si	3.086±0.018	3.340	0.580	−0.233
^{40}Ca	3.482±0.025	3.766	0.586	−0.161
^{88}Sr	4.17 ±0.02	4.83	0.496	
^{112}Cd	4.608±0.007	5.38	0.532	
^{148}Sm	4.989±0.037	5.771	0.596	
^{184}W	5.42 ±0.07	6.51	0.535	
^{206}Pb	5.509±0.029	6.61	0.545	
^{238}U	15.84	6.805	0.605	

A somewhat better description of the observed density is provided by a modified Fermi form with an additional parameter w,

$$\rho_{3pF}(r) = \rho_0 \frac{1 + w(r/c)^2}{1 + \exp\{(r - c)/z\}} \tag{4-23}$$

generally known as the three-parameter Fermi form. Other formulas, such as the three-parameter Gaussian form,

$$\rho_{3pG}(r) = \rho_0 \frac{1 + w(r/c)^2}{1 + \exp\{(r^2 - c^2)/z^2\}}$$

and the harmonic oscillator model form,

$$\rho_{HO}(r) = \rho_0 \Big\{1 + z\Big(\frac{r}{c}\Big)^2\Big\} e^{-(r/c)^2}$$

are also in use. Tabulated values of these parameters for various nuclei using these forms can be found, for example, in de Vries, de Jager, and de Vries [51].

4-3 Nucleon Form Factor

At sufficiently high energies, giga-electron-volts and above, the de Broglie wavelength of an electron becomes much shorter than the size of a typical nucleus. In such cases, the scattering result is dominated by the charge distributions within individual nucleons. The primary interest shifts to the structure of nucleons rather than that for the nucleus as a whole. In the place of nuclear form factors, we are concerned with the analogous quantities for nucleons. We shall return at the end of §4-4 to the question of whether there is any difference in scattering off bound nucleons instead of free ones.

Nucleon form factors. Since nucleons are spin-$\frac{1}{2}$ particles, both electric and magnetic scattering contribute to the cross section. For reasons that will soon become obvious, it is more convenient to use the Sachs form factors $G_E(q^2)$ and $G_M(q^2)$ rather than $F(q^2)$ and $F_T(q^2)$ of Eq. (4-6). This gives us the Rosenbluth formula for the differential cross section of electron scattering off nucleons,

$$\left(\frac{d\sigma}{d\Omega}\right)_{\text{lab.}} = \left(\frac{d\sigma}{d\Omega}\right)_{\text{point}} \left\{ \frac{G_E^2(q^2) + \zeta G_M^2(q^2)}{1+\zeta} + 2\zeta G_M^2(q^2) \tan^2\frac{\theta}{2} \right\} \tag{4-24}$$

where the dimensionless quantity ζ is given by

$$\zeta = \left(\frac{\hbar q}{2Mc}\right)^2$$

The relation between Sachs form factors and longitudinal and transverse form factors may be seen by comparing Eq. (4-24) with (4-6). The Sachs form factors have the property that, at zero momentum transfer,

$$G_E(0) = \begin{cases} 1 & \text{for a proton} \\ 0 & \text{for a neutron} \end{cases} \qquad G_M(0) = \begin{cases} \mu_p & \text{for a proton} \\ \mu_n & \text{for a neutron,} \end{cases} \tag{4-25}$$

where μ_p and μ_n are, respectively, the magnetic dipole moments of a proton and a neutron in units of nuclear magnetons (see Table 2-4).

In the place of $G_E(q^2)$ and $G_M(q^2)$, the scattering cross section may also be written in terms of Dirac and Pauli form factors $F_1(q^2)$ and $F_2(q^2)$, defined by

$$G_E(q^2) = F_1(q^2) - \zeta F_2(q^2) \qquad\qquad G_M(q^2) = F_1(q^2) + F_2(q^2)$$

The main difference between these two sets of form factors is that, instead of electric and magnetic scattering, $F_1(q^2)$ and $F_2(q^2)$ are distinguished according to helicity $\boldsymbol{\sigma}\cdot\boldsymbol{p}/|\boldsymbol{p}|$, the projection of electron intrinsic spin $\boldsymbol{\sigma}$ along its direction of motion $\boldsymbol{p}/|\boldsymbol{p}|$. The Dirac form factor $F_1(q^2)$ represents the helicity-preserving part of the scattering and the Pauli form factor $F_2(q^2)$ represents the helicity-flipping part. The Rosenbluth formula (4-24) is actually derived using a first Born approximation, involving only the exchange of one photon for the electron-nucleon interaction. In principle, corrections due to two or more photon exchanges are needed. However, comparisons with observations have shown that the formula works well to fairly high energies.

Asymptotic forms. There are two interesting points connected with nucleon form factors. The first is that, in the limit of large momentum transfer, the two proton form factors and the magnetic form factor of a neutron are identical to each other except for a *scaling factor*. The required factors can be deduced by examining Eq. (4-25),

$$G_E^p(q^2) = \frac{1}{\mu_p} G_M^p(q^2) = \frac{1}{|\mu_n|} G_M^n(q^2) = G(q^2) \tag{4-26}$$

The function $G(q^2)$ may be described by a *dipole form*,

$$G(q^2) = \frac{1}{\{1 + (q/q_0)^2\}^2}$$

Empirically, the parameter q_0 is found to be $\hbar q_0 = 0.84$ GeV/c.

Using the dipole form, the proton charge distribution becomes

$$\rho_{\rm ch}(r) = \rho_0 e^{-q_0 r}$$

From this, we obtain the square of the charge radius of a proton,

$$\langle r^2 \rangle = \frac{\int r^2 \rho_{\rm ch}(r)\, r^2\, dr}{\int \rho_{\rm ch}(r)\, r^2\, dr} = \frac{12}{q_0^2} = (0.81 \text{ fm})^2$$

Because of the scaling relation (4-26), the magnetic radius of a proton must also have the same value. Note that the value 0.81 fm for the rms radius of a nucleon is slightly smaller than the corresponding average value $\langle r^2 \rangle^{1/2}/A^{1/3} = 0.97 \pm 0.04$ for nucleons in nuclei. If we use a uniform density sphere to approximate the proton charge distribution, as we did in the previous section for nuclei, we obtain the radius of a proton to be around 1 fm.

The electric form factor of a neutron $G_E^n(q^2)$ is only known at small momentum transfers, $q < 10$ GeV/c, and is found to be much smaller than the corresponding magnetic form factor $G_M^n(q^2)$ at the same momentum transfer. In addition, there are two other reasons why measurements of neutron electric form factors are difficult at high q. The first is the increase in the value of ζ in Eq. (4-24) with q, and as a result, the scattering cross section at high q is dominated by the magnetic form factor. The second is the absence of a fixed neutron target, and all our experimental knowledge on neutrons must be deduced indirectly from scattering off such targets as deuteron and

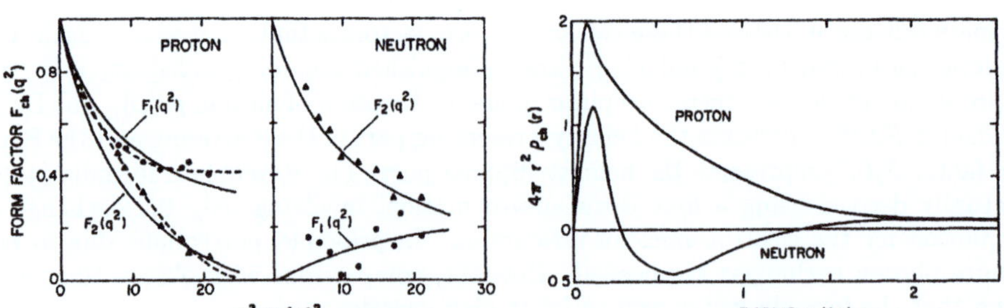

Figure 4-3: Dirac and Pauli form factors and charge distributions of proton and neutron. (Adapted from Refs. [85] and [98].)

^{3}He. Based on the information available, the charge distributions obtained by applying Eq. (4-5) to the measured form factors are shown in Fig. 4-3. Since a neutron has no net charge, it is not surprising to find the $F_1(q^2)$ to be much smaller than the corresponding values for a proton. On the other hand, the values may be important in understanding some of the details in nuclear charge distribution to be discussed later in §4-5.

4-4 High-Energy Lepton Scattering

Let us return to the question of electron scattering off nuclei. At low energies, where the electron wavelength is much larger than the nuclear size, the scattering cross section is essentially given by the Mott formula (4-8). As the energy of the incident electron is increased, the extended size of the nuclear charge distribution comes into play and the scattering cross section is modified by the nuclear form factors. Our interest is still confined to elastic scattering, with the energy of the scattered electron E_f differing from its incident value E_i only by the amount taken up by nuclear recoil. The energy transferred to the nucleus is then

$$\hbar\omega \equiv E_a - E_b = \frac{(\hbar q)^2}{2M} \tag{4-27}$$

The relation between ω and q^2 given above may be taken as a definition of elastic scattering.

Quasi-elastic scattering. As we increase the incident energy further, the electron wavelength eventually becomes short enough to be comparable with the size of a nucleon. At this point, coherence in the scattering from several nucleons at the same time is no longer important and the scattering takes place essentially from individual nucleons. However, the situation is different from scattering off free nucleons in that the energy transferred is given by

$$\hbar\omega = \frac{(\hbar q)^2}{2M_N}$$

The difference from Eq. (4-27) is that the nucleon mass M_N rather than M, the mass of the target nucleus, appears in the denominator. Since M_N is different from M, the scattering of an electron from a "bound" nucleon is no longer a true elastic scattering by the definition given in Eq. (4-27). It is, instead, a *quasi-elastic* scattering of electrons off individual nucleons.

Quasi-elastic scattering differs from elastic scattering off free nucleons also in that nucleons in a nucleus are not stationary with respect to the nuclear center of mass. The average momentum of a nucleon may be estimated from the uncertainty relation,

$$p_F \sim \frac{\hbar}{R}$$

where R is the size of the potential well that binds the nucleon. This is essentially the Fermi momentum of a nucleon inside a nucleus. Since R is of the order of the size of the nucleus, i.e., a few femtometers, p_F is of the order of 100 to 200 MeV/c. As a result, there is a spread of the order of 100 MeV in the energy transferred in quasi-elastic scattering, around 10% of the total.

Structure functions. At forward angles, the momentum transferred is small and the cross section is dominated by elastic scattering. Since form factors decrease in value very quickly with increasing momentum transfer, the elastic scattering cross section rapidly becomes very small as the momentum transfer is increased, and the presence of inelastic scattering processes becomes apparent. The reaction cross section now depends in general on the amount of energy $\hbar\omega$ as well as the momentum q transferred. The result is usually expressed as a double differential cross section,

$$\frac{d^2\sigma}{dq^2d\omega} = \frac{4\pi Z^2\alpha^2}{q^4}\frac{\hbar E_b}{E_a Mc^2}\left\{\frac{Mc^2}{\hbar\omega}F_2(q^2,\omega)\cos^2\frac{\theta}{2} + 2F_1(q^2,\omega)\sin^2\frac{\theta}{2}\right\} \qquad (4\text{-}28)$$

where the factor $4\pi Z^2\alpha^2/q^4$ is the familiar Rutherford scattering cross section off a point charge given in Eq. (4-15). The functions $F_1(q^2,\omega)$ and $F_2(q^2,\omega)$, related to the form factors defined earlier for the (single) differential cross section $d\sigma/d\Omega$, are usually referred to as the nucleon *structure functions,* as they express the difference of a nucleon from a point particle.

The definition of momentum transfer in an inelastic scattering remains the same as that given by Eq. (4-9); however, its relation to energy is slightly different. In the limit that the electron rest mass can be ignored, the final form of Eq. (4-11) is equivalent to

$$(\hbar cq)^2 = 4E_aE_b\sin^2\frac{\theta}{2} \qquad (4\text{-}29)$$

the same as that given by Eq. (4-13). In addition to the energy taken away by target recoil, some of the incident energy is also expended in promoting particles from ground to excited states.

Since nucleons are not "elementary" particles, quasi-elastic scattering off the constituent quarks and inelastic scattering involving excitation of nucleon internal degrees of freedom can take place in the same way as inelastic scattering off nuclear targets. The only difference between these two types of processes is that the energies involved in nucleon scattering are usually much higher. In high-energy electron scattering, it is customary to express the scattering in terms of the following two dimensionless quantities:

$$x = \frac{\hbar q^2}{2M\omega} \qquad\qquad y = \frac{\hbar\omega}{E_a} \qquad (4\text{-}30)$$

instead of q^2 and ω. Let us rewrite the double differential cross section for inelastic scattering in terms of these two variables.

From Eq. (4-29), we obtain the relation

$$\sin^2\frac{\theta}{2} = \frac{(\hbar cq)^2}{4E_aE_b} = \frac{Mc^2}{2E_b}xy \qquad (4\text{-}31)$$

and for high-energy scattering in the forward directions,

$$\cos^2\frac{\theta}{2} = 1 - \sin^2\frac{\theta}{2} = 1 - \frac{(\hbar cq)^2}{4E_aE_b} \approx 1 \qquad (4\text{-}32)$$

Instead of Eq. (4-29), we may also write q^2 in terms of x and y,

$$q^2 = \frac{2ME_a}{\hbar^2} xy$$

Using Eqs. (4-31) and (4-32), the angular dependence on the right-hand side of Eq. (4-28) may be expressed in terms of x and y,

$$\frac{d^2\sigma}{dq^2\, d\omega} = \frac{4\pi Z^2 \alpha^2}{q^4} \frac{1}{\omega} \Big\{ F_2(q^2, x)(1-y) + F_1(q^2, x) xy^2 \Big\}$$

where we have made use of the fact that

$$\frac{E_b}{E_a} = 1 - y$$

from the definitions of $\hbar\omega$ in Eq. (4-27) and y in (4-30).

In terms of x and y, the double differential cross section may be written in the form usually found in the literature,

$$\frac{d^2\sigma}{dx\, dy} = \frac{4\pi Z^2 \alpha^2}{q^4} \frac{2ME_a}{\hbar^2} \Big\{ F_2(q^2, x)(1-y) + F_1(q^2, x) xy^2 \Big\} \tag{4-33}$$

In the derivation, the only property of an electron used is that its rest mass may be ignored; the formula can therefore be applied at sufficiently high energies to describe the scattering of other charged leptons, such as muons. For neutrino scattering, however, one must replace the factor $4\pi\alpha^2/q^4$ for Rutherford scattering with $G_F^2/2\pi$, where G_F is the Fermi coupling constant for weak interactions (see §5-5). The details can be found in standard texts on particle physics, such as Perkins [115].

European Muon Collaboration group effect in deep-inelastic scattering. Many different types of final states can be reached in high-energy scattering. If the cross section includes all the possible final states, the process is called a *deep-inelastic*, or *inclusive*, scattering, in contrast to exclusive scattering to a particular final state. One of the interesting questions in high-energy, deep-inelastic lepton scattering off nuclei and nucleons concerns the quark substructure of nucleons. Indeed, it was the identification of point-like objects inside nucleons in lepton-nucleon scattering, known as partons at the time of discovery, that provided the early experimental evidence for the existence of quarks in hadrons.

The effect of quark substructure in lepton scattering may be formulated in terms of the nucleon structure functions introduced in Eq. (4-33). The relation applies equally well to lepton scattering off nucleons as well as nuclei. For the convenience of comparison between the measured results on nucleon and nuclear targets, we shall define the structure functions for nuclear targets in terms of their values for a single nucleon, $F_1^A(q^2, x)$ and $F_2^A(q^2, x)$, by taking out a constant factor A from $F_1(q^2, x)$ and $F_2(q^2, x)$ in Eq. (4-33). This gives us the result

$$\frac{d^2\sigma}{dx\, dy} = \frac{4\pi Z^2 \alpha^2}{q^4} \frac{2ME_a}{A\hbar^2} \left\{ F_2^A(q^2, x)(1-y) + F_1^A(q^2, x) xy^2 \right\}$$

Let us compare the structure functions per nucleon obtained from high-energy lepton scattering off free nucleons and off bound nucleons in nuclei.

From their differences in the charge distributions, we expect the structure functions for neutrons and protons to be different from each other and may, principle, be measured by separate experiments. However, for scattering off finite nuclei, the measured values will be an average between the contributions from neutrons and protons, without any easy way to distinguish between them. Let us assume for simplicity that we have $N = Z$, which is true for light nuclei. For such targets, the contributions from bound neutrons and protons are equal in weight. The same relation between neutrons and protons is obtained from scattering off a deuteron target. Since the deuteron is a loosely bound system, we can treat the two nucleons as essentially free, as we have done in the previous chapter for nucleon-nucleon scattering.

Again, for simplicity, we shall ignore $F_1(q^2, x)$, as it contributes only a small amount to the measured cross sections. If the quark substructure of a free nucleon is the same as that of a bound nucleon in a nucleus, we expect the ratio

$$R(x) = \frac{F_2^A(q^2, x)}{F_2^d(q^2, x)} \tag{4-34}$$

to be unity for all values of momentum transfer characterized by the dimensionless variable x. The experimental results, first reported by the European Muon Collaboration (EMC) group [12] and later confirmed by others, however, showed that not only $R(x)$ differs from unity but also the ratio changes as a function of x. Examples of such results are shown in Fig. 4-4. The apparent departure would, on the surface, imply that

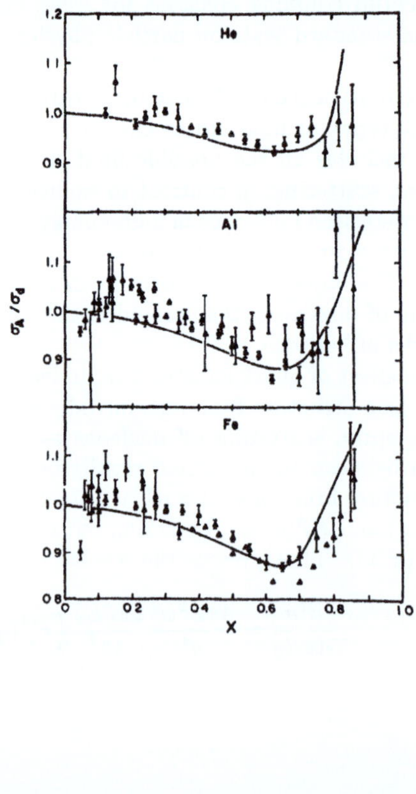

Figure 4-4: Examples of cross sections for high-energy lepton scattering off ^{4}He, ^{26}Al, and ^{56}Fe as ratios to that for deuteron, essentially the same quantities as $R(x)$ of Eq. (4-34). The points are the measured values taken from Sloan, Smadja, and Voss [130] and the smooth curves are the calculated results of Akulinichev et al. [5].

the internal structure of a bound nucleon, as shown by $F_2(q^2, x)$, is somewhat different from that of a free nucleon. Most of the departure from $R(x) = 1$ can, however, be explained by the binding energy of nucleons in a nucleus. The only exception is in the very small x region, where measurements are difficult to make. As a result, it remains to be one of the unsolved problems in high-energy nuclear physics.

4-5 Matter Density and Charge Density

We have seen in the previous sections that electron scattering is capable of mapping the charge density distribution in nuclei. The high precision achieved in the measurements enables us to ask a variety of interesting questions. Among these, the influence of neutrons on the charge distribution has been studied using *isotopic shift*, the difference in the charge distributions of nuclei with the same number of protons but a different number of neutrons. If charge distribution in a nucleus is independent of neutrons, we expect the isotopic difference to be negligible. The measured results indicate that, in general, the shifts are small but not zero. The same effect can also be observed in other measurements, such as the energy of x-rays from muonic atoms and scattering using pions and nucleons.

Isotopic shifts in calcium isotopes. Let us take the even calcium isotopes as an example. The isotopic shift data, obtained from electron scattering [68], are summarized in Table 4-2. In addition to the root-mean-square radius, the values of the surface thickness t, the distance between 90% and 10% of the peak density, and parameters c, z, and w of the three-parameter Fermi distribution defined in Eq. (4-23) are also given to provide a feeling of the large surface region in these nuclei.

Table 4-2: Charge distribution of calcium isotopes.

Nucleus	$\langle r^2\rangle^{1/2}$ (fm)	t (fm)	c (fm)	z (fm)	w
^{40}Ca	3.4869	2.681	3.6758	0.5851	−0.1017
^{42}Ca	3.5166	2.724	3.7278	0.5911	−0.1158
^{44}Ca	3.5149	2.630	3.7481	0.5715	−0.0948
^{48}Ca	3.4762	2.351	3.7444	0.5255	−0.03
^{48}Ti	3.5844	2.580	3.8551	0.5626	−0.0761

The differences, for example, in the root-mean-square radius $\langle r^2\rangle^{1/2}$ between the isotopes given in the table are quite small. However, the good accuracies achieved in the measured values indicate a genuine difference among them. Since the radius decreases by 0.01 fm in going from ^{40}Ca to ^{48}Ca, it means that the addition of neutrons to calcium isotopes reduces the size of the charge distribution of the same 20 protons when the neutron number is increased from 20 to 28. If we take the simple view that charges were distributed evenly throughout the nuclear volume, the charge radius should have increased by 6% based on the simple $R = r_0A^{1/3}$ relation. This is found to be true in the case of ^{48}Ti, a nucleus with two more protons and six more neutrons than ^{40}Ca. Here, the size of the charge distribution is increased by 0.1 fm for ^{48}Ti, not far from the expectation of an $A^{1/3}$-dependence, instead of decreasing for ^{48}Ca.

There are two possible explanations for the decrease in the charge radius with increasing neutron number among even calcium isotopes. The first is that the addition of neutrons makes the protons more tightly bound and, hence, the charge radius is smaller. This is, however, not true for nuclei in general and has led to the speculation that there are some other nuclear structure reasons for ^{48}Ca to be a more tightly bound nucleus than its neighbors.

A second explanation is based on the charge distribution within a neutron. The net charge of a neutron is zero; however, as we have seen in §4-3, the charge distribution inside a neutron does not vanish everywhere. One possible model for the charge distribution in a neutron is that the central part is positive and the region near the surface is negative, as shown in Fig. 4-3. The detailed charge distribution is not well known, because of the difficulty in measuring the small charge form factor. However, a small excess of negative charge in the surface region can produce about a third of the decrease in the charge radius in going from ^{40}Ca to ^{48}Ca, as suggested by Bertozzi et al. [26]. The other two-thirds may be attributed to the spin dependence (Darwin-Foldy term) in the interaction of protons with other nucleons in the nucleus.

Regardless of the exact cause of the isotopic shift among calcium isotopes, it is clear that neutrons have a definite influence on the measured charge distribution of a nucleus. Unfortunately data obtained based on electromagnetic interaction are not sufficient to provide us with the type of detailed information we need. In principle, strong interaction probes can be used to deduce the neutron distribution. The difficulty here lies in separating out the small effects due to neutrons from a multitude of others, including those involving aspects of strong interaction that are not yet very well known.

Muonic atom. A muon is a lepton with properties very similar to an electron. For this reason, it is possible to replace one of the electrons in an atom by a (negative) muon to form a *muonic atom*. However, since the mass of a muon is 207 times larger than that of an electron, the radii of the muonic orbits are much smaller than those of electrons.

Consider first the case of a simple, hydrogen-like atom with Z protons in the nucleus and only a single electron outside. Using the Bohr model for this hydrogen-like atom, the radius of the nth orbit is given by

$$r_n(e^-) = [4\pi\epsilon_0]\,\frac{n^2\hbar^2}{Zm_e e^2} = \frac{1}{\alpha\hbar c}\frac{n^2\hbar^2}{Zm_e} \tag{4-35}$$

where m_e is the mass of an electron, α is the fine structure constant, and the quantity inside the square brackets converts the formula from cgs to SI units. For a hydrogen atom ($Z = 1$), the ground state ($n = 1$) radius is the well-known Bohr radius,

$$a_0 = \frac{\hbar}{\alpha c m_e} = 5.29 \times 10^{-11}\ \text{m}$$

In arriving at this result, we have only made use of the electrostatic potential between a negatively charged electron and a positively charged nucleus. The charge distribution in this case may be considered to be concentrated at a point, as the electron is much further away than the extent of the nucleus. Since nothing special about the electron enters into the derivation, we can obtain the analogous results for a muonic atom by replacing m_e in Eq. (4-35) by m_μ, the mass of a muon,

$$r_n(\mu^-) = \frac{1}{\alpha\hbar c}\frac{n^2\hbar^2}{Zm_\mu} = a_0\frac{n^2m_e}{Zm_\mu} \tag{4-36}$$

Using similar arguments, we can arrive at the energy level of a muonic atom consisting of a single muon,

$$E_n = -\frac{m_\mu}{2}\frac{(Z\alpha c)^2}{n^2} \tag{4-37}$$

by starting with the energy levels of a hydrogen-like atom given in most standard textbooks on quantum mechanics.

The results of Eqs. (4-36) and (4-37) apply only to a hydrogen-like atom. For atoms with $Z > 1$, it is necessary that all the electrons except one are stripped off so as to remove the influence due other electrons. Such "screening" effects are, in general, quite difficult to calculate accurately. Fortunately, this is not a serious problem for muonic atoms with only a single electron replaced by a muon. Since the muonic orbits are so much smaller than the electronic orbits, there is very little chance of finding electrons between the muon and the nucleus, in particular for the low-lying orbits of interest to us. For this reason, the screening effect due to electrons in a muonic atom may be ignored here and, as a result, the energy levels of a muonic atom may be approximated by that of Eq. (4-37)

For a heavy nucleus, such as ^{208}Pb with $Z = 82$, the radius of the lowest muonic orbit from Eq. (4-36) is

$$r_1(\mu^-) \simeq a_0\frac{0.511}{82\times 106} = 3.1\times 10^{-15}\ \text{m}$$

or 3.1 fm, using a muon mass of 106 MeV/c^2. This is actually smaller than the value of 7.1 fm for the radius of ^{208}Pb, estimated using $R = r_0A^{1/3}$ with $r_0 = 1.2$ fm. A more elaborate calculation shows that the muon spends roughly 50% of the time inside a heavy nucleus. As a result, the actual muonic orbits differ from those given by Eqs. (4-36) and (4-37) for a point-charge nucleus. Being very close to the nuclear surface, the low-lying muonic orbits are sensitive to the detailed charge distribution. The resulting changes in the energy levels may be observed as shifts in the positions. This, in turn, changes the energy of x-rays emitted when the muonic atom decays from one level to another.

When a muon is captured by an atom, it is likely that the orbit is initially one of the higher ones. Being a different particle from the atomic electrons, there is no Pauli effect to prevent the muon from decaying to lower levels by emitting electromagnetic radiations Since the differences in the energy levels are larger than the corresponding electronic orbits because of the greater muonic mass, as can be seen in Eq. (4-37), the wavelengths of the radiation emitted are in the x-ray range. From the energies of these x-rays, we can deduce the muonic energy levels. The differences from the values given by Eq. (4-37) provide a measure of the charge distribution in the nucleus (for more details, see, e.g., Devons and Duerdoth [52]).

Pion-nucleus scattering. We have seen earlier in §2-6 that there is a strong pion-nucleon resonance in the spin-isospin $(T,S)=(\frac{3}{2},\frac{3}{2})$ channel at pion laboratory energy of 195 MeV (see Fig. 2-3). The dominance of the P_{33}-resonance provides us with a unique opportunity to examine the difference between neutron and proton density distributions in a nucleus.

There are six different possible reactions in pion-nucleon scattering with charged pions as the projectile:

$$\begin{array}{llll} (a) & \pi^+ + p \to \pi^+ + p & (b) & \pi^- + n \to \pi^- + n \\ (c) & \pi^+ + n \to \pi^+ + n & (d) & \pi^- + p \to \pi^- + p \\ (e) & \pi^+ + n \to \pi^0 + p & (f) & \pi^- + p \to \pi^0 + n \end{array} \tag{4-38}$$

The last two may be ignored here, as the scattered neutral pions are much harder to detect. Among the remaining four, only (a) and (b) have $|t_0|=\frac{3}{2}$ and must therefore take place entirely in the isospin-$\frac{3}{2}$ channel. For reactions (c) and (d), $|t_0|=\frac{1}{2}$, and the isospin is a mixture of $\frac{1}{2}$ and $\frac{3}{2}$. From simple Clebsch-Gordan coupling of isospin, we find that

$$|\pi^+n\rangle = \sum_t < 11\tfrac{1}{2},-\tfrac{1}{2}|t\,\tfrac{1}{2} > |t,\tfrac{1}{2}\rangle = \sqrt{\tfrac{1}{3}}\,|t{=}\tfrac{3}{2},t_0{=}\tfrac{1}{2}\rangle + \sqrt{\tfrac{2}{3}}\,|t{=}\tfrac{1}{2},t_0{=}\tfrac{1}{2}\rangle$$

$$|\pi^-p\rangle = \sum_t < 1,-1\tfrac{1}{2}\tfrac{1}{2}|t,-\tfrac{1}{2} > |t,-\tfrac{1}{2}\rangle = \sqrt{\tfrac{1}{3}}\,|t{=}\tfrac{3}{2},t_0{=}-\tfrac{1}{2}\rangle - \sqrt{\tfrac{2}{3}}\,|t{=}\tfrac{1}{2},t_0{=}-\tfrac{1}{2}\rangle$$

That is, only a third of the scattering amplitude of either one of these two reactions is in the isospin $t=\frac{3}{2}$ channel and the other two thirds are in the $t=\frac{1}{2}$ channel.

Since isospin is conserved in pion-nucleus reactions, the scattering amplitudes for the first four reactions in Eq. (4-38) may be decomposed in terms of isospin:

$$\begin{array}{llll} (a) & f_{\pi^+p}(\theta) = f_{t=3/2}(\theta) & (b) & f_{\pi^-n}(\theta) = f_{t=3/2}(\theta) \\ (c) & f_{\pi^+n}(\theta) = \frac{1}{3}f_{t=3/2}(\theta) + \frac{2}{3}f_{t=1/2}(\theta) & (d) & f_{\pi^-p}(\theta) = \frac{1}{3}f_{t=3/2}(\theta) + \frac{2}{3}f_{t=1/2}(\theta) \end{array}$$

At energies near the P_{33}-resonance, the scattering cross section in the isospin $t=\frac{1}{2}$ channel is much smaller than that in the $t=\frac{3}{2}$ channel. To simplify the argument, we shall ignore the $t=\frac{1}{2}$ contributions. In this approximation, we obtain the ratios

$$\frac{\sigma(\pi^+p)}{\sigma(\pi^+n)} \approx \frac{\sigma(\pi^-n)}{\sigma(\pi^-p)} \approx 9$$

by taking squares of the scattering amplitudes. Thus, the difference between the elastic π^+- and π^--scattering cross sections off a nucleus is, to ~10% uncertainty, the difference between neutron and proton density distributions. The method is particularly useful since pion scattering is sensitive mainly to the nuclear surface region where most of the differences in the neutron and proton densities are expected. This point has been verified in tests applied to a variety of nuclei.

For the even calcium isotopes discussed earlier, differential cross sections for charged pion scattering confirm the observation, made originally with electromagnetic probes, that the proton distribution is essentially unchanged as we add more neutrons. When the results of 180 MeV π^+-scattering off ^{40}Ca and ^{48}Ca are plotted on the same graph as shown in the left side of Fig. 4-5, very little difference can be detected. Since the energy is very close to the P_{33}-resonance and the cross sections are dominated by scattering off protons, the data give strong support to the similarity of proton distributions in ^{40}Ca and ^{48}Ca. On the other hand, a definite difference is found in the results of scattering of π^- off the same two nuclei at the same energy, as shown in the right side of Fig. 4-5. This is caused by the differences in the neutron distribution, a result expected from the eight additional neutrons in ^{48}Ca. When we examine the π^--scattering data off the same two nuclei at energies away from the P_{33}-resonance, the same type of difference is also observed except that the magnitudes are smaller, and the ratio between scattering off neutrons and protons is closer to unity. However, the precision that can be achieved in pion scattering is not yet as high as that with electrons. As a result, it is not possible to examine the detailed differences as we have done earlier with electron scattering results.

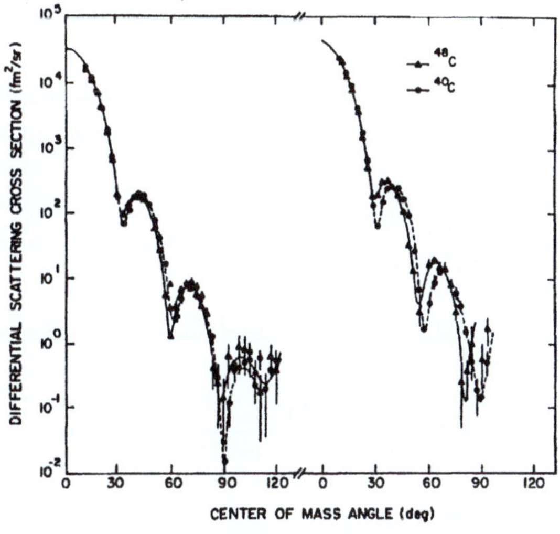

Figure 4-5: Angular distributions of 180 MeV π^+ (left) and π^- (right) scattering off ^{40}Ca (triangles) and ^{48}Ca (circles) targets. Similarity in the π^+ cross sections shows that proton distributions in the surface regions of the two nuclei are essentially the same. The differences in the π^- results demonstrate that the neutron distributions are different. (Adapted from Ref. [87].)

Nucleon-nucleus scattering. We shall see later in Chapter 8 that the time spent by an intermediate-energy (100 to 1000 MeV) proton in a nucleus is sufficiently short that it is unlikely to suffer multiple scattering, i.e., projectile scattered more than once inside a target nucleus. The incident proton has only the opportunity to interact with one of the nucleons in the target, and as a result, the scattering is sensitive to the density of nucleons in a direct way. The projectile proton can interact with either a proton or a neutron in the nucleus. From the small isospin dependence in the nucleon-nucleon interaction, we expect the result to be dependent on the differences in neutron and proton distributions. The isospin dependence of nuclear force is, however, not strong enough to clearly differentiate between neutrons and protons, as in pion-nucleon scattering at P_{33}-resonance. One way to proceed is to take it for granted that the proton density distribution in a nucleus is already known through electron scattering. Any difference observed in proton scattering may therefore be attributed to the presence of neutrons. Neutron density distributions obtained in this way depend a great deal on the model used to analyze the data, and the results are somewhat ambiguous compared with, for example, those obtained from charged pion scattering at the P_{33}-resonance.

4-6 Nuclear Shape and Electromagnetic Moments

Multipole expansion of charge density. In electromagnetism, the potential $\phi(\boldsymbol{r})$ at a point $\boldsymbol{r}$ due to a finite charge distribution $q(\boldsymbol{r}')$ is given by

$$\phi(\boldsymbol{r}) = \left[\frac{1}{4\pi\epsilon_0}\right] \int \frac{q(\boldsymbol{r}')}{|\boldsymbol{r} - \boldsymbol{r}'|}\, d\boldsymbol{r}'$$

where the quantity inside the square brackets converts the expression from cgs to SI units. In the region $r > r'$, the potential may be expressed as an infinite series in terms of spherical harmonics,

$$\phi(\boldsymbol{r}) = \left[\frac{1}{4\pi\epsilon_0}\right] \sum_{\lambda\mu} \int \frac{4\pi q(\boldsymbol{r}')}{2\lambda+1} \frac{(r')^{\lambda}}{r^{\lambda+1}} Y^*_{\lambda\mu}(\theta', \phi') Y_{\lambda\mu}(\theta, \phi)\, d\boldsymbol{r}'$$

where we have made use of the notation $\boldsymbol{r} = (r, \theta, \phi)$ and $\boldsymbol{r}' = (r', \theta', \phi')$. For a charge distribution $\rho_{\rm ch}(\boldsymbol{r}')$ in a nucleus

$$\phi(\boldsymbol{r}) = \left[\frac{1}{4\pi\epsilon_0}\right] \sum_{\lambda\mu} \frac{4\pi Z}{2\lambda+1} \frac{1}{r^{\lambda+1}} \mathcal{Q}_{\lambda\mu} Y_{\lambda\mu}(\theta, \phi) \tag{4-39}$$

where the multipole coefficients

$$\mathcal{Q}_{\lambda\mu} = \frac{1}{Z} \int e r'^{\lambda} Y^*_{\lambda\mu}(\theta', \phi') \rho_{\rm ch}(\boldsymbol{r}')\, d\boldsymbol{r}' \tag{4-40}$$

are quantities characterizing the distribution. Along the z-axis, we have $\cos\theta = 1$ and Eq. (4-39) reduces to the familiar form given in many texts on electromagnetism for the potential of an arbitrary, finite charge distribution,

$$\phi(r) = \left[\frac{1}{4\pi\epsilon_0}\right] \frac{1}{r} \sum_{\lambda\mu} \sqrt{\frac{4\pi}{2\lambda+1}} \frac{\mathcal{Q}_{\lambda\mu}}{r^{\lambda}} \delta_{\mu 0}$$

If the charge distribution is nearly spherical in shape, Eq. (4-39) is a fast convergent series and the importance of higher order terms decreases very rapidly with increasing λ. In fact, the potential of such a charge distribution can often be approximated by the contribution from the lowest nonvanishing order alone. We shall adopt the same philosophy here to describe the electromagnetic properties of a nucleus. Our interest is mainly in the lowest few multipoles, as the moments of these are the only ones that can be measured in practice.

Using the normalization for charge density given in Eq. (4-1), we can rewrite the right-hand side of Eq. (4-40) as an expectation value,

$$\mathcal{Q}_{\lambda\mu} = \langle \Psi(\boldsymbol{r}) | e r^{\lambda} Y^{*}_{\lambda\mu}(\theta,\phi) | \Psi(\boldsymbol{r}) \rangle$$

This allows us to identify the operator for the (λ, μ) electric multipole as

$$\boldsymbol{O}_{\lambda\mu}(E) = e r^{\lambda} Y^{*}_{\lambda\mu}(\theta,\phi)$$

The complex conjugation on the spherical harmonics is irrelevant in most of the subsequent discussions, as we shall be concerned mainly with the $\mu = 0$ component. If we adopt a model that the nuclear wave function is made of products of single-particle wave functions (see §7-1), $\boldsymbol{O}_{\lambda\mu}(E)$ may be expressed as a sum of operators, each one acting on an individual nucleon,

$$\boldsymbol{O}_{\lambda\mu}(E) = e \sum_{\text{protons}} r_i^{\lambda} Y^{*}_{\lambda\mu}(\theta_i,\phi_i) = \sum_{i=1}^{A} e(i) r_i^{\lambda} Y^{*}_{\lambda\mu}(\theta_i,\phi_i) \tag{4-41}$$

where, in the final form, we have introduced the symbol

$$e(i) = \begin{cases} e & \text{for a proton} \\ 0 & \text{for a neutron} \end{cases}$$

so that the summation may be taken over all A nucleons in the nucleus. Equation (4-41) is a general one, useful in discussing electric multipole transitions as well.

If the charge distribution of a nucleus is spherical in shape, only $\mathcal{Q}_{00}$ is different from zero. All higher order moments vanish, as can be seen from Eq. (4-40). Nonvanishing multipole coefficients, other than $\mathcal{Q}_{00}$, are therefore measures of "deformation," or departures from a spherical shape.

Multipole coefficients can also vanish for reasons of symmetry. For example, under an inversion of the coordinate system,

$$(r, \theta, \phi) \xrightarrow[P]{} (r, \pi - \theta, \pi + \phi)$$

as given in Eq. (A-2). On the other hand, spherical harmonics has the property

$$Y_{\lambda\mu}(\pi - \theta, \pi + \phi) = (-1)^{\lambda} Y_{\lambda\mu}(\theta, \phi)$$

As a result, we expect all odd electric multipole coefficients to vanish.

Since a nonvanishing odd electric multipole implies a breakdown of the symmetry under a parity inversion, it is of interest to find out if it happens in practice. The lowest order is dipole. The neutron turns out to be the best candidate for this purpose, as it is a neutral particle and, hence, $\mathcal{Q}_{00} = 0$. To have a nonzero electric dipole moment for a neutron, both time and parity invariance symmetries must be violated. Currently, the measured upper limit stands at 0.97×10^{-25} e-cm, consistent with zero. On the other hand, it does not rule out the possibility of a small symmetry-violating contribution either. For our purposes, we shall assume that both parity and time invariance are exact symmetries and only even-order electric multipole moments may be different from zero.

Angular momentum coupling imposes a restriction on the highest order multipole coefficients a state can have. The multipole operator $r^\lambda Y_{\lambda\mu}(\theta, \phi)$ is a spherical tensor of rank (λ, μ) and carries an angular momentum $\boldsymbol{\lambda}$. The expectation value of such an operator vanishes for a state with spin J unless $\boldsymbol{J}$, $\boldsymbol{\lambda}$, and $\boldsymbol{J}$ can be coupled together to form a closed triangle. This is the same as saying that only multipole coefficients of $\lambda \leq 2J$ can be nonzero. For this reason, a $J = 0$ state has no multipole moment except $\lambda = 0$. This may also be illustrated using the following argument.

Classically, we can "see" the shape of an object, for example, by taking a photograph. This is possible even for an object rotating at high angular velocities—all we need to do is to use an exposure time that is sufficiently short. For a quantum-mechanical object, we can also think in terms of taking a photograph of the object in order to find out its shape. The only difference here is that the Heisenberg uncertainty relation plays a role. Since $\Delta E\, \Delta t \geq \hbar$, small Δt implies large ΔE. For our "thought experiment" of taking a photograph, Δt is the exposure time. If it is short, the photograph cannot be that for an object in a definite energy state. Instead, it is a superposition of all the states in an energy interval $\Delta E = \hbar/\Delta t$. For a picture of an object in a particular eigenstate, we need good energy resolution and consequently long exposure time. As a result, the rotating object may appear, under certain circumstances, to be "spherical" in the sense that it looks to be the same regardless of the direction to view it. Quantum mechanically, such an object is in a $J = 0$ state. Instead of photography, scattering is used in practice to carry out the observation. The limitations imposed by the uncertainty principle on our thought experiment nevertheless apply.

The expectation value of an operator depends on the spin J as well as M, the projection of $\boldsymbol{J}$ on the quantization axis. The dependence on M is, however, a trivial one and given by a Clebsch-Gordan coefficient. Using the Wigner-Eckart theorem, we find that, from Eq. (A-15),

$$\langle JM|Q_{\lambda\mu}|JM\rangle = (-1)^{J-M} \begin{pmatrix} J & \lambda & J \\ -M & \mu & M \end{pmatrix} \langle J\|Q_\lambda\|J\rangle$$

where all the dependence of the matrix element on M is contained in the $3j$-coefficient $\left(\begin{smallmatrix} J & \lambda & J \\ -M & \mu & M \end{smallmatrix}\right)$. Since the reduced matrix element $\langle J\|Q_\lambda\|J\rangle$ is common to all the states differing only by their M-values, there is only a single independent quantity characterizing multipole coefficients of order λ for all $2J+1$ magnetic substates. For this reason, it is convenient to define the *multipole moment* as the expectation value in the state of maximum M, as we have done earlier for both the magnetic dipole moment of baryons and the electric quadrupole moment of the deuteron.

Electric quadrupole and hexadecapole moments. The lowest even-order electric multipole moment that can give us some idea of the "shape" of a nucleus is the quadrupole moment ($\lambda = 2$). The existence of a nonvanishing electric quadrupole moment implies that the charge distribution of the state is no longer a spherical one and the nucleus is said to be *deformed.* Usually nuclei near closed shells are more or less spherical in shape and have small absolute values for their quadrupole moments.

In contrast, nuclei in the middle of a major shell are often deformed and have large absolute values for their quadrupole moments. More detailed discussions of deformed nuclei are given later in §6-3 and §9-2.

The quadrupole moment is defined as the expectation value of the operator $r^2Y_{20}(\theta,\phi)$ in the substate of $M = J$,

$$\begin{aligned} Q &= \sqrt{\frac{16\pi}{5}}e\,\langle J, M{=}J|r^2Y_{20}(\theta,\phi)|J, M{=}J\rangle \\ &= e\,\langle J, M{=}J|(3z^2 - r^2)|J, M{=}J\rangle \end{aligned} \tag{4-42}$$

For a spherical nucleus, $\langle x^2\rangle = \langle y^2\rangle = \langle z^2\rangle = \frac{1}{3}\langle r^2\rangle$, and the quadrupole moment vanishes. For a deformed nucleus having an oblate shape, one with the polar axis shorter than the equatorial axis, Q is negative. On the other hand, for a prolate-shape nucleus, with polar axis longer than equatorial axis, the quadrupole moment is positive, as we have seen earlier for the deuteron.

The next higher order electric multipole is hexadecapole. Here, the spherical tensor rank of the operator is $\lambda = 4$ and the expectation value vanishes for states with $J < 2$. For such a state, the quadrupole moment is usually nonzero as well. As a result, it is not easy, in general, to measure the hexadecapole moment, as it is difficult to separate the contributions from those of the quadrupole in the observed results. Furthermore, since most nuclei are very nearly spherical in shape, any measured effect due to deformation tends to be dominated by the lowest order, the quadrupole here. The shortage of hexadecapole moment data comes also, in part, from the limitation that static moment measurements are far easier to carry out on ground states and there are only a few stable nuclei with ground state spin $J \geq 2$. Most of the known values of hexadecapole moments are for excited states, often deduced in a model-dependent way from measurements such as Coulomb excitation (see §8-1).

Magnetic moments. In addition to (electric) charge distribution, a deformed nucleus may also have a nonspherical "magnetic charge" distribution. Nuclear magnetism, as we have seen earlier, originates from a combination of two sources, the intrinsic magnetic dipole moment of individual nucleons and the orbital motion of protons. Analogous to an electric charge density distribution, we may define a *magnetic charge density* $\rho_{\rm m}(\boldsymbol{r})$ as the divergence of a magnetization density $\boldsymbol{M}(\boldsymbol{r})$,

$$\rho_{\rm m}(\boldsymbol{r}) = -\nabla\cdot\boldsymbol{M}(\boldsymbol{r})$$

The density can, in turn, be written in terms of a magnetization current density,

$$\mathcal{J}(\boldsymbol{r}) = \frac{c}{[c]}\nabla\times\boldsymbol{M}(\boldsymbol{r}) \tag{4-43}$$

For $\mathcal{J}(\boldsymbol{r})$, we can again adopt a model that nuclei are made of point nucleons having an intrinsic spin but no internal structure. A neutron is, then, a particle with magnetic dipole moment $\frac{1}{2}g_n$, and a proton, one having a magnetic dipole moment $\frac{1}{2}g_p$ as well as a unit of positive charge. In such a point-particle picture, the magnetization current density may be written as a sum of the contributions from all the nucleons in the nucleus in the following way:

$$\mathcal{J}(\boldsymbol{r}) = \sum_{i=1}^{A}\Big\{eg_\ell(i)\frac{\boldsymbol{p}(i)}{M_N} + \frac{e\hbar}{2M_N}g_s(i)\nabla\times\boldsymbol{s}(i)\Big\}\delta(\boldsymbol{r}-\boldsymbol{r}(i)) \tag{4-44}$$

where, in units of nuclear magneton μ_N,

$$g_\ell = \begin{cases} 1 & \text{for a proton} \\ 0 & \text{for a neutron}\end{cases} \qquad g_s = \begin{cases} 5.586 & \text{for a proton} \\ -3.826 & \text{for a neutron}\end{cases}$$

and M_N is the mass of a nucleon.

Similar to charge distributions, we may decompose a magnetization density distribution in terms of multipole coefficients given by the integral

$$\mathcal{M}_{\lambda\mu} = \int r^\lambda Y^*_{\lambda\mu}(\theta,\phi)\rho_{\rm m}(\boldsymbol{r})\,d\boldsymbol{r} = -\int r^\lambda Y^*_{\lambda\mu}(\theta,\phi)\,\nabla\cdot\boldsymbol{M}(\boldsymbol{r})\,d\boldsymbol{r} \tag{4-45}$$

Because of the divergence operator, the parity of the magnetic multipole operator of order λ is $(-1)^{\lambda+1}$, instead of $(-1)^\lambda$ in the case of the electric multipole moment. As a result, even-order magnetic multipole moments vanish for the same reason as odd-order electric multipole moments.

The lowest order nonvanishing magnetic multipole for a nucleus is the dipole. From Eq. (4-45), we see that the operator is proportional to $rY_{1\mu}$. Since $rY_{1\mu}$ is given by the μth component of the vector $\boldsymbol{r}$ in spherical coordinates, the expectation value of the magnetic dipole operator may be obtained in the following way. Using the definition given in Eq. (4-45), the magnetic dipole coefficient may be written as

$$\mathcal{M}_{1\mu} = -\int rY_{1\mu}\nabla\cdot\boldsymbol{M}(\boldsymbol{r})\,d\boldsymbol{r}$$

From this, we obtain the result

$$\mathcal{M}_{1\mu} = \int \boldsymbol{M}_\mu(\boldsymbol{r})\,d\boldsymbol{r} \tag{4-46}$$

using integration by parts.

The operator for magnetic dipole $\boldsymbol{O}_{1\mu}(M1)$ is given by the integrand of Eq. (4-46). Consider first the contribution from intrinsic spin alone. By comparing Eq. (4-43) with the second term of Eq. (4-44), we can express the intrinsic spin part of $\boldsymbol{M}_\mu(\boldsymbol{r})$ in terms of $\boldsymbol{s}_\mu(i)$. This gives us the result

$$\boldsymbol{O}_{1\mu}(M1,s) = \frac{e\hbar[c]}{2M_N c}\sum_{i=1}^{A} g_s(i)\,\boldsymbol{s}_\mu(i) \tag{4-47}$$

for the contribution from the nucleon intrinsic magnetic moment to the dipole operator. For orbital motion, we note that

$$\mathcal{M}_{1\mu}(\ell) \equiv \int \boldsymbol{M}_\mu(\ell)\,d\boldsymbol{r} = \frac{1}{2}\int\Big(\boldsymbol{r}\times(\nabla\times\boldsymbol{M}(\ell))\Big)_\mu d\boldsymbol{r} = \frac{[c]}{2c}\int(\boldsymbol{r}\times\mathcal{J}(\ell))_\mu\,d\boldsymbol{r}$$

obtained, again, with the help of integration by parts. For a proton, orbital motion contributes $(e/M_p)\boldsymbol{p}$ to the magnetization current $\mathcal{J}(\ell)$, where $\boldsymbol{p}$ is the linear momentum of the proton. Using the fact that orbital angular momentum is the vector product of $\boldsymbol{r}$ and $\boldsymbol{p}$,

$$\boldsymbol{\ell}\hbar = \boldsymbol{r} \times \boldsymbol{p}$$

we obtain the contribution due to proton orbital motion as

$$\boldsymbol{O}_{1\mu}(M1,\ell) = \frac{[c]}{2c}\left(\boldsymbol{r} \times \sum_{i=1}^{A} e g_\ell(i)\frac{\boldsymbol{p}_i}{M_N}\right)_\mu = \frac{e\hbar[c]}{2M_N c}\sum_{i=1}^{A} g_\ell(i)\,\boldsymbol{\ell}_\mu(i) \tag{4-48}$$

(see also Problem 2-6).

Combining Eqs. (4-47) and (4-48), we obtain the magnetic dipole operator in terms of orbital angular momentum $\boldsymbol{\ell}(i)$ and intrinsic spin $\boldsymbol{s}(i)$ of each nucleon,

$$\begin{aligned}\boldsymbol{O}_{1\mu}(M1) &= \boldsymbol{O}_{1\mu}(M1,\ell) + \boldsymbol{O}_{1\mu}(M1,s) \\ &= \frac{e\hbar[c]}{2M_N c}\sum_{i=1}^{A}\left\{g_\ell(i)\,\boldsymbol{\ell}_\mu(i) + g_s(i)\,\boldsymbol{s}_\mu(i)\right\} \\ &= \mu_N \sum_{i=1}^{A}\left\{g_\ell(i)\,\boldsymbol{\ell}_\mu(i) + g_s(i)\,\boldsymbol{s}_\mu(i)\right\}\end{aligned} \tag{4-49}$$

The general expression for a magnetic multipole operator of arbitrary order is

$$\boldsymbol{O}_{\lambda\mu}(M\lambda) = \mu_N \sum_{i=1}^{A}\left\{\frac{2}{\lambda+1}g_\ell(i)\boldsymbol{\ell}(i) + g_s(i)\boldsymbol{s}(i)\right\}\cdot\nabla_i(r_i^\lambda Y_{\lambda\mu}^*(\theta_i,\phi_i))$$

For a derivation, see, e.g., Bohr and Mottelson [34] or de Shalit and Talmi [50].

4-7 Magnetic Dipole Moment of Odd Nuclei

The magnetic dipole moment μ is defined as the expectation value of $\boldsymbol{O}_{1\mu}(M1)$ given by Eq. (4-49) in a state with $M = J$. In units of nuclear magnetons,

$$\begin{aligned}\mu &= \langle J, M{=}J|\boldsymbol{O}_{1\mu}(M1)|J, M{=}J\rangle \\ &= \sum_{i=1}^{A}\langle J, M{=}J|g_\ell(i)\boldsymbol{\ell}_0(i) + g_s(i)\boldsymbol{s}_0(i)|J, M{=}J\rangle\end{aligned} \tag{4-50}$$

Since $\boldsymbol{O}_{1\mu}(M1)$ carries one unit of angular momentum, the expectation value vanishes in states with spin $J < \frac{1}{2}$. By making a few simplifying assumptions on the wave function, it is possible to evaluate μ for a given state, in particular the ground state.

Single-particle model. As we shall see in the next section, pairs of identical nucleons in the ground state of a nucleus prefer to couple to angular momentum zero. For such *zero-coupled pairs,* contributions to the magnetic dipole moment vanish. This is easy to see both from angular momentum considerations and, more explicitly, from the following argument. For a zero-coupled pair of protons or a pair of neutrons, the total orbital angular momentum $\boldsymbol{L}$ and intrinsic spin $\boldsymbol{s}$ must be either the combination $(L, S) = (0, 0)$ or $(L, S) = (1, 1)$ to satisfy the Pauli principle. The $(L, S) = (1, 1)$ case has a higher energy and is of no interest in the ground state. To form an $S = 0$ state, the intrinsic spins of the two nucleons must be aligned antiparallel to each other. As a result, their contributions to the magnetic dipole moment cancel each other. Similarly, for a pair of protons coupled to $L = 0$, the net contribution from orbital motion must also vanish, as the two are moving in opposite directions.

In the limit that pairing completely dominates the nuclear ground state, all pairs of neutrons and protons are coupled to $J = 0$. Since these nucleons make no contribution to the magnetic dipole moment, we can leave them out for our present purpose. In the case of even-even nuclei, the magnetic dipole moment vanishes due to the fact that the spin J must be zero. This is observed to be true in practice for all stable even-even nuclei. For odd-mass nuclei, only one nucleon is outside zero-coupled pairs and Eq. (4-50) reduces to the expectation value of the unpaired nucleon alone,

$$\mu_{\text{s p.}} = \mu_N \langle j, m{=}j | g_\ell \ell_0 + g_s s_0 | j, m{=}j \rangle \tag{4-51}$$

where $|j, m\rangle$ is the single-particle wave function of the unpaired nucleon in a state with angular moment (j, m).

Using the Landé formula (A-20), as we have done earlier in the case of the deuteron magnetic moment in §3-2, we obtain

$$\mu_{\text{s.p}} = \frac{1}{j(j+1)} \langle j, m{=}j | (\boldsymbol{\mu} \cdot \boldsymbol{j}) j_0 | j, m{=}j \rangle \tag{4-52}$$

From Eq. (4-51), we identify the magnetic dipole operator for a single nucleon to be

$$\boldsymbol{\mu} = g_\ell \boldsymbol{\ell} + g_s \boldsymbol{s}$$

Thus, the product $\boldsymbol{\mu} \cdot \boldsymbol{j}$ in Eq. (4-52) may be expressed as a sum of $\boldsymbol{\ell} \cdot \boldsymbol{j}$ and $\boldsymbol{s} \cdot \boldsymbol{j}$. The expectation values of these operators can come from the relations

$$\begin{aligned} \boldsymbol{\ell} \cdot \boldsymbol{j} &= \boldsymbol{\ell} \cdot (\boldsymbol{\ell} + \boldsymbol{s}) = \boldsymbol{\ell}^2 + \tfrac{1}{2}(\boldsymbol{j}^2 - \boldsymbol{\ell}^2 - \boldsymbol{s}^2) \\ \boldsymbol{s} \cdot \boldsymbol{j} &= \boldsymbol{s} \cdot (\boldsymbol{\ell} + \boldsymbol{s}) = \boldsymbol{s}^2 + \tfrac{1}{2}(\boldsymbol{j}^2 - \boldsymbol{\ell}^2 - \boldsymbol{s}^2) \end{aligned}$$

This gives us the final result as

$$\mu_{\text{s.p}} = j\left\{ g_\ell \pm \frac{g_s - g_\ell}{2\ell + 1} \right\} \qquad \text{for} \qquad j = \ell \pm \tfrac{1}{2} \tag{4-53}$$

In this extreme single-particle picture, the magnetic dipole moment of an odd-mass nucleus is completely determined by the ℓ and j values of the unpaired nucleon.

Schmidt value. We can make use of the same single-particle model to deduce the spin and orbital angular momentum of the unpaired nucleon. If all the nucleons except one are members of zero-coupled pairs, the spin of the state is also given by that of the unpaired nucleon. Thus we have $j = J$, where J is the observed ground state spin. For a given j, there are two possible ℓ-values, $\ell = j \pm \frac{1}{2}$. The choice between them is determined by parity. Since all the other nucleons are grouped in pairs, with each pair occupying single-particle orbits having the same ℓ-value, the parity of the state is given by the orbital angular momentum of the unpaired nucleon, $\pi = (-1)^\ell$. In this way, the spin and parity of the state, together with whether the unpaired nucleon is a proton or a neutron, provide all the information we need to calculate the magnetic dipole moment using Eq. (4-53). For each j-value, the results obtained fall into two groups depending on the two possible values of ℓ. These are known as the Schmidt values and they are compared with observations in Fig. 4-6.

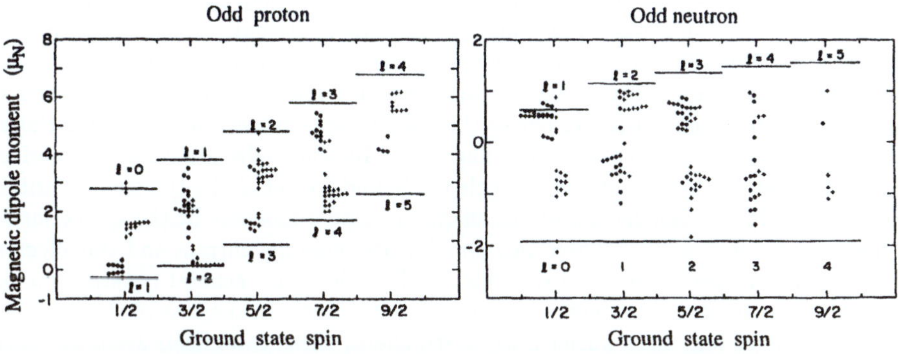

Figure 4-6: Magnetic dipole moment of odd-mass nuclei. The solid lines are calculated results using Eq. (4-53) and the observed values are taken from Ref. [95].

From the figure, we see that most of the observed magnetic dipole moments for odd-mass nuclei fall in between the two Schmidt values for $\ell = j \pm 1$. This is not surprising. The main approximation used in the model is that all the nucleons except one are tied in zero-coupled pairs. A more realistic ground state wave function includes other components as well. For our purpose here, we can characterize them by the number of "broken" zero-coupled pairs, pairs of nucleons coupled to $J > 0$. From considerations based on angular momentum coupling alone, it is unlikely that these non-zero-coupled pairs can contribute to the magnetic dipole moment of the nucleus in some coherent way to change the values given by $\mu_{\text{s p}}$ above by any substantial amount. More quantitative calculations also tend to support such an intuitive argument. Since configuration mixing of components with broken pairs decreases the weight of the single-particle component, the absolute value of the magnetic dipole moment is reduced from that given by the single-particle model. For this reason, the Schmidt values form, more or less, the limits of the possible ground state magnetic dipole moments.

Corrections to single-particle model. The model should work best for nuclei with one nucleon away from a closed shell (see §7-2). Consider the closed shell nuclei ^{4}He and ^{16}O. By removing one neutron from the former, we obtain ^{3}He, and removing one proton from the latter results in ^{15}N. Similarly, by adding a nucleon to ^{16}O, we obtain ^{17}O and ^{17}F. In such cases, the single-particle model is expected to be good, and as a result, we have the best chance to see the corrections required to such an extreme picture. The actual results found are, however, somewhat surprising. Corrections to the Schmidt values from nuclear structure considerations alone are found to be much larger than those required to account for the differences between the observed and Schmidt values. This implies that other factors are playing a role here.

In calculating the expectation values, we have assumed that each nucleon in the nucleus has the same properties as a free one. For example, the intrinsic magnetic dipole moment of a proton is taken to be $\frac{1}{2}g_p$ and that of a neutron $\frac{1}{2}g_n$. Such an assumption is sometimes referred to as the *impulse approximation*, a term originated from scattering studies (see §8-4). There are two ways by which the impulse approximation may fail. The first is the effect of mesons in nuclei. If charged mesons are exchanged between nucleons, their flow constitutes an electric current which may also contribute to the observed magnetic moments. The influence of such a *mesonic current* may also be responsible for the discrepancies in calculating other nuclear properties based on the impulse approximation. The second is that we have made the naive assumption in Eq. (4-44) that nucleons in nuclei behave like point particles carrying the same charge and magnetic dipole moments as free nucleons. Instead, effective values should be used to account for modifications of nucleons bound in nuclei, in the same spirit as we did earlier in our calculation of the magnetic dipole moment in the quark model for the baryon octet in §2-8. These two possibilities, mesonic current and the effective nuclear operator, are related to each other and to the more general question on the modifications a nucleon experiences in the nuclear medium. In the case of the magnetic dipole moment of odd-mass nuclei near closed shells, these two effects seem to cancel each other to a large extent, resulting in much closer agreement to the Schmidt values than expected from the size of either correction term alone.

4-8 Ground State Spin and Isospin

The ground state is the lowest one in energy for a nucleus. It is a special state of a system of N neutrons and Z protons by virtue of the fact that it is the most stable one. In addition, it is in general the most accessible and, as a result, often the best known and most extensively studied state in the nucleus.

The properties commonly observed are binding energy, spin, isospin, and static electromagnetic moments. We have already discussed the lowest order electromagnetic moments in the previous two sections, and we shall return to the question of binding energy in the next section. Other observables such as transition rates to and from ground states and reactions involving ground states will be covered in the later chapters. In this section, we shall concentrate on the possible values for the ground state spin and isospin of a nucleus.

Ground state spin. Since each nucleon has an intrinsic spin $\boldsymbol{s} = \frac{1}{2}$ and an (integer) orbital angular momentum $\boldsymbol{\ell}$, the total angular momentum or spin $\boldsymbol{j}$ carried by a nucleon is a half integer quantity. As a result, the total spin, the vector sum of the spins of all the nucleons in a nucleus,

$$\boldsymbol{J} = \sum_{i=1}^{A} \boldsymbol{j}_i$$

is half integer for odd-mass (A = odd) nuclei and integer otherwise. The same considerations apply to isospin as well,

$$\boldsymbol{T} = \sum_{i=1}^{A} \boldsymbol{t}_i$$

where $t_i = \frac{1}{2}$ is the isospin of the ith nucleon. Even-mass nuclei can be divided further into two categories: Those with both neutron and proton numbers even (N = even, Z = even) are called even-even nuclei and those with both neutron and proton numbers odd (N = odd, Z = odd) are called odd-odd nuclei.

For even-even nuclei, the ground spin is observed to be zero without any exception. This remarkable phenomenon reflects a fundamental property of nuclear interaction known as pairing. Since ground state spins of odd-odd nuclei are observed to be nonzero in general, we conclude further that pairing interaction is important only between two identical nucleons, two protons or two neutrons, but not between a neutron and a proton. For example, we have seen that the deuteron ground state spin is $J = 1$ (and isospin $T = 0$). If there were a strong pairing force between a neutron and a proton, the spin would have been $J = 0$ instead. In terms of isospin, we see that pairing force is present only in the $T = 1$ state of two nucleons but not in the $T = 0$ state. Because of antisymmetrization, a neutron and a proton occupying the same single-particle orbit and having relative angular momentum $\ell = 0$ form an isoscalar pair three-fourths of the time and an isovector pair one-fourth of the time (see §3-8). If the $T = 1$ pairing were strong enough to dominate over the $T = 0$ contribution, a neutron-proton pair would have preferred to be in a $T = 1$ state instead. Since this is not observed to be true, we have here another piece of information saying that the isospin dependence of nuclear force is not very prominent.

Because of pairing, the ground state spin of an odd-mass nucleus is given by the j-value of the unpaired nucleon. We have made use of this point already in the previous section. The basic idea here is that an odd-mass nucleus may be considered as a nucleon coupled to an even-even core consisting of neutrons and protons locked in zero-coupled pairs. The total angular momentum of such a core is zero and, as a result, the ground state spin of an odd-mass nucleus assumes the value of the unpaired nucleon. In Chapter 7 we shall see that the j-value of the unpaired nucleon may be found from the single-particle energy level spectrum, and as a result, the ground state spin of an odd-mass nucleus can often be deduced from its neutron and proton numbers.

For odd-odd nuclei, it is not easy to predict the ground state spin. An estimate may be made in the following way. As an extension of the idea used for deducing the ground state spin of odd-mass nuclei, we can treat an odd-odd nucleus as made of an even-even core plus a neutron and a proton outside. Again, the even-even core may be ignored here, as its lowest state must have spin zero due to pairing. If the spin of the unpaired proton is $\boldsymbol{j}_{\rm p}$ and that of the unpaired neutron is $\boldsymbol{j}_{\rm n}$, the total angular momentum of the neutron-proton pair outside the core is the vector sum of these two spins. The possible range of values is then

$$|j_{\rm p} - j_{\rm n}| \leq J \leq j_{\rm p} + j_{\rm n}$$

We can find the value of $j_{\rm p}$ from the ground state spin of the neighboring odd-mass nucleus with one less neutron. Similarly, the value of $j_{\rm n}$ may be obtained from the neighbor with one less proton. However, it is not possible to narrow down the possible J-values any further. Some guidance may be obtained from the empirical Nordheim rules:

Strong rule:	$J = \lvert j_{\rm p} - j_{\rm n}\rvert$	for $\eta = 0$
Weak rule:	$J =$ either $\lvert j_{\rm p} - j_{\rm n}\rvert$ or $j_{\rm p} + j_{\rm n}$	for $\eta = \pm 1$

Here, $\eta = j_p - \ell_p + j_n - \ell_n$. In practice, many exceptions are found and the rule can only provide a general guide to the likely value for the ground state spin of an odd-odd nucleus.

Ground state isospin. The possible isospin of a nucleus may be deduced from the proton and neutron numbers. For a system of Z protons and N neutrons, the projection of isospin on the quantization axis is

$$T_0 = \tfrac{1}{2}(Z - N)$$

The absolute value of T_0 gives the minimum of the possible isospin of a nucleus.

The maximum possible value is limited by the total number of nucleons. This may be seen from the following arguments. Isospin is related to the symmetry in interchanges between protons and neutrons. Since each nucleon has $|t_0| = \frac{1}{2}$, the maximum absolute value of T_0 for a system of A nucleons is $A/2$, attained when all the nucleons are either protons or neutrons. This must be the maximum value of T itself, as a larger value requires a larger $|T_0|$ and this is impossible. Together with the minimum value given above, the value of isospin is limited within the range

$$\tfrac{1}{2}|Z - N| \leq T \leq \tfrac{1}{2}(Z + N)$$

The isospin dependence of the nuclear force is not strong enough to put states belonging to different T into isolated groups in energy. However, except for odd-odd nuclei, the lowest member of each allowed T-value is well separated in energy from each other.

Based on the fact that there is a bound two-nucleon state for $T = 0$, the deuteron, but not for $T = 1$, we can infer that nuclear force favors the minimum value, $T = |Z - N|/2$, as the isospin for the ground state. For higher isospins, the lowest member of each group usually appears at successively higher energies. For example, in ^{16}O the ground state has $T = 0$ and the lowest $T = 1$ state occurs around 13 MeV excitation and the lowest $T = 2$ state at 23 MeV, as shown later in Fig. 4-8. States with $T \geq 3$ are expected to be at energies above 30 MeV. However, the density of states is too high at these excitation energies for individual states to be identified.

In odd-odd nuclei, the separation in energy between the lowest members of the two smallest possible isospins is often quite small. This is the result of competition between T- and J-dependence in nuclear force. Because of this, the ground state isospin is often a choice between the minimum value of $\frac{1}{2}|Z - N|$ and one larger. For example, the ground state of $^{26}_{13}$Al has $(J^\pi, T) = (5^+, 0)$ and the first excited state at 0.229 MeV has $(J^\pi, T) = (0^+, 1)$. In $^{42}_{21}$Sc, the ground state has $(J^\pi, T) = (0^+, 1)$ and the lowest state with $T = 0$ occurs at 0.6 MeV with $J^\pi = 7^+$.

Isospin mixing. We have assumed from the start that nuclear force depends only on the isospin and not the charge state of the interacting nucleons. This is, however, not the complete picture. In addition to strong interaction, we have also Coulomb interaction between protons. (We shall not be concerned with the much weaker magnetic dipole-dipole interaction between nucleons that depends on whether it is a proton pair, a neutron pair, or a proton-neutron pair.) Any charge-dependent term in the interaction violates the symmetry between proton-neutron exchange. Although at the nucleon level Coulomb force is much weaker than nuclear force, and may be ignored for most purposes, it is not true for the nucleus as a whole. As we have seen in §1-3, through short-range nuclear force a nucleon can only interact with a few of its neighbors. For this reason, its contribution in a many-body system increases, to a first-order approximation, only linearly with nucleon number. In contrast, the Coulomb term has a long range and its contribution to the binding energy, for example, increases quadratically with proton number. As a result, the effect of Coulomb interaction may be quite significant in heavy nuclei and isospin may no longer be a good quantum number.

We can investigate whether isospin is conserved by examining the amount of admixture of different isospin components in an eigenstate of a Hamiltonian containing such a symmetry-breaking term. Consider two eigenstates of the isospin-conserving part of the Hamiltonian, $|JTx\rangle$ and $|J'T'y\rangle$. Here, x and y are labels other than spin and isospin required to specify these two states. When we include also an isospin-breaking term, $|JTx\rangle$ and $|J'T'y\rangle$ are no longer eigenstates. If the symmetry-breaking term does not make connections to states outside our model space, we can find the new eigenstates $|\psi_1\rangle$ and $|\psi_2\rangle$ using $|JTx\rangle$ and $|J'T'y\rangle$ as the basis and solve the eigenvalue problem in this model space.

It is convenient to carry out this calculation using a matrix method, as we have done earlier in §3-4 for the deuteron D-state problem. The Hamiltonian matrix in the present case may be represented as

$$\{H\} = \begin{pmatrix} H_{xx} & H_{xy} \\ H_{yx} & H_{yy} \end{pmatrix}$$

The diagonal matrix elements

$$H_{xx} \equiv \langle JTx|H|JTx\rangle \qquad\qquad H_{yy} \equiv \langle J'T'y|H|J'T'y\rangle$$

are expected to be large, as they include contributions from nuclear interaction. In contrast, only isospin-breaking terms are effective in the off-diagonal elements

$$H_{xy} \equiv \langle JTx|H|J'T'y\rangle \qquad\qquad H_{yx} \equiv \langle J'T'y|H|JTx\rangle$$

in the basis we have chosen. They are, in general, much smaller than the diagonal elements. Furthermore, the Coulomb force preserves rotational symmetry and is invariant under time reversal. As a result, the Hamiltonian matrix is real and symmetric. We can therefore take

$$H_{xy} = H_{yx} = S\,\delta_{JJ'}$$

where S is the size of the off-diagonal matrix element. The eigenvectors are now linear combinations of the two basis states

$$\begin{aligned} |\psi_1\rangle &= \cos\theta|JTx\rangle + \sin\theta|J'T'y\rangle \\ |\psi_2\rangle &= -\sin\theta|JTx\rangle + \cos\theta|J'T'y\rangle \end{aligned}$$

where the angle θ is given by the relation

$$\tan 2\theta = \frac{2S}{H_{xx} - H_{yy}} \tag{4-54}$$

From this expression, we see that the amount of mixing among the two basis states depends on the size of the off-diagonal matrix element as well as the difference in the values of the diagonal matrix elements.

In a real nucleus, the number of states that can be admixed by isospin-breaking forces is likely to be larger. However, the general features are well illustrated by our simple, two-dimensional example. Several interesting problems may be studied using this model. The first is that the admixture is important only between states whose "unperturbed" locations, given essentially by the values of the diagonal matrix elements in the example above, are close to each other in energy, as can be seen by looking at the denominator on the right-hand side of Eq. (4-54). The second is that since the isospin-breaking term in the Hamiltonian is rotational invariant, off-diagonal matrix elements vanish between states having different J-values. As a result, isospin mixing can take place only between states of the same J.

We can go one step further by examining more closely the value of the off-diagonal matrix element. Since Coulomb force has a long range, the value is small unless the wave functions of the states involved are very similar to each other in every respect except isospin. This can be most easily seen in the limit that the isospin-breaking term has an infinite range, and as a result, the radial dependence may be approximated by a constant. In this case, the off-diagonal matrix elements vanish unless $x = y$. In other words, the two states must be identical except for isospin. In reality, we find that isospin mixing is important only between nearby states having the same spin and a large overlap between their spatial wave functions.

In light nuclei, the isospin purity of a state is preserved by a combination of two factors. First, the Coulomb force is relatively weak, as the number of protons is still small. As a result, the value of S in Eq. (4-54) is small in general. Second, the density of states is relatively low in the low-lying regions of interest. It is, therefore, rare to find two (unperturbed) states of different isospin near each other in energy and having a large overlap in their radial wave functions as well.

In heavy nuclei, isospin remains pure for the ground state and a few of the low-lying states nearby for a quite different set of reasons. Because of neutron excess, the Fermi energy for neutrons is much higher than that for protons. The lowest possible isospin is $T_{\text{min}} = |T_0| = \frac{1}{2}|Z - N|$, as we saw earlier, and the dominant component in the ground state wave function is given by the configuration with nucleons occupying the lowest available single-particle states. Admixtures of isospin will have to come from states with $T = T_{\text{min}} + 1$. The location of the lowest member of such a state may be found by estimating the excitation energy of the isobaric analogue (see below) to the ground state of a neighboring isobar with one more proton and one less neutron. The main configuration for the ground state of such a nucleus may be obtained by changing a proton in the nucleus with $T_0 = T_{\text{min}}$ to a neutron. Since the neutron and proton Fermi levels are quite different, it takes a large amount of energy to make the change. A crude estimate puts the amount to be the same as the neutron-proton Fermi energy difference and is therefore a function of the neutron excess, as can be seen from looking at Fig. 4-7. As a result, the location of the isobaric analogue to the ground state of the $T = T_{\text{min}} + 1$ nucleus is quite high for nuclei with a large neutron excess. This, in turn, means that it is difficult to have any significant isospin mixing in the low-lying states of heavy nuclei.

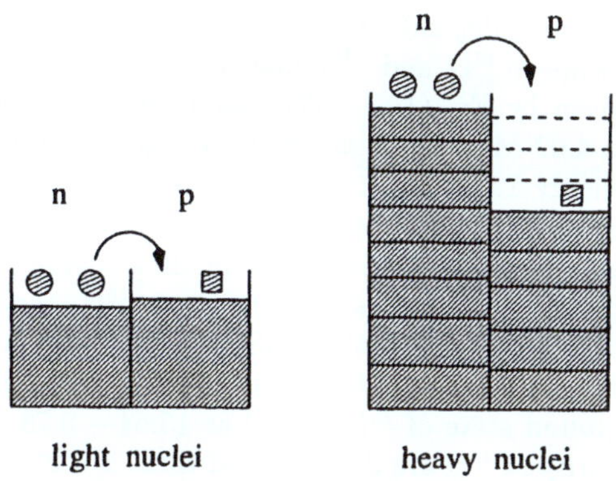

Figure 4-7: Occupancies of neutron and proton single-particle states. The location of the isobaric analogue state of an isobar with one more proton and one less neutron may be found by changing a valence neutron (circles) to a proton (squares). Because of neutron excess, such states are much higher in energy in heavy nuclei (right) than in light nuclei (left).

In a sense, isospin is no longer a meaningful symmetry in heavy nuclei, as the active neutrons and protons are occupying different single-particle states and are therefore distinguishable by the states they occupy. Furthermore, the argument for isospin purity given above applies only to the ground state region. For excited states, the strong Coulomb effect leads to large mixing, and isospin ceases to be a meaningful quantum number altogether.

Isobaric analogue state. The importance of isospin in light nuclei can also be seen by the similarities in the properties among members of an isobar, nuclei having the same A but different Z and N. An example was shown earlier in Fig. 3-1 for the low-lying states in $A = 11$ and $A = 21$ mirror nuclei. A more interesting one for our present purpose is the case of the $A = 16$ isobar shown in Fig. 4-8. Let us start with ^{16}N with $N = 9$ and $Z = 7$. The ground state isospin is $T = 1$, on account of the fact that $T_0 = -1$. If we apply an isospin-raising operator T_+ on the ground state wave function of ^{16}N, we obtain a state with $(T, T_0) = (1, 0)$. This is a state in ^{16}O ($N = Z = 8$). Since the ground state of ^{16}O is $T = 0$, the state produced by operating T_+ on the ground state of ^{16}N must be an excited state in ^{16}O. The isospin-raising operator does nothing other than change a neutron to a proton. Furthermore, since nuclear force is charge independent, the state produced must also be an eigenstate of the nuclear Hamiltonian and, hence, correspond to an actual state in ^{16}O. This state should be very similar in properties to the ground state of ^{16}N, since the wave functions are identical except for T_0. Two such states, one in ^{16}N and one in ^{16}O in our example here, related by isospin-lowering and isospin-raising operators, are called *isobaric analogue states* (IAS) of each other.

We can easily estimate where such an excited state in ^{16}O should lie. If the forces acting on a nucleon are completely charge independent, the excitation energy of the isobaric analogue to the ground state of ^{16}N in ^{16}O is given by the binding energy differences of these two nuclei. From a table of values, we find that

$$E_B(^{16}\text{O}) = 127.62 \text{ MeV} \qquad\qquad E_B(^{16}\text{N}) = 117.98 \text{ MeV}$$

The difference is 9.64 MeV.

Two corrections must be applied before we can compare the result with the observed value in ^{16}O. The first is that contributions of the Coulomb interaction to the binding energies depend on the number of protons. The difference in the Coulomb energy between ^{16}O and ^{16}N may be estimated using the uniformly charged sphere model given by Eq. (1-7). The result, calculated with a radius of $R = 1.2A^{1/3} = 3.02$ fm, is 4.00 MeV. This means that, instead of 9.64 MeV, the IAS should be at excitation energy 13.64 MeV. A second correction comes from the difference in mass between a neutron and a proton [together with an atomic electron to keep the atom neutral as required in the definition of binding energy given in Eq. (1-1)]. Since ^{16}O has one more proton and one less neutron than ^{16}N, the calculated value of the excitation energy of the IAS we obtained above is too large by an amount equal to the neutron-proton mass difference of 0.78 MeV. This puts the calculated excited energy of the IAS to the ground state of ^{16}N in ^{16}O at $13.64 - 0.78 = 12.86$ MeV. The ground state spin and parity of ^{16}N are $J^\pi = 2^-$ and there is a $2^-, T = 1$ state at 12.97 MeV in ^{16}O, as shown in Fig. 4-8. It is also known that this state is very similar in property to the ground state of ^{16}N.

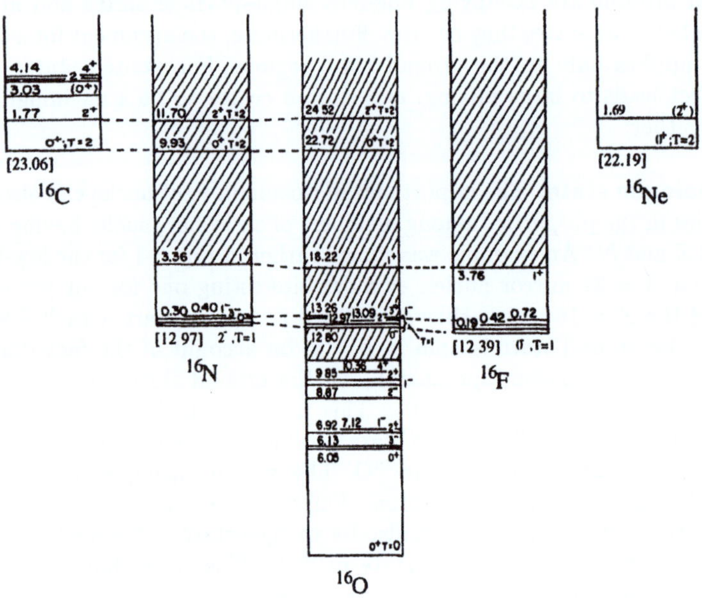

Figure 4-8: The $A = 16$ isobar with known isobaric analogue states connected by dashed lines. The energies of ^{16}C, ^{16}N, ^{16}F, ^{16}Ne relative to the ground state of ^{16}O, corrected for Coulomb effect and neutron-proton mass difference, are shown inside square brackets. (Plotted using data from Ref. [136].)

The difference of 0.11 MeV between our estimate and the observed excitation energy may be attributed to the crudeness of our Coulomb energy calculation and to a possible small difference in the wave function of the 12.97-MeV state in ^{16}O and that produced by applying T_+ to the ground state of ^{16}N. In fact, a more careful examination of Fig. 4-8 shows that there is a quartet of states with $J^\pi = 2^-$, 0^-, 3^-, and 1^- in the ground state region of ^{16}N. The same quartet of $T = 1$ states is also found in ^{16}O at around 12.9 MeV. However, the sequence of the four levels is different from what is observed in ^{16}N, showing that some small violation of the IAS idea is found in actual nuclei.

Besides $T = 1$ states, the IAS for $T = 2$ are also known in the $A = 16$ isobar. The ground state of $^{16}_{6}$C ($T_0 = -2$) is 0^+ and an excited 2^+ is found at 1.77 MeV. Similarly, the ground state of ^{16}Ne is $(J^\pi, T) = (0^+, 2)$ and a 2^+, $T = 2$ state is found at 1.69 MeV. The IAS of these two $T = 2$ states are found in ^{16}N at 9.93 MeV for the 0^+ state and at 11.90 MeV for the 2^+ state. Similarly, the same $T = 2$ pair is found in ^{16}O at 22.72 MeV (0^+) and 24.52 MeV (2^+). On the other hand, since the ground state of ^{16}F is not stable ($T_{1/2} \sim 10^{-19}$ s), the level structure is not known to sufficiently high excitation energies to identify the $T = 2$ states.

4-9 Semi-Empirical Mass Formulas

We saw in §1-3 that nuclear binding energy is more or less a smooth function of nucleon number and other "macroscopic" degrees of freedom. In fact, if we are willing to ignore small local departures, it is possible to develop simple formulas that express the binding energy $E_B(Z, N)$, or the equivalent quantity $M(Z, N)$, the mass of a nucleus, in terms of variables such as nucleon number A, proton number Z, and neutron number N. To keep the forms as simple as possible, it will not be possible to make a direct connection to the underlying nucleon-nucleon interactions responsible for the binding energies. We shall, instead, take a semi-empirical approach and determine the parameters involved by fitting appropriate observed values; hence the name *semi-empirical mass formulas.* The power of these formulas lies in providing us with some idea of the general trends in nuclear binding energies that is useful for a large variety of purpose.

Weizacker mass formula. One of the popular ways to find nuclear binding energy is based on the analogy of a nucleus with a drop of incompressible fluid. We have seen earlier that nuclear volume increases linearly with the number of nucleons, in support of such a "liquid drop" model. For binding energy, we saw in Fig. 1-2 that, to a first-order approximation, it is linearly proportional to the nucleon number,

$$E_B(Z, N) = \alpha_1 A$$

where α_1 is known as the volume energy parameter. Several corrections must be applied before the formula can provide us with useful results.

Similar to a drop of liquid, nucleons on the surface of a nucleus are less tightly bound, as there are fewer particles nearby with which to interact. As a result, we expect a decrease in the binding energy with increasing number of nucleons on the surface. Since the nuclear volume increases linearly with A, the surface area is proportional to $A^{2/3}$. With the correction, the expression for nuclear binding energy becomes

$$E_B(Z, N) = \alpha_1 A - \alpha_2 A^{2/3}$$

Here α_2 is the surface energy parameter.

The next correction term comes from the repulsive Coulomb interaction between protons. For a uniformly charged sphere, the Coulomb energy is given by Eq. (1-7).

Taking $R = 1.2A^{1/3}$ fm, the result is

$$E_c = \left[\frac{1}{4\pi\epsilon_0}\right]\frac{3}{5}\frac{Z(Z-1)e^2}{R} = \frac{3\alpha\hbar c}{5r_0}\frac{Z(Z-1)}{A^{1/3}} = 0.72\frac{Z(Z-1)}{A^{1/3}} \text{ MeV}$$

A real nucleus has a diffused surface region and its shape may not necessarily be spherical. For these reasons, we do not expect the uniformly charged sphere value to be completely correct. On the other hand, the expression captures the essence of the dependence on A and Z. We shall therefore adopt the form of the Coulomb correction given above but leave the strength as an adjustable parameter. The binding energy now takes on the form

$$E_B(Z,N) = \alpha_1 A - \alpha_2 A^{2/3} - \alpha_3\frac{Z(Z-1)}{A^{1/3}}$$

with α_3 as the Coulomb energy parameter.

So far purely classical ideas have been used. The first quantum-mechanical consideration is the isospin dependence of nuclear force. We have seen in earlier sections that, apart from Coulomb repulsion between protons, stable nuclei prefer to have $N \approx Z$. Such an effect may be expressed by a quadratic dependence on $N - Z$, and the binding energy formula expands to

$$E_B(Z,N) = \alpha_1 A - \alpha_2 A^{2/3} - \alpha_3\frac{Z(Z-1)}{A^{1/3}} - \alpha_4\frac{(N-Z)^2}{A} \tag{4-55}$$

The new contribution is sometimes known as *symmetry energy,* with α_4 as the adjustable parameter. The factor A in the denominator is used to compensate in part for the observed fact that the increase in neutron excess $(N - Z)$ with A is much faster than that for the isospin effect.

We have also seen earlier that, because of pairing force, even-even nuclei are more tightly bound than their odd-odd counterparts with the same A, and odd-mass nuclei have intermediate values between them. To account for pairing, a factor Δ is included. The complete binding formula now has the form

$$E_B(Z,N) = \alpha_1 A - \alpha_2 A^{2/3} - \alpha_3\frac{Z(Z-1)}{A^{1/3}} - \alpha_4\frac{(N-Z)^2}{A} + \Delta \tag{4-56}$$

where the pairing energy parameter

$$\Delta = \begin{cases} \delta & \text{for even-even nuclei} \\ 0 & \text{for odd-mass nuclei} \\ -\delta & \text{for odd-odd nuclei} \end{cases}$$

with δ taken as a parameter to be fitted to known data.

The final form, Eq. (4-56), is known as the Weizacker mass formula. The values of the five parameters α_1, α_2, α_3, α_4, and δ depend somewhat on the binding energies used to find their values. A commonly used set,

$$\begin{array}{lll} \alpha_1 = 16 \text{ MeV} & \alpha_2 = 17 \text{ MeV} & \alpha_3 = 0.6 \text{ MeV} \\ \alpha_4 = 25 \text{ MeV} & \delta = \frac{25}{A} \text{ MeV} & \end{array} \tag{4-57}$$

provides some idea of the magnitude of each of the terms. The value of the pairing parameter δ is, perhaps, the least well determined quantity among the five and is found to be much larger, for example, if we restrict the fit to heavy nuclei alone (see Problem 4-11).

The usefulness of such a mass formula lies, for example, in obtaining the energies involved when a heavy nucleus undergoes fission into lighter fragments. The calculated results give a correct overall picture of the dependence on A, Z, and N values. However, there are local increases in the binding energy for nuclei near closed shells, known as the *shell effect,* that are important for a number of applications. Corrections may be included in the ways suggested by Myers and Swiatecki [108] and Strutinsky [134], as we shall see later in §9-2.

Kelson-Garvey mass formula. The primary aim of the Weizacker mass formula is to obtain global agreement with observed values. For this reason, it does not always give the best result for the binding energy difference between neighboring nuclei. In applications where such values are important, the Kelson-Garvey approach [88] is more useful. Instead of a liquid drop, a microscopic model is used as the starting point. Nuclear binding energy is considered to be a sum of one- and two-nucleon interaction terms. The values of these terms may vary from one mass region to another, but in a small region differing only by a few nucleons, they must be essentially constant and their values may be found from neighboring nuclei.

Let us illustrate the procedure by considering first an example consisting of one-body terms alone. This is not a realistic approximation and is used here only for its simplicity. In this limit, the binding energy of a nucleus made of Z protons and N neutrons is given by

$$E_B(Z, N) = \alpha N + \beta Z$$

where parameters α and β represent, respectively, the average values of the interaction of a neutron and a proton with the rest of the nucleons in the nucleus.

We can deduce the values of α and β for a small region in mass from the differences in the known binding energies in, for example, the following way:

$$\alpha = E_B(Z, N+1) - E_B(Z, N) \qquad \beta = E_B(Z+1, N) - E_B(Z, N)$$

In terms of these parameters, the difference between $E_B(Z+1, N+1)$ and $E_B(Z, N)$ is given by

$$\begin{aligned} E_B(Z+1, N+1) - E_B(Z, N) &= \alpha + \beta \\ &= E_B(Z, N+1) - E_B(Z, N) + E_B(Z+1, N) - E_B(Z, N) \end{aligned}$$

This may be rewritten as a difference equation relating the binding energies of four nearby nuclei,

$$E_B(Z+1, N+1) + E_B(Z, N) - E_B(Z, N+1) - E_B(Z+1, N) = 0$$

In other words, the binding energy of any one of the four nuclei may be deduced from the known values of the other three. Such a simple relation is, of course, the result of our one-body model and is correct only if the binding energy is predominantly one body in nature. We can test whether the premise is correct by calculating the differences between four such nuclei using known binding energies,

$$\Delta E_B = E_B(Z+1, N+1) + E_B(Z, N) - E_B(Z, N+1) - E_B(Z+1, N)$$

In practice, we find that the values of ΔE_B are rather large, reflecting the shortcomings in ignoring two-body terms.

There are, in general, three types of two-body contributions to be included: proton-proton, neutron-neutron, and neutron-proton interactions. For a nucleus of N neutrons and Z protons, the number of pairs for each type is

$$\text{Number of pairs} = \begin{cases} \frac{1}{2}Z(Z-1) & \text{proton-proton} \\ \frac{1}{2}N(N-1) & \text{neutron-proton} \\ NZ & \text{neutron-proton} \end{cases}$$

Using these results, we can express a model for the binding energy of a nucleus in terms of one- and two-body contributions as

$$E_B(Z, N) = aN + bZ + cN(N-1) + dZ(Z-1) + eNZ \tag{4-58}$$

with five parameters, a, b, c, d, and e, to be determined from known binding energies in neighboring nuclei.

We can follow the same approach as used earlier in the one-body model and rewrite Eq. (4-58) as a difference equation. One of the several possible forms is

$$\begin{aligned} &E_B(Z+1, N-1) + E_B(Z-1, N) + E_B(Z, N+1) \\ &-E_B(Z, N-1) - E_B(Z+1, N) - E_B(Z-1, N+1) = 0 \end{aligned} \tag{4-59}$$

To derive this relation, we can, for example, rearrange the six binding energies into three groups, each one consisting of the difference between a pair of nuclei having the same neutron number. In this way, terms depending only on neutron numbers do not enter and the difference equation reduces to one involving only three parameters: b, d, and e.

We shall do this for two different pairs of nuclei. First consider a pair with neutron number $N-1$ and proton numbers Z and $Z+1$. For such a choice, the purely neutron terms in Eq. (4-58) can be ignored and we obtain the difference in the binding energy as

$$\begin{array}{lllll} E_B(Z+1, N-1) = & b(Z+1) & +d\,Z(Z+1) & +e\,(N-1)(Z+1) & +f(a,c,N) \\ -\ E_B(Z, N-1) = & bZ & +d\,Z(Z-1) & +e\,(N-1)Z & +f(a,c,N) \\ \hline E_B(Z+1, N-1) & & & & \\ \quad -E_B(Z, N-1) = & b & +2d\,Z & +e\,(N-1) & \end{array}$$

Similarly, the difference for a second pair of nuclei with neutron number $N+1$ and proton numbers Z and $Z+1$ is given by

$$\begin{array}{lllll} E_B(Z, N+1) = & bZ & +d\,Z(Z-1) & +e\,(N+1)Z & +g(a,c,N) \\ -\ E_B(Z-1, N+1) = & b(Z-1) & +d\,(Z-1)(Z-2) & +e\,(N+1)(Z-1) & +g(a,c,N) \\ \hline E_B(Z, N+1) & & & & \\ \quad -E_B(Z-1, N+1) = & b & +2d\,(Z+1) & +e\,(N+1) & \end{array}$$

The sum of these two differences is the same as the difference between a pair with N neutrons and proton number differing by 2,

$$\begin{array}{rl} E_B(Z+1,N) = & b(Z+1) + dZ(Z+1) \qquad + eN(Z+1) + h(a,c,N) \\ -E_B(Z-1,N) = & b(Z-1) + d(Z-1)(Z-2) + eN(Z-1) + h(a,c,N) \\ \hline E_B(Z+1,N) - E_B(Z-1,N) = & 2b \qquad + 2d(2Z+1) \qquad + 2eN \end{array}$$

By eliminating b, d, and e from these three equations, we obtain the result given in Eq. (4-59). The same relation can also be derived by considering three pairs of differences with each pair having the same proton number.

We can check how well Eq. (4-58) works in practice by calculating, from six known binding energies, the difference

$$\begin{aligned} \Delta E_B &= E_B(Z+1,N-1) + E_B(Z-1,N) + E_B(Z,N+1) \\ &\quad -E_B(Z,N-1) - E_B(Z+1,N) - E_B(Z-1,N+1) \end{aligned}$$

in the same way as we did earlier when we had only one-body terms. We do not expect ΔE_B to be exactly zero for an arbitrary group of six neighboring nuclei even with two-body terms included. However, if the assumptions that went into constructing Eq. (4-58) are essentially correct, ΔE_B should be much smaller here and distributed randomly around zero. This was found to be true and the standard deviation for a sample of N such differences turns out to be

$$\frac{1}{N}\left\{\sum_{i=1}^{N}(\Delta E_B)^2\right\}^{1/2} \approx 100 \text{ keV}$$

We can take this result to mean that Eq. (4-59) may be used to calculate the binding energy for any one of six nuclei from the other five with an uncertainty of about 100 keV on average.

The usefulness of a relation represented by Eq. (4-58) is not limited to extrapolating binding energies for nuclei one nucleon away from known masses. To make estimates for nuclei further away, we can use the Kelson-Garvey mass formula to calculate the values of the unknown intermediate ones and include them as a part of the input. The uncertainties in the prediction, however, increase roughly as the square root of the number of steps.

One can, in principle, include contributions other than one- and two-body effects to reflect higher order terms that may occur in an effective nucleon-nucleon interaction. The difference equation, in this case, involves more than six nuclei. Since binding energies are known for a large number of nuclei, it is possible to consider such a higher order approach for nuclei far away from the valley of stability.

4-10 Alpha-Particle Decay

In light nuclei, the threshold for α-particle decay is comparable with that for nucleon emission, as can be seen from the plot of separation energies given later in Fig. 7-2. For this reason, α-particle decay is not energetically favored until about $A > 150$. Even for heavy nuclei, the lifetimes are long by strong interaction time scales. Furthermore, the energy of α-particles emitted tends to be confined in a narrow range of 5 to 9 MeV whereas the half-lives vary by several orders of magnitude. As mentioned in Chapter 1, these observations led to the discovery of quantum-mechanical tunneling, a milestone in the development of modern physics.

Barrier for α-particle emission. It is perhaps easier to visualize the barrier an α-particle must tunnel through by considering first the inverse process of a low-energy α-particle approaching a heavy nucleus from large distances away. For simplicity, we shall take the nucleus to be a uniform density sphere of radius $R = r_0 A^{1/3}$. Outside the nuclear surface, $r > R$, the interaction is purely Coulomb and the repulsive potential may be represented as

$$V_c(r) = \left[\frac{1}{4\pi\epsilon_0}\right]\frac{2Ze^2}{r} = \frac{2Z\alpha\hbar c}{r} = \frac{2.88Z}{r}\ \text{MeV} \tag{4-60}$$

where Z is the charge number of the nucleus and r is measured in femtometers in the final equality.

Once the α-particle is inside the nuclear surface, short-range nuclear force becomes effective. The fact that it is bound at distances $r < R$ means that the combination of Coulomb and nuclear forces must be attractive. For simplicity, we shall take the potential for this part to be a square well, as shown schematically in Fig. 4-9. In this approximation, the height of the barrier that retains the α-particle inside the potential well may be estimated from the amount of work required to overcome Coulomb repulsion in bringing an α-particle to the surface of a heavy nucleus, such as $^{238}_{92}$U,

$$E_c = \left[\frac{1}{4\pi\epsilon_0}\right]\frac{2Ze^2}{R} = \frac{2Z\alpha\hbar c}{r_0 A^{1/3}} \approx 35\ \text{MeV}$$

Better estimates put the Coulomb energy of an α-particle in this region of mass number to be just below 30 MeV.

Classically, an α-cluster inside a heavy nucleus must acquire enough energy to reach the top of the potential barrier before it can escape. Furthermore, the kinetic energy

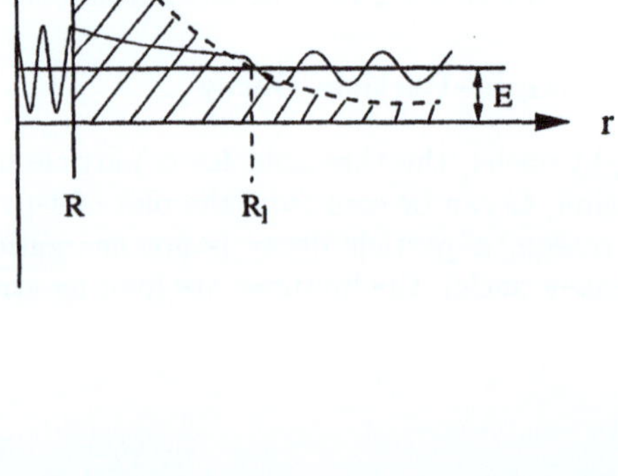

Figure 4-9: Schematic diagram showing the wave function of an α-particle tunneling through a potential barrier. Inside the potential well, $r \leq R$, the particle is bound and the wave function is sinusoidal. In the region $R < r < R_1$, the energy of the particle is less than the barrier height and the amplitude of the wave decreases more or less exponentially. Once outside, $r > R_1$, the α-particle is essentially free except for Coulomb interaction with the residual nucleus.

of the particle emitted is expected to be as least as large as the barrier height, as the α-particle carries with it all the energy it acquired to reach the top of the barrier. The observed values, however, are much smaller and, consequently, some other mechanism must be operating here. The quantum-mechanical explanation is that the α-particle does not have to go over the top of the potential barrier before being emitted. Instead, it tunnels through the barrier. The basic reason comes from the fact that the α-particle wave function does not vanish inside a small region of (finite) repulsive potential. In quantum mechanics, it is well known that the amplitude of a wave function, to a first-order approximation, decays exponentially inside such a barrier, and as a result, there is a finite probability to find the particle outside, as shown schematically in Fig. 4-9.

The reason that α-particle emission is an important channel of decay for heavy nuclei comes from a combination of two reasons. The first is the saturation of nuclear force which favors a pair of neutrons and a pair of protons inside a nucleus to form an α-cluster. As we have seen earlier, the average binding energy between α-clusters is much less than the corresponding value between nucleons within the same nucleus. This makes it energetically more favorable to emit an α-particle rather than a nucleon. The second reason is the increase in Coulomb repulsion in heavy nuclei due to the larger number of protons present. The combined effect of both reasons enables the Q-value for α-emission to become positive (negative in terms of the separation energy) for $A > 150$, as can be seen from Fig. 7-2. Examples of Q-values for some of the heavy nuclei as measured by the kinetic energy of the α-particles emitted are given in Table 4-3, together with the half-life associated with each case.

Table 4-3: Half-lives and energies of α-particle decay in typical heavy nuclei.

α-Emitter	E_α (MeV)	$\tau_{1/2}$	α-Emitter	E_α (MeV)	$\tau_{1/2}$
^{206}Po	5.22	8.8 days	^{236}U	4.49	2.39×10^7 yr
^{208}Po	5.11	2.90 yr	^{238}U	4.20	4.51×10^9 yr
^{210}Po	5.31	138 days	^{238}Pu	5.50	86 yr
^{212}Po	8.78	0.30 μs	^{240}Pu	5.17	6.58×10^3 yr
^{214}Po	7.68	164 μs	^{242}Pu	4.90	3.79×10^5 yr
^{216}Po	6.78	0.15 s	^{244}Pu	4.66	8×10^7 yr
^{228}U	6.69	9.1 months	^{240}Cm	6.29	26.8 days
^{230}U	5.89	20.8 days	^{242}Cm	6.12	163 days
^{232}U	5.32	72 yr	^{244}Cm	5.80	17.6 yr
^{234}U	4.77	2.47×10^5 yr	^{246}Cm	5.39	5.5×10^3 yr

Decay probability. Even before there were any theoretical models, it was found empirically that the large range of α-decay half-lives, from 10^{-6} s to 10^{17} years, may be related to the square root of the kinetic energy E_α of the α-particles emitted in the following way:

$$\log_{10} \mathcal{W} = C - \frac{D}{\sqrt{E_\alpha}} \tag{4-61}$$

where $\mathcal{W}$ is the decay probability. The parameters C and D are weakly dependent on Z but not on the neutron number, as can be seen from the plot in Fig. 4-10. This is known as the Geiger-Nuttall law. A simple, one-body theory of α-particle decay described below provides the foundation for this observed relation.

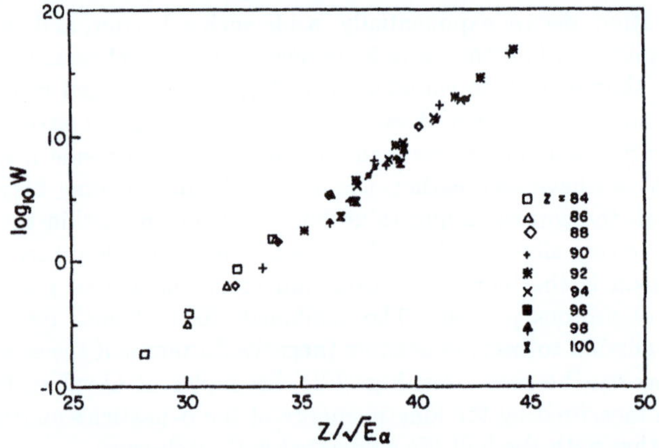

Figure 4-10: Semi-log plot of the transition rate $\mathcal{W}$ as a function of the ratio $Z/\sqrt{E_\alpha}$ for α-decay of heavy nuclei with proton number Z and α-particle kinetic energy E_α. The clustering of data along a straight line, described by Eq. (4-61), is known as the Geiger-Nuttall law. (Plotted using data from Ref. [95].)

The probability $\mathcal{W}$ for α-particle emission from a heavy nucleus by tunneling may be separated into a product of three factors. The first is the probability p_α to find an α-particle inside the nucleus. In a heavy nucleus, there is a good chance for two protons and two neutrons to form an α-like entity. We shall call such an object an α-cluster. However, this is only one of the many possible components of the wave function for such a nucleus. As a result, it is not easy to make an estimate for the value of p_α. A crude way is to say that it must be essentially of the same order of magnitude for all heavy nuclei, as there are only small fractional differences in their masses and we shall take $p_\alpha \sim 0.1$ as a rough guide.

Once an α-cluster is formed inside the nucleus, it must come to the surface before it can tunnel through the barrier. The frequency ν with which it appears at the edge of the potential well depends on the velocity v it travels and the size of the potential well. A reasonable way to estimate ν is to take the well size as twice the nuclear radius R. With this assumption we obtain the result,

$$\nu = \frac{v}{2R} = \frac{\sqrt{2K/M_\alpha}}{2R}$$

where K is the kinetic energy of the α-cluster inside the well and M_α its mass. The precise value of K depends on the depth of the potential well and is not well known.

It is reasonable to expect that K is of the same order of magnitude as E_α, the kinetic energy of the α-particle emitted. For our simple model, we shall take $K = E_\alpha$. This leads to the result

$$\nu = \frac{\sqrt{2E_\alpha/M_\alpha}}{2R} = \frac{\sqrt{E_\alpha}}{A^{1/3}} \times 2.9 \times 10^{21}\ \mathrm{s}^{-1}$$

where E_α is measured in mega-electron-volts and, as usual, we have taken $R = r_0 A^{1/3}$ with $r_0 = 1.2$ fm. From this we obtain, for example, a value of $\nu = 10^{21}\ \mathrm{s}^{-1}$ for ^{238}U with $E_\alpha = 5.6$ MeV. It is about an order of magnitude larger than the best values deduced from measurements. Part of the reason for the poor agreement comes from the fact that heavy nuclei do not have the simple spherical shape assumed here. Furthermore, the replacement of K by E_α may also have cost some loss of accuracy.

Transmission coefficient. Once at the barrier, the probability for an α-particle to tunnel through it is given by the transmission coefficient $\mathcal{T}$. For a one-dimensional square potential barrier of height V_0 and width b, the value is given in standard quantum mechanics texts as

$$\mathcal{T} = \left\{1 + \frac{V_0^2}{4E_\alpha(V_0 - E_\alpha)} \sinh^2 \kappa b\right\}^{-1} \tag{4-62}$$

Here

$$\kappa = \frac{1}{\hbar}\sqrt{2m(V_0 - E_\alpha)}$$

E_α is the kinetic energy ($E_\alpha < V_0$), and m is the mass of the particle. Outside the potential barrier, the particle is free and the wave function is sinusoidal. Inside the barrier, the wave function decays exponentially, as the kinetic energy of the particle is less than the barrier height. An extreme situation with $V_0 \gg E$ is particularly simple to calculate. In this limit, $\kappa b \to \infty$, and $\sinh \kappa b \longrightarrow e^{\kappa b}$. The transmission coefficient in Eq. (4-62) simplifies to the form

$$\mathcal{T} \longrightarrow e^{-2\kappa b} \tag{4-63}$$

The factor $e^{-\kappa b}$ expresses the attenuation of the amplitude of the wave in going through the barrier, and it is quite reasonable to expect that the transmission coefficient is essentially given by the square of this factor. For our case of $V_0 \approx 30$ MeV and E_α in the range 4 to 9 MeV, the condition of $V_0 \gg E_\alpha$ is adequately satisfied for the accuracies we need. As a result, we can use the approximate form of $\mathcal{T}$ given in Eq. (4-63) for the rest of the discussion.

The true potential barrier experienced by an α-particle in heavy nuclei is more complicated than the square-well example in the previous paragraph. However, the results of the one-dimensional treatment remains valid on the whole. As long as the potential well is spherically symmetric, the radial part of the Schrödinger equation has the form

$$\frac{d^2u(r)}{dr^2} + \frac{2\mu}{\hbar^2}\left\{\left(E_\alpha - V(r)\right) - \frac{\ell(\ell+1)\hbar^2}{2\mu r^2}\right\}u(r) = 0$$

where μ is the reduced mass of the α-particle inside the barrier and $u(r)$ is the radial wave function divided by r.

Since we are interested here in the region outside the range of nuclear force, the potential $V(r)$ may be taken as purely Coulomb, having the form given by Eq. (4-60). All the angular dependence is contained in the $\ell(\ell+1)/r^2$ term and may be taken as a part of the potential barrier. For this purpose, we shall define an "effective" potential barrier,

$$V_b(r) = \left[\frac{1}{4\pi\epsilon_0}\right]\frac{zZe^2}{r} + \frac{\ell(\ell+1)\hbar^2}{2\mu r^2} = \frac{zZ\alpha\hbar c}{r} + \frac{\ell(\ell+1)\hbar^2}{2\mu r^2}$$

The radial equation reduces to the form

$$\frac{d^2u(r)}{dr^2} + \frac{2\mu}{\hbar^2}\left\{E_\alpha - V_b(r)\right\}u(r) = 0$$

This result has essentially the same form as that for a square well used for our simple model above, except that the barrier height is now a function of r. The equation must now be solved by better techniques, such as the Wentzel-Kramers-Brillouin (WKB) method.

The form of the solution, however, remains very similar to that given in Eq. (4-63) if we make the replacement

$$\kappa b \longrightarrow \int_R^{R_1} \sqrt{\tfrac{2\mu}{\hbar^2}}\left\{V_b(r) - E_\alpha\right\}^{1/2} dr$$

The integration is taken from $r = R$ at the nuclear surface to $r = R_1$, where R_1 is the classical turning radius given by the relation

$$V_b(R_1) - E_\alpha = 0$$

For $\ell = 0$, we have

$$R_1 = \left[\frac{1}{4\pi\epsilon_0}\right]\frac{zZe^2}{E_\alpha} = \frac{zZ\alpha\hbar c}{E_\alpha}$$

In this case, the integral can be carried out explicitly, and the result for the transmission coefficient becomes

$$\ln \mathcal{T} = -2\kappa b = -\frac{2R_1}{\hbar}\sqrt{2\mu E_\alpha}\left\{\cos^{-1}\sqrt{\frac{R}{R_1}} - \sqrt{\frac{R}{R_1}\left(1 - \frac{R}{R_1}\right)}\right\}$$

Since we can take $R \ll R_1$ for the accuracies we need, the arc cosine term may be approximated by $(\pi/2 - \sqrt{R/R_1})$ and R/R_1 may be dropped in the last term. The result simplifies to

$$\ln \mathcal{T} = -\frac{2R_1}{\hbar}\sqrt{2\mu E_\alpha}\left(\frac{\pi}{2} - 2\sqrt{\frac{R}{R_1}}\right) = -\pi zZ\alpha c\sqrt{\frac{2\mu}{E_\alpha}} + 8\sqrt{\frac{\mu r_0 \alpha c}{\hbar}}\sqrt{ZA^{1/3}}$$

At low energies, such as those encountered in stellar evolution in astrophysics (see §10-2), the first term dominates. In this limit, the transmission coefficient is generally referred to as the Coulomb penetration factor and often written as

$$\mathcal{T} = e^{-2\pi\eta} = \exp\left\{-2\pi\frac{zZ\alpha c}{v}\right\} \tag{4-64}$$

with η known as the Sommerfeld number.

Energy and mass dependence. Using p_α, ν, and $\mathcal{T}$ obtained above, we can write the transition probability as

$$\mathcal{W} = p_\alpha \nu \mathcal{T}$$

To put this expression into a form so that it can be compared with the Geiger-Nuttall law of Eq. (4-61), we take the logarithm in the base 10 for both sides and obtain the result

$$\begin{aligned}\log_{10}\mathcal{W} &= \log_{10} p_\alpha + \log_{10}\nu + \log_{10}\mathcal{T} \\ &= 20.46 + \log_{10}\frac{\sqrt{E_\alpha}}{A^{1/3}} + 1.42\sqrt{ZA^{1/3}} - 1.72\frac{Z}{\sqrt{E_\alpha}} \end{aligned} \tag{4-65}$$

The dominant energy dependence comes from the last term, in agreement with the empirical result of the Geiger-Nuttall law.

In addition to energy, there is also a dependence of $\log_{10}\mathcal{W}$ in Eq. (4-65) on Z and, to a lesser extent, on A. To show the Z-dependence, we can plot $\log_{10}\mathcal{W}$ as a function of $\sqrt{E_\alpha}$ separately for each element. The observed values are now found to cluster much closer to straight lines than we have seen earlier in Fig. 4-10. The curves belonging to different Z-values are running, more or less, parallel to each other, as shown in Fig. 4-11. The remaining deviations in the distribution of data points from straight lines, usually too small to be noticed in such a plot, are due to nuclear structure effects that can only be explained by a proper account of p_α. For our purposes, we shall be satisfied with the success of a relatively simple theory to explain almost 30 orders of magnitude difference in the half-lives.

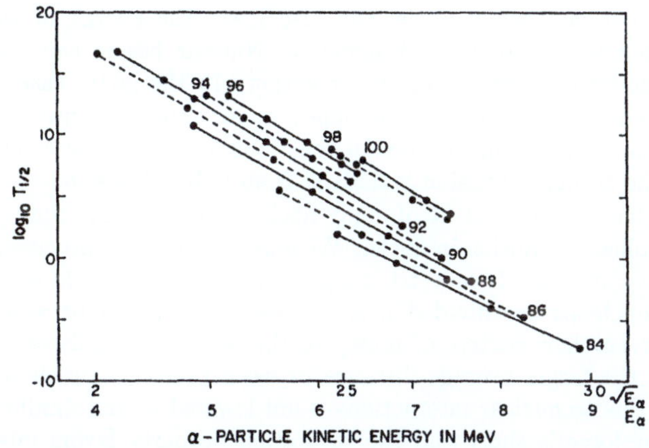

Figure 4-11: Semi-log plot of the transition rate $\mathcal{W}$ for α-decay as a function of the square root of E_α, the kinetic energy of the α-particle emitted. The observed values for different isotopes of each element, labeled by proton number Z, are closer to straight lines than those given in Fig. 4-10. (Plotted using data from Ref. [95].)

We have implicitly assumed in the above discussion that there is only a single kinetic energy E_α for all the α-particles emerging from a nucleus. In fact, it is common to find several different groups of α-particles emitted by the same parent nucleus. The example of ^{212}Bi shown in Fig. 4-12 has more than 14 different decays known and each one leaves the residual nucleus in a different state.

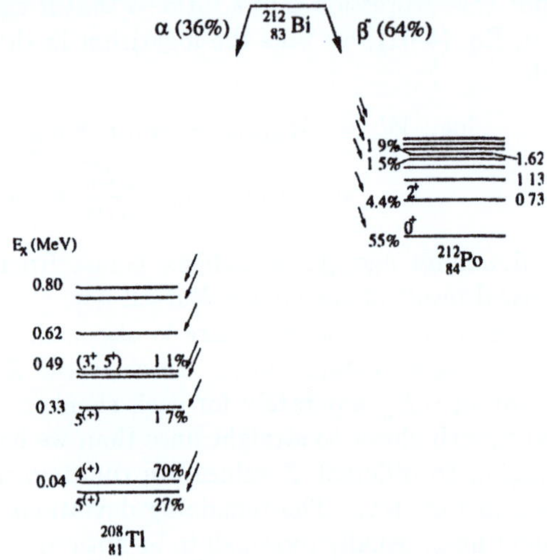

Figure 4-12: Energy level diagram showing the decay of ^{212}Bi. The Q-value for α-decay is 6.206 MeV leading to the ground state of ^{208}Tl and 2.246 MeV for β-decay to the ground state of ^{212}Po. (Plotted using data from Ref. [95].)

4-11 Nuclear Fission

In addition to α-decay, heavy nuclei can also lower the energy of the system as a whole by fission into two or more fragments. Nuclear fission can either take place spontaneously or be induced by another reaction. In the latter case, the reaction is the result of stimulation from energy supplied by an external source, such as neutron capture. In general, spontaneous fission reactions are rare events, and most of our knowledge on the subject of fission is derived from induced fission.

Fission involves the movement of many nucleons at the same time and is therefore an example of collective motion in nuclei. We may visualize the process in the following way. For a heavy nucleus, it may be energetically more favorable to assume a shape such that the nucleons are divided into two overlapping groups separated by some distance d between their centers of mass. In this way, the repulsive Coulomb energy is decreased by the larger average distance between protons. On the other hand, the binding energy due to nuclear interactions is not lowered in any significant way, as the range of nuclear force is short and the saturation property favors interaction among a few nearby nucleons. In terms of the Weizacker mass formula, the reduction comes mainly from a small increase in the surface area as a result of deformation. By sacrificing some surface energy, the binding energy of the system is increased by a larger decrease in Coulomb repulsion and the equilibrium shape of heavy nuclei becomes deformed.

For such nuclei, *spontaneous fission* is possible in principle, as the system may gain even more binding energy by splitting into two completely separated pieces. The only thing preventing spontaneous fission from happening more often is the fission barrier through which the fragments must tunnel. We shall return to the source of this barrier later. Since the condition for large deformation required for fission is found only in extremely heavy nuclei, spontaneous fission does not become an important decay channel until $A > 240$. For example, the partial half-life for spontaneous fission of ^{232}U is around 10^{14} years, whereas the value for α-decay is only 2.85 years. On the other hand, by the time we get to $^{254}_{98}$Cf, the half-life reduces to 60 days with a branching ratio of 99% for spontaneous fission.

Induced fission. Nuclear fission was actually first found through *induced fission* before the discovery of spontaneous fission. Induced fission may be defined as any reaction of the type $A(a, b)B$ with final products b and B being nuclei of roughly comparable mass. The best-known example is the fission of $^{235}_{92}$U induced by thermal neutrons. A compound nucleus ^{236}U* (where the asterisk in the superscript indicates that the nucleus is in an excited state) constitutes the intermediate state for the reaction. For all practical purposes, we may regard the kinetic energy of the incident thermal neutron to be zero. In this limit, the excitation energy of the compound nucleus is simply the excess in binding energy for ^{235}U plus a "free" neutron over that for (the ground state of) ^{236}U. The value found from a table of binding energies is 6.5 MeV. The ground state of ^{236}U is unstable toward α-particle emission with a half-life of 2.4×10^7 years. However, the excitation energy brought along by the neutron capture sets the entire nucleus into vibration, and as a result, fission into a number of different products becomes possible. A typical channel is the reaction

$$^{235}_{92}\text{U} + n \to {}^{92}_{36}\text{Kr} + {}^{142}_{56}\text{Ba} + 2n \tag{4-66}$$

The energy liberated in this example is around 180 MeV. The two neutrons emitted in the process are called *prompt neutrons*, since they are released as a part of the fission process. However, both ^{92}Kr and ^{142}Ba are neutron unstable, as the heaviest stable isotopes are, respectively, ^{86}Kr and ^{138}Ba. Among other possibilities, neutrons are released either directly from ^{92}Kr and ^{142}Ba or from unstable daughter nuclei derived from their decays. These neutrons are the *delayed neutrons* since they emerge after some delays—as, for example, due to intervening β- and γ-ray emissions.

It is often found that there is a preference for asymmetric fission, fission into two fragments of unequal mass. When a heavy nucleus with large neutron excess undergoes fission, for instance, into two medium-weight nuclei with nucleon number roughly $A/2$ each, the fragments would have to be nuclei far away from the valley of stability. As a result, many prompt neutrons would have to be emitted, greatly increasing the number of independent entities in the final state. Phase space considerations, on the other hand, strongly favor exit channels with a minimum number of products. Fission involving one heavier fragment, together with one lighter counterpart, reduces the number of prompt neutrons that have to be emitted. The net result, shown schematically in Fig. 4-13, is that the fission fragments have a bimodal distribution as a function of mass. Besides binary fission involving two final nuclei (plus prompt neutrons), ternary fission consisting of three final nuclei is also commonly observed. In principle, larger numbers of fragments are also possible, but, again, phase space considerations reduce their probability.

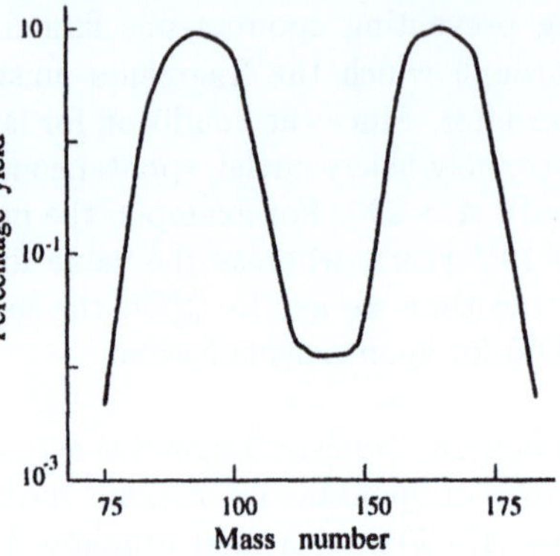

Figure 4-13: Schematic diagram showing the distribution of fission fragments as a function of nucleon number. The reason for a bimodal distribution comes from the larger ratio of neutron to proton numbers for the heavy parent than those for medium-weight nuclei making up the fragments.

The reason for using thermal neutrons to induce ^{235}U fission in most nuclear reactors is related to the neutron absorption cross section. As can be seen from Fig. 4-14, the probability at low energies for ^{235}U to form a compound nucleus decreases almost exponentially with increasing neutron energy. The only exceptions are small resonances, related to the single-particle states of ^{236}U at neutron threshold energy. For our purpose here, we may ignore them, as their combined contribution to the fission cross section is not significant. In contrast, the induced fission cross section for ^{238}U, the dominant component in uranium ore, is negligible until the neutron energy is above 1 MeV, or 10^{10} K in terms of temperature. Consequently, fast neutrons are more suitable to induce ^{238}U fission, the principle behind breeder reactors.

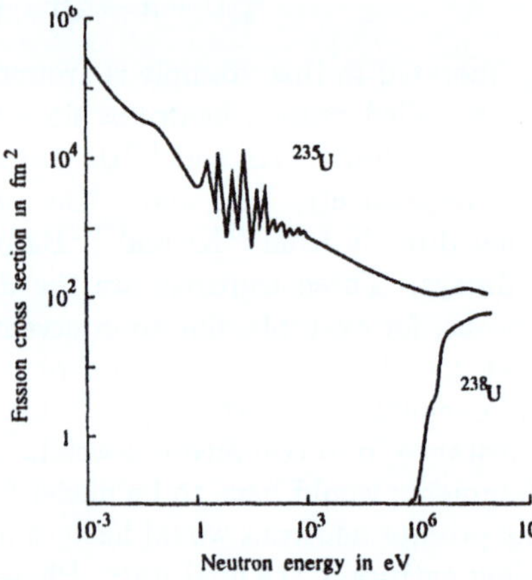

Figure 4-14: Schematic diagram showing the cross sections for neutron-induced fission on ^{235}U and ^{238}U as functions of neutron energy. Large values for ^{235}U at low energies favor thermal neutron fission. In contrast, neutron induced fission for ^{238}U is insignificant until the energy is above 1 MeV.

Fission barrier. A crude simulation of the barrier that inhibits fission may be constructed from a liquid drop model using the Weizacker mass formula given in Eq. (4-56) as the starting point. For simplicity, we shall take for our calculation a hypothetical nucleus of $A = 300$ and $Z = 100$ and symmetric fission into two equal fragments of $A = 150$ and $Z = 50$ each. The volume energy term in Eq. (4-56) is unchanged when the nucleus is split into two separate pieces and may be omitted from our considerations here. Similarly, changes in the pairing and symmetry energy terms are too small for us to be concerned with here. Only two terms, the surface energy $\alpha_2 A^{2/3}$ and the Coulomb energy $\alpha_3 Z(Z-1)/A^{1/3}$, remain and they form the main components of the fission barrier. Since both terms enter the binding energy equation with negative signs, a decrease in either term will increase the binding energy of the system as a whole.

When the hypothetical nucleus undergoes a transformation from $A = 300$ to two fragments of $A = 150$ each, the loss in binding energy due to changes in the surface energy is given by

$$E_{\text{surface}} = \alpha_2 \left\{ 2 \times \left(\tfrac{1}{2}A\right)^{2/3} - A^{2/3} \right\} = 0.26\alpha_2 A^{2/3} \approx 200 \text{ MeV}$$

where we have used the value $\alpha_2 = 17$ MeV from Eq. (4-57). The amount is the decrease in binding energy of the system when the two clusters are sufficiently far away from each other that there is no more nuclear interaction remaining between them. Let us use d, the distance between the two centers of mass, as a measure of the separation between the two fragments. Before fission, we shall assume $d = 0$, and after the reaction is completed, d is greater than twice the radius of each of the fragments. For simplicity, we shall take the radius of each fragment to be $R_f = r_0(150)^{1/3} \approx 6.4$ fm. It is not easy to calculate the contribution from the surface energy term at distances between $d = 0$ and $d = 2R_f \approx 13$ fm. Schematically, it may take on a form as shown by the dashed line in Fig. 4-15(a).

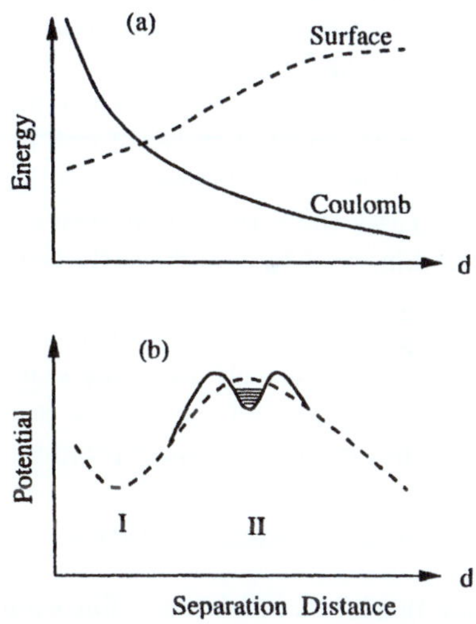

Figure 4-15: Schematic diagram of fission barrier as a function of the distance d between two fragments. Competition between surface energy, represented by the dashed curve in (a), and Coulomb energy, the solid curve in (a), produces a potential shown by the dashed curve in (b). Realistic potential barriers can be more complicated as, for example, shown by the solid curve in (b), with a second minimum, indicated as II in the diagram, in addition to the primary well at I.

Let us now consider the contribution to fission due the Coulomb energy term. For $A = 300$, the Coulomb energy from Eq. (4-56) is

$$\alpha_3 \frac{Z(Z-1)}{A^{1/3}} \approx 900 \text{ MeV}$$

using $\alpha_3 = 0.6$ MeV. For each of the fragments, the value is found to be roughly 275 MeV by the same method. There is therefore a gain of 350 MeV in binding energy from the Coulomb term as a result of splitting the nucleus into two. The amount is obtained under the assumption that the two fragments are separated by an infinite distance without any Coulomb interaction between them. As the distance decreases, the Coulomb energy increases with a $1/r$ dependence if we ignore any possible shape changes in the fragments. At $d < 13$ fm, the value cannot be estimated easily. Again, we can take it to be something like that shown by the solid curve in Fig. 4-15(a).

The net change in binding energy as a result of fission in our simple model is the difference between the gain in the Coulomb energy and the loss in the surface energy outlined above. The value $350 - 200 = 150$ MeV is of the correct order of magnitude as, for example, compared with the energy released through the reaction given by Eq. (4-66). The fission barrier may therefore be schematically represented by a sum of the two curves shown in Fig. 4-15(a) and plotted as the dashed curve in Fig. 4-15(b). For lighter nuclei, the difference between the contributions from surface and Coulomb terms is not as clearly marked as in our example. The resulting barrier is much broader and spontaneous fission is suppressed. For extremely heavy nuclei, the opposite is true and the nucleus becomes unstable toward fission.

Deformation, however, complicates the shape of fission barriers, even though the main features are given correctly by our simple model. The most interesting aspect due to deformation and other finer considerations is that the detailed shape of the fission barrier may be somewhat different, as shown schematically by the solid line in Fig. 4-15(b). Sometimes a local minimum in the potential, marked as II in the figure, may develop. The evidence for such a secondary potential well is obtained from narrow resonances found in the fission cross section that cannot come from levels in the main well, marked as I in the figure. Such a barrier is often called a double-hump potential for obvious reasons. With such a potential, it is possible that the nucleus may be trapped in the excited states of well II. Such states, especially the low-lying ones, may prefer to de-excite by fission rather than returning to the main well by γ-ray emission. In some cases the rate of spontaneous fission for some of the low-lying states in well II may be hindered by a sufficiently large factor that a fission isomer may develop as a result.

Partly because of its importance to applied work, fission of heavy nuclei has been investigated extensively by many experiments. Theoretical studies to understand the equilibrium shape and the process leading to fission are, however, complicated, and it is somewhat difficult to make progress.

4-12 Infinite Nuclear Matter

We saw in §4-2 and §4-5 that a finite nucleus has a large surface region where the density drops gradually to zero with radial distance. Even for a heavy nucleus, only a small fraction of the nucleons are in the central region where the density may be considered to be constant (see Problem 4-4). For many theoretical investigations, it is much easier if the density is uniform throughout the nuclear volume. For this reason, infinite nuclear matter is created as an idealized system of bound nucleons with an uniform density that approximates the interior of a heavy nucleus. To make the situation even simpler, we shall assume that the neutron number is the same as the proton number. Such a system is convenient for testing nucleon-nucleon interaction as well as techniques for solving many-body problems. Furthermore, being an infinite system, we do not have complications caused by motion of the center of mass, as in the case of finite nuclei. Electromagnetic interactions are usually ignored in such studies, as the primary interest is nuclear.

There is obviously no observed data on such an idealized system. A neutron star is as close to an infinite system of nuclear matter as we can imagine; however, experimental measurements on neutron stars of direct interest to nuclear physics may not be forthcoming for a while. As a result, most information concerning infinite nuclear matter must be inferred from our knowledge on finite nuclei.

Energy and density. Let us first try to deduce the binding energy in nuclear matter, for instance, using the Weizacker mass formula given in §4-9. For a finite nucleus, the volume is proportional to nucleon number A and the surface area to $A^{2/3}$. The ratio of the surface term to the volume term therefore varies as $A^{-1/3}$. For infinite nuclear matter, we can ignore the surface term in Eq. (4-56), as $A^{-1/3} \to 0$. Contributions from Coulomb repulsion between protons can also be put to zero, as we do not wish to consider any electromagnetic effect here. Similarly, the symmetry energy vanishes with the assumption of $N = Z$. Pairing effect may also be ignored because A is infinite. This leaves only the volume term in the binding energy. From studies made on finite nuclei, we have the result

$$\frac{E_B}{A} = 16 \pm 1 \text{ MeV} \tag{4-67}$$

for the binding energy per nucleon in infinite nuclear matter, the same value as α_1 given in Eq. (4-57). The uncertainty of 1 MeV here, in part, reflects variations in the values obtained by different ways of determining the parameter.

The density of infinite nuclear matter can be inferred from the maximum or *saturation* density in finite nuclei. The value commonly used is

$$\rho_0 = 0.16 \pm 0.02 \text{ nucleons/fm}^3 \tag{4-68}$$

It is slightly higher than the average density of $3/(4\pi r_0^3) = 0.14$ nucleons/fm^3 obtained using $r_0 = 1.2$ fm. The difference comes from the absence of a surface region here.

Fermi gas model. We can also relate ρ_0 to the Fermi momentum of nucleons in the following way. If the excitation energy is not very large—we shall come back at the end of this derivation to give an estimate of what is considered to be large—most of the low-lying single-particle states are occupied. As a result, the Pauli exclusion principle plays a more important role than nucleon-nucleon interaction in determining the motion of nucleons inside infinite nuclear matter. For such a system, we can adopt a degenerate Fermi gas model to study the momentum distribution of the nucleons. In this approximation, nucleons are treated as noninteracting fermions, with the ground state formed by filling up all the available low-lying single-particle states.

For a free particle in a cubic box of length L on each side, the wave function may be represented by a plane wave,

$$\psi(\boldsymbol{r}) = \frac{1}{L^{3/2}} e^{\imath \boldsymbol{k}\cdot\boldsymbol{r}}$$

In the absence of spin and isospin degrees of freedom, the allowed values for the wave numbers $\boldsymbol{k} = (k_x, k_y, k_z)$ are given by the condition that the wave function vanishes at the boundary of the box. As a result,

$$k_x = \frac{2\pi}{L} n_x \qquad k_y = \frac{2\pi}{L} n_y \qquad k_z = \frac{2\pi}{L} n_z$$

where n_x, n_y, and n_z are integers 0, ±1, ±2, ... The number of allowed plane wave states in a volume element d^3k is then

$$dn = 4\left(\frac{L}{2\pi}\right)^3 d^3k \tag{4-69}$$

where the factor of 4 comes from the fact that there are equal numbers of protons and neutrons and that each of them can be in a state with either spin up or spin down.

We can now relate nuclear density to the Fermi momentum. Since the total number of nucleons is A and, in the ground state, they fill all the low-lying states up to Fermi momentum k_F, this gives us the relation

$$A = \int_0^{\epsilon_F} dn = \int_0^{k_F} 4\left(\frac{L}{2\pi}\right)^3 d^3k = 4\left(\frac{L}{2\pi}\right)^3 \frac{4\pi}{3} k_F^3$$

The nucleon density is found from the number of nucleons in volume L^3,

$$\rho_0 = \frac{A}{L^3} = \frac{2}{3\pi^2} k_F^3$$

On inverting the relation, we obtain the Fermi momentum in terms of the density of infinite nuclear matter,

$$k_F = \left(\frac{3\pi^2}{2}\rho_0\right)^{1/3} = 1.33 \pm 0.05 \text{ fm}^{-1} \tag{4-70}$$

where the final result is calculated using the value of ρ_0 given in Eq. (4-68).

The average kinetic energy of nucleons in infinite nuclear matter may be found from Eq. (4-69) using the fact that the value for a nucleon is $(\hbar k)^2/2M_N$, with M_N as the mass of a nucleon. On averaging over all the nucleons, we obtain the result

$$\bar{\epsilon} = \frac{1}{A}\int_0^{k_F} \frac{(\hbar k)^2}{2M_N} 4\left(\frac{L}{2\pi}\right)^3 d^3k = \frac{3}{5}\frac{(\hbar k_F)^2}{2M_N} = \frac{3}{5}\epsilon_F \tag{4-71}$$

The value of the Fermi energy ϵ_F can be found using k_F given in Eq. (4-70),

$$\epsilon_F = \frac{(\hbar k_F)^2}{2M_N} \approx 37 \text{ MeV} \tag{4-72}$$

The average energy of a nucleon is then $\bar{\epsilon} \approx 22$ MeV. From this result, we see that only a small fraction of the nucleons in a nucleus of nucleon number A can be excited unless the energy involved is comparable to $A\epsilon_F$, of the order of 10^3 MeV for a medium-weight nucleus. This in turn justifies the use of the Fermi gas model above.

Compression modulus. The state of minimum energy is an equilibrium for infinite nuclear matter, stable against small variations of the density. Variations of the binding energy per particle due to changes in Fermi momentum must therefore vanish,

$$\frac{d}{dk_F}\left(\frac{E_B}{A}\right) = 0$$

The second-order derivative of E_B/A depends on the difficulty, or *stiffness*, of nuclear matter against changes in the density. This is measured by the compression modulus,

$$\mathcal{K} = k_F^2 \frac{d^2}{dk_F^2}\left(\frac{E_B}{A}\right)_{\rho=\rho_0} \tag{4-73}$$

which is the slope in the variation of binding energy per nucleon as a function of k_F. Since it is evaluated at the energy minimum, it is a positive quantity. The compression modulus is the equivalent of the bulk modulus in mechanics that characterizes variations of the volume of a material as a function of applied pressure (see Problem 4-19).

For nuclear matter, generally

$$\mathcal{K} \approx 200 \text{ MeV} \tag{4-74}$$

estimated, for example, from the energy required to excite a nucleus without changing its shape. This is called a *breathing mode* of excitation (see §6-1) and is most easily recognized in even-even nuclei, in particular, closed shell nuclei. Since the ground state spin and parity of such a nucleus is 0^+, one way to excite to another 0^+ state is by a change in the density.

Nuclear matter calculations. One of the aims of infinite nuclear matter calculations is to reproduce the values of the three pieces of known "data" deduced from finite

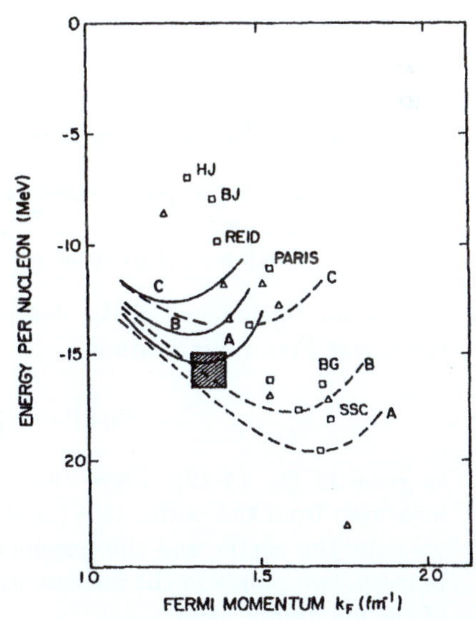

Figure 4-16: Binding energy per nucleon as a function of nucleon Fermi momentum in infinite nuclear matter. Shaded area represents the possible values extrapolated from finite nuclei and small squares are calculated results using potentials shown (HJ, Hamada-Johnston; BJ, Bethe-Johnson; BG, Bryan-Gersten; and SSC, super-soft-core). The solid lines are the results of different Dirac-Brueckner calculations and dashed lines are more conventional Brueckner calculation. (Adapted from Ref. [100].)

nuclei: the binding energy per nucleon of Eq. (4-67), the saturation density in terms of the Fermi momentum of Eq. (4-70), and (not always carried out) the compression modulus of Eq. (4-74). The hope is that the simple geometry offered by the idealized system provides us with a direct and meaningful test for nucleon-nucleon interactions and techniques for many-body problem. The test is a nontrivial one. For example, most of the properties of finite nuclei are not very sensitive to the hard core in nucleon-nucleon interactions. In contrast, the value of this term is crucial in reproducing the saturation density in infinite nuclear matter. Without the short distance repulsion, infinite nuclear matter can gain binding energy by increasing its density. In fact, many calculations failed to produce the correct binding energy per nucleon at the correct saturation density. The situation is more promising for some of the relativistic approaches. Results, obtained using a version of the Bonn potential for nucleon-nucleon interaction and a relativistic version of the most fruitful many-body technique, seem to produce the correct answers for both quantities, as shown in Fig. 4-16. We shall not go any further into these calculations, as it would require a lengthy discussion of the various techniques required to solve the many-body problem involved here.

Problems

4-1. The differential cross section for Rutherford scattering is proportional to $\sin^{-4}(\theta/2)$ where θ is the scattering angle. Explain why, in reality, experimental differential cross sections remain finite as $\theta \to 0$.

4-2. Derive an expression for the form factor $F(q^2)$, assuming that the nuclear density is given by a uniform sphere of radius R.

4-3. Show that, for high-energy inelastic scattering where the projectile rest mass may be ignored, the momentum transfer is given by

$$(\hbar c q)^2 = 4EE' \sin^2 \frac{\theta}{2}$$

of Eq. (4-29), where E is the energy of the projectile, E' the energy of the scattered particle, and θ the scattering angle.

4-4. Use the values of the parameters given in Table 4-1 to calculate the number of nucleons in the surface region of ^{206}Pb, i.e., the region where the density is 90% or less of the value at the center of the nucleus.

4-5. The radial dependence of the density of a nucleus may be described by a two-parameter Fermi distribution,

$$\rho_{2\text{pF}}(r) = \frac{\rho_0}{1 + \exp\{(r - c)/z\}}$$

as given in Eq. (4-22). Show that the parameter c is the radius of the nucleus measured from the center to a point where the density falls to roughly half of its value at the center and the parameter $z \approx t/4.4$, where t is the radial distance between two points in the nucleus whose densities are, respectively, 10% and 90% of the maximum value.

4-6. A muon (μ^-) can take the place of an electron in an atom and form a muonic atom. Since the muon is 207 times more massive than an electron, the radii of its orbits in an atom are much smaller. For a heavy atom, the low-lying muonic orbits may have a substantial overlap with the nuclear wave function as a result. Calculate the fraction of time a muon is inside a spherical nucleus of radius $R = 1.2A^{1/3}$ fm, when the muonic atom is in the lowest energy state. Take $A = 200$ and $Z = 70$.

4-7. Use a table of binding energies (e.g., from http://www.nndc.bnl.gov) to:

(a) Calculate the energy released in fusing two free protons and two free neutrons into an α-particle.

(b) Show that ^{238}U is unstable toward α-decay and calculate the kinetic energy of the α-particle emitted.

(c) Calculate the maximum possible energy released in the fission of ^{235}U induced by thermal neutrons. Make a reasonable assumption of the fission fragments released in the process.

4-8. The radius of a nucleus may be approximated as $R = r_0 A^{1/3}$. From the binding energy differences between odd-A mirror nuclei (i.e., a pair of nuclei having the same A but with neutron and proton numbers interchanged) in the mass range $A = 11$ to $A = 17$, estimate the value of r_0 assuming a uniform spherical charge distribution for each nucleus and ignoring effects other than Coulomb.

4-9. From the binding energies of members of the $A = 16$ isobar, ^{16}C, ^{16}N, ^{16}O, ^{16}F, and ^{16}Ne, calculate the position of the $T = 1$ isobaric analogue state to the ground state of ^{16}F in ^{16}O. Use a uniformly charged sphere of radius $R = 1.2A^{1/3}$ femtometers to find the Coulomb energy difference between the two nuclei. Repeat the calculation starting from the ground state of ^{16}N and compare the results obtained. Deduce also the positions of the lowest $J^\pi = 0^+$, $T = 2$ states in ^{16}N, ^{16}O, and ^{16}F from the binding energies of ^{16}C and ^{16}Ne. Compare the results with the observed values shown in Fig. 4-8.

4-10. From a table of binding energies show that the ground state of ^{8}Be is stable against β-decay and nucleon emission but unstable against decay into two α-particles. Calculate the energy released in the decay $^8\text{Be} \rightarrow \alpha + \alpha$.

4-11. Obtain the masses of members of the $A = 135$ isobar from a table of binding energies and plot the results as a function of Z. From the results deduce the value of the symmetry energy parameter α_4 in Eq. (4-56) for the Weizacker mass formula. Carry out the same calculation for members of the $A = 136$ isobar and estimate the value of the pairing parameter δ from the results.

4-12. The Weizacker mass formula is useful to obtain global distribution of binding energies as a function of A, Z, and N. Use this formula to show that the Q-value, i.e., kinetic energy released, in fission is positive only for heavy nuclei.

4-13. Use the Kelson-Garvey mass relation to find the binding energy of ^{12}O from the values of nearby nuclei. Estimate the uncertainties in the value deduced.

Using the value of binding energy obtained and a minimum amount of additional experimental data, find the locations of isobaric analogue states to the ground state of ^{12}O in ^{12}N and ^{12}C.

4-14. Calculate the average density of ^{238}U assuming it is a uniform sphere of radius $R = 1.2A^{1/3}$ femtometers. If, instead, R is the radius of ^{238}U measured from the center to a point on the surface where the density is half of that in the central region, and the nuclear density is given by the two-parameter Fermi form given in Eq. (4-22) with $z = 0.6$ fm, what is the central density of the nucleus? Compare the value obtained for the average density.

4-15. Determine the Coulomb and symmetry energy parameters of the Weizacker mass formula from the binding energies of three members of the $A = 27$ isobar, $(A, Z) = (27, 12), (27, 13), (27, 14)$.

4-16. The barrier penetration factor for an α-particle to pass through the Coulomb field of the residual nucleus may be estimated using a triangular-shaped, one-dimensional barrier as shown in the figure on the right. For a heavy nucleus with $Z = 72$, we can use $R = 9$ fm. The height of barrier V_0 may be taken as the Coulomb energy at distance R and the distance R_1 may be estimated using the point where the Coulomb energy is equal to the kinetic energy E_α of the α-particle. Find the probability for an α-particle of $E_\alpha = 5$ MeV to escape from the nucleus.

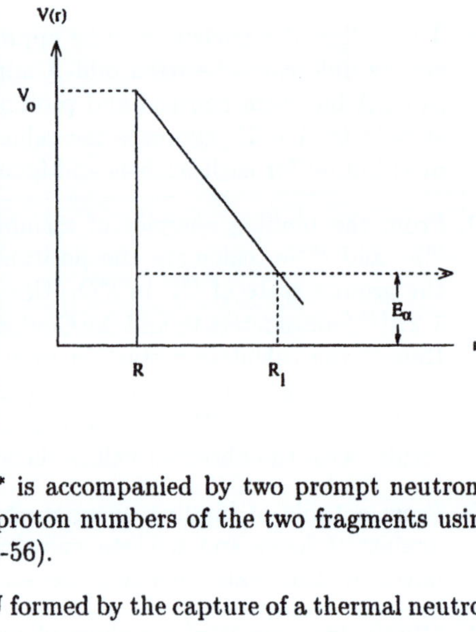

4-17. Assume that a binary fission of ^{236}U* is accompanied by two prompt neutrons. Calculate the optimum neutron and proton numbers of the two fragments using the Weizacker mass formula of Eq. (4-56).

4-18. Calculate the excitation energy of ^{236}U formed by the capture of a thermal neutron to the ground state of ^{235}U.

4-19. In the study of the property of ordinary matter, the bulk modulus B is defined as the negative of the ratio of the change in pressure applied, Δp, to the fractional change in the volume, $\Delta V/V$,

$$B = -V\frac{\Delta p}{\Delta V}$$

where the negative sign simply means that volume shrinks when the pressure is increased. Find the relation of B to the compression modulus $\mathcal{K}$ defined in Eq. (4-73).

Chapter 5

Electromagnetic and Weak Interaction

The wave functions of nuclear states are usually taken as the eigenvectors of a Hamiltonian in which the only interaction is nuclear. Electromagnetic and weak interactions are treated as perturbations, inducing transitions from one state to another. In this chapter we shall examine the rates of these transitions and see what they can tell us about the atomic nucleus. Other processes, such as nucleon and α-particle emission, also change the state a nucleus. They are examined in Chapters 4 and 8.

5-1 Nuclear Transition Matrix Element

Transition probability. If we have a sample of N radioactive nuclei, the probability for any one of them to decay at a given time is independent of the status of other members in the sample. The number of decays taking place in a given time interval is therefore proportional to $N(t)$, the number of radioactive nuclei present at time t,

$$\frac{dN}{dt} = -\mathcal{W}N(t) \tag{5-1}$$

The constant of proportionality, $\mathcal{W}$, is the *transition probability* or *decay constant*, and its value depends on the nature of the perturbation that causes the decay as well as the properties of the initial and final states involved. For this reason, the decay constant is the quantity of central interest in examining nuclear transitions.

From Eq. (5-1), we obtain the familiar exponential decay law,

$$N(t) = N_0 e^{-\mathcal{W}t}$$

where N_0 is the number of radioactive nuclei at time $t = 0$. The half-life, $T_{1/2}$, is the amount of time it takes for the activity of a sample to be reduced by half and is inversely proportional to the transition rate,

$$T_{1/2} = \frac{\ln 2}{\mathcal{W}} = \frac{0.693}{\mathcal{W}}$$

The lifetime, or mean life, of an excited state is the average amount of time it takes for a radioactive nucleus to decay. It is connected to the transition probability and half-life by the relation

$$\overline{T} = \frac{\int_0^\infty te^{-\mathcal{W}t}\,dt}{\int_0^\infty e^{-\mathcal{W}t}\,dt} = \frac{1}{\mathcal{W}} = \frac{T_{1/2}}{0.693}$$

Transition probability, half-life, and lifetime are three different ways to characterize the same physical observable.

Width. If a nucleus is in an excited state, it must discard the excess energy it has by undergoing a decay. It is, however, impossible to predict when the decay will actually take place. As a result, there is an uncertainty in time $\Delta t = \overline{T}$ associated with the existence of the excited state. Because of the limited lifetime, it is not possible for us to measure its energy to infinite precision. This is independent of the instrumental accuracy in the measurement. In fact, for our purpose here, we can regard uncertainties introduced by the measuring apparatus to be sufficiently small that they may be ignored.

In quantum mechanics, the expectation value corresponding to an observable is interpreted as the average over measurements for a large number of identically prepared samples. In other words, if we carry out the energy measurement for N nuclei in the same excited state, there will be a distribution of the values obtained. If the value for the ith excited nucleus is E_i, the average $\langle E\rangle$ is given by

$$\langle E\rangle = \frac{1}{N}\sum_{i=1}^{N} E_i$$

An idea of the spread in the measured values is provided by the square root of the variance,

$$\Gamma = \left\{\frac{1}{N}\sum_{i=1}^{N}\left(E_i^2 - \langle E\rangle^2\right)\right\}^{1/2}$$

The Heisenberg uncertainty principle says that the product of Γ and $\overline{T}$ is equal to $\hbar$ under the best circumstances, or

$$\Gamma = \frac{\hbar}{\overline{T}} = \hbar\mathcal{W} \tag{5-2}$$

The quantity Γ is known as the natural line width, or width for short, of a state, as we saw earlier in §2-6. It is also a way to indicate the transition probability of a state and is proportional to (the inverse of) the lifetime and half-life. Since $\hbar = 6.6 \times 10^{-22}$ MeV-s, a width of 1 MeV corresponds to a lifetime of 6.6×10^{-22} s.

One can also relate Γ to the probability of finding the excited state at a specific energy E. In terms of the wave function, the decay constant $\mathcal{W}$ may be defined in the following way:

$$|\Psi(\boldsymbol{r},t)|^2 = |\Psi(\boldsymbol{r},t{=}0)|^2 e^{-\mathcal{W}t} \tag{5-3}$$

For a stationary state, the time-dependent wave function may be separated into a product of spatial and temporal parts,

$$\Psi(\boldsymbol{r},t) = \psi(\boldsymbol{r})e^{-iEt/\hbar}$$

To carry such a separation over to a decaying (excited) state, the energy E must be changed into a complex quantity,

$$E \longrightarrow \langle E \rangle - \tfrac{1}{2} i\hbar \mathcal{W}$$

The time-dependent wave function now takes on the form

$$\Psi(\boldsymbol{r}, t) = \psi(\boldsymbol{r}) e^{-i\langle E \rangle t/\hbar - \mathcal{W}t/2} \tag{5-4}$$

so as to satisfy Eq. (5-3).

Alternatively we can make use of the fact that the excited state is one without a definite energy and write the wave function as a superposition of components having different energies,

$$\Psi(\boldsymbol{r}, t) = \psi(\boldsymbol{r}) \int a(E) e^{-iEt/\hbar}\, dE$$

where $a(E)$ is the probability amplitude for finding the state at energy E. Comparing this form with Eq. (5-4), we arrive at a relation between the decay constant $\mathcal{W}$ and probability amplitude $a(E)$,

$$e^{-\mathcal{W}t/2} = \int a(E) e^{-i(E - \langle E \rangle)t/\hbar}\, dE$$

That is, $e^{-\mathcal{W}t/2}$ is the Fourier transform of $a(E)$.

The relation can be inverted and we obtain the result

$$a(E) = \frac{1}{2\pi\hbar} \int_0^\infty e^{\{i(E - \langle E \rangle)/\hbar - \mathcal{W}/2\}t}\, dt = \frac{i}{2\pi} \frac{1}{(E - \langle E \rangle) + i\frac{1}{2}\hbar\mathcal{W}}$$

The probability for finding the excited state at energy E is given by the absolute square of the amplitude,

$$|a(E)|^2 = \frac{1}{4\pi^2} \frac{1}{(E - \langle E \rangle)^2 + (\frac{1}{2}\Gamma)^2}$$

where we have replaced $\hbar\mathcal{W}$ with Γ using Eq. (5-2). The shape of such a distribution is *Lorentzian* and the width Γ may be interpreted as the full width at half maximum (FWHM) of such a distribution. Since the question of instrumental uncertainty does not enter here, the width is the "natural line width" of the distribution in energy of the excited state.

Branching ratio. A given excited state may decay to several final states. If the transition probability to the ith final state is $\mathcal{W}(i)$, the total transition probability for the initial state is the sum of the probabilities to all final states,

$$\mathcal{W} = \sum_i \mathcal{W}(i) \tag{5-5}$$

Similarly, the total width Γ is the sum of all the partial widths,

$$\Gamma = \sum_i \Gamma(i)$$

The relation between half-life $T_{1/2}$ and partial half-lives $T_{1/2}(i) = \ln 2/\mathcal{W}(i)$ is given by

$$\frac{1}{T_{1/2}} = \sum_i \frac{1}{T_{1/2}(i)}$$

as evident from Eq. (5-5).

The *branching ratio* gives the partial transition probability to a particular final state as a fraction of the total from a specific initial state. For example, the mean lifetime of a π^0-meson is 8.4×10^{-17} s and decays 98.8% of the time to two γ-rays, 1.17% of the time to a γ-ray together with an electron-positron pair, and 2×10^{-7} of the time to an electron-positron pair alone. The branching ratios to these three decay channels are therefore 98.8%, 1.17%, and 2×10^{-7}, respectively. Among the radioactive nuclei, the ground state of the odd-odd nucleus $^{226}_{89}$Ac at the start of the actinide series has a half-life of 29 h and can decay by emitting an electron to $^{226}_{90}$Th, transform to $^{226}_{99}$Ra by capturing one of the atomic electrons, or decay by α-particle emission to $^{222}_{87}$Fr with branching ratios 83%, 10%, and 0.06%, respectively.

Transition matrix element. The transition probability is proportional to the square of the nuclear matrix element,

$$\mathcal{M}_{fi}(M_f, M_i) = \langle J_f M_f \xi | \boldsymbol{O}_{\lambda\mu} | J_i M_i \zeta \rangle \tag{5-6}$$

where $| J_i M_i \zeta \rangle$ and $| J_f M_f \xi \rangle$ are, respectively, the wave functions of the initial and final states and $\boldsymbol{O}_{\lambda\mu}$ is the nuclear part of the transition operator (see §5-3 and §5-6) with spherical tensor rank (λ, μ). The labels ζ and ξ here denote quantum numbers other than J and M associated with angular momentum that are required to specify the nuclear states uniquely. Since the transition may also involve the emission of a particle such as an electron or a nucleon, the initial and final states are not necessarily in the same nucleus. The transition may also be induced by the interaction of the nucleus with an external field. For this reason, the exact relation between the transition probability $\mathcal{W}$ and the nuclear matrix element $\mathcal{M}_{fi}$ depends also on factors related to the external field. We shall treat each type of transition separately in later sections.

The dependence of the matrix element $\mathcal{M}_{fi}$ on M_i and M_f, respectively, projections of the initial and final total angular momentum on the quantization axis, may be factored out using the Wigner-Eckart theorem,

$$\mathcal{M}_{fi}(M_f, M_i) = (-1)^{J_f - M_f} \begin{pmatrix} J_f & \lambda & J_i \\ -M_f & \mu & M_i \end{pmatrix} \langle J_f \xi \| \boldsymbol{O}_\lambda \| J_i \zeta \rangle$$

where $\begin{pmatrix} J_f & \lambda & J_i \\ -M_f & \mu & M_i \end{pmatrix}$ is the $3j$-symbol and $\langle J_f \xi \| \boldsymbol{O}_\lambda \| J_i \zeta \rangle$ is the reduced matrix element defined in Eq. (A-15). Our main interest will be in $\langle J_f \xi \| \boldsymbol{O}_\lambda \| J_i \zeta \rangle$, as it is invariant under a rotation of the coordinate system.

If the measurement is not sensitive to the spin orientation of the final state, the transition includes all the possible final states differing only by the value of M_f. Furthermore, if the operator is not restricted to any specific direction in space, all the allowed values of μ must be included in considering the transition. Under these conditions, the square of the transition matrix element reduces to

$$\begin{aligned}
|\mathcal{M}_{fi}|^2 &= \sum_{\mu M_f} \left| (-1)^{J_f - M_f} \begin{pmatrix} J_f & \lambda & J_i \\ -M_f & \mu & M_i \end{pmatrix} \langle J_f \xi \| \boldsymbol{O}_\lambda \| J_i \zeta \rangle \right|^2 \\
&= |\langle J_f \xi \| \boldsymbol{O}_\lambda \| J_i \zeta \rangle|^2 \sum_{\mu M_f} \left| \begin{pmatrix} J_f & \lambda & J_i \\ -M_f & \mu & M_i \end{pmatrix} \right|^2 \\
&= \frac{\Delta(J_f, \lambda, J_i)}{2J_i + 1} |\langle J_f \xi \| \boldsymbol{O}_\lambda \| J_i \zeta \rangle|^2 \qquad (5\text{-}7)
\end{aligned}$$

In arriving at the final result, we have made use of the orthogonality relation between the $3j$-symbols given by Eq. (A-12),

$$\sum_{m_1 m_2} \begin{pmatrix} j_1 & j_2 & j_3 \\ m_1 & m_2 & m_3 \end{pmatrix} \begin{pmatrix} j_1 & j_2 & j_3' \\ m_1 & m_2 & m_3' \end{pmatrix} = \frac{\Delta(j_1, j_2, j_3)}{2j_3 + 1} \delta_{j_3, j_3'} \delta_{m_3, m_3'}$$

The factor

$$\Delta(J_f, \lambda, J_i) = \begin{cases} 1 & \text{for } \boldsymbol{J}_f = \boldsymbol{\lambda} + \boldsymbol{J}_i \\ 0 & \text{otherwise} \end{cases}$$

expresses the angular momentum selection rule that forbids transitions where the triangular relation among the three angular momentum vectors $\boldsymbol{J}_f$, $\boldsymbol{\lambda}$, and $\boldsymbol{J}_i$ is not satisfied. Note also that $|\mathcal{M}_{fi}|^2$ defined in Eq. (5-7) is independent of M_i.

5-2 Transition Probability in Time-Dependent Perturbation Theory

A connection between the transition probability and the transition matrix element may be established using time-dependent perturbation theory. Consider a time-dependent Hamiltonian

$$H(t) = H_0 + H'(t) \qquad (5\text{-}8)$$

with H_0 independent of time. All the time dependence is contained in $H'(t)$. In particular, we are interested here in the case where the strength of $H'(t)$ is sufficiently weak that it may be considered as a perturbation to H_0.

Let $\phi_n(\boldsymbol{r})$ represent the eigenfunction of H_0,

$$H_0 \phi_n(\boldsymbol{r}) = E_n \phi_n(\boldsymbol{r})$$

We shall assume that all $\phi_n(\boldsymbol{r})$ together form a complete set of orthonormal functions. Again, we have suppressed any indications of possible dependence of $\phi_n(\boldsymbol{r})$ on spin, isospin, and other variables so as to simplify the notation. The eigenfunction $\psi(\boldsymbol{r}, t)$ for H_0 alone is the solution of the time-dependent Schrödinger equation,

$$i\hbar \frac{\partial \psi(\boldsymbol{r}, t)}{\partial t} = H_0 \psi(\boldsymbol{r}, t)$$

and may be expressed in terms of $\phi_n(\boldsymbol{r})$,

$$\psi(\boldsymbol{r}, t) = \sum_n c_n \phi(\boldsymbol{r}) e^{-iE_n t/\hbar}$$

Here, the expansion coefficients,

$$c_n = \int \phi_n^*(\boldsymbol{r}) e^{iE_n t/\hbar} \psi(\boldsymbol{r}, t)\, dV$$

are independent of time, since we have not yet included $H'(t)$.

For the complete Hamiltonian, the eigenfunctions $\Psi(\boldsymbol{r}, t)$ may still be expressed in terms of $\phi_n(\boldsymbol{r})$ except that the expansion coefficients are now time dependent,

$$\Psi(\boldsymbol{r}, t) = \sum_n c_n(t) \phi_n(\boldsymbol{r}) e^{-iE_n t/\hbar} \tag{5-9}$$

The coefficient $c_n(t)$ may be interpreted as the probability amplitude for finding the system in the unperturbed state n at time t. On substituting the results of Eq. (5-9) into the time-dependent Schrödinger equation for $H(t)$,

$$i\hbar \frac{\partial \Psi(\boldsymbol{r}, t)}{\partial t} = \{H_0 + H'(t)\} \Psi(\boldsymbol{r}, t) \tag{5-10}$$

we obtain the equation governing coefficients $c_n(t)$,

$$i\hbar \sum_n \left\{ \frac{dc_n(t)}{dt} - c_n(t) i \frac{E_n}{\hbar} \right\} \phi_n(\boldsymbol{r})\, e^{-iE_n t/\hbar} = \{H_0 + H'(t)\} \sum_n c_n(t) \phi(\boldsymbol{r}) e^{-iE_n t/\hbar}$$

By taking products with $\phi_k^*(\boldsymbol{r}) \exp\{iE_k t/\hbar\}$ on both sides of the equation and integrating over all the independent variables except t, we obtain the result

$$\begin{aligned} i\hbar \sum_n \left\{ \frac{dc_n(t)}{dt} - ic_n(t) \frac{E_n}{\hbar} \right\} e^{i(E_k - E_n)t/\hbar} \langle \phi_k(\boldsymbol{r}) | \phi_n(\boldsymbol{r}) \rangle \\ = \sum_n c_n(t) \Big\{ \langle \phi_k(\boldsymbol{r}) | H_0 | \phi_n(\boldsymbol{r}) \rangle + \langle \phi_k(\boldsymbol{r}) | H'(t) | \phi_n(\boldsymbol{r}) \rangle \Big\} e^{i(E_k - E_n)t/\hbar} \end{aligned} \tag{5-11}$$

Since $\phi_n(\boldsymbol{r})$ is a member of an orthonormal set of eigenfunctions for H_0, we have the conditions

$$\langle \phi_k(\boldsymbol{r}) | \phi_n(\boldsymbol{r}) \rangle = \delta_{kn} \qquad\qquad \langle \phi_k(\boldsymbol{r}) | H_0 | \phi_n(\boldsymbol{r}) \rangle = E_n \delta_{kn}$$

On inserting these results into Eq. (5-11), we obtain a differential equation for $c_k(t)$,

$$i\hbar \frac{dc_k(t)}{dt} = \sum_n \langle \phi_k | H'(t) | \phi_n(t) \rangle\, e^{i\omega_{kn} t}$$

where $\omega_{kn} \equiv (E_k - E_n)/\hbar$.

As initial conditions, let us assume that at $t = 0$ the system is in state $\phi_0(\boldsymbol{r})$. That is,

$$c_n(0) = \begin{cases} 1 & \text{for } n = 0 \\ 0 & \text{for } n \neq 0 \end{cases}$$

If the perturbation is sufficiently weak, we expect that

$$c_k(t) \approx \begin{cases} 1 & \text{for } k = 0 \\ 0 & \text{for } k \neq 0 \end{cases}$$

for all time t of interest. As a result, we can approximate Eq. (5-10) by retaining only the $n = 0$ term in the sum on the right-hand side. This gives us the result

$$i\hbar\frac{dc_k(t)}{dt} = \langle\phi_k|H'(t)|\phi_0(t)\rangle\, e^{i\omega_{k0}t} \tag{5-12}$$

Furthermore, if the time variation of $H'(t)$ is slow compared with $\exp\{i\omega_{k0}t\}$, we may take H' to be a constant. In this approximation, Eq. (5-12) can be solved explicitly and the result may be expressed as

$$c_k(t) = \frac{\langle\phi_k|H'|\phi_0(t)\rangle}{E_k - E_0}(1 - e^{i\omega_{k0}t})$$

From this, we obtain

$$|c_k(t)|^2 = 2|\langle\phi_k(\boldsymbol{r})|H'|\phi_0(\boldsymbol{r})\rangle|^2\frac{1-\cos\omega_{k0}t}{(E_k - E_0)^2}$$

as the probability for finding the system in state k at time t if it started from state 0 at time $t = 0$.

The total probability to a group of states within some interval labeled by f is given by a summation over all the final states k in the interval,

$$\begin{aligned}\sum_{k\in f}\left|c_k(t)\right|^2 &= 2\sum_{k\in f}\left|\langle\phi_k(\boldsymbol{r})|H'|\phi_0(\boldsymbol{r})\rangle\right|^2\frac{1-\cos\omega_{k0}t}{(E_k - E_0)^2}\\ &= \frac{2}{\hbar^2}\int|\langle\phi_k(\boldsymbol{r})|H'|\phi_0(\boldsymbol{r})\rangle|^2\frac{1-\cos\omega_{k0}t}{\omega_{k0}^2}\rho(E_k)\,dE_k\end{aligned}$$

In the last step, the summation over all possible final states is replaced by an integration over energy multiplied by the density of final states $\rho(E_k)$ for reasons that will soon become clear.

The transition probability per unit time, $\mathcal{W}$, corresponds to the rate of finding the system in the group of final states labeled by f and may be expressed as

$$\mathcal{W} = \frac{d}{dt}\sum_{k\in f}|c_k(t)|^2 = \frac{2}{\hbar^2}\int|\langle\phi_k(\boldsymbol{r})|H'|\phi_0(\boldsymbol{r})\rangle|^2\frac{\sin\omega_{k0}t}{\omega_{k0}}\rho(E_k)\,dE_k$$

Since the function $\sin\omega_{k0}t/\omega_{k0}$ oscillates very quickly except where $\omega_{k0} \approx 0$, only a small region around $E_k = E_0$ can contribute to the integral. In this small energy interval we may regard the matrix element $\langle\phi_k(\boldsymbol{r})|H'|\phi_0(\boldsymbol{r})\rangle$ and the state density $\rho(E_k) = \rho(E_f)$ to be constant and may be taken outside the integral. Furthermore, the limits of integration over E_k may be replaced by $\pm\infty$ under these conditions without sacrificing too much accuracy. The final form of the transition probability per unit time becomes

$$\mathcal{W} = \frac{2\pi}{\hbar}|\langle\phi_f(\boldsymbol{r})|H'|\phi_0(\boldsymbol{r})\rangle|^2\rho(E_f) \tag{5-13}$$

where we have made use of the fact that

$$\int_{-\infty}^{+\infty}\frac{\sin\omega_{k0}t}{\omega_{k0}}\,d\omega_{k0} = \pi$$

This is the starting point for our calculations of transition probabilities in the next two sections. Since Fermi called it the "golden rule of time-dependent perturbation theory," it is often referred to as *Fermi's golden rule*.

5-3 Electromagnetic Transition

In this section we shall deal mainly with nuclear decay through the emission of a γ-ray. The transition is caused by the interaction of the nucleus with an external electromagnetic field. For our purpose here, we may regard the nucleus as made of point nucleons, each carrying a magnetic dipole moment and, in the case of protons, a net charge as well. The charge distribution couples with the external field, causing "electric" transitions. At the same time, interaction with the intrinsic magnetism of each nucleon and the magnetism generated by current loops due to proton orbital motion induces "magnetic" transitions.

Electromagnetic transitions form the dominant mode of decay for low-lying excited states in nuclei, particularly for the light ones. The main reason is that nucleon emission, a much faster process than γ-decay, is not possible until the excitation energy is above nucleon separation energies. As we can see later in Fig. 7-2, these are of the order of 8 to 10 MeV for neutrons and somewhat lower for protons because of Coulomb repulsion. Other possible modes are β-decay, α-particle emission, and fission. Generally speaking, these are slower processes than γ-decays.

Besides γ-ray emission, electromagnetic perturbation can also induce nuclear decay through internal conversion whereby one of the atomic electrons is ejected. This is usually more important for heavy nuclei, where the nuclear electromagnetic fields are strong and the orbits of the inner shell electrons are close to the nucleus. Similarly, the decay can also proceed by creating an electron-positron pair. The probability for such internal pair creation processes is, in general, much smaller than γ-ray emission and becomes important, for example, where γ-ray emission is forbidden by angular momentum considerations. This happens, for example, in the case of transitions from an initial state with $J^\pi = 0^+$ to a final state that is also 0^+.

The first step in a discussion on electromagnetic transitions is to establish a connection between the transition probability $\mathcal{W}$ of Eq. (5-1) and the nuclear matrix element $\mathcal{M}_{if}$ of Eq. (5-6) using first-order time-dependent perturbation given by Eq. (5-13). The perturbation H' comes from coupling between nuclear and electromagnetic fields, and the density of final states $\rho(E_f)$ is a product of the number of nuclear and electromagnetic states per energy interval at E_f. Similarly, the initial and final wave functions $\phi_0(\boldsymbol{r})$ and $\phi_k(\boldsymbol{r})$ of Eq. (5-13) are products of nuclear and electromagnetic parts. In keeping with the custom used by most workers in the field, all the electromagnetic calculations in this section will be carried out only in cgs units.

Coupling to electromagnetic field. Our primary interest is in the nuclear part of the matrix element of H'. For this purpose, we shall first separate H' into a product of two operators, one acting only on the nuclear wave function and the other on the external electromagnetic field. Since each nuclear state involved has a definite angular momentum, it is necessary that the external electromagnetic field is also quantized and decomposed by a multipole expansion into components with definite spherical tensor ranks. As we shall see later, the decomposition is an important one, as the lowest order multipole tends to dominate the transition. Both quantization and multipole expansion of the electromagnetic field are fairly straightforward but tedious procedures. We shall attempt here only a brief outline of the steps involved and leave the proper derivations to more advanced textbooks, such as Blatt and Weisskopf [32] and Sakurai [121].

We can visualize the form of electromagnetic perturbation in a nucleus in the following way. Consider first the simple case of a point particle carrying a charge q and, for the time being, no magnetic moment. In the absence of any external electromagnetic field, the particle is free and the Hamiltonian consists of only the kinetic energy term,

$$H_0 = \frac{1}{2m}\boldsymbol{p}^2 \tag{5-14}$$

where $\boldsymbol{p}$ is the momentum. In the presence of an electromagnetic field, the momentum conjugate to $\boldsymbol{r}$ is modified from that for a free particle by the following transformation:

$$\boldsymbol{p} \to \boldsymbol{p} + \frac{q}{c}\boldsymbol{A}$$

where $\boldsymbol{A}$ is the vector potential for the electromagnetic field. The Hamiltonian for the charged particle now takes on the form

$$H = \frac{1}{2m}\Big(\boldsymbol{p} - \frac{q}{c}\boldsymbol{A}\Big)^2 \tag{5-15}$$

For simplicity we have not included in the Hamiltonian terms pertaining solely to the external electromagnetic field or the effect of any electrostatic potential that may be present to interact with the charged particles. We shall return to this point later.

The Hamiltonian of Eq. (5-15) may be written in the form of Eq. (5-8) as a sum of two terms: a free-particle term H_0 given by Eq. (5-14) and a perturbation term H' expressing the coupling with external electromagnetic field. Comparing Eq. (5-15) with (5-14), we can make the identification

$$\begin{aligned} H' &= -\frac{q}{2mc}\big(\boldsymbol{p}\cdot\boldsymbol{A} + \boldsymbol{A}\cdot\boldsymbol{p}\big) + \frac{q^2}{2mc^2}\boldsymbol{A}\cdot\boldsymbol{A} \\ &= -\frac{q}{mc}\boldsymbol{A}\cdot\boldsymbol{p} + \frac{q^2}{2mc^2}\boldsymbol{A}\cdot\boldsymbol{A} \\ &\longrightarrow -\frac{q}{mc}\boldsymbol{A}\cdot\boldsymbol{p} \end{aligned} \tag{5-16}$$

where we have replaced $\boldsymbol{p}\cdot\boldsymbol{A}$ by $\boldsymbol{A}\cdot\boldsymbol{p}$ using the requirement that an electromagnetic field can have only transverse components, i.e., $\boldsymbol{p} = -i\hbar\nabla$ and $\nabla\cdot\boldsymbol{A} = 0$, generally known as the transversality condition. The quadratic term in $\boldsymbol{A}$ involves two photons at the same time and may therefore be ignored in the lowest order consideration we are interested in here.

In general, H' may be written in a more convenient form by expressing the momentum of the charged particle in terms of a current density,

$$\mathcal{J} = q\boldsymbol{v} = q\frac{\boldsymbol{p}}{m} \tag{5-17}$$

The first-order term in Eq. (5-16) becomes

$$H' = -\frac{1}{c}\boldsymbol{A}\cdot\mathcal{J} \tag{5-18}$$

In addition to the electric charge carried by protons in a nucleus, the intrinsic magnetic dipole moment of the nucleon can also interact with the external electromagnetic field. We have seen earlier, in Eq. (4-44), in the discussion of static magnetic moments, that the intrinsic magnetic dipole moment of a nucleon may also be expressed in the form of a current. Consequently, it is unnecessary to change the form of H' given in Eq. (5-18) to include the effect of nucleon magnetic moments, except that we need a new definition for the current density $\mathcal{J}$ than what is given in Eq. (5-17). The most general form of H' must also include the possibility for the charge distribution to interact with an external electrostatic field. Such a perturbing Hamiltonian is most conveniently written in four-component notation,

$$H' = -\frac{1}{c}\sum_{\mu=1}^{4} A_\mu J_\mu \tag{5-19}$$

where $A_\mu = (\boldsymbol{A}, iV)$ includes a scalar potential V and $J_\mu = (\mathcal{J}, i\rho c)$ a charge distribution ρ. Contributions from the fourth components are usually not important in nuclear transitions.

External electromagnetic field. The electromagnetic field is given by the solution to Maxwell's equations. In a region outside any charge and current distributions, the four-potential is obtained from the time-dependent partial differential equation,

$$\left(\nabla^2 - \frac{1}{c^2}\frac{\partial^2}{\partial t^2}\right)A_\mu(\boldsymbol{r}, t) = 0$$

Our primary interest is in the vector potential $\boldsymbol{A}(\boldsymbol{r}, t)$, the first three components of A_μ. The time dependence may be removed from the equation by expanding $\boldsymbol{A}(\boldsymbol{r}, t)$ in terms of components with definite wave number $\boldsymbol{k}$,

$$\boldsymbol{A}(\boldsymbol{r}, t) = \sum_{\boldsymbol{k}} \boldsymbol{A}_{\boldsymbol{k}}(\boldsymbol{r})e^{-i\omega t} \tag{5-20}$$

where $\omega = kc$, with k being the magnitude of $\boldsymbol{k}$. The spatial dependence of $\boldsymbol{A}$ is given by the equation

$$(\nabla^2 + k^2)\boldsymbol{A}_{\boldsymbol{k}}(\boldsymbol{r}) = 0 \tag{5-21}$$

The solution of this second-order differential equation has the familiar form

$$\boldsymbol{A}_{\boldsymbol{k}}(\boldsymbol{r}) = \boldsymbol{B}_{\boldsymbol{k}}e^{i\boldsymbol{k}\cdot\boldsymbol{r}} + \boldsymbol{C}_{\boldsymbol{k}}e^{-i\boldsymbol{k}\cdot\boldsymbol{r}}$$

where $\boldsymbol{B}_{\boldsymbol{k}}$ and $\boldsymbol{C}_{\boldsymbol{k}}$ are constants to be determined by boundary conditions. Among others, these conditions must simulate the source of the electromagnetic field. Substituting the spatial dependence of the vector potential into Eq. (5-20), we obtain the expression

$$\boldsymbol{A}(\boldsymbol{r}, t) = \frac{1}{N}\sum_{\boldsymbol{k}}\sum_{\eta}\left\{b_{\boldsymbol{k}\eta}\boldsymbol{\epsilon}_\eta e^{i(\boldsymbol{k}\cdot\boldsymbol{r}-\omega t)} + b^\dagger_{\boldsymbol{k}\eta}\boldsymbol{\epsilon}_\eta e^{-i(\boldsymbol{k}\cdot\boldsymbol{r}+\omega t)}\right\} \tag{5-22}$$

where N is a normalization constant. Being the quantum of a vector field, each γ-ray carries one unit of angular momentum. However, because of the transversality condition $\nabla\cdot\boldsymbol{A} = 0$, only two of the three components of the vector field are independent quantities. These two components may be identified by two unit vectors $\boldsymbol{\epsilon}_\eta$, with $\eta = 1$,

2. This is similar to expressing ordinary light waves as a linear combination of horizontal and vertical polarizations. At this stage, the factors $b_{k\eta}$ and $b^{\dagger}_{k\eta}$ remain as constants related to B_k and C_k and must be determined by boundary conditions.

Up to now the solution we have obtained for Maxwell's equation is purely classical in nature. Since the form of Eq. (5-21) is identical to the equation for a harmonic oscillator, we may think of the electromagnetic field as a collection of harmonic oscillators, one for each frequency (or wave number $\boldsymbol{k}$) and polarization direction η. The separation in energy between harmonic oscillator states is in units of $\hbar\omega = \hbar ck$. It is now quite straightforward to have a quantum-mechanical description of the electromagnetic field. The quantity $\hbar\omega$ may be taken as the energy of a field quantum for a given wave number and the electromagnetic field is now characterized by the number of quanta for each $\boldsymbol{k}$ and η. We may also interpret $b^{\dagger}_{k\eta}$ and $b_{k\eta}$ in Eq. (5-22) as the creation and annihilation operators, respectively, of a photon with labels $(\boldsymbol{k}, \eta)$. In this way we see that the lowest order term of H', given by Eq. (5-19), involves a linear combination of $b^{\dagger}_{k\eta}$ and $b_{k\eta}$. The physical interpretation of H' is that coupling between nuclear and electromagnetic fields makes it possible for the nucleus to create a photon when it decays from a higher state to a lower one and to absorb a photon when it is excited to a higher state. A proper derivation of the quantized electromagnetic field will require us to demonstrate that $b^{\dagger}_{k\eta}$ and $b_{k\eta}$ are indeed creation and annihilation operators of photons by showing that they have the correct commutation relations, and so on. We shall dispense with this important step here to keep the discussion focused on the concerns of nuclear physics.

Multipole expansion of the electromagnetic field. The expansion of $\boldsymbol{A}(\boldsymbol{r}, t)$ in Eq. (5-22) is carried out, implicitly, in Cartesian coordinates. For applications to problems with rotational symmetry, it is more convenient to express $\boldsymbol{A}(\boldsymbol{r}, t)$ in terms of operators having definite spherical tensor ranks, as we have done on several earlier occasions. The advantage here is quite obvious. We have already seen that the nuclear current density $\mathcal{J}$ may be written as a sum over terms, each carrying a definite amount of angular momentum. Since H' is a scalar, only multipoles of the same order in both $\mathcal{J}$ and $\boldsymbol{A}$ can be coupled together. For this purpose, we shall first rewrite the radiation field in terms of the eigenfunctions of angular momentum operators,

$$\boldsymbol{A}(\boldsymbol{r}, t) = \sum_{\lambda\mu} \boldsymbol{A}_{\lambda\mu}(\boldsymbol{r}, t)$$

where the vector function $\boldsymbol{A}_{\lambda\mu}(\boldsymbol{r}, t)$ with spherical tensor rank (λ, μ) satisfies the following relations as the eigenfunction of angular momentum operators $\boldsymbol{J}^2$ and $\boldsymbol{J}_0$:

$$\boldsymbol{J}^2 \boldsymbol{A}_{\lambda\mu}(\boldsymbol{r}, t) = \lambda(\lambda+1)\boldsymbol{A}_{\lambda\mu}(\boldsymbol{r}, t) \qquad \boldsymbol{J}_0 \boldsymbol{A}_{\lambda\mu}(\boldsymbol{r}, t) = \mu\, \boldsymbol{A}_{\lambda\mu}(\boldsymbol{r}, t) \tag{5-23}$$

The time dependence, given by Eq. (5-20), is sufficiently simple that we shall drop it from now on so as to simplify the notation. The functions $\boldsymbol{A}_{\lambda\mu}(\boldsymbol{r}, t)$ are, however, different from spherical harmonics in that they are vector functions. They may be expressed in terms of *vector spherical harmonics*, vector functions constructed from (scalar) spherical harmonics $Y_{\ell m}(\theta, \phi)$.

Instead of the two polarization directions allowed for $\boldsymbol{A}(\boldsymbol{r}, t)$ in Eq. (5-22), we have now two different types of multipole fields satisfying Eq. (5-21). The decays induced by them are called *electric* multipole transitions, indicated hereafter as $E\lambda$, and *magnetic* multipole transitions, labeled $M\lambda$. In terms of spherical harmonics, they may be written in the following ways:

$$\begin{aligned} \boldsymbol{A}_{\lambda\mu}(E\lambda,\boldsymbol{r}) &= \frac{-i}{k}\nabla\times(\boldsymbol{r}\times\nabla)\big(j_\lambda(kr)Y_{\lambda\mu}(\theta,\phi)\big) \\ \boldsymbol{A}_{\lambda\mu}(M\lambda,\boldsymbol{r}) &= (\boldsymbol{r}\times\nabla)\big(j_\lambda(kr)Y_{\lambda\mu}(\theta,\phi)\big) \end{aligned} \tag{5-24}$$

where $j_\lambda(kr)$ is a spherical Bessel function of order λ. A general solution of $\boldsymbol{A}(\boldsymbol{r},t)$ is a linear combination of both types of terms, with time dependence given by Eq. (5-20). Again, we shall refer the reader to standard references, such as Morse and Feshbach [106], for a demonstration that $\boldsymbol{A}_{\lambda\mu}(E\lambda,\boldsymbol{r})$ and $\boldsymbol{A}_{\lambda\mu}(M\lambda,\boldsymbol{r})$ satisfy Eq. (5-23).

Electromagnetic multipole transition operators. Using the operators given in Eq. (5-24), we can write the (λ,μ) multipole part of the perturbing Hamiltonian H' of Eq. (5-19) as

$$\begin{aligned} \boldsymbol{O}_{\lambda\mu}(E\lambda) &= -\frac{i(2\lambda+1)!!}{ck^{\lambda+1}(\lambda+1)}\mathcal{J}(\boldsymbol{r})\cdot\nabla\times(\boldsymbol{r}\times\nabla)(j_\lambda(kr)Y_{\lambda\mu}(\theta,\phi)) \\ \boldsymbol{O}_{\lambda\mu}(M\lambda) &= -\frac{(2\lambda+1)!!}{ck^{\lambda}(\lambda+1)}\mathcal{J}(\boldsymbol{r})\cdot(\boldsymbol{r}\times\nabla)(j_\lambda(kr)Y_{\lambda\mu}(\theta,\phi)) \end{aligned} \tag{5-25}$$

where $(2\lambda+1)!!=1\cdot3\cdot5\cdots(2\lambda+1)$. The normalizations used in the definitions of these operators are such that they reduce to those for the static moments of Eqs. (4-41) and (5-43) in the limit $k\to0$. For now, we shall not be concerned with the multipole expansion of the nuclear current density $\mathcal{J}(\boldsymbol{r})$. Since both $\boldsymbol{O}_{\lambda\mu}(E\lambda)$ and $\boldsymbol{O}_{\lambda\mu}(M\lambda)$ are scalar operators in the combined nuclear and electromagnetic fields, only the $(\lambda,-\mu)$ multipole part of $\mathcal{J}(\boldsymbol{r})$ can make a nonvanishing contribution in the transition.

The spherical Bessel function in Eq. (5-25) may expanded in a power series,

$$j_\lambda(kr)\approx\frac{(kr)^\lambda}{(2\lambda+1)!!}\Big(1-\frac{1}{2}\frac{(kr)^2}{2\lambda+3}+\cdots\Big) \tag{5-26}$$

The typical γ-rays involved in nuclear transitions have energies E_γ less than 10 MeV, corresponding to wave numbers of the order $k=E_\gamma/\hbar c\approx1/20\ \text{fm}^{-1}$ or less. Since the multipole operators act on the nuclear wave function, they cannot have contributions coming from regions outside the nucleus. The dimension of a nucleus is characterized by the nuclear radius R, and even for a heavy nucleus, such as ^{208}Pb ($R=r_0A^{1/3}\sim7$ fm), it is less than 10 fm. As a result, the dimensionless argument kr of the spherical Bessel function is much less than unity for typical γ-rays involved in a nuclear decay. The series given by Eq. (5-26) is, then, a fast convergent one and $j_\lambda(kr)$ may be approximated by its first term in the expansion alone. This is called the *long-wavelength* limit. Physically, it comes from the observation that, for γ-rays at these energies, the wavelength is $2\pi\hbar c/E_\gamma$, of the order of 10^2 fm, much larger than nuclear dimension. As a result, these γ-rays cannot be sensitive to the details of the nuclear radial wave functions. Under such conditions, the expectation value of $j_\lambda(kr)$ is simply proportional to that for its leading-order term $(kr)^\lambda$.

We are now in a position to calculate the contribution of each multipole order to the transition probability given in Eq. (5-13). On inserting the density of final states and using the multipole operator given in Eq. (5-25) for H', we can express the transition probability for multipole λ from an initial nuclear state $| J_i M_i \zeta \rangle$ to a final nuclear state $| J_f M_f \xi \rangle$ as

$$\mathcal{W}(\lambda; J_i\zeta \rightarrow J_f\xi) = \frac{8\pi(\lambda+1)}{\lambda[(2\lambda+1)!!]^2}\frac{k^{2\lambda+1}}{\hbar}B(\lambda; J_i\zeta \rightarrow J_f\xi) \tag{5-27}$$

where the reduced transition probability $B(\lambda; J_i\zeta \rightarrow J_f\xi)$ may be written in terms of the reduced matrix element of the multipole operator for either electric or magnetic transition, in the same way as we did in Eq. (5-7),

$$B(\lambda; J_i\zeta \rightarrow J_f\xi) = \sum_{\mu M_f} |\langle J_f M_f \xi | O_{\lambda\mu} | J_i M_i \zeta\rangle|^2 = \frac{1}{2J_i+1}|\langle J_f\xi \| O_\lambda \| J_i\zeta\rangle|^2 \tag{5-28}$$

It is worth noting here that the reduced transition probabilities are quantities with dimensions. For electric transitions, the units are $e^2\text{fm}^{2\lambda}$, and for magnetic transitions, $\mu_N^2\text{fm}^{2\lambda-2}$. The transition rate $\mathcal{W}$ is the number of decays per unit time. In relating the numerical values of $\mathcal{W}$ and reduced transition probability, one must be careful with the factors e^2 in electric transitions and μ_N^2 for magnetic transitions. For example, the values in Table 5-1 are obtained using the following relations:

$$\mathcal{W}(\lambda) = \begin{cases} \alpha\hbar c\dfrac{8\pi(\lambda+1)}{\lambda[(2\lambda+1)!!]^2}\dfrac{1}{\hbar}\Big(\dfrac{1}{\hbar c}\Big)^{2\lambda+1}E_\gamma^{2\lambda+1}\,B(E\lambda \text{ in } e^2\text{fm}^{2\lambda}) \\[2ex] \alpha\hbar c(\dfrac{\hbar c}{2M_pc^2})^2\dfrac{8\pi(\lambda+1)}{\lambda[(2\lambda+1)!!]^2}\dfrac{1}{\hbar}\Big(\dfrac{1}{\hbar c}\Big)^{2\lambda+1}E_\gamma^{2\lambda+1}\,B(M\lambda \text{ in } \mu_N^2\text{fm}^{2\lambda-2}) \end{cases}$$

where we have used short-hand notation $B(E\lambda)$ for reduced electric transition probability and $B(M\lambda)$ for reduced magnetic transition probability. The numerical values for e^2 and μ_N^2 may be obtained using the relation $e^2 = \alpha\hbar c$ in cgs units.

Table 5-1: Electromagnetic transition probabilities for the lowest four multipoles.

$\mathcal{W}(E1)=1.59\times10^{15}\,E_\gamma^3\times B(E1)$	$\mathcal{W}(M1)=1.76\times10^{13}\,E_\gamma^3\times B(M1)$
$\mathcal{W}(E2)=1.23\times10^{9}\,E_\gamma^5\times B(E2)$	$\mathcal{W}(M2)=1.35\times10^{7}\,E_\gamma^5\times B(M2)$
$\mathcal{W}(E3)=5.71\times10^{2}\,E_\gamma^7\times B(E3)$	$\mathcal{W}(M3)=6.31\times10^{0}\,E_\gamma^7\times B(M3)$
$\mathcal{W}(E4)=1.70\times10^{-4}E_\gamma^9\times B(E4)$	$\mathcal{W}(M4)=1.88\times10^{-6}E_\gamma^9\times B(M4)$

E_γ are in MeV, $B(E\lambda)$ in $e^2\text{fm}^{2\lambda}$, and $B(M\lambda)$ in $\mu_N^2\text{fm}^{(2\lambda-2)}$.

If we take that the electric charge in a nucleus consists of point charges carried by individual protons and the magnetization currents are due to the magnetic dipole moments of individual nucleons and the orbital motion of protons, the electric and magnetic multipole operators in the long-wavelength limit simplify to the following forms:

$$O_{\lambda\mu}(E\lambda) = \sum_{i=1}^{A} e(i) r_i^{\lambda} Y_{\lambda\mu}(\theta_i, \phi_i) \tag{5-29}$$

$$\begin{aligned} O_{\lambda\mu}(M\lambda) &= \sum_{i=1}^{A} \Big\{ g_s(i)\boldsymbol{s}_i + g_\ell(i)\frac{2\boldsymbol{\ell}_i}{\lambda+1}\Big\} \cdot \nabla_i(r_i^{\lambda} Y_{\lambda\mu}(\theta_i, \phi_i)) \\ &= \sqrt{\lambda(2\lambda+1)} \sum_{i=1}^{A} r_i^{(\lambda-1)} \Big\{ \Big(g_s(i) - \frac{2g_\ell(i)}{\lambda+1}\Big)(Y_{\lambda-1}(\theta_i, \phi_i) \times \boldsymbol{s}_i)_{\lambda\mu} \\ &\qquad + \frac{2g_\ell(i)}{\lambda+1}(Y_{\lambda-1}(\theta_i, \phi_i) \times \boldsymbol{j}_i)_{\lambda\mu} \Big\} \end{aligned} \tag{5-30}$$

Here $\boldsymbol{j}_i = \boldsymbol{\ell}_i + \boldsymbol{s}_i$ and

$$e(i) = \begin{cases} 1e \\ 0 \end{cases} \qquad g_\ell(i) = \begin{cases} 1\mu_N \\ 0 \end{cases} \qquad g_s(i) = \begin{cases} 5.586\mu_N & \text{for a proton} \\ -3.826\mu_N & \text{for a neutron} \end{cases}$$

the same values as given in §4-6. The multiplication symbols here indicate the angular momentum coupled products defined in Eq. (A-10). We have omitted the derivation from Eqs. (5-25) to (5-27), (5-29), and (5-30), as it involves a large amount of angular momentum recoupling and properties of vector spherical harmonics. For a proper treatment, the reader is directed to references such as Blatt and Weisskopf [32]. Note that, for historical reasons, the definitions of the operators for the static moments of a state differ from those for $O_{\lambda\mu}$ here by constant factors, for example, as in the case of the quadrupole moment operator given in Eq. (4-42).

Dimensional check. It may be helpful to make a dimensional analysis of some of the results obtained so far in this section. From Eq. (5-1), we see that $\mathcal{W}$ is measured in number of transitions per unit time interval. For electric multipole transitions, the operator $O_{\lambda\mu}(E\lambda)$ given in Eq. (5-29) is proportional er^{λ}. This gives the reduced transition probability $B(E\lambda)$ of multipolarity λ in units of charge squared times length to the power 2λ. It is customary to measure charge in units of e, the magnitude of the charge on a proton, and length in units of femtometer (fm). Electric multipole operators are therefore in units of $e\text{fm}^{\lambda}$. For magnetic transitions, we have the nuclear magneton μ_N in the place of electric charge e. Because of the gradient operator, the power for length is reduced by one compared with the corresponding electric transition of the same order. As a result, operator $O_{\lambda\mu}(M\lambda)$ in Eq. (5-30) is in units of $\mu_N\text{fm}^{\lambda-1}$. From Eq. (5-28), we find that the units for the reduced transition probabilities are as follows:

Electric multipole λ	$B(E\lambda)$:	$e^2\text{fm}^{2\lambda}$
Magnetic multipole λ	$B(M\lambda)$:	$\mu_N^2\text{fm}^{2\lambda-2}$

Note that since the units for reduced transition probabilities involve e^2 or μ_N^2, the explicit values of these two quantities in conventional units must be put into the expressions in order to evaluate the transition probability $\mathcal{W}$.

Besides the reduced transition probability, the only other dimensioned quantity in the expression for $\mathcal{W}(\lambda)$ given in Eq. (5-27) is the factor $k^{2\lambda+1}/\hbar$. Since k is in units of inverse length and $\hbar$ is in units of energy multiplied by time, the units of electric transition probability is

$$\mathcal{W}(E\lambda): \quad \frac{[\mathrm{fm}^{-1}]^{2\lambda+1}}{[\mathrm{MeV\text{-}s}]} \times e^2[\mathrm{fm}]^{2\lambda}$$

Using the relation $e^2 = \alpha\hbar c$ and the fact that the fine structure constant α is a dimensionless quantity and $\hbar c$ is in units MeV-fm, we obtained the correct result,

$$\mathcal{W}(E\lambda): \quad \frac{[\mathrm{fm}^{-1}]^{2\lambda+1}}{[\mathrm{MeV\text{-}s}]} \times [\mathrm{MeV\text{-}fm}][\mathrm{fm}]^{2\lambda} = [\mathrm{s}]^{-1}$$

as expected. For magnetic multipole transitions, we need only to examine the difference between μ_N^2 and e^2. Since

$$\mu_N = \frac{e\hbar}{2M_pc}$$

in cgs units, the dimension of μ_N^2 is

$$\mu_N^2: \quad e^2[\mathrm{fm}]^2$$

Since $B(M\lambda)$ is measured in units of $\mu_N^2\mathrm{fm}^{2\lambda-2}$, we see that $\mathcal{W}(M\lambda)$ also has the correct dimension of inverse time.

Selection rules. We have stated earlier that, for γ-rays with energy of the order of a few mega-electron-volts, transitions of different multipolarities have quite different rates. We can also see this from the energy dependence of $\mathcal{W}(\lambda)$ given by the factor $k^{2\lambda+1}$. In terms of the ratio between two multipole transitions λ and $\lambda+1$, we find that

$$\mathcal{R} = \frac{\mathcal{W}(\lambda+1)}{\mathcal{W}(\lambda)} \sim (kr)^2$$

A length factor r is included in the calculation so as to make the ratio a dimensionless quantity. Since nuclear size is of the order of a few femtometers, we can take r to be 1 fm for the purpose of making an estimate. This gives us the result that, for a 1-MeV γ-ray, $\mathcal{R}$ is of the order 3×10^{-5}. A more precise calculation of the factor relating transition probabilities with reduced probabilities for both electric and magnetic transitions produces the results shown in Table 5-1. Note that the difference in numerical factor between electric and magnetic transitions of the same multipolarity is $(\hbar/2M_pc)^2$ femtometers squared, coming from the difference in the units for $B(E\lambda)$ and $B(M\lambda)$.

Because of the large reduction in probability with increasing multipolarity order, the transition between an initial nuclear state with spin-parity $J_i^{\pi_i}$ and a final state $J_f^{\pi_f}$ is usually dominated by the lowest order allowed by angular momentum and parity selection rules. For transition of order λ, the operator carries λ units of angular momentum. As a result, the transition vanishes unless $\boldsymbol{J}_f = \boldsymbol{\lambda} + \boldsymbol{J}_i$. The angular momentum selection rule for the λth multipole electromagnetic transition is therefore given by

$$|J_f - J_i| \leq \lambda \leq J_f + J_i$$

The same condition is expressed also by the factor $\Delta(J_f, \lambda, J_i)$ in Eq. (5-7) and is implicit in the reduced matrix element of Eq. (5-28). Together with the energy dependence in $\mathcal{W}(\lambda)$, we have the multipolarity selection rule that, for allowed values of λ,

$$\mathcal{W}(E\lambda) \gg \mathcal{W}(E(\lambda+1)) \qquad\qquad \mathcal{W}(M\lambda) \gg \mathcal{W}(M(\lambda+1))$$

We shall not make a comparison of electric with magnetic transitions until later, as the nature of these two operators is quite different.

The operator for an $E\lambda$-transition is proportional to spherical harmonics $Y_{\lambda\mu}(\theta, \phi)$, as can be seen from Eq. (5-29). Under an inversion of the coordinate system, the transformation property of $\boldsymbol{O}_{\lambda\mu}(E\lambda)$ follows that for spherical harmonics of order (λ, μ). This gives us the transformation property under parity for $E\lambda$-transitions,

$$\boldsymbol{O}_{\lambda\mu}(E\lambda) \xrightarrow[P]{} (-1)^{\lambda}\boldsymbol{O}_{\lambda\mu}(E\lambda) \tag{5-31}$$

The magnetic operator, on the other hand, is related to $\nabla(r^{\lambda}Y_{\lambda\mu}(\theta, \phi))$. The presence of the ∇-operator introduces an "extra" minus sign under a parity transformation, and the net result is

$$\boldsymbol{O}_{\lambda\mu}(M\lambda) \xrightarrow[P]{} (-1)^{\lambda+1}\boldsymbol{O}_{\lambda\mu}(M\lambda) \tag{5-32}$$

Equations (5-31) and (5-32) give us the parity selection rule,

$$\pi_i\pi_f = (-1)^{\lambda} \quad \text{for} \quad E\lambda \qquad\qquad \pi_i\pi_f = (-1)^{\lambda+1} \quad \text{for} \quad M\lambda$$

where π_i and π_f are, respectively, the parities of the initial and final states. Because of this selection rule, $E\lambda$- and $M\lambda$-transitions of the same multipolarity cannot occur between the same pair of nuclear states. For example, in a $2^+ \rightarrow 0^+$ transition, only an $E2$-transition can take place, whereas in a $2^- \rightarrow 0^+$ transition, only $M2$ is allowed.

The difference in the nature of electric and magnetic transition operators also plays a role in determining the dominant mode of transition between a pair of nuclear states. In general we find that magnetic transitions are weaker than electric ones. It is not easy to make a direct comparison here, as electric and magnetic transitions of the same multipolarity cannot both occur between a given pair of nuclear states because of parity selection rules. The alternative is the following. For a given pair of nuclear states, if both $E\lambda$ and $M(\lambda+1)$ are allowed by angular momentum and parity selection rules, the $E\lambda$-mode usually dominates the transition by a large factor. On the other hand, if both $M(\lambda)$ and $E(\lambda+1)$ transitions are allowed, the higher multipole order electric transition may be competitive, as far as transition rates are concerned, with the magnetic transition. This is true in spite of the hindrance factor due to energy dependence. In fact, it is common to find both types of transitions between a pair of states, such as in the case of 2^+ to 1^+, where both $M1$- and $E2$-transitions are allowed. The mixture is characterized by the *mixing ratio* δ, defined by the relation

$$\delta^2 = \frac{\mathcal{W}(E(\lambda+1); J_i\zeta \rightarrow J_f\xi)}{\mathcal{W}(M\lambda; J_i\zeta \rightarrow J_f\xi)}$$

The sign of δ is given by the relative sign of the reduced matrix elements of the two transitions where such a sign can be determined.

The reason for the dominance of electric over magnetic transitions may be inferred from the fact that the operators for magnetic transitions differ from the corresponding ones for electric transitions by a gradient operator ∇. Since it is a differential operator, it tends to reduce the size of the matrix element. Similar differences are also found in other electromagnetic processes.

Internal conversion and internal pair production. Besides γ-ray emission, electromagnetic decay can also take place through internal conversion and internal pair production. In *internal conversion*, an atomic electron is ejected instead of a γ-ray. The kinetic energy of the electron is equal to the de-excitation energy $E_i - E_f(= E_\gamma)$ minus the (atomic) binding energy of the electron. As a result, electrons emitted from internal conversion processes have discrete energies and can therefore be distinguished from the continuous spectrum of electrons emitted in β^--decays to be discussed later in §5-6. Since both types of decay can take place from the same excited state in medium and heavy nuclei, the difference makes it possible to distinguish between them.

The process of internal conversion may be visualized in the following way. When a nucleus de-excites, for example, either by a nucleon jumping from one single-particle orbit to another or by a change in the rotational motion of the nucleus as a whole, a sudden disturbance is sent to the surrounding electromagnetic field. Atomic electrons, especially those in the innermost orbits, such as the K- and L-orbits, spend a large fraction of the time in the vicinity of the nucleus, the source of the electromagnetic field of interest here. It is therefore probable for the disturbance in the electromagnetic field to transfer the excess energy in the nucleus to one of the electrons and eject it from the atomic orbit. This is similar to the atomic *Auger* effect where, instead of emitting a photon when an atomic electron de-excites from a higher to a lower energy orbit, one of the atomic electrons is ejected.

Internal conversion is important in heavier nuclei for two reasons. First, the radii of atomic electron orbits are smaller because of the strong Coulomb fields provided by heavy nuclei. The probability of transition is increased as a result of the larger overlap between the wave functions of the nucleus and the inner shell atomic electrons. For this reason, the electrons ejected come mainly from the innermost shells. Second, the stronger Coulomb field in heavy nuclei exerts a larger influence on the surrounding. For these reasons, the importance of internal conversion increases roughly as Z^3 and becomes competitive with γ-ray emission for medium and heavy nuclei.

In *internal pair production*, an electron-positron pair is emitted in the place of a γ-ray when an excited nucleus decays through electromagnetic processes. As long as the energy of decay is greater than $2m_ec^2 \approx 1.02$ MeV, pair production is possible, at least in principle. However, the process is not an efficient one and is usually several orders of magnitude retarded compared with allowed γ-ray emissions. Pair production therefore becomes important only when γ-ray emission is forbidden. For example, a $0^+ \rightarrow 0^+$ transition is not allowed by γ-ray emission, as a γ-ray must carry away at least one unit of angular momentum. In such cases, pair production (and internal conversion for heavy nuclei) becomes the dominant decay mode. Other possibilities include such second-order processes as emitting a pair of γ-rays, which we have excluded from our discussions.

The inverse of γ-ray emission is Coulomb excitation. Here, the nucleus is excited to a higher state as a result of changes in the surrounding electromagnetic field. This can take place, for example, as the result of a charged particle passing nearby. We shall see in §8-1 that the nuclear transition matrix elements involved are identical to γ-ray transitions we have been discussing here.

5-4 Single-Particle Value

It is useful to have some feeling of the magnitude of reduced transition probability between a pair of states. For this purpose, we shall make an estimate of the sizes of $B(E\lambda)$ and $B(M\lambda)$ that can be expected on the average. A second motivation for doing this is to have a basis with which we can form some judgment on observed values. As we shall see later, this is an important function. Since transition probability $\mathcal{W}(\lambda)$ is dominated by the energy-dependent factor $k^{2\lambda+1}$, it is difficult to obtain an idea of the size of the nuclear matrix element involved by looking at the numerical value of $\mathcal{W}(\lambda)$. The common practice is to compare these values with Weisskopf estimates. In fact, these estimates are so widely used that they are often regarded as the "standard," or units, for measuring transition rates.

***$E\lambda$*-transition.** A calculation of the reduced transition probability requires a knowledge of both the initial and final wave functions. As the first step toward establishing an average, we shall make some assumptions about these wave functions so that a reasonable estimate may be made without reference to the specific states involved in a transition. Again, we shall adopt an extreme independent particle picture and consider nuclear transitions to be taking place by a nucleon moving from one single-particle orbit to another without affecting the rest of the nucleus. In the case of $E\lambda$-transitions, this means that a proton moves from an initial single-particle state $|\,j_im_i\,\rangle$ to a final one $|\,j_fm_f\,\rangle$. In this limit, the matrix element of operator $\boldsymbol{O}_{\lambda\mu}(E\lambda)$ between many-body nuclear wave functions $|\,J_iM_i\zeta\,\rangle$ and $|\,J_fM_f\xi\,\rangle$ reduces to a single-particle matrix element,

$$\langle J_fM_f\xi|\sum_{k=1}^{A} e(k)r_k^\lambda Y_{\lambda\mu}(\theta_k,\phi_k)|J_iM_i\zeta\rangle = \langle j_fm_f|er^\lambda Y_{\lambda\mu}(\theta,\phi)|j_im_i\rangle \tag{5-33}$$

where we have made use of the explicit form of $\boldsymbol{O}_{\lambda\mu}(E\lambda)$ given in Eq. (5-29) to express the $E\lambda$-transition operator as a sum of single-particle operators.

A single-particle wave function may be decomposed into a product of three parts: a radial wave function $R_{n\ell}(r)$, an orbital angular momentum part given by spherical harmonics $Y_{\ell m}(\theta,\phi)$, and an intrinsic spin part $\chi_{1/2}$. By coupling $Y_{\ell m}(\theta,\phi)$ with $\chi_{1/2}$ to angular momentum (j,m), we have the result

$$|jm\rangle = R_{n\ell}(r)\{Y_\ell(\theta,\phi)\times\chi_{1/2}\}_{jm}$$

where n is the principal quantum number. The single-particle matrix element on the right-hand side of Eq. (5-33) can now be written in terms of a product of a radial integral and a matrix element in angular momentum space,

$$\begin{aligned}\langle j_fm_f|er^\lambda Y_{\lambda\mu}|j_im_i\rangle = &\int_0^\infty R^*_{n_f\ell_f}(r)\,r^\lambda R_{n_i\ell_i}(r)\,r^2\,dr\\ &\times\langle\{Y_{\ell_f}(\theta,\phi)\times\chi_{1/2}\}_{j_fm_f}|Y_{\lambda\mu}(\theta,\phi)|\{Y_{\ell_i}(\theta,\phi)\times\chi_{1/2}\}_{j_im_i}\rangle\end{aligned} \tag{5-34}$$

We shall now make some further simplifying assumptions so that the averages of both quantities on the right-hand side may be evaluated without having to specify explicitly the single-particle states involved.

The radial dependence of Eq. (5-34) is contained in the integral

$$\langle r^\lambda \rangle = \int R^*_{n_f \ell_f}(r)\, r^\lambda R_{n_i \ell_i}(r)\, r^2\, dr$$

where $R_{n_f \ell_f}(r)$ and $R_{n_i \ell_i}(r)$ are, respectively, the normalized radial wave functions of the initial and final single-particle states. The exact value of the integral depends on the radial shapes. However, to a first approximation, it is determined by the power λ and the size of the nucleus. For the purpose of an estimate, we can simplify the situation greatly by assuming that the nucleus is a sphere of uniform density with the radius $R = r_0 A^{1/3}$. In this approximation, the radial integral reduces to

$$\langle r^\lambda \rangle = \frac{3}{\lambda + 3} r_0^\lambda A^{\lambda/3} \tag{5-35}$$

For r_0, we can use the value of 1.2 fm from electron scattering. (See Problem 5-5 for values obtained with the more realistic harmonic oscillator radial wave functions.)

Using Eq. (5-28), an estimate for the reduced transition probability of an $E\lambda$-transition may be put into the form

$$\begin{aligned} B_{\text{est.}}(E\lambda) &= \sum_{\mu M_f} |\langle J_f M_f \xi | O_{\lambda\mu}(E\lambda) | J_i M_i \zeta \rangle|^2 \\ &= e^2 \langle r^\lambda \rangle^2 \sum_{m_f \mu} \langle (Y_{\ell_f}(\theta,\phi) \times \chi_{1/2})_{j_f m_f} | Y_{\lambda\mu}(\theta,\phi) | (Y_{\ell_i}(\theta,\phi) \times \chi_{1/2})_{j_i m_i} \rangle^2 \end{aligned} \tag{5-36}$$

The only matrix element remaining involves angular momentum and can therefore be evaluated using standard techniques of spherical tensors. However, there is very little point to do this. Since the total solid angle about a point is 4π steradians, an average of any angular dependence must be around the value $1/4\pi$. Hence, for the purpose of an estimate, we can take

$$B_{\text{est}}(E\lambda) = e^2 \langle r^\lambda \rangle^2 \frac{1}{4\pi}$$

On substituting the value of the radial integral given in Eq. (5-35) and using 1.2 fm for r_0, we obtain the *Weisskopf single-particle estimate* for the λth multipole reduced electric transition probability,

$$B_{\text{W}}(E\lambda) = \frac{1}{4\pi}\Big(\frac{3}{\lambda+3}\Big)^2 (1.2)^{2\lambda} A^{2\lambda/3}\ e^2\text{fm}^{2\lambda}$$

As mentioned earlier, this value is often used as the unit for $E\lambda$-transition and is also called the Weisskopf unit for reduced transition probability for this reason.

***M*λ-transition.** Estimates for magnetic transitions are slightly more complicated, as we have contributions from both nucleon intrinsic spin and proton orbital motion. We may proceed essentially along the same line as we have done above for electric transitions. This involves adopting an extreme single-particle model and making use of the last form of operator $O_{\lambda\mu}(M\lambda)$ given in Eq. (5-30) to reduce the nuclear matrix element to a single-particle matrix element. In parallel with Eq. (5-36), we have the expression

$$\begin{aligned}
B_{\text{est.}}(M\lambda) &= \sum_{\mu M_f} |\langle J_f M_f \xi | O_{\lambda\mu}(E\lambda) | J_i M_i \zeta\rangle|^2 \\
&= \lambda(2\lambda+1)\langle r^{(\lambda-1)}\rangle^2 \\
&\quad \times \sum_{m_f \mu} \Big\{ \Big(g_s - \frac{2g_\ell}{\lambda+1}\Big) \\
&\qquad \times \langle (Y_{\ell_f}(\theta,\phi) \times \chi_{1/2})_{j_f m_f} | (Y_{(\lambda-1)}(\theta,\phi) \times \boldsymbol{s})_{\lambda\mu} | (Y_{\ell_i}(\theta,\phi) \times \chi_{1/2})_{j_i m_i} \rangle \\
&\qquad + \frac{2g_\ell}{\lambda+1} \langle (Y_{\ell_f}(\theta,\phi) \times \chi_{1/2})_{j_f m_f} | (Y_{(\lambda-1)}(\theta,\phi) \times \boldsymbol{j})_{\lambda\mu} | (Y_{\ell_i}(\theta,\phi) \times \chi_{1/2})_{j_i m_i} \rangle \Big\}^2
\end{aligned} \tag{5-37}$$

Again, we shall take the nucleus to be a constant-density sphere of radius R and average over the angular dependence. The result of the radial integral may be taken to be

$$\langle r^{(\lambda-1)}\rangle = \frac{3}{\lambda+3} r_0^{(\lambda-1)} A^{(\lambda-1)/3}$$

the same as we have done to arrive at Eq. (5-35). [Strictly speaking, the factor in front on the right-hand side should instead be $3/(\lambda+2)$ but is kept in the form given to conform with that for $E\lambda$-transitions.]

For the purpose of an estimate, it is adequate to evaluate either one of the two terms inside the curly brackets in Eq. (5-37) and multiply the result by 2 (before taking the square). Factors related to the gyromagnetic ratios in the first term may be replaced by a reasonable average value, and this is generally taken to be

$$\lambda(2\lambda+1)\Big(g_s - \frac{2g_\ell}{\lambda+1}\Big)^2 \approx 10$$

For the average of the square of the angular part, we can again take the value $1/4\pi$ used earlier for $E\lambda$-transitions. This gives us the result

$$B_{\text{W}}(M\lambda) = \frac{10}{\pi}\Big(\frac{3}{\lambda+3}\Big)^2 (1.2)^{2\lambda-2} A^{(2\lambda-2)/3}\ \mu_N^2 \text{fm}^{2\lambda-2} \tag{5-38}$$

as the final form of the Weisskopf estimate for reduced magnetic multipole transition probability.

The results of Eqs. (5-36) and (5-38) may be substituted into (5-27) to produce the Weisskopf units for transition probability:

$$\mathcal{W}_{\text{W}}(E\lambda) = \alpha\hbar c \frac{8\pi(\lambda+1)}{\lambda[(2\lambda+1)!!]^2} \frac{1}{\hbar} \Big(\frac{1}{\hbar c}\Big)^{2\lambda+1} \frac{1}{4\pi} \Big(\frac{3}{\lambda+3}\Big)^2 (1.2)^{2\lambda} A^{2\lambda/3} E^{2\lambda+1}$$

$$\mathcal{W}_{\text{W}}(M\lambda) = \alpha\hbar c \Big(\frac{\hbar}{2M_p c}\Big)^2 \frac{8\pi(\lambda+1)}{\lambda[(2\lambda+1)!!]^2} \frac{1}{\hbar} \Big(\frac{1}{\hbar c}\Big)^{2\lambda+1} \frac{10}{\pi} \Big(\frac{3}{\lambda+3}\Big)^2 (1.2)^{2\lambda-2} A^{(2\lambda-2)/3} E^{2\lambda+1}$$

Explicit values in terms of nucleon number A and transition energy E_γ are listed in Table 5-2.

Table 5-2: Weisskopf single-particle estimates for $E\lambda$- and $M\lambda$-transition probabilities and widths.

Order	$E\lambda$			$M\lambda$		
λ	$\mathcal{W}$ (s^{-1})	Γ (MeV)		$\mathcal{W}$ (s^{-1})	Γ (MeV)	
1	1.02×10^{14}	6.75×10^{-8}	$A^{2/3}E_\gamma^3$	3.15×10^{13}	2.07×10^{-8}	E_γ^3
2	7.28×10^{7}	4.79×10^{-14}	$A^{4/3}E_\gamma^5$	2.24×10^{7}	1.47×10^{-14}	$A^{2/3}E_\gamma^5$
3	3.39×10	2.23×10^{-20}	$A^{2}E_\gamma^7$	1.04×10	6.85×10^{-21}	$A^{4/3}E_\gamma^7$
4	1.07×10^{-5}	7.02×10^{-27}	$A^{8/3}E_\gamma^9$	3.27×10^{-6}	2.16×10^{-27}	$A^{2}E_\gamma^9$
5	2.40×10^{-12}	1.58×10^{-33}	$A^{10/3}E_\gamma^{11}$	7.36×10^{-13}	4.84×10^{-34}	$A^{8/3}E_\gamma^{11}$

E_γ in MeV. The E_γ- and A-dependent factors are common to both $\mathcal{W}$ and Γ.

In terms of Weisskopf units, the measured reduced rates are observed to vary by several orders of magnitude, sometimes even for transitions within a single nucleus. This shows the richness of physics contained in electromagnetic transitions between nuclear states. For a transition to be enhanced by an order of magnitude or more over the single-particle values we have estimated, many nucleons must be acting together in a coherent manner. As we shall see in Chapter 6, this leads to the concept of collective motion in the form of nuclear vibrations and rotations.

5-5 Weak Interaction and Beta Decay

Nuclear β-decay is one of the many facets of weak interaction. In addition to transitions between nuclear states, a variety of other phenomena involving hadrons and leptons share the same origin. Being slower by several orders of magnitude, these weak processes cannot be observed if there are competing reactions induced by electromagnetic and strong interactions. For this reason, weak interactions can be studied only in cases where these faster processes are either forbidden or hindered by selection rules.

The basic reaction involving weak interaction in nuclei may be characterized by the decay of a neutron and a (bound) proton,

$$n \rightarrow p + e^- + \bar{\nu}_e \tag{5-39}$$

$$p_{\text{bound}} \rightarrow n + e^+ + \nu_e \tag{5-40}$$

introduced earlier in Eqs. (2-1) and (2-2). These transitions are examples of a general class of decay taking place in other hadrons as well. For example,

$$\pi^+ \rightarrow \begin{cases} \mu^+ + \nu_\mu \\ e^+ + \nu_e \end{cases} \tag{5-41}$$

$$\pi^- \rightarrow \begin{cases} \mu^- + \bar{\nu}_\mu \\ e^- + \bar{\nu}_e \end{cases} \tag{5-42}$$

$$\Sigma^- \rightarrow n + \pi^- \tag{5-43}$$

$$K^+ \rightarrow \begin{cases} \pi^+ + \pi^0 \\ \pi^+ + \pi^- + \pi^+ \end{cases} \tag{5-44}$$

Reactions such as those given by Eqs. (5-39) to (5-42) are referred to as semi-leptonic processes, since both hadrons and leptons are involved. Some weak interaction processes do not involve any leptons at all and are called non-leptonic processes, such as those given by Eqs. (5-43) and (5-44). There are also purely leptonic processes, such as the decay of a muon,

$$\mu^- \rightarrow e^- + \overline{\nu}_e + \nu_\mu \tag{5-45}$$

Our main concern will be the semi-leptonic mode, as nuclear β-decay is a part of it.

Universal weak interaction. Weak interaction processes are often said to be *universal,* as the strength of the basic process is the same for all three different types of reactions described in the previous paragraph. This point is illustrated by the fact that the coupling constant G_F, generally known as the Fermi coupling constant, has the same value,

$$G_F = 1.43572(3) \times 10^{-62}\ \text{J-m}^3 = 1.16639(2) \times 10^{-11}(\hbar c)^3\ \text{MeV}^{-2} \tag{5-46}$$

regardless of whether it is measured through superallowed β-decay in nuclei (see next section and Problem 5-2), the decay of muons shown in Eq. (5-45), or other weak interaction processes.

Weak interactions are mediated by vector bosons $W^\pm$ and Z^0, in the same way as electromagnetic interactions are carried by photons. However, because of their large masses,

$$m_W c^2 = 80.36 \pm 0.12\ \text{GeV} \qquad\qquad m_Z c^2 = 91.187 \pm 0.007\ \text{GeV}$$

the range of weak interactions is extremely short ($r_0 = \hbar/mc \sim 10^{-3}$ fm), about three orders of magnitude smaller than the long-range part of nuclear force. For this reason, weak interactions may be considered as zero-range, or "contact," interactions for all practical purposes in nuclear physics.

Since bosons $W^\pm$ carry net charges, they change the charge state of a particle as, for example, in the reactions given in Eqs. (5-39) and (5-40). Pictorially, these two processes may be represented by the diagrams shown in Fig. 5-1. Most weak decays

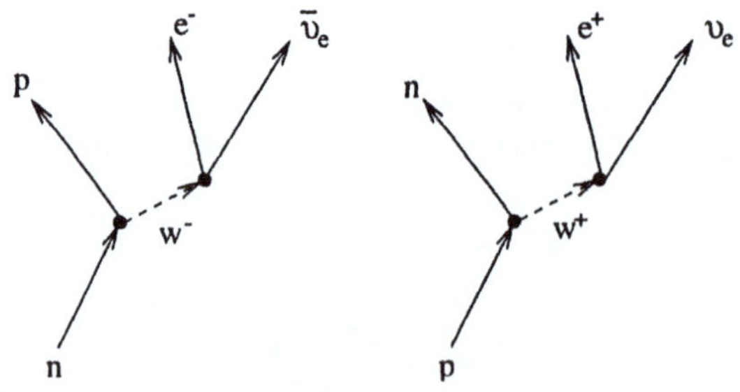

Figure 5-1: Diagram showing β^--decay of a neutron into a proton by emitting a W^--boson and β^+-decay of a (bound) proton into a neutron by emitting a W^+-boson. In both cases, the W-boson decays into a pair of leptons.

are mediated by charged bosons, as illustrated by the examples given in Eq. (5-41) to (5-45). The neutral boson Z^0 is the source of *neutral weak current* and is responsible for reactions such as neutrino-electron scattering:

$$\nu + e^- \rightarrow \nu + e^-$$

In spite of its small cross section, such processes are important, for example, in trapping the energy inside the outer shell of a supernova, preventing large amounts being carried away by neutrinos immediately after the explosion (see §10-4).

On a more fundamental level, β-decay of hadrons may be viewed as the transformation of one type of quark to another through the exchange of charged weak currents. As we have seen in Chapter 2, the flavor of quarks is conserved in strong interactions. However, through weak interactions, it is possible for quarks to change flavor, for example, by transforming from a d-quark to a u-quark,

$$d \rightarrow u + e^- + \overline{\nu}_e \tag{5-47}$$

This is what takes place in the β^--decay of a neutron. In terms of quarks, Eq. (5-39) may be written as

$$(udd) \rightarrow (uud) + e^- + \overline{\nu}_e$$

Similarly, the β^+-decay of a bound proton to a neutron involves the transformation of a u-quark to a d-quark,

$$u \rightarrow d + e^+ + \nu_e \tag{5-48}$$

Diagrammatically, the processes given by Eqs. (5-47) and (5-48) may be represented by Fig. 5-2(a) and (b). The other weak transitions given in Eq. (5-41) to (5-45) are represented by diagrams such as those shown in Fig. 5-2(c) to (e).

When a quark decays, it does not necessarily have to result in a quark of definite flavor. For the simple case of weak decay among the four lightest quarks, u, d, s, and c, the flavor mixing in the decay product may be expressed in terms of a single parameter, the Cabibbo angle θ_c,

$$\begin{aligned} u \longrightarrow d' &= d\cos\theta_c + s\sin\theta_c \\ c \longrightarrow s' &= -d\sin\theta_c + s\cos\theta_c \end{aligned} \tag{5-49}$$

This is reminiscent of what we saw in §2-7 on SU_3(flavor) symmetry mixing. There, the observed pairs of neutral mesons, (η, η') and (ρ, ω), are mixtures of the SU_3(flavor) symmetry-conserving pairs (η_0, η_8) and (ϕ_0, ϕ_8), respectively.

The observed weak transitions are, however, between quarks of definite flavor, for example, u- and d-quarks, as in the decay of a neutron to a proton. The relation given in Eq. (5-49) implies that the observed β-decay strength in reactions is not the fundamental weak interaction coupling constant G_F itself, but a value modified by the mixing angle.

It is customary to express the transformation given by Eq. (5-49) in the form of a charged current,

$$\mathcal{J}^+_{\text{weak}} = (\overline{u}\ \overline{c}) \begin{pmatrix} \cos\theta_c & \sin\theta_c \\ -\sin\theta_c & \cos\theta_c \end{pmatrix} \begin{pmatrix} d \\ s \end{pmatrix}$$

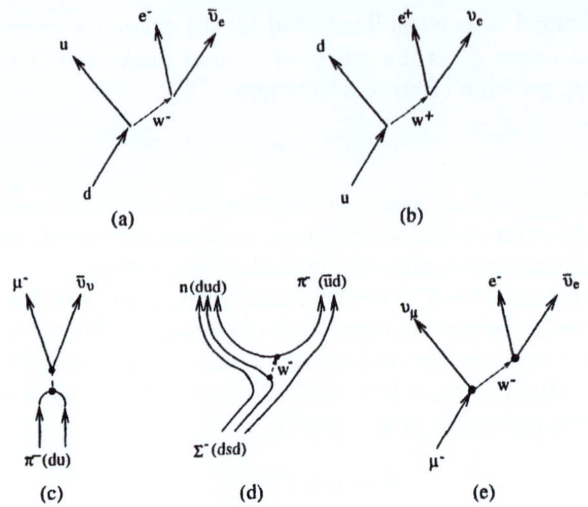

Figure 5-2: Weak decay of quarks: (*a*) a d-quark becomes a u-quark by emitting a W^--boson, similar to a neutron decaying into a proton; (*b*) a u-quark is changed into a d-quark similar to a bound proton decay; (*c*) an example of semi-leptonic decay given by Eq. (5-42); (*d*) an example of nonleptonic weak decay in Eq. (5-43); and (*e*) an example of purely leptonic decay given by Eq. (5-45).

In terms of such a current, the more general case involving all six quarks may be written as

$$\mathcal{J}^{+}_{\text{weak}} = (\overline{u}\;\overline{c}\;\overline{t})\begin{pmatrix} M_{11} & M_{12} & M_{13} \\ M_{21} & M_{22} & M_{23} \\ M_{31} & M_{32} & M_{33} \end{pmatrix}\begin{pmatrix} d \\ s \\ b \end{pmatrix}$$

where the 3×3 matrix is known as the Cabibbo-Kobayashi-Maskawa matrix. The nine matrix elements are functions of three mixing angles and a phase factor. A complete determination of all the independent matrix elements involves weak decays of the heavy quarks as well.

For nuclear β-decay, we are mainly concerned with the transition between u- and d-quarks. As a result, only the product between the Fermi coupling constant G_F and $\cos\theta_c$ enters into the process. The mixing angle is sufficiently small that we can ignore it for most of our purposes. In order to simplify the notation and to avoid any possible confusion, we shall use the symbol G_V, the vector coupling constant, to represent the product and omit the explicit presence of the mixing angle. However, we must be aware of the difference, for example, when we compare the value of G_F with the measured strength of weak decays in nuclei, as done in Problem 5-2.

Parity nonconservation. One of the more remarkable properties of weak decay is that parity is not conserved. We need to go into this point somewhat, since it is closely linked with the nature of nuclear β-decay operator. As described in §A-1, parity transformation is the operation that inverts the spatial coordinates,

$$(x, y, z) \xrightarrow[P]{} (-x, -y, -z)$$

It is often described in terms of taking a mirror image of the coordinate system, as can be seen from Fig. 5-3. Under a parity operation a scalar (S) is unchanged, but a vector (V) of the type we normally use changes sign. Where needed, we shall refer to such vectors as *polar vectors*, to distinguish them from axial vectors, to be defined next. Examples of polar vectors are spatial location $\boldsymbol{r}$ and momentum $\boldsymbol{p}$.

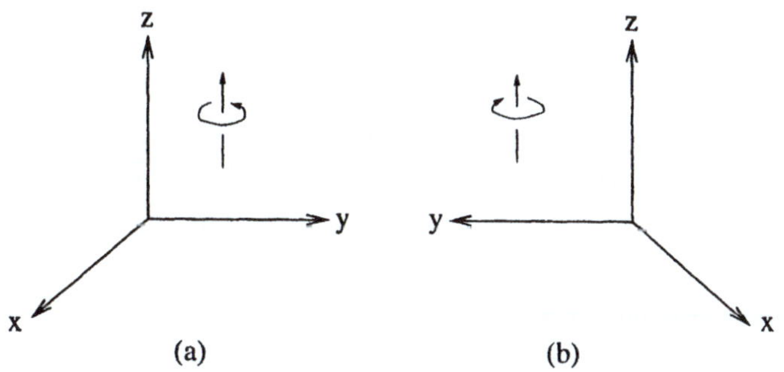

Figure 5-3: Parity and direction of spin. Under a parity transformation, a right-handed rotation, shown in (a), is changed into its mirror image, a left-handed rotation, shown in (b). The transformation in the coordinate axes is accomplished by $(x, y, z) \rightarrow (-x, -y, -z)$ followed by a rotation of 180° around the y-axis.

We can also construct vectors that do not change sign under a parity transformation. For example, the angular momentum vector $\boldsymbol{\ell}\hbar = \boldsymbol{r} \times \boldsymbol{p}$ does not change sign under a parity operation, as both $\boldsymbol{r}$ and $\boldsymbol{p}$ reverse signs. Vectors that do not change sign under an inversion of the coordinate system are called *axial vectors* (A). All angular momentum operators, including intrinsic spin operators, are axial vectors. The scalar product of an axial vector and a polar vector is a scalar that changes sign under a parity operation. Such scalars are called *pseudoscalars* (P). There is also a fifth category of quantities called *tensors* (not to be confused with spherical tensors, which we use for angular momentum algebra) that behave differently from S, V, A, and P under a parity transformation, but we shall not be concerned with them here.

An operator made of a linear combination of scalars and pseudoscalars, or vectors and axial vectors, does not have a definite parity, and as a result, parity is not conserved under its action. For strong and electromagnetic interactions, parity is strictly conserved. That is, all such processes are invariant under spatial inversion. However, this is not true for weak interaction.

The suspicion of parity violation in weak interaction originated from observations made on the decay of a K^+-meson. Two different modes have been observed, one having two pions in the final state and the other having three pions, as shown in Eq. (5-44). Since these two modes have different parities, parity nonconservation in weak decays was proposed as the resolution. The confirmation of this suggestion came from observing the β^--decay of ^{60}Co [154],

$$^{60}\text{Co} \to {}^{60}\text{Ni} + e^- + \overline{\nu}_e$$

The ground state of odd-odd nucleus $^{60}_{27}$Co has spin-parity $J^\pi = 5^+$, as shown in Fig. 5-4. A nonzero spin is essential here so that the nucleus can be polarized (that is, have the nuclear spin aligned) along the direction of an external magnetic field. The ground state decays predominantly (99% of the time) to the $J^\pi = 4^+$ excited state of $^{60}_{28}$Ni at excitation energy 2.51 MeV. The decay is purely of the Gamow-Teller type (see next section).

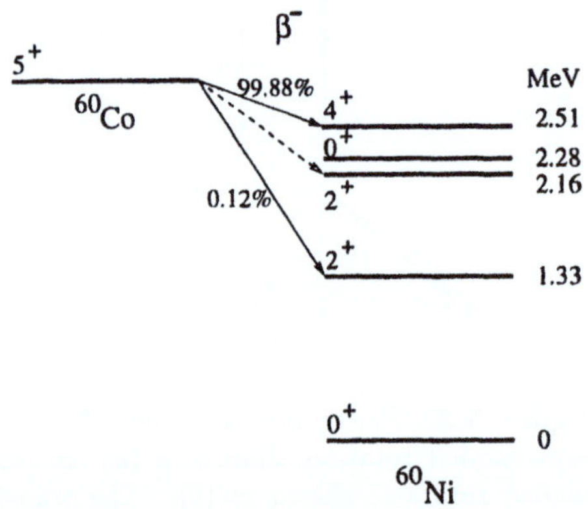

Figure 5-4: Decay scheme of ^{60}Co to ^{60}Ni. The main branch leading to the 4^+-state at 2.51 MeV in ^{60}Ni is a Gamow-Teller transition and is used in one of the first experiments to demonstrate parity nonconservation in β-decay. (Plotted using data from Ref. [95].)

If the spins of all the ^{60}Co nuclei are aligned, we have a fixed direction in space that is defined in a natural way by the experimental setup. This direction may be indicated by a unit vector $\boldsymbol{\sigma}$ parallel to the alignment of the ^{60}Co ground state spin $\boldsymbol{J}$. The angular distribution of electrons emitted with momentum $\boldsymbol{p}$ and energy E may be expressed in the following form (see, for example, p. 67 of Morita [105] or p. 290 of Eisenberg and Greiner [55] for a derivation),

$$W(\theta) \propto 1 + a\frac{\boldsymbol{\sigma} \cdot \boldsymbol{p}c}{E} = 1 + a\frac{v}{c}\cos\theta \tag{5-50}$$

where θ is the angle with respect to $\boldsymbol{J}$ by which the electron is emitted, E is the total (relativistic) energy of the electron, and the parameter a gives the intensity of angular dependence. Under a parity operation $\boldsymbol{p}$, being a polar vector, changes sign and $\boldsymbol{\sigma}$, being an axial vector, does not change sign. The product $\boldsymbol{\sigma} \cdot \boldsymbol{p}$ in the second term of Eq. (5-50) is then a pseudoscalar and changes sign under an inversion of the spatial coordinate system. On the other hand, the first term (unity), being a scalar, remains invariant. If parity is conserved in the decay, the second term of Eq. (5-50) must vanish on account of the fact that $\boldsymbol{\sigma} \cdot \boldsymbol{p} \to_P -\boldsymbol{\sigma} \cdot \boldsymbol{p}$. As a result, we expect $a = 0$ and the angular distribution of electrons emitted to be isotropic. Experimentally, a turned out to be -1, indicating a maximum degree of parity violation. The same conclusion is later confirmed by other measurements, such as the decay of pions and muons.

The result $a = -1$ may also be examined from the point of view of the helicities of the leptons involved. The *helicity* of a particle is defined as the projection of $\boldsymbol{\sigma}$, which we shall take here as twice the intrinsic spin $\boldsymbol{s}$ along its direction of motion

$$h = \frac{\boldsymbol{\sigma} \cdot \boldsymbol{p}}{|\boldsymbol{p}|} \tag{5-51}$$

For a massless particle, the eigenvalues of h can only be ± 1. An example of a particle with only two possible orientations is provided by the photon, which, as we have seen earlier, can have only two linearly independent polarization directions. For electrons and other particles with nonzero rest mass, the helicity takes on values $\pm v/c$. Particles with positive helicity are often referred to as "right-handed" particles and negative-helicity particles are often called "left-handed" particles. If the neutrinos are massless, they should behave in ways similar to photons and h can have values ± 1. Experimentally the helicity of neutrinos was first determined by Goldhaber, Grodzins, and Sunyan [75] through electron capture (see next section) in the 0^+ ground state of ^{152}Eu leading to the 1^- excited state of ^{152}Sm at 963 keV,

$$e^- + {}^{152}\mathrm{Eu} \longrightarrow \nu_e + {}^{152}\mathrm{Sm}^*(1^-) \longrightarrow \nu_e + {}^{152}\mathrm{Sm}(0^+) + \gamma$$

By measuring the polarization of the γ-ray emerging from the decay of ^{152}Sm to its 0^+ ground state, the helicity of ν_e emitted in the electron capture process was determined to be -1. Other nuclear β-decays put the helicity of $\overline{\nu}_e$ to be $+1$, as well as $h = -v/c$ for the electrons and $h = +v/c$ for the positrons emitted.

Two important consequences follow from these experimental observations. The first comes from the fact that all the leptons emitted in β-decays are observed to be left-handed ($h < 0$) and all antilepton right-handed ($h > 0$). For reasons we shall not go into here, operators that are scalars, pseudoscalars, and tensors produce leptons (as well as antileptons) of both helicities under a parity transformation. Only vector operators V and axial-vector operators A can accommodate the observed result that all leptons are of one helicity and antileptons are of the other value. Furthermore, since V and A are of different parity, a linear combination of V and A is required as the operator for β-decay. This leads to the $V - A$ theory of β-decay. [The minus sign is related to the fact that $a = -1$ in Eq. (5-50) rather than $+1$.]

A second consequence of the observation that neutrinos are found only with helicity $h = -1$ and antineutrinos with $h = +1$ is that neutrinos may be described by a two-component wave function. In Dirac theory, wave functions of spin-$\frac{1}{2}$ particles have four components so as to describe both particle and antiparticle, each with projections of spin $\pm\frac{1}{2}$ along the quantization axis. If neutrinos are always of one helicity and antineutrinos always of the opposite helicity, then a two-component theory will be adequate, as particles have $h = -1$ and antiparticles $h = +1$. However, such a simplification also implies that neutrinos are massless, and an experimental determination whether a neutrino has a nonzero rest mass is of fundamental interest here as well.

Fermi and Gamow-Teller operators. Since β-decay contains both a vector part and an axial-vector part, we expect that there are two independent operators, each with its own strength and its own radial dependence. As far as nuclear β-decay is concerned, the situation is somewhat simpler for two reasons. The first comes from the fact that weak interaction has very short range, much smaller than nuclear dimensions, as we saw earlier. For this reason, the radial dependence of the operators may be approximated by a delta function. This leaves only the strengths, or coupling constants, of each of the two operators to be specified.

The two coupling constants for nuclear β-decay may be put in the form of a vector coupling constant G_V for the vector part of the operator and a Gamow-Teller coupling constant G_A for the axial-vector part. We have already seen that G_V is related to G_F, the coupling constant underlying all weak interaction processes. The second simplifying feature of nuclear β-decay processes is that G_A is related to G_V. This results from the belief that the difference between G_A and G_V is only because of modifications of the axial-vector operator in the presence of strong interaction. The vector current, which may be indicated by a four-vector V_μ, is known to be a conserved quantity, i.e.,

$$\sum_{\mu=1}^{4} \frac{\partial V_\mu}{\partial x_\mu} = 0 \tag{5-52}$$

This is generally referred to as the *conserved vector current* (CVC) hypothesis and is analogous to the continuity equation in electromagnetism.

On the other hand, the axial-vector current A_μ does not have such a relation, i.e., the divergence of A_μ does not vanish. (This is related to the decay of pions, which are pseudoscalar particles.) Since A_μ is an axial vector, its divergence is a pseudoscalar. As we have seen in §2-7, the pion is a pseudoscalar particle and therefore is described by a pseudoscalar field. This leads to the *partially conserved axial-vector current* (PCAC) hypothesis,

$$\sum_{\mu=1}^{4} \frac{\partial A_\mu}{\partial x_\mu} = a\phi_\pi \tag{5-53}$$

where ϕ_π represents the pion field and a is a constant. In other words, the axial current is not conserved, but its divergence is proportional to the pion field ϕ_π. The weak axial-vector current is now related to a strong interaction field through a PCAC. (For further details, see, e.g., de Shalit and Feshbach [49] and Lee [96].)

A connection between the two weak coupling constants G_A and G_V can be made in a similar way. This is known as the Goldberger-Trieman relation, which, for our purpose here, may be stated in the form of the ratio between G_A and G_V,

$$g_A \equiv \frac{G_A}{G_V} = \frac{f_\pi\, g_{\pi N}}{M_N c^2} \tag{5-54}$$

where M_N is the nucleon mass. The quantity $f_\pi = F_\pi/\sqrt{2}$ is measured to be ~93 MeV and F_π is known as the pion decay constant. The quantity $g_{\pi N}$ is the pion-nucleon coupling constant and its value is known empirically to be

$$\frac{|g_{\pi N}|^2}{4\pi} \approx 14$$

From these values, Eq. (5-54) gives the result

$$|g_A| \approx 1.31$$

The measured value from nuclear β-decay is $g_A = -1.259 \pm 0.004$ (see also Problem 5-2), in agreement with the result of the Goldberger-Trieman relation. This in turn confirms the PCAC.

5-6 Nuclear Beta Decay

Nuclear β-decay is the process by which a nucleus made of Z protons and N neutrons decays to a nucleus of the same nucleon number A but with $(Z\pm1, N\mp1)$. A β^--decay,

$$A(Z,N) \longrightarrow A(Z+1,N-1)+e^- +\overline{\nu}_e \tag{5-55}$$

may be regarded as the transformation of one of the neutrons in the nucleus to a proton, and a β^+-decay,

$$A(Z,N) \longrightarrow A(Z-1,N+1)+e^+ +\nu_e$$

as that one of the protons to a neutron.

Analogous to internal conversion in electromagnetic decays, an atomic electron may be captured by the nucleus instead of emitting a positron in β^+-decay,

$$e^- + A(Z,N) \longrightarrow A(Z-1,N+1)+\nu_e$$

Except for a small difference in the energies involved, which we shall return to later, such an *electron capture* process has the same selection rule as β^+-decay and is usually in competition with it. The probability of electron capture increases as Z^3, again, because of increased strength of the nuclear Coulomb field and decreased radii of electronic orbits in atoms with increasing proton number.

Q-values. Some care is needed in calculating the Q-value for nuclear β-decay and electron capture. The Q-value of a reaction is defined as the difference in the total kinetic energies of the system before and after a reaction,

$$Q = T_f - T_i$$

For a nuclear β-decay, the parent nucleus may be assumed to be at rest in the laboratory, and the initial kinetic energy T_i in the system is zero. For the decay to take place, the total kinetic energies in the final state T_f and, hence, the Q-value must be positive. Since either an electron or a positron is emitted in the process, the Q-value is not simply the difference between the energies of the initial and final nuclear states. (The neutrino mass is too small, if nonzero, to play a significant role in the considerations here.)

A further complication comes from the fact that mass and binding energy of a nucleus are defined in terms of those for the corresponding neutral atom, as we have seen earlier. That is, the mass difference between the parent and daughter nuclei in a β-decay,

$$\Delta M_{\beta\pm} = M(Z,N) - M(Z\mp1,N\pm1)$$

includes the mass and the binding energy of an atomic electron as well. For this reason, the Q-value of β^--decay is given by the expression

$$Q_{\beta^-} = (M(Z,N) - M(Z+1,N-1))c^2$$

as the electron emitted in the decay may be used, as far as energy calculations are concerned, to compensate for the additional electron required to make the daughter atom neutral. On the other hand, the Q-value for β^+-decay is given by

$$Q_{\beta^+} = (M(Z,N) - M(Z-1,N+1))c^2 - 2m_ec^2$$

The "additional" amount of $2m_ec^2$ is required to create the positron emitted and the atomic electron that must be ejected in going from a neutral atom of Z electrons to one with $Z-1$ electrons.

In contrast, for electron capture, we have the relation

$$Q_{\rm EC} = (M(Z,N) - M(Z-1,N+1))c^2 - B_e$$

where B_e is the ionization energy of the atomic electron captured. Since B_e is of the order of 10 eV, we may ignore it unless we are concerned with accuracies of such order as, for example, in the case of neutrino mass measurements. The difference of $2m_ec^2$ in the Q-values between β^+-decay and electron capture is, however, important. For example, the mass difference between ^{7}Be and ^{7}Li is 0.86 MeV/c^2, less than $2m_ec^2$. As a result, β^+-decay from ^{7}Be to ^{7}Li is impossible, and the transition goes purely by electron capture with a half-life of 53.4 days. Only a neutrino emerges from an electron capture. Because of the difficulty in detecting neutrinos, the most prominent signature of electron capture processes is the x-ray emitted when atomic electrons in higher orbitals decay to the lower orbitals left empty when an inner shell electron is absorbed by the nucleus.

In terms of binding energies, the Q-values above correspond to the following expressions:

$$\begin{aligned} Q_{\beta^-} &= E_B(Z+1,N-1) - E_B(Z,N) + 0.782 \text{ MeV} \\ Q_{\beta^+} &= E_B(Z-1,N+1) - E_B(Z,N) - 2m_ec^2 - 0.782 \text{ MeV} \\ Q_{\rm EC} &= E_B(Z-1,N+1) - E_B(Z,N) - B_e - 0.782 \text{ MeV} \end{aligned} \qquad (5\text{-}56)$$

where the amount of 0.782 MeV comes from the mass difference between a neutron and a neutral hydrogen atom.

Transition rates for β-decay. To relate the transition probability $\mathcal{W}$ for a β-decay with the nuclear matrix element involved, we shall follow a procedure that closely resembles the one used earlier in electromagnetic transitions. To simplify the discussion, we shall ignore electron capture. Again, we start with Fermi's golden rule,

$$\mathcal{W} = \frac{2\pi}{\hbar}|\langle\phi_k(\boldsymbol{r})|H'|\phi_0(\boldsymbol{r})\rangle|^2\rho(E_f) \qquad (5\text{-}57)$$

given earlier by Eq. (5-13). The initial state is simple, involving only the parent nucleus,

$$|\phi_0(\boldsymbol{r})\rangle = |J_iM_i\zeta\rangle \qquad (5\text{-}58)$$

and we shall assume it to be stationary in the laboratory.

The final state consists of three particles, a neutral lepton, a charged lepton, and the daughter nucleus. For simplicity, we shall begin by ignoring any Coulomb effect between the charged lepton and the daughter nucleus. In this limit, both leptons are free particles and are described by plane waves traveling with wave numbers $\boldsymbol{k}_e$ and $\boldsymbol{k}_\nu$, respectively. The final state wave function is then a product of three parts,

$$|\phi_k(\boldsymbol{r})\rangle = \frac{1}{\sqrt{V}}e^{i\boldsymbol{k}_e\cdot\boldsymbol{r}}\frac{1}{\sqrt{V}}e^{i\boldsymbol{k}_\nu\cdot\boldsymbol{r}}|J_fM_f\xi\rangle$$

where $|J_fM_f\xi\rangle$ is the wave function of the daughter nuclear state. The two factors of $V^{-1/2}$ are required to normalize the two lepton wave functions. We may expand the product of the two plane waves in terms of spherical harmonics, as done in Eq. (B-11),

$$e^{i\boldsymbol{k}\cdot\boldsymbol{r}} = \sum_{\lambda=0}^{\infty} \sqrt{4\pi(2\lambda+1)}\, i^\lambda j_\lambda(kr)\, Y_{\lambda 0}(\theta, 0)$$

where $k = |\boldsymbol{k}| = |\boldsymbol{k}_e + \boldsymbol{k}_\nu|$ and θ is the angle between $\boldsymbol{k}$ and $\boldsymbol{r}$. The spherical harmonics $Y_{\lambda\mu}(\theta, \phi)$ are independent of the azimuthal angle ϕ for $\mu = 0$.

We can, again, make use of the long-wavelength approximation, as the Q-value of the transition is typically of the order of a few mega-electron-volts. In this limit, we only need to retain the first term in the expansion of the spherical Bessel function $j_\lambda(kr)$, as we did earlier in Eq. (5-26),

$$j_\lambda(kr) \approx \frac{(kr)^\lambda}{(2\lambda+1)!!}$$

The final state wave function may now be written as

$$|\phi_{\boldsymbol{k}}(\boldsymbol{r})\rangle = \frac{1}{V}\Big\{1 + i\sqrt{\frac{4\pi}{3}}\,(kr)Y_{10}(\theta,0) \; + O((kr)^2)\Big\}|J_fM_f\xi\rangle \tag{5-59}$$

This is very similar to what we have done earlier in the multipole expansion of electromagnetic transition matrix elements. The only difference, as we shall see later, is that the higher order terms are retarded by even larger factors in β-decay than in the corresponding reduction between successive higher orders in electromagnetic transitions.

Nuclear transition matrix elements. Let us examine first the possible forms of the nuclear part of the β-decay operator before proceeding to find the transition matrix element. Since a neutron is transformed into a proton in β^--decay and the other way around in β^+-decay, the nuclear operator must be one body in nature, i.e., only one nucleon is involved at a time, and must involve the single-particle isospin-raising or isospin-lowering operator $\tau_\pm$. Furthermore, according to the $V - A$ theory, there are two terms in the weak interaction, a polar vector part with coupling constant G_V and an axial-vector part with coupling constant G_A. In the nonrelativistic limit, the vector part may be represented by the unity operator times $\tau_\pm$ and the axial-vector part by a product of the intrinsic spin operator $\boldsymbol{\sigma}$ and $\tau_\pm$. A proper derivation of this result requires manipulations with Dirac wave functions and γ-matrices. We shall not carry out the discussions here, as they can be found in standard references such as Morita [105].

Putting this result for the operator together with those obtained earlier for the wave functions in Eqs. (5-58) and (5-59), we can write the $\beta^\pm$-decay transition matrix element of Eq. (5-57) as

$$\begin{aligned}\langle\phi_{\boldsymbol{k}}(\boldsymbol{r})|H'|\phi_0(\boldsymbol{r})\rangle &= \frac{1}{V}\langle J_fM_f\xi|\sum_{j=1}^{A}\big\{G_V\tau_\mp(j) + G_A\boldsymbol{\sigma}(j)\tau_\mp(j)\big\}\\ &\quad\times\Big\{1 - i\sqrt{\frac{4\pi}{3}}(kr)Y_{10}(\theta,0) + O((kr)^2)\Big\}|J_iM_i\zeta\rangle\end{aligned} \tag{5-60}$$

For the moment, we shall be mainly concerned with the two leading-order terms in the expression, generally known as the operators for "allowed transitions." The higher order terms involve spherical harmonics of orders greater than zero, and these induce the "forbidden decays."

For the allowed decays, the nuclear part of the $\beta^{\pm}$-decay operator has the form

$$O_{\lambda\mu}(\beta) = G_V \sum_{j=1}^{A} \tau_{\mp}(j) + G_A \sum_{j=1}^{A} \boldsymbol{\sigma}(j)\tau_{\mp}(j) \tag{5-61}$$

The angular momentum carried by the first term is $\lambda = 0$ and the second term is $\lambda = 1$. The transition matrix element for this β-decay operator is then

$$\begin{aligned}\langle\phi_{\boldsymbol{k}}(\boldsymbol{r})|H'|\phi_0(\boldsymbol{r})\rangle \approx & \frac{G_V}{V}\sum_{\mu M_f}\Big\{\langle J_f M_f \xi|\sum_{j=1}^{A}\tau_{\mp}(j)|J_i M_i \zeta\rangle \\ & + g_A\langle J_f M_f \xi|\sum_{j=1}^{A}\boldsymbol{\sigma}(j)\tau_{\mp}(j)|J_i M_i \zeta\rangle\Big\}\end{aligned} \tag{5-62}$$

where $g_A = G_A/G_V$. The first term here is usually referred to as *Fermi decay* and the second term as *Gamow-Teller decay*. Transitions matrix elements of operators with $\lambda > 1$ are usually much smaller in value, as they come from the higher order terms in Eq. (5-60). Their contributions are important only in cases where the two lowest order terms are forbidden by angular momentum and parity selection rules.

Density of final states. The density of states in Eq. (5-57) is complicated somewhat by the three-body final state in nuclear β-decay . Because of conservation laws, the energy and momentum of one particle are limited in value in a two-body situation by the amounts taken up by the other. For this reason, the two-body problem simplifies to an equivalent one-body one. In a nuclear β-decay, the available kinetic energy, after taking care of the nuclear recoil, is shared between the neutrino and the electron (or positron). As a result, continuous energy spectra of the charged lepton and the neutrino are produced, subject only to the condition that their sum, together with the nuclear recoil, satisfies energy-momentum conservation for the decay. Furthermore, the charged lepton is emitted in the Coulomb field of the daughter nucleus and its wave function is "distorted" as a result of electromagnetic interaction. This also has an effect on the density of final states available to the charged lepton.

Since a neutrino hardly interacts with its surroundings, it may be considered as a free particle once it is created. For such a particle, the number of states with momentum p_ν $(= \hbar k_\nu)$, without any regard to the direction in which the particle is moving, is given in statistical mechanics to be

$$dn_\nu = \frac{V}{2\pi^2\hbar^3} p_\nu^2\, dp_\nu \tag{5-63}$$

where V is the same volume as that used for normalizing a three-dimensional plane wave in Eq. (5-58). If the rest mass of the neutrino is m_ν, its total energy is given by the relativistic relation

$$E_\nu^2 = (m_\nu c^2)^2 + p_\nu^2 c^2 \tag{5-64}$$

The amount E_ν is a part of the energy released by the nucleus in going from the initial to the final state. The rest of the energy is taken up by the charged lepton and the daughter nucleus.

Instead of the Q-value, it is customary to express the energies involved in a β-decay in terms of the maximum kinetic energy of the charged lepton emitted. The reason for this is a practical one, as the electron (or positron) energy is a quantity that can be observed directly. The maximum value, E_0, is generally referred to as the *end-point* energy in the sense that it is the point, in a plot of the number of charged leptons observed as a function of the kinetic energy, beyond which no more particle is detected. In terms of the end-point energy, we have the relation

$$E_\nu = E_0 - E_e$$

where E_e is the kinetic energy of the charged lepton. For simplicity, we have ignored variations of the end-point energy due to small differences in the recoil energy of the daughter nucleus in a three-body final state. Since nuclear mass is much larger than those of leptons, the nuclear recoil needs to be accounted for only where high precision is required. In terms of E_0 and E_e, the density of neutrino states in Eq. (5-63) may be expressed as

$$dn_\nu = \frac{V}{2\pi^2\hbar^3}\frac{(E_0 - E_e)}{c^3}\{(E_0 - E_e)^2 - (m_\nu c^2)^2\}^{1/2}\, dE_e \tag{5-65}$$

where we have made use of the result $p_\nu c = \sqrt{E_\nu^2 - (m_\nu c^2)^2}$ from Eq. (5-64).

The charged leptons emitted cannot be treated as free particles, as the decay takes place in the Coulomb field of the daughter nucleus. A good approximation may be obtained by starting from a free particle and folding in a distortion factor $F(Z, E_e)$ to correct for Coulomb effects. Analogous to Eq. (5-63), the density of charged-lepton states may be written as

$$dn_e = \frac{V}{2\pi^2\hbar^3}F(Z, E_e)p_e^2\, dp_e \tag{5-66}$$

The correction factor $F(Z, E_e)$ is known as the Fermi function. In nonrelativistic limit, with the velocity of the charged lepton $v \ll c$, the function is related to the absolute square of the Coulomb wave function at the origin [see also Eq. (B-50)] and has the approximate form

$$F(Z, E_e) = \frac{x}{1 - e^{-x}} \tag{5-67}$$

where $x = \mp 2\pi\alpha Zc/v$ for $\beta^\pm$-decay, with α as the fine structure constant. The general form of the function is much more complicated and does not have a simple analytical representation. Extensive tabulated values are available and they are the ones usually used in practical applications. A full discussion of the Fermi function can be found in Morita [105].

The results of Eqs. (5-62), (5-65), and (5-66) can now be put into (5-57), and we arrive at the transition probability for an electron or positron emitted with momentum $p_e(= |\boldsymbol{p}_e|)$,

$$\begin{aligned}\mathcal{W}(p_e) &= \frac{1}{2\pi^3\hbar^7 c^3}\sum_{\mu M_f}\left|\langle J_f M_f \xi | \boldsymbol{O}_{\lambda\mu}(\beta) | J_i M_i \zeta\rangle\right|^2 \\ &\qquad \times F(Z, E_e)p_e^2(E_0 - E_e)\{(E_0 - E_e)^2 - (m_\nu c^2)^2\}^{1/2}\end{aligned}$$

where all factors related to V, the (arbitrary) normalization volume, cancel each other, and the summation over M_f takes care of the requirement to include all the possible nuclear final states.

Let us ignore the possible small neutrino mass for the time being. The expression for $\mathcal{W}(p_e)$ simplifies to

$$\mathcal{W}(p_e) = \frac{1}{2\pi^3\hbar^7c^3} \sum_{\mu M_f} \left|\langle J_f M_f \xi | O_{\lambda\mu}(\beta) | J_i M_i \zeta \rangle\right|^2 F(Z, E_e) p_e^2 (E_0 - E_e)^2 \qquad (5\text{-}68)$$

The approximation affects mainly the region where E_e is very close to the end-point energy E_0 and the influence of m_ν is most evident. From Eq. (5-68), we see that $\{\mathcal{W}(p_e)/p_e^2F(Z, E_e)\}^{1/2}$ is proportional to $E_0 - E_e$. A plot of the former quantity as a function of the latter produces a straight line (except in the region of the end-point energy) with a slope proportional to the nuclear matrix element. Such a graph, represented schematically in Fig. 5-5, is called a Kurie plot.

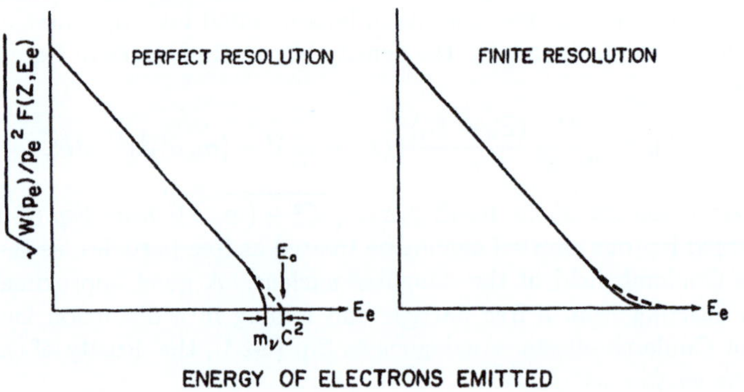

Figure 5-5: Schematic diagram of a Kurie plot. The solid lines are for the case of finite neutrino mass m_ν and the dashed lines are for $m_\nu = 0$. With perfect resolution the plot, as shown in (*a*), is a straight line intersecting the horizontal axis at the end-point energy E_0 if $m_\nu = 0$. Finite resolution of the detector modifies the region near the end-point energy, as shown in (*b*).

Neutrino mass measurement. It is convenient to digress here into a few remarks on the measurement of m_ν, the mass of an electron neutrino. All evidence to date indicates that m_ν is small, of the order of electron-volts, although the possibility of $m_\nu = 0$ is not ruled out either. The masses of the other two neutrinos, ν_μ and ν_τ, are expected to be larger; the present upper limits are, respectively, 0.5 and 70 MeV/c^2.

Most direct measurements of m_ν make use of the β^--decay of the triton, consisting of one proton and two neutrons. There are many reasons for favoring this reaction. The decay

$$t \longrightarrow {}^3\text{He} + e^- + \overline{\nu}_e$$

has a half-life of 12.3 years. The Q-value is low, 18.6 keV, so that the influence of a small m_ν stands out more prominently than otherwise, for example, in a Kurie plot.

The excited states of the daughter nucleus ^{3}He are very high in energy, and as a result, the ground state is the only possible final nuclear state for the decay. In addition, the radioactive tritium source is relatively easy to prepare. There are, however, several difficulties associated with the measurement that are hard to overcome. The first one arises mainly from the low counting rate near the end point, a common problem in all nuclear β-decays. The second one comes from the fact that the expected rest mass energy of the neutrino is comparable to the excitation energies in atoms. As a result, atomic effects, which are seldom a problem in nuclear measurements, become an important issue here. For example, there are two possible final atomic states for ^{3}He, and the relative probability of forming them must be known fairly well in order to obtain a reliable final answer on m_ν. Currently, there is still no agreement between the measured values from different laboratories. One of the quoted values, ~ 30 eV/c^2, came from the first measurement of Lubimov et al. [99], and the results are shown in Fig. 5-6. Observations of neutrinos from the supernova SN 1987a, atmospheric neutrinos produced from cosmic rays, and those obtained from reactors and accelerators put the possible value to be much lower, somewhere around a few electron-volts divided by c^2. However, the possibility of zero is by no means ruled out.

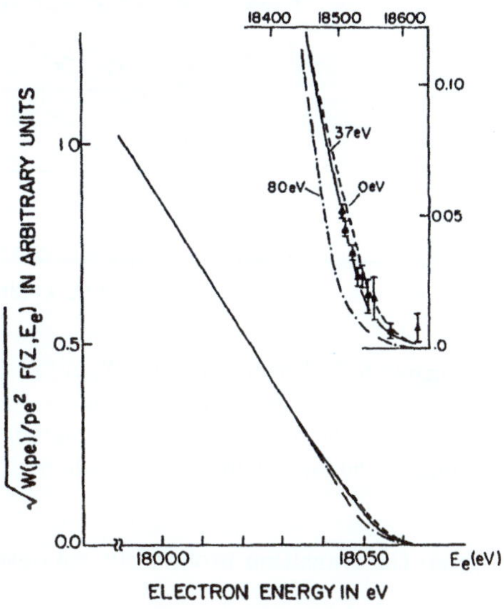

Figure 5-6: Kurie plot near the end-point energy for tritium β-decay, used for neutrino mass measurements. The dashed curve is for $m_\nu = 0$ and the dash-dot curve for $m_\nu = 80$ eV/c^2. The best-fit value of $m_\nu = 37$ eV/c^2 is given by the solid curve. Adapted from Ref. [99].

Total transition probability. Let us return to the question of nuclear β-decay. If we are not interested in the distribution of charged leptons emitted as a function of E_e, we can integrate $\mathcal{W}(p_e)$, given by Eq. (5-68), over all possible values of momentum p_e and obtain the total transition probability $\mathcal{W}$,

$$\mathcal{W} = \int \mathcal{W}(p_e)\,dp_e = \frac{m_e^5 c^4}{2\pi^3\hbar^7} f(Z, E_0) \Big| \sum_{\mu M_f} \langle J_f M_f \xi | \boldsymbol{O}_{\lambda\mu}(\beta) | J_i M_i \zeta \rangle \Big|^2$$

where the dimensionless function

$$f(Z,E_0) \equiv \int F(Z,E_e)\Big(\frac{p_e}{m_e c}\Big)^2\Big(\frac{E_0-E_e}{m_e c^2}\Big)^2 \frac{dp_e}{m_e c}$$
$$= \frac{1}{m_e^5 c^7}\int F(Z,E_e)\, p_e^2 (E_0-E_e)^2\, dp_e \qquad (5\text{-}69)$$

is known as the Fermi integral. Except in the trivial case of $Z = 1$ for the daughter nucleus, the integral must be evaluated numerically. Extensive tables of calculated values are available [60], and some of the typical values are plotted in Fig. 5-7.

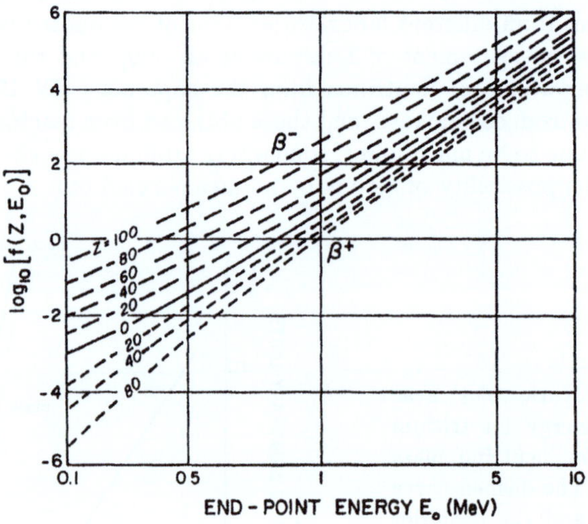

Figure 5-7: Fermi integral $f(Z, E_0)$ of Eq. (5-69) as a function of the end-point energy E_0 for different proton number Z in the daughter nucleus. The long dashed curves in the upper half of the diagram are for β^--decay and the short dashed ones in the lower half are for β^+-decay. (Adapted from Refs. [59, 61].)

From the transition probability, we obtain the expression for β-decay half-life,

$$T_{1/2} = \frac{\ln 2}{\mathcal{W}} = \frac{1}{f(Z,E_0)} \frac{2\pi^3\hbar^7}{m_e^5 c^4} \frac{\ln 2}{|\sum_{\mu M_f}\langle J_f M_f \xi | \boldsymbol{O}_{\lambda\mu}(\beta) | J_i M_i \zeta\rangle|^2}$$

Instead of half-lives, nuclear β-decay rates are often quoted in terms of ft-values, the product of the Fermi integral $f(Z, E_0)$ and $T_{1/2}$,

$$ft \equiv f(Z,E_0)T_{1/2} = \frac{2\pi^3\hbar^7}{m_e^5 c^4} \frac{\ln 2}{|\sum_{\mu M_f}\langle J_f M_f \xi | \boldsymbol{O}_{\lambda\mu}(\beta) | J_i M_i \zeta\rangle|^2} \qquad (5\text{-}70)$$

As we can see from the definition, the ft-value is a more meaningful physical quantity in nuclear β-decay studies, as it is directly related to the square of the nuclear transition matrix element. The half-life, on the other hand, involves $f(Z, E_0)$, which depends in a complicated way on the proton number of the daughter nucleus and the end-point energy. We have already encountered a similar problem in the study of electromagnetic decays. There, half-lives are dominated by energy dependence and the quantities more directly related to nuclear physics are the reduced transition probabilities $B(\lambda)$ and transition rates measured in Weisskopf units. The ft-value in β-decay plays a similar role as $B(\lambda)$ in electromagnetic transitions.

The measured ft-values are found to vary over many orders of magnitude, especially when we consider both allowed and forbidden decays. For this reason, it is often more convenient to use $\log ft$ values, the logarithm to the base 10 of the ft-values. The distribution of measured $\log ft$ values is shown in Fig. 5-8.

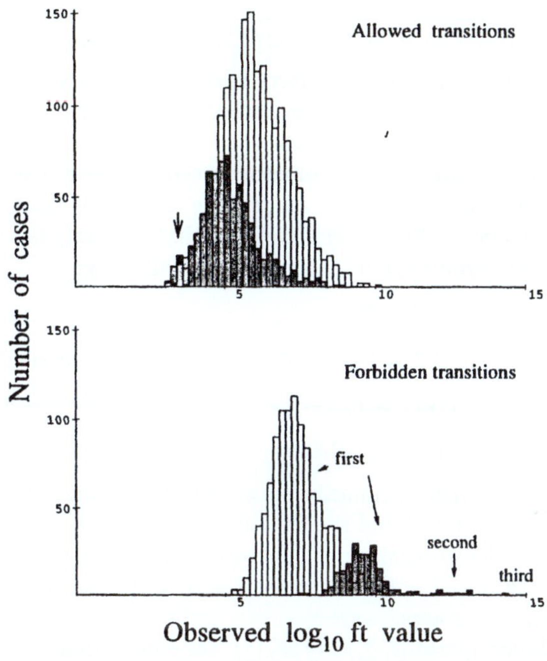

Figure 5-8: Systematics of observed $\log ft$ values. The grey area in the upper panel shows 718 cases of $0^+ \rightleftharpoons 1^+$ allowed transitions, and the remaining 1741 cases of other allowed decays are shown by the white histogram. The peak of the distribution for the 24 cases of $0^+ \to 0^+$ superallowed decay is indicated by the arrow. The 216 first-forbidden unique transitions ($|J_i - J_f| = 2$) are shown by the shaded part in the lower panel and the 1086 cases of other first-forbidden transitions by the white histogram. Only 37 second-forbidden and 3 third-forbidden cases are known. The four reported cases of the highest order, fourth, have $\log ft$ values above 20. (Based on data in Ref. [79], selected by Singh et al. [128].)

Allowed β-decay. Let us return to the operator for allowed decays, given earlier in Eq. (5-61). The Fermi term involves only the isospin-raising or isospin-lowering operator. As a result, it is possible to carry out the summation over all the nucleons explicitly,

$$\sum_{j=1}^{A} \tau_{\mp}(j) = \boldsymbol{T}_{\mp}$$

where $\boldsymbol{T}_{\mp}$ lowers or raises the third component of the isospin for the nucleus as a whole. If isospin is an exact quantum number, the matrix element for the Fermi operator can be evaluated without having to know explicitly the wave functions involved,

$$\langle J_f M_f T_f T_{0f} | \sum_{j=1}^{A} \tau_{\mp}(j) | J_i M_i T_i T_{0i} \rangle$$
$$= \sqrt{T_i(T_i+1) - T_{0i}(T_{0i} \mp 1)}\, \delta_{J_f J_i} \delta_{M_f M_i} \delta_{T_f T_i} \delta_{T_{0f}(T_{0i} \mp 1)} \tag{5-71}$$

As we have seen earlier, both the Coulomb force and the difference in the mass between charged and neutral pions violate isospin symmetry and, consequently, affect the actual value of the Fermi matrix element. In practice, it is found that, for light nuclei, the necessary correction factors are quite small and, in certain cases, can be evaluated to sufficient accuracy so that the final results are reliable to an uncertainty of 0.1% or less.

From the results given by Eq. (5-71), we find that the angular momentum and isospin selection rules of Fermi-type $\beta^{\pm}$-decay are the following:

$$\begin{array}{ll} J_f = J_i & (\Delta J = 0) \\ T_f = T_i \neq 0 & (\Delta T = 0, \text{ but } T_i = 0 \rightarrow T_f = 0 \text{ forbidden}) \\ T_{0f} = T_{0i} \mp 1 & (\Delta T_0 = 1) \\ \Delta\pi = 0 & \text{no parity change} \end{array} \tag{5-72}$$

In other words, Fermi decay goes primarily between isobaric analogue states where the only difference between the initial and the final states is the replacement of a proton by a neutron or vice versa. However, since the operator has isospin rank unity, the matrix element vanishes if both initial and final states has $T = 0$. For this reason, the decay is forbidden between $T = 0$ states.

For the Gamow-Teller operator, $\sum_j \boldsymbol{\sigma}(j)\tau_{\mp}(j)$, the summation over nucleons cannot be carried out explicitly, as both spin and isospin of a nucleon are acted upon at the same time. Unlike Fermi decays, matrix elements for the Gamow-Teller operator, in general, cannot be evaluated unless both the initial and the final wave functions are known. The angular momentum and isospin selection rules, however, can be deduced from the properties of the operator itself. Since the spherical tensor ranks are unity in both spin and isospin spaces, it is necessary that the initial and final states are related in the following ways:

$$\begin{array}{ll} \Delta J = 0, 1 & \text{but } J_i = 0 \rightarrow J_f = 0 \text{ forbidden} \\ \Delta T = 0, 1 & \text{but } T_i = 0 \rightarrow T_f = 0 \text{ forbidden} \\ T_{0f} = T_{0i} \mp 1 & (\Delta T_0 = 1) \\ \Delta\pi = 0 & \text{no parity change} \end{array} \tag{5-73}$$

The last point, on the parity selection rule, comes from the fact that $\boldsymbol{\sigma}$ is an axial-vector operator and, consequently, cannot change the parity between initial and final states. The absolute values of Gamow-Teller matrix elements are generally smaller than those for Fermi transitions, as both spin and isospin are involved.

For allowed decays, the square of the nuclear transition matrix element may be written as

$$\begin{aligned}
&\sum_{\mu M_f} \left|\langle J_f M_f \xi | \boldsymbol{O}_{\lambda\mu}(\beta) | J_i M_i \zeta\rangle\right|^2 \\
&\quad = G_V^2 \Big\{ \sum_{M_f} \left|\langle J_f M_f \xi | T_{\mp} | J_i M_i \zeta\rangle\right|^2 + g_A^2 \sum_{\mu M_f} \Big|\langle J_f M_f \xi | \sum_{j=1}^{A} \sigma_\mu(j) \tau_{\mp}(j) | J_i M_i \zeta\rangle\Big|^2 \Big\} \\
&\quad \equiv G_V^2 \{\langle F\rangle^2 + g_A^2 \langle GT\rangle^2\}
\end{aligned} \tag{5-74}$$

There is no cross term between the Fermi and Gamow-Teller operators, as the matrix element vanishes on summing over all the possible projections on the quantization axis. To simplify the notation, we shall use the abbreviations for the matrix elements adopted in the final form of Eq. (5-74).

For allowed β-decays, the ft-value of Eq. (5-70) may be written as

$$ft = \frac{K}{\langle F\rangle^2 + g_A^2 \langle GT\rangle^2}$$

where the vector coupling constant G_V, as well as other universal constants, is absorbed into the definition of the constant,

$$K = \frac{2\pi^3 \hbar^7 \ln 2}{m_e^5 c^4 G_V^2} = 6141.2 \pm 3.2 \text{ s}$$

Among the factors included, the value of the vector coupling constant G_V is perhaps least well known. A determination of K is then one way to deduce the absolute value of G_V. The best measured value of K is currently 6141.2 ± 3.2 s, obtained from superallowed decays after applying such corrections as the finite size and the charge distribution of the nucleus [129]. The value of the vector coupling constant deduced in this way is $G_V = 1.41556(74) \times 10^{-62}$ J-m^3 or, in its more commonly quoted form, $G_V/(\hbar c)^3 = 1.14984(60) \times 10^{-11}$ MeV^{-2}. In order to obtain the Fermi coupling constant of $G_F/(\hbar c)^3 = 1.16639(2) \times 10^{-11}$ MeV^{-2}, further correction factors are required (see, e.g., Ref. [33]).

Superallowed β-decay. Transitions from an initial nuclear state with $J_i^\pi = 0^+$ to a final state with $J_f^\pi = 0^+$ with the same isospin T form a special class of β-decay, as the Gamow-Teller term does not contribute. These are known as *superallowed* β-decays. The transitions are purely Fermi and, as a result, are least sensitive to the details of nuclear wave functions. Such decays are useful, for example, in determining the value of K, and hence G_V, as mentioned in the previous paragraph. Light nuclei are preferred here, as isospin-breaking effects are smaller than in heavier ones. Superallowed β^--decay is often forbidden by Q-value considerations, as Coulomb energies are higher for daughter nuclei with one more proton than the parent. Most examples of superallowed β-decay are positron emitters for this reason. An example is the case of

$$^{14}\text{O} \rightarrow {}^{14}\text{N} + e^{+} + \nu_e$$

leading to the 0^{+} first excited state of ^{14}N at 2.311 MeV. The half-life of ^{14}O is 74 s and the Q-value of the reaction is 1.12 MeV. The ft-value is 3037.7 s, among the smallest known. If the initial and final nuclear states are truly isobaric analogue states of each other, the value of the Fermi matrix element may be obtained using Eq. (5-71) without referring explicitly to the nuclear wave functions.

The determination of g_A, the ratio between axial-vector and vector coupling constants, requires nuclear Gamow-Teller decay, as can be seen from Eq. (5-74). The best-known value is

$$g_A = \frac{G_A}{G_V} = -1.259 \pm 0.004$$

In principle, β^{-}-decay of free neutrons is the ideal reaction to use for the purpose, as only the nucleon intrinsic spin wave functions enter into the calculation. In this case, the Gamow-Teller matrix element can be evaluated using the relation

$$\sum_{\mu m_f} |\langle \chi_{m_f} | \sigma_\mu | \chi_{m_i} \rangle|^2 = 3$$

However, we are limited here by our knowledge of the half-life of neutrons. The value that is quoted nowadays is $T_{1/2} = 614.6 \pm 1.3$ s. Unfortunately, it belongs to the class of data that change with time, as newer and better measurements are carried out.

Forbidden decay. From the selection rules given in Eqs. (5-72) and (5-73) and summarized in Table 5-3, we see that, for allowed β-decays, the spins of the initial and final states can be different at most by unity and the parities must be same. Transitions between states of different parity and $\Delta J > 1$ are also known to take place, albeit with much larger ft-values (i.e., smaller probabilities). These are referred to as *forbidden decays.* As we can see from Eq. (5-59), the operators for forbidden decays involve spherical harmonics of order greater than zero.

Forbidden decays are classified into different groups by the ℓ-value of the spherical harmonics involved. For a given order ℓ, the possible operators with definite spherical

Table 5-3: Selection rules and observed range of $\log ft$ values for nuclear β-decay.

Decay type	ΔJ	ΔT	$\Delta\pi$	$\log_{10} ft_{1/2}$
Superallowed	$0^{+} \rightarrow 0^{+}$	0	no	3.1–3.6
Allowed	0, 1	0, 1	no	2.9–10
First forbidden	0, 1, 2	0, 1	yes	5–19
Second forbidden	1, 2, 3	0, 1	no	10–18
Third forbidden	2, 3, 4	0, 1	yes	17–22
Fourth forbidden	3, 4, 5	0, 1	no	22–24

tensor ranks are $Y_{\ell m}(\theta,\phi)$ and $(Y_\ell(\theta,\phi)\times\boldsymbol{\sigma})_{\lambda\mu}$. The angular momentum and parity selection rules for the ℓth-order forbidden transition are then

$$\Delta J = \ell \text{ or } \ell \pm 1 \qquad\qquad \Delta\pi = (-1)^\ell$$

The isospin selection rule remains the same as for allowed decays,

$$\Delta T_0 = 1 \qquad \Delta T = 0,\ 1 \qquad \text{but } T_i = 0 \to T_f = 0 \text{ transitions forbidden}$$

as nothing is different in the isospin structure of the operators between forbidden and allowed decays. Thus, for first-order forbidden transitions, the operators are $r\,Y_{1\mu}(\theta,\phi)$ (proportional to $\boldsymbol{r}$) and $(\boldsymbol{\sigma}\times rY_1(\theta,\phi))_{\lambda\mu}$ with $\lambda = 0,\ 1,\ 2$. Since the parity of $Y_{1\mu}(\theta,\phi)$ is -1, a parity change between the initial and final states is necessary.

The reason for the large ft-values in forbidden β-decays comes from the angular momentum barrier that inhibits lepton emission when $\ell > 0$. This results in a reduction in the size of the nuclear transition matrix element and, hence, an increase in the ft-values. Typical values for the various order decays are given in Table 5-3. Distributions of observed $\log ft$ values for different orders are shown Fig. 5-8. In general, it is quite difficult to calculate the nuclear matrix elements for forbidden β-decays with any reliability, and as a result, relatively few theoretical investigations are found in the literature.

Charge exchange reactions. Charge exchange reactions can also replace a proton in a nucleus by a neutron or the other way around. Although the process involves primarily nuclear interaction, the nuclear matrix element that enters into the reaction rate is essentially the same as that in a β-decay induced by weak interaction. Relations between β-decay and charge exchange reactions are therefore of interest both from the point of making a connection between these two types of interaction and in studying the nuclear matrix elements involved.

A typical charge exchange may be illustrated by a (p,n) or (n,p) reaction. In the former case, a nucleus A is bombarded by a beam of protons. Among the different possible reactions, we are interested here in the one where the proton is absorbed by the nucleus and a neutron is emitted in exchange. The nuclear structure part of this reaction bears strong resemblance to β^--decay. Apart from the dynamics of scattering, the main difference between them is that the (p,n) reaction is not restricted by Q-value considerations leading only to final states that are lower in energy, as given by Eq. (5-56) for the β^--decay.

Analogous to the relation between the (p,n) reaction and β^--decay, we have the (n,p) reaction,

$$A(Z,N) + n \longrightarrow A(Z-1,N+1) + p$$

as the complement of β^+-decay. From a practical point of view, the reaction is difficult to study because of the scarcity of energetic neutron beams. A combination of (p,n) and (n,p) reactions, however, allows a whole range of interesting questions to be investigated.

Charge exchange processes may also be induced by reactions involving nuclei as the projectile and scattered particles, such as (^{3}He,t), (^{6}Li,^{6}He), and their inverses. The use of ^{3}He and heavier ions has the complication that both the incident and the scattered particle, as well as the target nucleus, may be excited in the process. In addition, (π^+, π^0) and (π^-, π^0) reactions are also used to examine charge exchange processes, as we shall see later in §8-6.

With pions and heavy ions, it is also possible to initiate double-charge exchange reactions, such as (π^-, π^+) and (π^+, π^-). In such reactions, a pair of nucleons change their charge states at the same time. These reactions are sensitive to two-body correlations in nuclei, a question of importance in nuclear structure studies. The information is closely related to that obtained through two-nucleon transfer reactions discussed in §8-2 and is also related to double β-decay described below.

Double β-decay. Double β-decay is the process by which two electrons or two positrons are emitted,

$$\begin{aligned} A(Z,N) &\rightarrow A(Z+2,N-2)+2e^-+2\overline{\nu}_e \\ A(Z,N) &\rightarrow A(Z-2,N+2)+2e^++2\nu_e \end{aligned}$$

These reactions are the result of second-order perturbations induced by weak interaction and are far slower than normal β-decays in which only a single charged lepton is emitted. As a result, double β-decays are expected to be long-lived, with typical half-lives of the order of 10^{20} years. Processes with such long half-lives may be observed only in nuclei where ordinary β-decay and other faster reactions are forbidden by Q-value considerations. A number of such cases are known and they can be identified by comparing the binding energies of neighboring nuclei. For example, ${}^{82}_{34}$Se is stable against β^--decay to ${}^{82}_{35}$Br, as the Q-value is -0.90 MeV. However, it is unstable against double β^--decay to ${}^{82}_{36}$Kr, with a Q-value of $+3.00$ MeV.

It is not surprising that a number of nuclei can, in principle, undergo double β-decay. In general, these are even-even nuclei with large neutron excess near, but not at, the bottom of the valley of stability. Because of pairing energy, they are more tightly bound compared with neighboring odd-odd nuclei (see §4-9). On the other hand, a neighboring even-even nucleus with two more protons and two less neutrons may be more tightly bound if the symmetry energy is larger. As we have seen earlier in the discussion of binding energies, this term is proportional to $(N-Z)^2$. Since most nuclei in the medium to heavy range have a large neutron excess, an isobar with two neutrons less can often be more tightly bound as a result. For this reason, more nuclei are known to be capable of double β^--decay than those for double β^+-decay. Double β^+-decay is possible, for example, in the case of ${}^{106}_{48}$Cd to ${}^{106}_{46}$Pd; the Q-value of 0.7 MeV is, however, smaller than the typical double β^--decay values of 2 to 3 MeV.

One of the interests in nuclear double β-decay is the question of whether the reaction can take place without emitting neutrinos. If neutrinos are Majorana fermions, with no distinction between particles and antiparticles, we can imagine that the neutrino from the first β-decay in a double β-decay process is absorbed in the intermediate state and that this absorption induces the emission of the second charged lepton. In such cases, no neutrinos emerge from the decay. On the other hand, such "neutrinoless" double β-decay processes are strictly forbidden if the neutrinos are Dirac particles, with particles distinct from their antiparticles.

The fact that neutrinos and antineutrinos are different particles has been confirmed in an experiment using the reaction

$$\text{neutrino} + {}^{37}\text{Cl} \rightarrow e^- + {}^{37}\text{Ar}$$

The source of the neutrino for this classic experiment by Davis [48] in 1955 was a reactor which produces mainly $\overline{\nu}_e$. The observed cross section for this reaction was much smaller than one expected for Majorana neutrinos. The result therefore constitutes a proof that neutrinos are Dirac particles.

However, in a neutrinoless double β-decay, the neutrinos are virtual particles and may be different from the real neutrinos observed in the experiment of Davis. If virtual neutrinos are Majorana particles, then double β-decay can take place without emitting any physical neutrinos and can, therefore, proceed on a much faster scale, perhaps by as much as six orders of magnitude. An important factor in support of the faster rate is that the phase space available for the final states of a neutrinoless double β-decay is much larger than the competing two-neutrino mode.

One way to distinguish between the two possible types of double β-decay is the spectrum of the electrons emitted. If no neutrinos are emitted, the sum of the energies of the two electrons is equal to the Q-value of the decay (again ignoring the small amount of energy taken away by nuclear recoil). On the other hand, if two neutrinos are also emitted, the sum of the energies of the two electrons has a continuous distribution given by energy-momentum conservation of the five-body final state. One recent measurement of the double β^--decay of ^{82}Se to ^{82}Kr by Elliot, Hahn, and Moe [57] gives a limit of the half-life of the decay to be 4.4×10^{20} years and an energy spectrum of the two electrons emitted consistent with the two-neutrino mode.

Long lifetimes alone do not necessarily rule out the possibility of neutrinoless double β-decay. As we have seen in the case of single β-decay, there is a large spread in the $\log ft$ values, even among the allowed decays. Such a divergence in the rate is due, primarily, to the wide range of possible values for the nuclear transition matrix element involved. The same may also be true for double β-decays. If the nuclear matrix elements in double β-decays are much smaller than expected, the lifetimes of 10^{20} years could even be an underestimate of the rate for the two-neutrino mode. Consequently, long measured half-lives by themselves do not rule out the neutrinoless mode. In this sense, the energy spectrum of the experiment of Elliot, Hahn, and Moe is a more conclusive evidence against the neutrinoless mode than lifetime measurements.

Since we are considering very slow processes, there are also other possibilities for double β-decay in addition to two-neutrino and neutrinoless modes. One is the weak decay of a Δ-particle to a nucleon with the emission of two charged leptons. The normal decay mode of Δ is to a pion plus a nucleon via strong interaction. However, a weak branch involving leptons cannot be ruled out, especially when the Δ-particle is a part of a nucleus. The other possibility is that, instead of two neutrinos, a boson, given the name "Majaron," may be emitted. The detection of any such events requires measurements involving half-lives of the order of 10^{20} year or longer, and they are not easy experiments to carry out.

Problems

5-1. The first excited state of ^{20}Ne at 1.634 MeV has $J^\pi = 2^+$ and decays to the 0^+ ground state with a half-life of 0.655 ps. Find the reduced transition probability $B(E2)$ in units of e^2fm^4 and the transition rate $\mathcal{W}$ in Weisskopf units.

5-2. The following list of corrected superallowed ft-values in seconds are taken from Sirlin [129]:

^{14}O $\rightarrow^{14}$N	3074.0 ± 3.9	^{26m}Al$\rightarrow^{26}$Mg	3068.1 ± 3.7
^{34}Cl $\rightarrow^{34}$S	3069.0 ± 4.7	^{38m}K $\rightarrow^{38}$Ar	3066.6 ± 4.6
^{42}Sc $\rightarrow^{42}$Ca	3077.5 ± 7.5	^{46}V $\rightarrow^{46}$Ti	3074.7 ± 4.3
^{50}Mn$\rightarrow^{50}$Cr	3069.6 ± 4.4	^{54}Co $\rightarrow^{54}$Fe	3069.0 ± 1.6

Find the vector coupling constant G_V from this list of results. From the value G_V obtained, find the ratio $|G_A/G_V|$ using the value of 615 s for the half-life of neutron β^--decay and an estimate of the value of $f(Z, E_0)$ from Fig. 5-7.

5-3. Since the end-point energy E_0 is difficult to measure precisely, the Q-value of a nuclear β-decay is often determined from the corresponding (p,n) or (n,p) reaction. Calculate the Q-value for the superallowed β^+-decay of ^{26m}Al ($E_x = 0.229 \pm 0.003$ MeV) to the ground state of ^{26}Mg, given that the measured Q-value for the ^{26}Mg$(p,n)^{26}$Al reaction leading to the ground state of ^{26}Al is -4.786 ± 0.002 MeV.

5-4. The $7/2^+$ state at 1.72 MeV in ^{21}Ne has a half-life of 48 fs (1 fs $= 10^{-15}$ s) and decays 94% of the time to the 0.33-MeV $5/2^+$ state with a mixing ratio $\delta = 0.14 \pm 0.02$ and 6% of the time to the $3/2^+$ ground state. Find the $B(E2)$ and $B(M1)$ values for the transitions involved.

5-5. A more realistic radial wave function for nucleons than the uniform-density-sphere model used in §5-4 to calculate the Weisskopf single-particle estimates of electromagnetic transitions is the spherical harmonic oscillator radial wave function. Use the explicit forms given in Table 7-1 to evaluate the matrix element $\langle r^2 \rangle$ and compare the results with those given by Eq. (5-35).

5-6. The orbital angular momentum part of a single-particle wave function is given by spherical harmonics $Y_{\ell m}(\theta, \phi)$. Use this together with the radial integrals evaluated above in Problem 5-5 to calculate the single-particle values for $E2$-transitions.

5-7. The nucleus $^{12}_{7}$N decays to $^{12}_{6}$C with a Q-value of 16.38 MeV. Calculate the maximum recoil energy of the daughter nucleus. If the probability of emitting leptons with momentum p_ν, up to $p \sim Q/c$, is $\sim p^2\,dp$, given purely by phase space conditions, calculate the distribution of the number of positrons emitted as a function of energy. Ignore Coulomb corrections to the charged lepton emitted.

Chapter 6

Nuclear Collective Motion

The experimental observations outlined in the previous two chapters on energy level positions, static moments, transition rates, and reaction cross sections provide us with the basis for nuclear structure studies. Many of the observed properties of a nucleus involve the motion of many nucleons "collectively." For these phenomena, it is more appropriate to describe them using a Hamiltonian expressed in terms of the bulk or *macroscopic* coordinates of the system, such as mass, radius, and volume.

6-1 Vibrational Model

We have seen earlier in the discussion of nuclear binding energies in §1-3 and §4-9 that, in many ways, the nucleus may be looked upon as a drop of fluid. A large number of the observed properties can be understood from the interplay between the surface tension and the volume energy of the drop. In this section, we shall take the same approach to examine nuclear excitation due to *vibrational* motion.

For simplicity we shall take that, at equilibrium, the shape of a nucleus is spherical, i.e., the potential energy is minimum when the nucleus assumes a spherical shape. This is purely an assumption of convenience for our discussion here. It is made, in part, for the reason that spherical nuclei do not have rotational degrees of freedom, and as a result, vibrational motion stands out clearly, without complications due to rotation. In practice, the most stable shape for many nuclei is deformed, as we shall see later in §6-3, and vibrational motions built upon deformed shapes are also commonly observed.

Breathing mode. When a nucleus acquires an excess of energy, for example, from Coulomb excitation due to a charged particle passing nearby, it can be set into vibration around its equilibrium shape. We can envisage several different types of vibration. For example, the nucleus may change its size without changing its shape, as shown in Fig. 6-1(a). Since the volume is now changing while the total amount of nuclear matter remains constant, the motion involves an oscillation in the density. Such a density vibration is similar to the motion involved in respiration and, for this reason, is called a *breathing* mode vibration.

For an even-even, spherical nucleus, the ground state spin and parity are 0^+. To preserve the nuclear shape, breathing mode excitation in this case generates states that are also $J^\pi = 0^+$. In Fig. 6-2, we see that, in the case of doubly magic nuclei of ^{16}O,

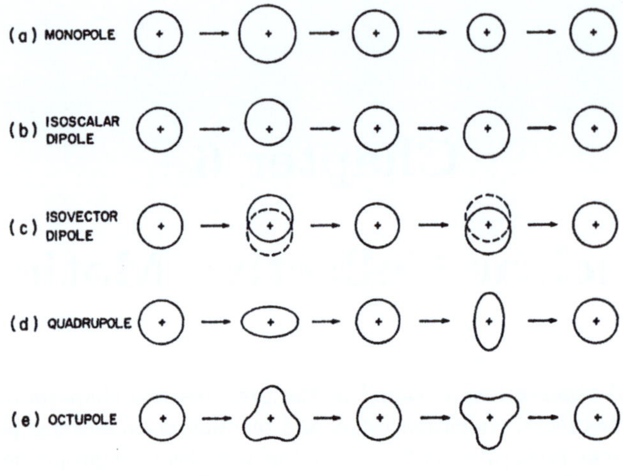

Figure 6-1: Time evolution of low-order vibrational modes. The monopole oscillation in (*a*) involves variations in the size without changing the overall shape. The nucleus moves as a whole in an isoscalar dipole vibration shown in (*b*). In contrast, an isovector dipole vibration consists of neutrons and protons oscillating in opposite phase, as in (*c*). In quadrupole vibrations the nucleus changes from prolate to oblate and back again, as in (*d*). Octupole vibrations are shown in (*e*).

^{40}Ca, ^{90}Zr, and ^{208}Pb, a low-lying $J^\pi = 0^+$ state is found among the first few excited states. Such low-energy states are often the result of collective excitation and may be identified as breathing mode states. On the other hand, nuclear matter is rather stiff against compression, and one expects the main part of the breathing mode strength to be much higher in energy. The observed value depends on the number of nucleons in the nucleus, and the peak location is usually found at around $80A^{-1/3}$ mega-electron-volts. The energy of breathing mode excitation is one of the few ways to find out something about the stiffness of nuclear matter that are important in understanding, for an example, the state of a star just before a supernova explosion (see §10-6) and in the study of infinite nuclear matter (§4-12).

Shape vibration. The more common type of vibration involves oscillations in the shape of the nucleus without changing the density. This is very similar to a drop of liquid suspended from a water faucet. If the drop is disturbed very gently, it starts to vibrate. Since the amount of energy is usually too small to compress the liquid, the motion simply involves an oscillation in the shape.

For a drop of fluid, departures from spherical shape without density change may be described in terms of a set of *shape* parameters $\alpha_{\lambda\mu}(t)$ defined in the following way:

$$R(\theta,\phi;t) = R_0\Big\{1 + \sum_{\lambda\mu} \alpha_{\lambda\mu}(t) Y_{\lambda\mu}(\theta,\phi)\Big\} \tag{6-1}$$

where $R(\theta,\phi;t)$ is the distance from the center of the nucleus to the surface at angles

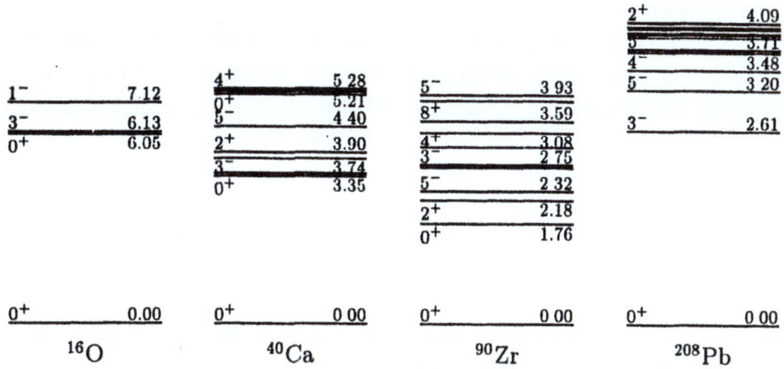

Figure 6-2: Observed low-lying energy level structure of doubly magic nuclei ^{16}O, ^{40}Ca, ^{90}Zr, and ^{208}Pb, showing the location of 0^+ breathing mode and 3^- octupole vibrational states. (Plotted using data from Ref. [95].)

(θ, ϕ) and time t. The equilibrium radius R_0 here is that for a sphere having the same volume. Each mode of order λ has, in general, $2\lambda + 1$ parameters, corresponding to $\mu = -\lambda, -\lambda + 1, \ldots, \lambda$. However, symmetry requirements reduce the number of independent ones to be somewhat smaller. For example, since

$$Y^*_{\lambda\mu}(\theta, \phi) = (-1)^\mu Y_{\lambda,-\mu}(\theta, \phi)$$

it is necessary for

$$\alpha_{\lambda\mu}(t) = (-1)^\mu \alpha_{\lambda,-\mu}(t)$$

to keep $R(\theta, \phi; t)$ real. Furthermore, rotational and other invariance requirements also impose a set of conditions on $\alpha_{\lambda\mu}(t)$. We shall see an example for quadrupole deformation later in §6-3.

The $\lambda = 1$ mode corresponds to an oscillation around some fixed point in the laboratory, as shown in Fig. 6-1(b). If all the nucleons are moving together as a group without any changes in the internal structure of the nucleus, the vibration corresponds to a motion of the center of mass of the nucleus. This is known as the isoscalar ($T = 0$) dipole mode and is of no interest if our wish is to study the internal dynamics of a nucleus. On the other hand, the corresponding isovector ($T = 1$) mode, as we shall see in the next section, corresponds to a dipole oscillation of neutrons and protons in opposite directions, as shown in Fig. 6-1(c). This is the cause of giant dipole resonances observed in a number of nuclei. The $\lambda = 2$ mode describes a quadrupole oscillation of the nucleus. A positive quadrupole deformation means that the nuclear shape is a prolate one, with polar radius longer than equatorial radius. On the other hand, a negative quadrupole deformation is one in which the nucleus has an oblate shape, with equatorial radius longer than polar one. A quadrupole vibration corresponds to the situation that the nucleus changes its shape back and forth, from spherical to prolate, back to spherical and then to oblate, and then back again to spherical, as shown in Fig. 6-1(d). Similarly, an octupole ($\lambda = 3$) vibration is depicted in Fig. 6-1(e).

The energy associated with vibrational motion may be discussed in terms of the variations in the shape parameters $\alpha_{\lambda\mu}(t)$ as functions of time. When a nucleus changes its shape, nucleons are moved from one location to another. This constitutes the kinetic energy in the vibration. At the same time, when a nucleus moves away from its equilibrium shape, the potential energy is increased, the same as a spring is compressed or stretched. Unless constrained, it will return to its lowest potential energy state. The amount of energy involved in each case is related to the nuclear shape and, as a result, the shape parameters become the appropriate canonical variables to describe the motion (rather than, for example, coordinates specifying the position of each nucleon in the nucleus).

For small-amplitude vibrations, the kinetic energy may be expressed in terms of the rate of change in the shape parameters,

$$T = \frac{1}{2}\sum_{\lambda\mu} D_\lambda \left|\frac{d\alpha_{\lambda\mu}}{dt}\right|^2$$

where D_λ is a quantity playing an equivalent role as mass in ordinary (nonrelativistic) kinetic energy in mechanics. For a classical irrotational flow, D_λ is related to the mass density ρ and equilibrium radius R_0 of the nucleus in a liquid drop model,

$$D_\lambda = \frac{\rho R_0^5}{\lambda}$$

Similarly, the potential energy may be expressed as

$$V = \frac{1}{2}\sum_{\lambda\mu} C_\lambda |\alpha_{\lambda\mu}(t)|^2 \tag{6-2}$$

Such a form follows naturally from the fact that we have assumed the equilibrium shape to be spherical and, as a result, the minimum in the potential energy lies at $\alpha_{\lambda\mu}(t) = 0$. In this case, there is no linear dependence of V on $\alpha_{\lambda\mu}(t)$ and the leading order is the quadratic term. For small-amplitude vibrations, terms depending on the higher powers of $\alpha_{\lambda\mu}$ may be ignored and we are led to Eq. (6-2). The quantity C_λ may be related to the surface and Coulomb energies of the fluid in a liquid drop model for the nucleus (see p. 660 of Ref. [35]),

$$C_\lambda = \frac{1}{4\pi}(\lambda-1)(\lambda+2)\,\alpha_2 A^{2/3} - \frac{3}{2\pi}\frac{\lambda-1}{2\lambda+1}\alpha_3\frac{Z(Z-1)}{A^{1/3}}$$

where α_2 and α_3 are the surface and Coulomb energy parameters defined in Eq. (4-56).

In terms of C_λ and D_λ, the Hamiltonian for vibrational excitation of order λ may be written as

$$H_\lambda = \frac{1}{2}C_\lambda\sum_\mu |\alpha_{\lambda\mu}(t)|^2 + \frac{1}{2}D_\lambda\sum_\mu\left|\frac{d\alpha_{\lambda\mu}}{dt}\right|^2 \tag{6-3}$$

If different modes of excitation are decoupled from each other, and with any other degrees of freedom the nucleus may have, H_λ, C_λ, and D_λ are constants of motion. Under these conditions, we can differentiate Eq. (6-3) with respect to time and obtain the equation of motion,

$$D_\lambda\frac{d^2\alpha_{\lambda\mu}}{dt^2} + C_\lambda\alpha_{\lambda\mu}(t) = 0$$

Comparing with the expression for an harmonic oscillator,

$$\frac{d^2x}{dt^2} + \omega^2 x = 0$$

we obtain the result that, for small oscillations, the amplitude $\alpha_{\lambda\mu}(t)$ undergoes harmonic oscillation with frequency

$$\omega_\lambda = \left(\frac{C_\lambda}{D_\lambda}\right)^{1/2}$$

with $\hbar\omega_\lambda$ as a quantum of vibrational energy for multipole λ.

Quadrupole and octupole vibrations. A vibrational quantum of energy is called a *phonon*, as it is a form of "mechanical" energy, reminiscent of the way sound wave propagates through a medium. Each phonon is a boson carrying $\lambda\hbar$ units of angular momentum and parity $\pi = (-1)^\lambda$. Consider the example of vibrations built upon the ground state of an even-even nucleus. In this case, the 0^+ ground state constitutes the zero-phonon state. The lowest vibrational state has $J = \lambda$ and $\pi = (-1)^\lambda$, obtained by coupling the angular momentum of the phonon to that of the ground state. Examples of one-phonon octupole excitations are found in the form of a low-lying 3^- state in all the closed shell nuclei from ^{16}O to ^{208}Pb, as shown earlier in Fig. 6-2. In terms of the single-particle picture discussed in the next chapter, excited states may be produced by promoting, for example, a particle from an occupied orbit below the Fermi surface to an empty one above. Since orbits below and above the Fermi surface near a closed shell have, in general, opposite parities (see §7-2), negative-parity states are formed from such one-particle, one-hole excitations. We shall see later in §7-2 that the typical energy involved in such cases is around $41A^{-1/3}$ mega-electron-volts, about 16 MeV in ^{16}O and 7 MeV in ^{208}Pb. As can be seen in Fig. 6-2, the observed 3^- vibrational states are much lower than this value. One way to lower the excitation energy in this case is to have the nucleons acting in a coherent or "collective" manner.

In nuclei such as the even-even cadmium (Cd) and tin (Sn) isotopes, the first excited state above the $J^\pi = 0^+$ ground state is inevitably a 2^+ state and, at roughly twice the excitation energy, there is often a triplet of states with $J^\pi = 0^+, 2^+, 4^+$. Such behavior is typical of nuclei undergoing quadrupole vibration. The first excited state is the one-phonon state, having $J^\pi = 2^+$ of a quadrupole phonon. The two-phonon states are expected at $2\hbar\omega_\lambda$ in excitation energy, twice that for the one-phonon state. The possible range of spin is from 0 to 4 ($=2\lambda$). However, symmetry requirements between the two identical phonons excludes coupling to 1^+ and 3^+ states (see Problem 6-1), and the only allowed ones are $J^\pi = 0^+, 2^+, 4^+$. If vibration is the only term in the nuclear Hamiltonian, we expect the three two-phonon states to be degenerate in energy. In practice, they are observed to be separated from each other by an amount generally much smaller than $\hbar\omega_\lambda$. We can take this as the evidence that forces in addition to vibration are also playing a role in forming these states. The fact that the order among these three levels is different in different nuclei implies that the nature of the J-dependence may be a complicated one.

With three quadrupole-phonons, there are five allowed levels, 0^+, 2^+, 3^+, 4^+, and 6^+. Since these states lie high in excitation energy, where the density of states is large, admixture with states formed by other excitation modes becomes important. As a result, it is not always easy to identify a complete set of three-phonon excited states. One such example, shown in Fig. 6-3, is found in ^{118}Cd.

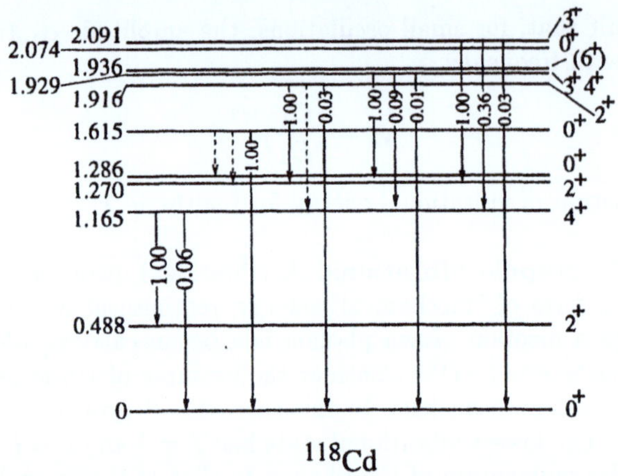

Figure 6-3: Observed low-lying energy levels of ^{118}Cd, showing quadrupole vibrational states up to three-phonon excitations. The spin and parity of the 1.929-MeV state may be either 3^+ or 4^+ and of the 1.936-MeV state, 5^+ or 6^+, with the possibility of 4^+ not ruled out. The 0^+ state at 1.615 MeV may not be a member of the vibrational spectrum. Vertical arrows indicate $B(E2)$ values relative to the observed strongest transition from each state and the dashed lines indicate transitions with only upper limits known. (Based on data from Refs. [8, 79].)

Electromagnetic transitions. Besides energy level positions, the vibrational model also predicts the electromagnetic transition rates between states having different numbers of excitation phonons. Since vibrational states have the same structure as those for an harmonic oscillator, we can make use of the result that the transition from an n-phonon state to an $(n-1)$-phonon state takes place by emitting one quantum of energy. If nuclear vibrations are purely harmonic in nature, the electric transition operator $\boldsymbol{O}_{\lambda\mu}(E\lambda)$ for a vibrational mode of order λ must be proportional to the annihilation operator $\boldsymbol{b}_{\lambda\mu}$ for a phonon of multipolarity (λ, μ),

$$\boldsymbol{O}_{\lambda\mu}(E\lambda) \propto \boldsymbol{b}_{\lambda\mu}$$

Because of its collective nature, nuclear excitations induced by quadrupole vibrations have large $E2$-transition rates between states differing in excitation energy by one phonon, compared with Weisskopf single-particle estimates given in §5-4. Similarly, strong $E3$-transition strengths to the ground states are also observed from octupole vibrational states.

The matrix element of a phonon annihilation operator $\boldsymbol{b}$ between two harmonic oscillator states is given by

$$\langle n'|\boldsymbol{b}|n\rangle = \sqrt{n}\,\delta_{n',n-1}$$

Since the reduced transition probability is proportional to the square of the transition matrix element, we find that its value between n- and $(n-1)$-phonon states is proportional to n, the number of phonons in the initial state of the decay,

$$B(E\lambda, n \rightarrow n-1) \propto n$$

Because of this relation, we expect the transition probability from a two-phonon state to a one-phonon state to be enhanced in comparison with single-particle estimates and roughly twice the value from a one-phonon state to a zero-phonon state in the same nucleus. Transitions between states differing by more than one phonon are higher in order, as they involve simultaneous emission of two or more phonons. The probability for such processes is much lower than that for single-phonon emission, and the corresponding transition rates are expected to be small. Both points are observed to be essentially correct in vibrational nuclei, as can be seen from the examples given in Table 6-1.

Table 6-1: Quadrupole moment and $B(E2)$ values of vibrational nuclei.

Nucleus	$B(E2;2_1^+\rightarrow 0_1^+)$ $10^2\,e^2\text{fm}^4$	$B(E2;2_1^+\rightarrow 0_1^+)$ W.u.	$B(E2;4_1^+\rightarrow 2_1^+)$ $10^2\,e^2\text{fm}^4$	$B(E2;4_1^+\rightarrow 2_1^+)$ W.u.	$\frac{B(E2;4^+\rightarrow 2_1^+)}{B(E2;2_1^+\rightarrow 0^+)}$	$B(E2;2_2^+\rightarrow 0_1^+)$ $10^2 e^2\text{fm}^4$	$B(E2;2_2^+\rightarrow 0_1^+)$ W.u.	Q $e\text{fm}^2$
^{60}Ni	1.88	13.5	2.3	16	1.2	0.03	0.22	3.0
^{62}Ni	1.8	12	2.6	18	1.5	0.09	0.6	8.8
^{102}Ru	12.4	43.9	19	66	1.5	0.31	1.09	−68
^{110}Cd	8.58	27.4	14	46	1.7	0.42	1.34	−39
^{112}Cd	9.69	30.2	20.0	62.4	2.1	0.21	0.65	−37
^{114}Cd	10	31	19	58	1.9	1.8	5.4	−36
^{116}Cd	10.6	31.6	19.4	57.8	1.8	0.37	1.1	−42
Vibrational model	large		large		2.0	small		0*

Note: W.u.=Weisskopf unit. *Spherical nuclei.

Implicit in our discussion is the assumption that the vibration is an axially symmetric one; i.e., variations along the x- and y-directions are equal to each other, only their ratio to that along the z-axis is changing as a function of time. This type of vibration is generally known as β-vibration. More generally, we can also have γ-vibrations, in which the nucleus changes into an ellipsoidal shape in the equatorial direction. In other words, a section of the nucleus in the xy-plane at any instant of time is an ellipse rather than a circle, as in the case of β-vibration. (The definitions for parameters β and γ are given later in Eq. 6-11.) In addition to purely harmonic vibrational motion, anharmonic terms may be present in a nucleus. Furthermore, vibrations may also be coupled to other modes of excitation in realistic situations.

If the amplitude of vibration is large, the above treatment no longer applies. In fact, if the vibration is energetic enough, a "drop" of nuclear matter may dissociate into two or more droplets. Such ideas are used with success in fission studies. However, in order for a nucleus to develop toward a shape for splitting into two or more fragments, there must be a superposition of many different vibrational modes. Furthermore, the various modes must be strongly coupled to each other so that energy can flow from one mode to another. The mathematical problem involved here is not simple, but the basic physical idea is a sound one. However, we shall not examine this topic here.

6-2 Giant Resonance

Giant resonance is a term used to describe the observed concentration of excitation strength at energies tens of mega-electron-volts above the ground state. Both the total values and distribution widths are found to be much larger than typical resonances built upon single-particle (noncollective) excitations. In the energy region where such resonances appear, the density of states is sufficiently high and the number of open decay channels sufficiently large that states in a narrow energy region cannot be very different from each other in character. As a result, only smooth variations are expected in the reaction cross sections, as can be seen from the example of the ^{208}Pb$(p,p')^{208}$Pb* reaction shown in Fig. 6-4. The concentration of strength localized in the region of a few mega-electron-volts is interesting, as it must be related to some special feature of the nuclear system particular to the energy region.

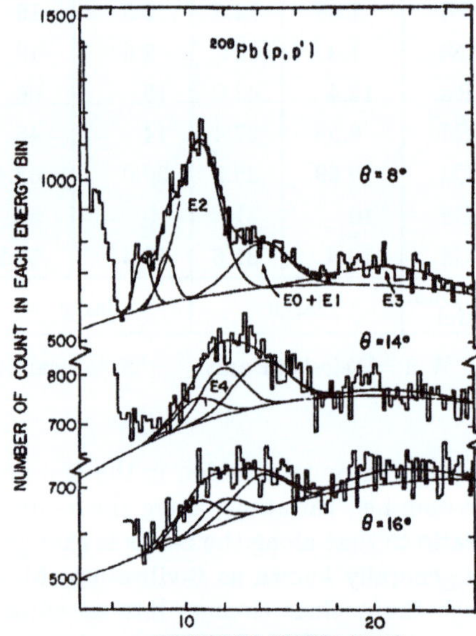

Figure 6-4: Differential cross section of ^{208}Pb(p,p') reaction with 200-MeV protons at different scattering angles, showing the angular dependence of giant resonances excited in the reaction. (Taken from Ref. [28].)

For most giant resonances, the strength is found to be essentially independent of the probe used to excite the nucleus, γ-rays, electrons, protons, α-particles, or heavy ions. Furthermore, both the width and peak of strength distribution vary smoothly with nucleon number A, without any significant dependence on the structure of the individual nucleus involved. For example, the location of the isovector giant dipole resonance in different nuclei is well described by the relation

$$E_1 \approx 78A^{-1/3} \tag{6-4}$$

Prominent dipole resonances, as well as other types of giant resonances, have been observed in almost all the nuclei studied, from ^{16}O to ^{208}Pb, as can be seen later in Figs. 6-5 and 6-6.

Giant resonances come from collective excitation of nucleons. As we shall see later in §7-2, the energy gap between two adjacent major shells, is roughly $41A^{-1/3}$ mega-electron-volts and the parity of states produced by $1p1h$-excitations up one major shell is negative in general. To a first approximation, this is the cause of negative-parity giant resonances. For positive-parity excitations there are two possibilities, rearranging the particles in the same major shell ($0\hbar\omega$-excitation) or elevating a particle up two major shells ($2\hbar\omega$-excitation). Other possibilities, such as excitations by four major shells ($4\hbar\omega$-excitation) for positive-parity resonances and three major shells ($3\hbar\omega$-excitation) for negative-parity resonances are less likely because of the higher energies involved.

Giant dipole resonance. Isovector giant dipole resonances have been studied since the late 1940s. They are the $J^\pi = 1^-$ excitation strength when nucleons are promoted up one major shell. In light nuclei, the observed peaks of strength occur around 25 MeV in energy and, in heavy nuclei, the values are lower, just below 14 MeV in ^{208}Pb. The variation with nucleon number A, as can be seen in Fig. 6-5(a), is fairly well described by the relation given by Eq. (6-4). The peak position is higher than that expected of a simple $1\hbar\omega$-excitation process of $41A^{-1/3}$ mega-electron-volts. The difference is caused by the residual interaction between nucleons which pushes isovector excitations to higher energies. The width of the resonance is found to be around 6 MeV without any noticeable dependence on the nucleon number, as can be seen in Fig. 6-5(b).

An explanation of giant dipole resonance is provided by the Goldhaber-Teller model, based on the collective motion of nucleons. Here, neutrons and protons act as two

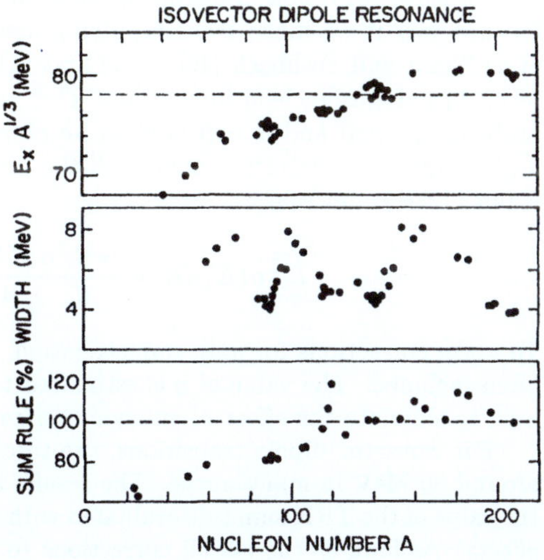

Figure 6-5: Variations of the observed peak location (a), width (b), and total strength (c) of isovector giant dipole resonance as functions of nucleon number. Dashed line in (c) is the value of the Thomas-Reiche-Kuhn (TRK) sum rule with $\eta = 0$. (Taken from Ref. [27].)

separate groups of particles and excitation comes from the motion of one group with respect to the other, with little or no excitations within each group. In the dipole mode, the neutrons are moving in one direction along some axis while the protons are going in the opposite direction, as shown schematically in Fig. 6-1(*c*). The opposite phase keeps the center of mass of the entire nucleus stationary. Since neutrons and protons are moving "out of phase" with respect to each other, it is an isovector mode of excitation. In contrast, if the neutrons and protons move in phase, it is an isoscalar dipole vibration, with all the nucleons moving in the same direction at any given time. The net result, in this case, is that the entire nucleus is oscillating around some equilibrium position in the laboratory. Such a motion constitutes a "spurious" state and is of no interest to the study of the nucleus, as it does not correspond to an excited state of the nucleus involving nuclear degrees of freedom.

Sum rule quantities. One question of interest in giant resonance studies is to find the fraction of total transition strength represented by the observed cross section. The amount may be estimated by calculating the corresponding sum rule quantity. The simplest one is the transition strength of a given multipolarity to all the possible final states. The starting state is usually chosen to be the ground state, as this is the only type that can be measured directly. The *non-energy-weighted* sum of the reaction cross section is then

$$S = \int_0^\infty \sigma(E)\, dE \tag{6-5}$$

where $\sigma(E)$ is the cross section at excitation energy E. Since an integration is carried out over all the final states, the resulting quantity is a function of the initial state only. For transitions originating from the ground state, S is the ground expectation value of an operator related to the transition. An example is given later for the case of Gamow-Teller giant resonance. Other sum rule quantities, such as energy-weighted ones, have also been studied; we shall, however, restrict ourselves to the simplest one defined in Eq. (6-5).

For isovector dipole transitions, the total strength S can be evaluated in a straightforward way if we make two simplifying assumptions (see, for example, pp. 709–713 of de Shalit and Feshbach [49]). The first is to ignore any possible velocity-dependent terms in nucleon-nucleon interaction. This has been done in a variety of other nuclear problems as well and is expected to be of very little consequence. The second is to neglect antisymmetrization among all the nucleons. The result is the Thomas-Reiche-Kuhn (TRK) sum rule,

$$\int_0^\infty \sigma(E)\, dE = \frac{2\pi^2\hbar^2\alpha}{M_p}\frac{NZ}{A} \approx 6.0\frac{NZ}{A}\ \text{MeV-fm}^2$$

To make corrections for antisymmetrization, an overall multiplicative factor $(1+\eta)$ is often included. The value of η is estimated to be around 0.5, depending on the model used to simulate the effect of antisymmetrization.

For isovector dipole transitions, the total strength is known experimentally up to around 30 MeV in many nuclei. The results are compared in Fig. 6-5(*c*) together with the value of the TRK sum rule evaluated with $\eta = 0$, i.e., no correction for antisymmetry effects. As long as the actual corrections to the TRK sum rules are not too different from the generally accepted value of $\eta \sim 0.5$, we see that the measured giant dipole cross sections exhaust most of the total possible strengths. Furthermore, the result is essentially independent of the particular nucleus from which the strength sum is taken. The large variety of nuclei included in Fig. 6-5 represents a wide spectrum of ground state wave functions. The fact that the value of S is essentially given by the TRK sum rule, without any specific reference to the ground state wave function of any of the nuclei involved, may be taken as another evidence of the collective nature of the excitation process itself.

Besides isovector dipole excitations, other giant resonances have also been observed in recent years. Both giant quadrupole ($E2$) and giant octupole ($E3$) resonances have been extensively studied in a variety of nuclei. The results for the former are shown in Fig. 6-6 as an example.

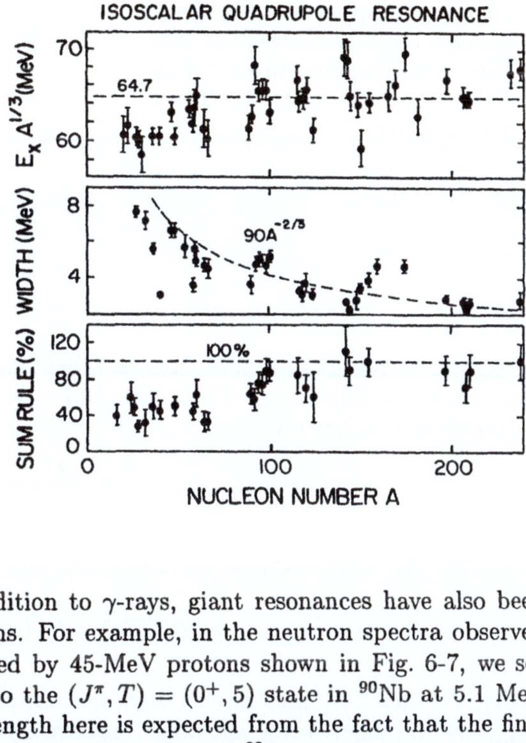

Figure 6-6: Variations of the observed peak location (a), width (b), and total strength (c) of isoscalar giant quadrupole resonances as functions of nucleon number. (Taken from Ref. [27].)

Gamow-Teller resonance. In addition to γ-rays, giant resonances have also been observed in charge exchange reactions. For example, in the neutron spectra observed in the ^{90}Zr$(p,n)^{90}$Nb reaction induced by 45-MeV protons shown in Fig. 6-7, we see that a sharp peak is found leading to the $(J^\pi, T) = (0^+, 5)$ state in ^{90}Nb at 5.1 MeV excitation. The concentration of strength here is expected from the fact that the final state in ^{90}Nb is the isobaric analogue to the ground state of ^{90}Zr. The operator involved in the reaction is similar to that in Fermi β-decay, namely, the isospin-raising operator $\boldsymbol{T}_+$. However, since the strength is concentrated in a single state, the distribution is essentially a delta function. The Fermi type of charge exchange strength, therefore, does not fit into the category of a giant resonance.

Unlike the Fermi case, the Gamow-Teller strength is shared by a number of states. However, in β-decay, the transition is allowed only if the initial state is higher in energy than the final state. As a result, only a small part of the total strength is actually observed. The main portion usually lies higher in excitation energy and is observed in charge exchange reactions. For example, in the case of the $^{90}Zr(p,n)^{90}Nb$ reaction, part of the strength appears as a "giant resonance" in the neutron spectra, as shown in Fig. 6-7, at energies just below the isobaric analogue strength peak.

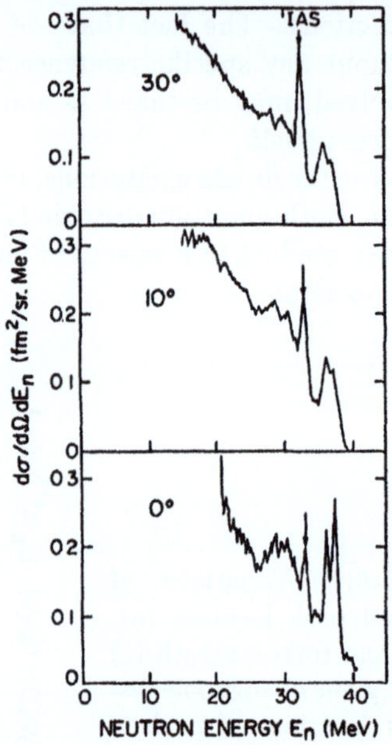

Figure 6-7: Neutron spectra at different scattering angles from the reaction $^{90}Zr(p,n)^{90}Nb$ induced by 45-MeV protons. The results give the angular dependence of the giant Gamow-Teller resonance and isobaric analogue strength excited in a charge exchange reaction. (Taken from Ref. [71].)

Let us evaluate the sum of Gamow-Teller transition strength in a charge exchange reaction as an illustration. From Eq. (5-61), we find that the operator for the axial-vector transition has the form

$$\boldsymbol{O}_{\rm GT}(\beta^\pm) = G_A \sum_{k=1}^{A} \sum_{\mu} \boldsymbol{\sigma}_\mu(k) \boldsymbol{\tau}_\mp(k)$$

Following Eq. (6-5), we may define the sum rule strength in the following way:

$$S_\pm = G_A^{-2} \sum_f |\langle f|\boldsymbol{O}_{\rm GT}(\beta^\pm)|i\rangle|^2$$

where $|i\rangle$ and $|f\rangle$ are, respectively, the initial and final nuclear states. We have removed the axial-vector coupling constant G_A from the definition of the operator itself so as to simplify the appearance of the final result. Since we are summing over all the final states, $S_\pm$ may be transformed into an expectation value using a closure relation,

$$\begin{aligned} S_\pm &= G_A^{-2} \sum_f \langle f|\boldsymbol{O}_{\rm GT}(\beta^\pm)|i\rangle^* \langle f|\boldsymbol{O}_{\rm GT}(\beta^\pm)|i\rangle \\ &= G_A^{-2} \sum_f \langle i|\boldsymbol{O}^\dagger_{\rm GT}(\beta^\pm)|f\rangle \langle f|\boldsymbol{O}_{\rm GT}(\beta^\pm)|i\rangle \\ &= G_A^{-2} \langle i|\boldsymbol{O}^\dagger_{\rm GT}(\beta^\pm)\boldsymbol{O}_{\rm GT}(\beta^\pm)|i\rangle \end{aligned} \tag{6-6}$$

Components of the operators involved here have the following properties:

$$\sigma_\mu^\dagger = (-1)^\mu \sigma_{-\mu} \qquad\qquad \tau_\mp^\dagger = \tau_\pm$$

as can be seen for σ_μ in Eq. (3-31) and for $\tau_\pm$ in Eq. (2-20). On substituting the explicit form of the Gamow-Teller operator into Eq. (6-6), we obtain the strength sum for β^+-transitions,

$$\begin{aligned} S_+ &= \langle i|\sum_{k=1}^{A}\sum_{\mu}(-1)^\mu \sigma_{-\mu}(k)\tau_+(k)\sigma_\mu(k)\tau_-(k)|i\rangle \\ &= \langle i|\sum_{k=1}^{A}\sigma^2(k)\tau_+(k)\tau_-(k)|i\rangle \end{aligned}$$

where we have made use of Eq. (A-19) to obtain σ^2 from $\sum(-1)^\mu\sigma_{-\mu}\sigma_\mu$. Similarly, for β^--transitions,

$$S_- = \langle i|\sum_{k=1}^{A}\sigma^2(k)\tau_-(k)\tau_+(k)|i\rangle \tag{6-7}$$

Since $\tau_+|n\rangle = |p\rangle$, $\tau_-|p\rangle = |n\rangle$, and $\tau_+|p\rangle = \tau_-|n\rangle = 0$, where $|p\rangle$ is the wave function of a proton and $|n\rangle$ is that for a neutron, we have the results

$$\begin{aligned} \tau_+\tau_-|p\rangle &= |p\rangle \qquad\qquad \tau_+\tau_-|n\rangle = 0 \\ \tau_-\tau_+|n\rangle &= |n\rangle \qquad\qquad \tau_-\tau_+|p\rangle = 0 \end{aligned}$$

In other words, we can treat $\tau_+\tau_-$ as the projection operator for protons and $\tau_-\tau_+$ as the corresponding quantity for neutrons.

Using these results, we can write

$$S_+ = \langle i|\sum_{k=1}^{Z}\sigma^2(k)|i\rangle = 3Z \tag{6-8}$$

where Z is the number of protons in the initial state. The summation is restricted to protons in the target nucleus, because of the projection operator $\tau_+\tau_-$. In obtaining the final result, we have made use of the fact that, for a single nucleon, $\boldsymbol{s} = \frac{1}{2}\boldsymbol{\sigma}$ and the expectation value of σ^2 is 3. By the same token, Eq. (6-7) can be simplified to

$$S_- = 3N \tag{6-9}$$

where N is the neutron number for the target.

Equations (6-8) and (6-9) are not very useful sum rules, as they represent, respectively, the total strength if all the protons and all the neutrons are excited by the reaction. Such processes involve extremely high energy components and cannot be achieved in practice. Experimentally, only nucleons near the Fermi surface are affected, and there is no easy way to estimate the numbers of such nucleons. However, the difference between the two sum rules

$$S_- - S_+ = 3(N - Z) \tag{6-10}$$

may not depend on how high in energy the excitation strengths are measured and may therefore be tested against observations.

A departure from Eq. (6-10) may also indicate the presence of particles other than nucleons in the nucleus, such as Δ-particles, resulting from exciting the internal degrees of freedom in nucleons. Such a component in the intermediate state has been conjectured as a possibility in many other reactions. For this reason, there is a great amount of interest in measuring the difference in strength between (p,n) and (n,p) reactions. However, the experiments are difficult to carry out and, at this moment, the results are still too preliminary to draw any conclusion.

The strength of Gamow-Teller excitation is related to the spin-isospin term in the nucleon-nucleon interaction, $V_{\sigma\tau}(r)\boldsymbol{\sigma}(1)\cdot\boldsymbol{\sigma}(2)\boldsymbol{\tau}(1)\cdot\boldsymbol{\tau}(2)$. A good knowledge of the giant Gamow-Teller resonance will therefore also shed light on this important term in the interaction between nucleons inside a nucleus. The same is true of other giant resonances as well, as each may be shown to be dependent predominantly on a particular term in the interaction.

6-3 Rotational Model

Deformation. In the previous two sections we have assumed, for the convenience of discussion, that the basic shape of a nucleus is spherical and excitations are built upon such an equilibrium configuration in the form of small vibrations. There is no compelling reason why the nuclear shape cannot be different. The interplay between short-range nuclear force, long-range repulsive Coulomb force, and centrifugal stretching due to rotation may well favor a nonspherical or *deformed* equilibrium shape.

In general, spherical nuclei are found around closed shells. This is easy to understand. As we shall see later in §7-2, the single-particle spectrum for nucleons is not uniform. Instead, the states are separated into groups, with energy differences between states within a group smaller than those between groups. This makes it more favorable for nucleons to fill up each group, or *shells*, before occupying those in the next one. A closed shell nucleus is formed when all the single-particle states in a group are fully occupied. When this condition is met, the total M-value, the projection of spin along the quantization axis, of the nuclear state is zero. Such an object is then invariant under a rotation of the coordinate system and must, therefore, be spherical in shape.

On the other hand, for nuclei in regions between closed shells, many single-particle states are available. In this case, it may be more favorable for a nucleus to minimize its energy by going to a deformed shape. In general, the nuclear shape tends to be prolate, i.e., elongated along the z-axis, at the beginning of a major shell and oblate, i.e., flattened at the poles, toward the end. This comes from a preference, arising from the pairing term in nuclear force, for nucleons to occupy single-particle states with the largest absolute m-values, starting from $m = \pm j$. As a result, there is an increase in the probability at the beginning of a shell to find nucleons in the polar regions. For example, among the light nuclei in the ds-shell, we find that the deformation is positive for ^{19}Ne and ^{19}Na, with three nucleons outside the closed shell at ^{16}O. At the middle of the major shell, around ^{28}Si, the deformation changes sign, as can be seen from the negative quadrupole moment for most of the nuclei in the ds-shell with $A > 28$.

For stable nuclei, departure from spherical equilibrium shape is generally small in the ground state region. Relatively large deformations are found, for example, in medium-heavy nuclei with $150 \lesssim A \lesssim 180$ and heavy nuclei with $220 \lesssim A \lesssim 250$, as shown in Fig. 6-8. The largest deformations, or "superdeformations," as we shall see later in §9-2, are observed in the excited configurations of medium-weight nuclei, created when two heavy ions are fused together into a single entity.

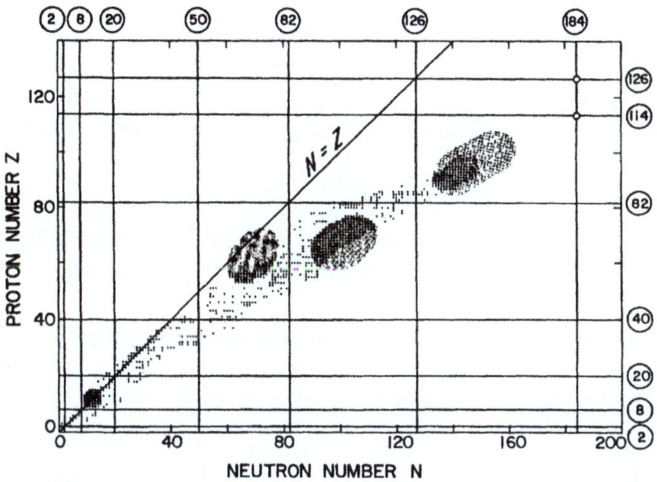

Figure 6-8: Regions of deformation. Deformed nuclei, indicated by the shaded areas, lie in regions between closed shells and among very heavy nuclei beyond $^{208}_{82}$Pb.

Quantum mechanically, there cannot be a rotational degree of freedom associated with a spherical object. For a sphere, the square of its wave function is, by definition, independent of angles—it appears to be the same from all directions. As a result, there is no way to distinguish the wave functions before and after a rotation. Rotation is therefore not a quantity that can be observed in this case and, consequently, cannot correspond to a degree of freedom in the system with energy associated with it. In contrast, rotational motion of a deformed object, such as an ellipsoid, may be detected, for example, by observing the changes in the orientation of the axis of symmetry with time.

Quadrupole deformation and Hill-Wheeler variables. The simplest and most commonly occurring type of deformation in nuclei is quadrupole. To simplify the discussion, let us assume that the nuclear density is constant throughout the volume and drops off sharply to zero at the surface. In this case, the surface radius $R(\theta, \phi; t)$ of Eq. (6-1) reduces to

$$R(\theta, \phi; t) = R_0 \left\{ 1 + \sum_{\mu=-2}^{2} \alpha_{2\mu}(t) Y_{2\mu}(\theta, \phi) \right\}$$

There are five shape parameters, $\alpha_{2\mu}(t)$ for $\mu = -2$ to $\mu = +2$.

The orientation of a nucleus in space is specified by three parameters, for example, the Euler angles $(\omega_\alpha, \omega_\beta, \omega_\gamma)$. Since the orientation is immaterial, as far as the intrinsic nuclear shape is concerned, we can regard the Euler angles as three "conditions" to be imposed on the five parameters. This may be expressed formally by transforming the coordinate system to one fixed with the nucleus,

$$\alpha_{\lambda\mu} = \sum_{\nu=-2}^{2} a_{\mu\nu} \mathcal{D}^J_{MM'}(\omega_\alpha, \omega_\beta, \omega_\gamma)$$

where $\mathcal{D}^J_{MM'}(\omega_\alpha, \omega_\beta, \omega_\gamma)$ is the rotation matrix defined by Eq. (A-5). Since there are only two degrees of freedom left, the body-fixed shape parameters $a_{\mu\nu}$ have the following properties

$$a_{2,-1} = a_{2,1} = 0 \qquad\qquad a_{2,-2} = a_{2,2}$$

Instead of $a_{2,0}$ and $a_{2,2}$, the two parameters remaining, it is common practice to use the Hill-Wheeler variables β and γ. They are defined by the relations

$$a_{20} = \beta \cos\gamma \qquad\qquad a_{2,-2} = a_{22} = \frac{\beta}{\sqrt{2}} \sin\gamma \tag{6-11}$$

Using β and γ, the surface radius may be written as

$$R(\theta, \phi) = R_0\Big\{1 + \beta\sqrt{\frac{5}{16\pi}}\Big(\cos\gamma(3\cos^2\theta - 1) + \sqrt{3}\sin\gamma\sin^2\theta\cos 2\phi\Big)\Big\} \tag{6-12}$$

Since we are mostly interested in fixed, permanently deformed shapes here, the surface radius in the body-fixed coordinate system is independent of time. The same is true for parameters β and α. From the definitions given by Eq. (6-11), we see that the parameter β provides a measure of the extent of deformation and γ, the departure from axial symmetry. A negative value of β indicates that the nucleus is oblate in shape while a positive value describes a prolate shape. This is illustrated in Fig. 6-9 for the axially symmetric case ($\gamma = 0$).

Figure 6-9: Quadrupole-deformed shapes for axially symmetric nuclei. On the left, the oblate shape has $\beta = -0.4$, and on the right, the prolate shape has $\beta = +0.4$.

We have two different sets of coordinate systems here. The intrinsic coordinate system, with frame of reference fixed to the rotating body, is convenient for describing the symmetry of the object itself. On the other hand, the nucleus is rotating in the laboratory and the motion is more conveniently described by a coordinate system that is fixed in the laboratory. Each system is better suited for a different purpose, and we shall make use of both of them in our discussions. Following general convention, the intrinsic coordinate axes are labeled by subscripts 1, 2, and 3 to distinguish them from the laboratory coordinates, labeled by subscripts x, y, and z.

We can also see from Eq. (6-12) that there is a certain degree of redundancy in the values of β and γ. For example, with positive values of β, we have prolate shapes for $\gamma = 0°, 120°, 240°$. However, the symmetry axis is a different one in each case: 3 for $\gamma = 0°$, 1 for $\gamma = 120°$, and 2 for $\gamma = 240°$. Similarly, the corresponding oblate shapes are found for $\gamma = 180°, 300°, 60°$. For this reason, most people follow the (Lund) convention in which $\beta \geq 0$ and $0° \leq \gamma \leq 60°$ if the rotation is around the smallest axis. If the rotation is around the largest axis, $-120° \leq \gamma \leq -60°$, and if around the intermediate axis, $-60° \leq \gamma \leq 0°$.

Rotational Hamiltonian. Classically, the angular momentum $\boldsymbol{J}$ of a rotating object is proportional to its angular velocity $\boldsymbol{\omega}$,

$$\boldsymbol{J} = \mathcal{I}\boldsymbol{\omega} \tag{6-13}$$

The ratio between $\boldsymbol{J}$ and $\boldsymbol{\omega}$ is the moment of inertia $\mathcal{I}$. The rotational energy E_J is given by the square of the angular frequency and is proportional to $\boldsymbol{J}^2$ as a result,

$$E_J = \frac{1}{2}\mathcal{I}\omega^2 = \frac{1}{2\mathcal{I}}J^2$$

By analogy, we can write the rotational Hamiltonian in quantum mechanics as

$$H = \sum_{i=1}^{3} \frac{\hbar^2}{2\mathcal{I}_i} J_i^2$$

where $\mathcal{I}_i$ is the moment of inertia along the ith axis. For an axially symmetric object with 3 as the symmetry axis, the moment of inertia along a body-fixed, or *intrinsic,* set of coordinate axes 1, 2, and 3 has the property

$$\mathcal{I}_1 = \mathcal{I}_2 \equiv \mathcal{I}$$

(and $\mathcal{I}_3 \neq \mathcal{I}$, or else it is spherical). The Hamiltonian in this case may be written as

$$H = \frac{\hbar^2}{2\mathcal{I}}(\boldsymbol{J}^2 - \boldsymbol{J}_3^2) + \frac{\hbar^2}{2\mathcal{I}_3}\boldsymbol{J}_3^2 \tag{6-14}$$

If we use K to represent the projection of $\boldsymbol{J}$ along the symmetry axis in the intrinsic frame, the expectation value of the Hamiltonian in the body-fixed system is then a function of $J(J+1)$, the expectation value of $\boldsymbol{J}^2$, and K, that of $\boldsymbol{J}_3$.

In classical mechanics, a rotating body requires three Euler angles (α, β, γ) to specify its orientation in space. In quantum mechanics, the analogous quantities may be taken as three independent labels, or quantum numbers, describing the rotational state. For two of these three labels we can use the constants of motion J, related to the eigenvalue of $\boldsymbol{J}^2$, and M, the projection of $\boldsymbol{J}$ along the quantization axis in the laboratory. For the third label, we can use K.

Rotational wave function. For the convenience of discussing rotational motion, we shall divide the wave function of a nuclear state into two parts, an intrinsic part describing the shape and other properties pertaining to the structure of the state and a rotational part describing the motion of the nucleus as a whole in the laboratory. Our main concern for the moment is in the rotational part, labeled by J, M, and K. Since it is a function of the Euler angles only, it must be given by $\mathcal{D}^J_{MK}(\alpha,\beta,\gamma)$ of Eq. (A-8), which relates the wave functions of an object in two coordinate systems rotated with respect to each other by Euler angles (α,β,γ). In terms of spherical harmonics, the function $\mathcal{D}^J_{MK}(\alpha,\beta,\gamma)$ may be defined by the relation

$$Y_{JK}(\theta',\phi') = \sum_M \mathcal{D}^J_{MK}(\alpha,\beta,\gamma)\, Y_{JM}(\theta,\phi)$$

where $Y_{JK}(\theta',\phi')$ are spherical harmonics of order J in a coordinate system rotated by Euler angles α,β,γ with respect to the unprimed system.

The transformation property of the $\mathcal{D}$-function under an inversion of the coordinate system (i.e., parity transformation) is given by

$$\mathcal{D}^J_{MK}(\alpha,\beta,\gamma) \xrightarrow[P]{} (-1)^{J+K}\mathcal{D}^J_{M-K}(\alpha,\beta,\gamma)$$

An arbitrary $\mathcal{D}$-function, therefore, does not have a definite parity since, in addition to the phase factor, the sign of label K is also changed. To construct a wave function of definite parity, a linear combination of $\mathcal{D}$-functions, with both positive and negative K, is required. As a result, the rotational wave function takes on the form

$$|JMK\rangle_{\text{rot.}} = \sqrt{\frac{2J+1}{16\pi^2(1+\delta_{K0})}}\left\{\mathcal{D}^J_{MK}(\alpha,\beta,\gamma) \pm (-1)^{J+K}\mathcal{D}^J_{M-K}(\alpha,\beta,\gamma)\right\} \tag{6-15}$$

where the plus sign is for positive parity and the minus sign for negative parity. Since both $+K$ and $-K$ appear on the right-hand side of Eq. (6-15), only $K \geq 0$ can be used to label a rotational wave function. The value K itself is no longer a good quantum number, but the absolute value of K remains a constant of motion for axially symmetric nuclei. In the more general tri-axial case with $\mathcal{I}_1 \neq \mathcal{I}_2 \neq \mathcal{I}_3$, a linear combination of $|JMK\rangle$ with different K values is required to describe nuclear rotation. In such cases, only J and M remain as good quantum numbers.

To complete the wave function for an observed nuclear state, we must also give the intrinsic part. Depending on the energy and other parameters involved, a nucleus can take on different shapes, and as a result, there can be more than one rotational band, each described by a different intrinsic wave function, in a nucleus. For the axial symmetric case, the constant of motion K is often used as a label to identify a particular intrinsic state.

Rotational band. A nucleus in a given intrinsic state can rotate with different angular velocities in the laboratory. A group of states, each with a different total angular momentum J but sharing the same intrinsic state, forms a rotational band. Since the only difference between these states is in their rotational motion, members of a band are related to each other in energy, static moments, and electromagnetic transition rates. In fact, a rotational band is identified by these relations.

The parity of a rotational state is given by Eq. (6-15). Because of the phase factor, the wave function for a positive-parity $K = 0$ state vanishes if the J-value is odd. As a result, only states with even J-values are allowed for a $K = 0^+$ band. Similarly, there are no states with even J-values in a $K = 0^-$ band. The results may summarized as

$$J = \begin{cases} 0,\ 2,\ 4, \ldots & \text{for } K^\pi = 0^+ \\ 1,\ 3,\ 5, \ldots & \text{for } K^\pi = 0^- \end{cases}$$

For $K > 0$, the only restriction on the allowed spin in a band is $J \geq K$, arising from the fact that K is the projection of J on the body-fixed quantization axis, the 3-axis. The possible spins are then

$$J = K,\ K+1,\ K+2, \ldots \quad \text{for} \quad K > 0$$

For the rotational Hamiltonian given in Eq. (6-14), the energy of a state is given by

$$E_J = \frac{\hbar^2}{2\mathcal{I}} J(J+1) + E_K \tag{6-16}$$

where E_K represents contributions from the intrinsic part of the wave function. An example of such a band is shown in Fig. 6-10 for ^{170}Hf.

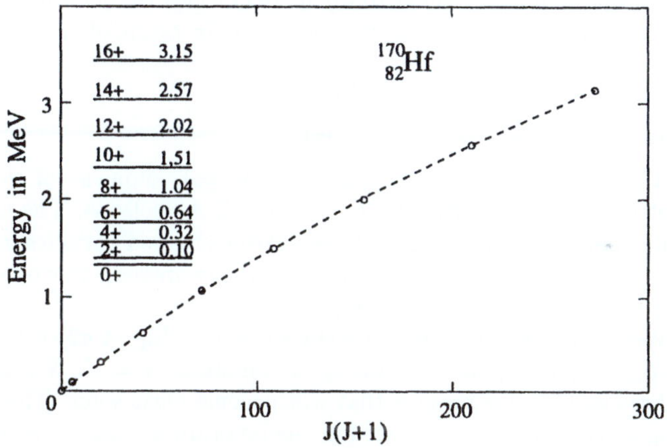

Figure 6-10: Rotational levels in $^{170}_{72}$Hf. For a simple rotor, the relation between E_J and $J(J+1)$ is a curve with constant slope. The small curvature found in the plot indicates that $\mathcal{I}$ increases slightly with large J, a result of centrifugal stretching of the nucleus with increasing angular velocity. (Plotted using data from Ref. [95].)

From Eq. (6-16) we see that the energy of a member of a rotational band is proportional to $J(J+1)$, with the constant of proportionality related to the momentum of inertia $\mathcal{I}$. The quantity E_K enters only in the location of the band head, the position where the band starts. Different bands are distinguished by their moments of inertia and by the positions of their band heads. Both features, in turn, depend on the structure of the intrinsic state assumed by the deformed nucleus that is rotating in the laboratory frame of reference.

Quadrupole moment. Besides energy level positions, the static moments of members of a band and the transition rates between them are also given by the rotational model. The discussions below depend on the property that all members share the same intrinsic state and differ only in their rotational motion. Let us start with the quadrupole moment given by the integral,

$$Q_0 = \int (3z^2 - r^2)\rho(\boldsymbol{r})\, dV \tag{6-17}$$

where $\rho(\boldsymbol{r})$ is the nuclear density distribution. Since it is related to the shape of the intrinsic state, Q_0 is known as the intrinsic quadrupole moment. For an axially symmetric object, it is related to the difference in the polar and equatorial radii, characterized by the parameter

$$\delta = \frac{3}{2}\frac{R_3^2 - R_\perp^2}{R_3^2 + 2R_\perp^2} \approx \frac{R_3 - R_\perp}{R} \tag{6-18}$$

where R_3 is the radius of the nucleus along the body-fixed symmetric (3-) axis, $R_\perp$ is the radius in the direction perpendicular to it, and R is the mean value. To the lowest order, δ is approximately equal to $3\sqrt{5/(16\pi)}$ times the parameter β defined in Eq. (6-11) for small, axially symmetric deformations. In terms of δ,

$$Q_0 = \frac{4}{3}\Big\langle \sum_{i=1}^{A} r_i^2 \Big\rangle \delta$$

The quantity Q_0 defined here is the "mass" quadrupole moment of the nucleus, as the density distribution $\rho(\boldsymbol{r})$ in Eq. (6-17) involves all A nucleons. The usual quantity measured in an experiment, for example, by scattering charged particles from a nucleus, is the "charge" quadrupole moment, differing from the expression above by the fact that the summation is restricted to protons only.

The observed quadrupole moment of a state given by Eq. (4-42) is the expectation value of the electric quadrupole operator $\boldsymbol{Q}$ in the state $M = J$. We shall represent this quantity here as Q_{JK} for reasons that will become clear soon. The value of Q_{JK} differs from Q_0, as the former is measured in the laboratory frame of reference and the latter in the body-fixed frame. The relation between them is given by a transformation from the intrinsic coordinate system to the laboratory system. Since this requires a $\mathcal{D}$-function, the result depends on both J and K. Inserting the explicit value of the $\mathcal{D}$-function for the $M = J$ case, we obtain the relation

$$Q_{JK} = \frac{3K^2 - J(J+1)}{(J+1)(2J+3)} Q_0 \tag{6-19}$$

In practice, direct measurements of quadrupole moments are possible, in most cases, only for the ground state of nuclei. For excited states, the quadrupole moment can sometimes be deduced indirectly through reactions such as Coulomb excitation (see §8-1).

To compare the values calculated using Eq. (6-19) with experimental data, we need a knowledge of the intrinsic quadrupole moment Q_0 as well as the value of K for the band. The latter may be found from the minimum J-value for the band. For Q_0, one way is to make use of the measured value of Q_{JK} for another member. If the values deduced in this way are available for several members of a band, they can be also used as a consistency check of the model. Unfortunately, it is difficult to measure the quadrupole moment for more than one member of a band. The alternative is make use of electric quadrupole transition rates, as we shall see next.

Electromagnetic transitions. In the rotational model, electromagnetic transitions between two members of a band can take place by a change in the rotational frequency and, hence, the spin J, without any modifications to the intrinsic state. We shall concentrate here on electric quadrupole ($E2$) and magnetic dipole ($M1$) transitions, as these are the most commonly observed intraband transitions. A change in the rotational frequency in such cases is described by the angular momentum recoupling coefficient. There are three angular momenta involved, the spin of the initial state J_i, the spin of the final state J_f, and the angular momentum rank of the transition operator λ. The recoupling is given by Clebsch-Gordan coefficients (see §A-3). For quadrupole deformations, the size of the $E2$-transition matrix element is also related to the deformation of the intrinsic state, characterized by Q_0. The reduced transition probability is given by

$$B(E2; J_i \to J_f) = \frac{5}{16\pi} e^2 Q_0^2 \langle J_i K 20 | J_f K \rangle^2 \tag{6-20}$$

(For a derivation see, e.g., Bohr and Mottelson [35].) For $K = 0$, $J_i = J$, and $J_f = J-2$, the square of the Clebsch-Gordan coefficient simplifies to

$$\langle J020|(J\text{-}2)\,0\rangle^2 = \frac{3J(J-1)}{2(2J+1)(2J-1)}$$

with the help of the identities given in Table A-1. The reduced transition rate for decay between adjacent members of a $K = 0$ band becomes

$$B(E2; J \to J\text{-}2) = \frac{15}{32\pi} e^2 Q_0^2 \frac{J(J-1)}{(2J+1)(2J-1)} \tag{6-21}$$

Alternatively, for electromagnetic excitation from J to $J + 2$,

$$B(E2; J \to J\text{+}2) = \frac{15}{32\pi} e^2 Q_0^2 \frac{(J+1)(J+2)}{(2J+1)(2J+3)} \tag{6-22}$$

a form more useful, for example, in Coulomb excitation. From the values of $B(E2)$ deduced from a measurement of the transition rates, we can again calculate the value of Q_0. The intrinsic quadrupole moment obtained this way may be different from that of Eq. (6-19), as it involves two members of a band. For this reason, it is useful to distinguish the value obtained from $B(E2)$ by calling it *transition* quadrupole moment and that from Eq. (6-19), by calling it *static* quadrupole moment.

Magnetic dipole transitions may be studied in the same way. The $K = 0$ bands are not suitable for our purpose here, as the J-values of the members differ by at least two units and $M1$-transitions are forbidden by angular momentum selection rule. The magnetic transition operator defined in Eq. (5-30) is given in terms of single-nucleon gyromagnetic ratios g_ℓ for orbital angular momentum and g_s for intrinsic spin. Here, we are dealing with collective degrees of freedom. Instead of g_ℓ and g_s, it is more appropriate to use g_R and g_K, respectively, the gyromagnetic ratio for rotational motion and the intrinsic state of a deformed nucleus. In terms of these two quantities, the magnetic dipole operator for $K > \frac{1}{2}$ bands remains to have a simple form, similar to that given by Eq. (4-49),

$$\boldsymbol{\mu} = g_R \boldsymbol{J} + (g_K - g_R)\frac{\boldsymbol{K}^2}{J+1}$$

where $\boldsymbol{K} = \boldsymbol{J}_3$, the operator measuring the projection of $\boldsymbol{J}$ on the 3-axis in the intrinsic frame. For a symmetric rotor, the expectation value is K, as we saw earlier.

In the same spirit as Eq. (6-20) for $E2$-transitions, the $B(M1)$ value in the rotational model is given by

$$B(M1, J_i \to J_f{=}J_i{\pm}1) = \frac{3}{4\pi}(g_K - g_R)^2 K^2 \langle J_i K 1 0 | J_f K\rangle^2 \tag{6-23}$$

in units of μ_N^2, the nuclear magneton squared. From Eqs. (6-20) and (6-23) the mixing ratio between $E2$- and $M1$-transition rates between two adjacent members of a $K > 0$ band can be calculated. The quantity relates the intrinsic quadrupole moment Q_0 with gyromagnetic ratios g_R and g_K and provides another check of the model against experimental data.

Transitions between members of different rotation bands, or interband transitions, involve changes in the intrinsic shape of a nucleus in addition to the angular momentum recoupling discussed above for intraband transitions. The main interest of interband transitions concerns the intrinsic wave function. However, we shall not be going into this more complicated subject here.

Corrections to the basic model. On closer examinations, the energy level positions of the members of a rotational band often differ from the simple $J(J+1)$ dependence given by Eq. (6-16). Similarly, the relations between transition rates are not governed exactly by those of Eqs. (6-20) and (6-23). There are many possible reasons for deviations from a simple rotational model. The main ones may be summarized as:

- We have seen that K is a constant of motion for a symmetric rotor. However, rotational wave functions require linear combinations of both $+K$ and $-K$ components in order to be invariant under a parity transformation. It is therefore possible to have a term in the Hamiltonian that couples between $\pm K$, analogous to the Coriolis force in classical rotation. The size of the coupling may depend on both J and K in general but is observed to be negligible except for $K = \frac{1}{2}$. This gives rise to the decoupling term in $K = \frac{1}{2}$ bands to be described later.

- The moment of inertia, which gives the slope in a plot of E_J versus $J(J+1)$, may not be a constant for states of different J. This is expected on the ground that the nucleus is not a rigid body and centrifugal force generated by the rotation can modify slightly the intrinsic shape when the angular velocities are high. Centrifugal stretching is observed at the higher J end of many rotational bands. In general, such small and gradual changes in the moment of inertia may be accounted for empirically by adding a $J^2(J+1)^2$-dependent term in the rotational Hamiltonian.

- Rotational bands have been observed with members having very high spin values, for example, $J \approx 40\hbar$ and beyond. Such high-spin states occur quite high in energy with respect to the ground state of the nucleus. As a result, it may be energetically more favorable for the underlying intrinsic shape to adjust itself slightly and change to a different stable configuration as the excitation energy is increased. Such changes are likely to be quite sudden, reminiscent of a phase change in chemical reactions. Compared with the smooth variation in centrifugal stretching, readjustment of the intrinsic shape takes place within a region of a few adjacent members of a rotational band. This gives rise to the phenomenon of "backbending," to be discussed later in §9-2.

In practice, departures from a $J(J+1)$ spectrum are small, except in the case of $K = \frac{1}{2}$ bands because of the decoupling term. As a result, the $J(J+1)$-level spacing remains, for most purposes, a signature of rotational band.

Decoupling parameter. For odd-mass nuclei, rotational bands have half-integer K-values. In the case of $K = \frac{1}{2}$, the band starts with $J = \frac{1}{2}$ and has additional members with $J = \frac{3}{2}, \frac{5}{2}, \frac{7}{2}, \ldots$. If the energy level positions of the band members are given by the simple rotational Hamiltonian of Eq. (6-14), we expect, for example, the $J = \frac{5}{2}$ member to be above the $J = \frac{3}{2}$ member in energy by an amount larger than the difference between the $J = \frac{3}{2}$ and $J = \frac{1}{2}$ members. The observed level sequence, however, can be quite different and, in many cases, is more similar to the example of ^{169}Tm shown in Fig. 6-11. Instead of a simple $J(J+1)$ sequence, we find the $J = \frac{3}{2}$ member of the band is depressed in energy and is located just above the $J = \frac{1}{2}$ member, the $J = \frac{7}{2}$ member is just above the $J = \frac{5}{2}$ member, and so on.

The special case of $K = \frac{1}{2}$ bands can be understood by adding an extra term $H'(\Delta K)$ to the basic rotational Hamiltonian given in Eq. (6-14). The term connects two components of a rotational wave function different in K by ΔK for $K \neq 0$. The contribution of this term to the rotational energy may be represented, to a first approximation, by the expectation value of $H'(\Delta K)$ with the wave function of Eq. (6-15),

$$\begin{aligned}\langle JMK|H'(\Delta K)|JMK\rangle_{\text{rot.}} \\ \longrightarrow \Big\{ & \langle \mathcal{D}^J_{MK}|H'(\Delta K)|\mathcal{D}^J_{MK}\rangle + \langle \mathcal{D}^J_{M-K}|H'(\Delta K)|\mathcal{D}^J_{M-K}\rangle \\ & + \langle \mathcal{D}^J_{MK}|H'(\Delta K)|\mathcal{D}^J_{M-K}\rangle + \langle \mathcal{D}^J_{M-K}|H'(\Delta K)|\mathcal{D}^J_{MK}\rangle \Big\}\end{aligned}$$

The first two terms on the right-hand side vanish since, by definition, $H'(\Delta K)$ cannot connect two wave functions having the same K-value. For $\Delta K = 1$, the last two terms are nonzero only for $K = \frac{1}{2}$.

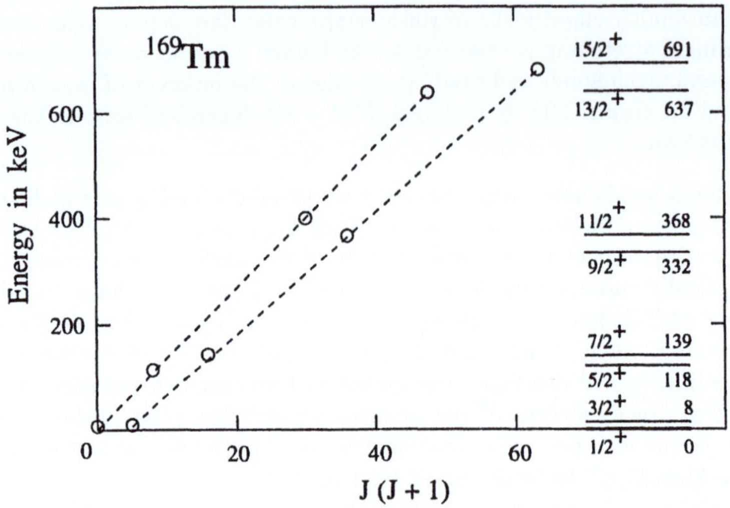

Figure 6-11: Rotational spectrum for the $K^\pi = \frac{1}{2}^+$ band in ^{169}Tm, showing the effect of the decoupling term of Eq. (6-25). (Plotted using data from Ref. [95].)

A term in the Hamiltonian that operates only between wave functions different in K-value by unity may be written as

$$H'(\Delta K = 1) \sim \omega_1 J_1 = \tfrac{1}{2}\omega_1(J_+ + J_-) \tag{6-24}$$

where J_1 is the component of the angular momentum operator J along the body-fixed 1-axis and ω_1 the corresponding angular frequency. The analogue of such a contribution is the Coriolis force in classical mechanics responsible, for example, for deflecting movement of air mass from polar to equatorial regions on Earth to a counterclockwise direction in the Northern Hemisphere and clockwise in the Southern Hemisphere as a result of Earth spinning on its own axis. Since a $K = \frac{1}{2}$ band is associated with an odd-mass nucleus, we can view the situation as a single nucleon moving in the average potential of an even-even core. Since the core is rotating, an additional force is felt by the nucleon, and the interaction does not preserve the sign of K in the intrinsic frame of reference.

The decoupling term given in this way is effective only for the $K = \frac{1}{2}$ band. Because of $H'(\Delta K)$, the rotational energy of a member of the $K = \frac{1}{2}$ band becomes

$$E_{J(K=1/2)} = \frac{\hbar^2}{2\mathcal{I}}\big\{J(J+1) + a(-1)^{J+1/2}(J+\tfrac{1}{2})\big\} + E_K \tag{6-25}$$

where a is the strength of the decoupling term. Instead of a $J(J+1)$ spectrum, each level is now moved up or down from its location given by Eq. (6-16) for an amount depending on whether $J + \frac{1}{2}$ of the level is even or odd. In cases where the absolute value of the decoupling parameter a is large, a higher spin level may appear below one with spin one unit less, as seen in the ^{19}F example in Problem 6-3. The signature of a rotational band can still be recognized by the fact that one-half of the members, $J = \frac{1}{2}, \frac{5}{2}, \frac{9}{2}, \ldots$, possess a E_J versus $J(J+1)$ relation with one (almost constant) slope, and the other half with a different slope, as can be seen from Eq. (6-25).

The basic concept behind rotational models is the classical rotor. Quantum mechanics enters in two places, a trivial one in the discrete (rather than continuous) distribution of energy and angular momentum and a more important one in evaluating the moment of inertia. The latter is a complicated and interesting question, as illustrated by the following consideration.

The equilibrium shape of a nucleus may be deduced from such measurements as the quadrupole moment. At the same time, the moment of inertia can be calculated, for example, by considering the nucleus as a rigid body,

$$\mathcal{I}_{\text{rig}} = \tfrac{2}{5}MR_0^2(1 + \tfrac{1}{3}\delta) \tag{6-26}$$

where M is the mass of the nucleus and R_0 its mean radius. The quantity δ may be expressed in terms of Q_0 using Eq. (6-18). Compared with observations, the rigid-body value turns out to be roughly a factor of 2 too large. Furthermore, the observed value of $\mathcal{I}$ for different nuclei changes systematically from being fairly small near closed shell nuclei, increasing toward the region in between, and decreasing once again toward the next set of magic numbers. An understanding of this question requires a knowledge of the equilibrium shape of nuclei under rotation. We shall discuss this point further in §9-2.

6-4 Interacting Boson Approximation

We have seen the importance of pairing and quadrupole terms in nuclear interaction in a number of nuclear properties examined earlier. For many states, the main features are often given by these two terms alone. In fact, it is possible to build a model for nuclear structure based on this approximation. One of the advantages in such an approach is that analytical solutions are possible under certain conditions. We shall examine only one representative model in this category, the interacting boson approximation (IBA).

Boson operators. A good starting point for IBA is to follow the philosophy behind vibrational models and treat the principal excitation modes in the model as canonical variables. Here, two types of excitation quanta, or *bosons*, can be constructed: a $J = 0$ quantum, or s-boson, and a $J = 2$ quantum, or d-boson. Both types may be thought to be made of pairs of identical nucleons coupled to $J = 0$ and $J = 2$, respectively. Such a realization of the bosons in terms of nucleons is important if one wishes to establish a microscopic foundation for the model. However, it is not essential for us if we only wish to see how the model accounts for the observed nuclear properties through very simple calculations.

Let $\boldsymbol{s}^\dagger$ be the operator that creates an s-boson and $\boldsymbol{d}_\mu^\dagger$ be the corresponding operator for a d-boson. Since a d-boson carries two units of angular momentum, it has five components, distinguished by the projections of the angular momentum on the quantization axis, $\mu = -2, -1, 0, 1, 2$. Corresponding to each of these boson creation operators, we have the conjugate annihilation operators $\boldsymbol{s}$ and $\boldsymbol{d}_\mu$. To complete the definition of these operators, we need to specify the commutation relations between them:

$$\begin{aligned}
&[\boldsymbol{s}^\dagger, \boldsymbol{s}] = 1 \qquad && [\boldsymbol{s}^\dagger, \boldsymbol{s}^\dagger] = [\boldsymbol{s}, \boldsymbol{s}] = 0 \\
&[\boldsymbol{d}_\mu^\dagger, \boldsymbol{d}_\nu] = \delta_{\mu\nu} && [\boldsymbol{d}^\dagger, \boldsymbol{d}^\dagger] = [\boldsymbol{d}, \boldsymbol{d}] = 0 \\
&[\boldsymbol{s}^\dagger, \boldsymbol{d}_\mu] = [\boldsymbol{s}, \boldsymbol{d}_\mu^\dagger] = [\boldsymbol{s}, \boldsymbol{d}_\mu] = [\boldsymbol{s}^\dagger, \boldsymbol{d}_\mu^\dagger] = 0
\end{aligned} \tag{6-27}$$

All other operators necessary to calculate nuclear properties in the model are expressed in terms of these operators.

Using $\boldsymbol{s}^\dagger$, $\boldsymbol{s}$, $\boldsymbol{d}_\mu^\dagger$, and $\boldsymbol{d}_\mu$, the number operators for s- and d-bosons are, respectively,

$$\boldsymbol{n}_s = \boldsymbol{s}^\dagger \cdot \overline{\boldsymbol{s}}, \qquad \boldsymbol{n}_d = \boldsymbol{d}^\dagger \cdot \overline{\boldsymbol{d}} = \sum_\mu (-1)^\mu \boldsymbol{d}_\mu^\dagger \overline{\boldsymbol{d}}_{-\mu} = \sum_\mu \boldsymbol{d}_\mu^\dagger \boldsymbol{d}_\mu$$

where the bar on top indicates the (spherical tensor) adjoint of $\boldsymbol{d}$ and $\boldsymbol{s}$, with

$$\overline{\boldsymbol{d}}_\mu = (-1)^{2+\mu} \boldsymbol{d}_{-\mu} \qquad \overline{\boldsymbol{s}} = \boldsymbol{s}$$

as shown in Eq. (A-9). In addition, we can construct five irreducible spherical tensors made of products of two boson operators

$$\begin{aligned}
&\boldsymbol{P} = \tfrac{1}{2}\{\overline{\boldsymbol{d}} \cdot \overline{\boldsymbol{d}} - \overline{\boldsymbol{s}} \cdot \overline{\boldsymbol{s}}\} \qquad && \boldsymbol{L} = \sqrt{10}(\boldsymbol{d}^\dagger \times \overline{\boldsymbol{d}})_1 \\
&\boldsymbol{Q} = (\boldsymbol{d}^\dagger \times \overline{\boldsymbol{s}})_2 + (\boldsymbol{s}^\dagger \times \overline{\boldsymbol{d}})_2 - \sqrt{\tfrac{7}{4}}(\boldsymbol{d}^\dagger \times \overline{\boldsymbol{d}})_2 && \\
&\boldsymbol{T}_3 = (\boldsymbol{d}^\dagger \times \overline{\boldsymbol{d}})_3 && \boldsymbol{T}_4 = (\boldsymbol{d}^\dagger \times \overline{\boldsymbol{d}})_4
\end{aligned} \tag{6-28}$$

where the multiplication symbol $\times$ stands for the angular momentum coupled product

$$(\boldsymbol{A}_r \times \boldsymbol{B}_s)_{tm} = \sum_{pq} \langle rpsq|tm\rangle \boldsymbol{A}_{rp} \boldsymbol{B}_{sq}$$

defined in Eq. (A-10).

The simple model. If we restrict ourselves to the simple case of having either active neutrons or protons, the most general IBA Hamiltonian we can construct may be expressed as a linear combination of the five operators given in Eq. (6-28) together with the boson number operators. This is generally referred to as IBA-1. In this limit, there are six parameters in the Hamiltonian,

$$H_{\text{IBA}-1} = \epsilon \boldsymbol{n}_d + a_0 \boldsymbol{P}^\dagger \cdot \boldsymbol{P} + a_1 \boldsymbol{L} \cdot \boldsymbol{L} + a_2 \boldsymbol{Q} \cdot \boldsymbol{Q} + a_3 \boldsymbol{T}_3 \cdot \boldsymbol{T}_3 + a_4 \boldsymbol{T}_4 \cdot \boldsymbol{T}_4 \tag{6-29}$$

where ϵ is the energy difference between a d- and an s-boson and a_J for $J = 0$ to $J = 4$ are the strengths of the other five components in the expression. The dot between two spherical tensor operators in Eq. (6-28) represents a scalar product, angular momentum coupled product with final tensorial rank zero. The number operator $\boldsymbol{n}_s$ for s-bosons does not enter into the expression, as the energy associated with it is taken to be zero and is absorbed into the definition of the energy scale. In the absence of a microscopic connection to the nucleon degrees of freedom, these six parameters must be found, for example, by fitting results calculated with the Hamiltonian to known data.

In addition to energy, operators corresponding to other observables in the space span by the s- and d-bosons can also be expressed in terms of tensor products of the boson creation and annihilation operators. For example, the possible electromagnetic transition operators in the space are

$$O_0(E0) = \beta_0(\boldsymbol{d}^\dagger \times \tilde{\boldsymbol{d}})_0 + \gamma_0(\boldsymbol{s}^\dagger \times \tilde{\boldsymbol{s}})_0 \qquad O_{1\mu}(M1) = \beta_1(\boldsymbol{d}^\dagger \times \tilde{\boldsymbol{d}})_{1\mu}$$
$$O_{2\mu}(E2) = \alpha_2\{(\boldsymbol{s}^\dagger \times \tilde{\boldsymbol{d}})_{2\mu} + (\boldsymbol{d}^\dagger \times \tilde{\boldsymbol{s}})_{2\mu}\} + \beta_2(\boldsymbol{d}^\dagger \times \tilde{\boldsymbol{d}})_{2\mu}$$
$$O_{3\mu}(M3) = \beta_3(\boldsymbol{d}^\dagger \times \tilde{\boldsymbol{d}})_{3\mu} \qquad O_{4\mu}(E4) = \beta_4(\boldsymbol{d}^\dagger \times \tilde{\boldsymbol{d}})_{4\mu}$$

where α_2, γ_0, and the β's are, again, adjustable parameters.

One of the interesting features of IBA-1 is that it has an underlying group structure, and as a result, powerful mathematical techniques may be applied to find the solutions. The communication relations among the boson creation and annihilation operators expressed in Eq. (6-27) imply that the operators form a group, the U_6 group, a unitary group in six dimensions. The energy of a state corresponding to one of the irreducible representations of this group may be expressed as a function of the six parameters in the Hamiltonian. Once the values of these parameters are determined, a large number of energy levels can be calculated. Examples of results for energy level positions obtained with IBA-1 are shown in Fig. 6-12.

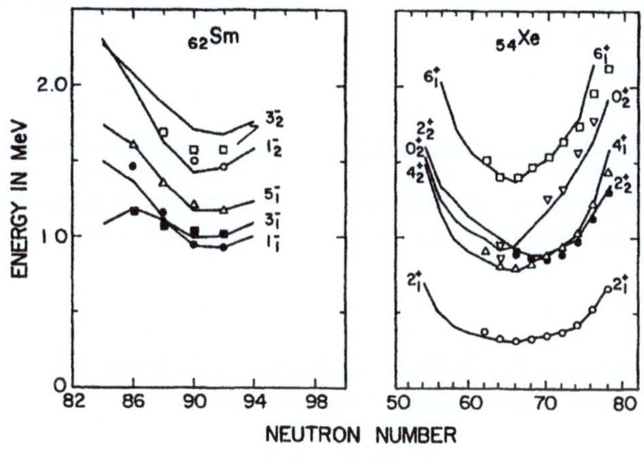

Figure 6-12: Comparison of experimental (squares, circles and triangles) and calculated level spectra (lines) in the IBA for octupole states in samarium (left) and xenon isotopes (right). (Taken from Ref. [9].)

The underlying group structure of IBA-1 lends itself also to three limiting cases that are of interest in nuclear structure. The U_6 group may be decomposed into a variety of subgroups. Among these, we shall limit ourselves to cases where the chain of reduction contains the three-dimensional rotational group as one of the subgroups. This is necessary if angular momentum is to be retained as a constant of motion.

If d-bosons are completely decoupled from the system, the Hamiltonian may be written in terms of s-boson operators alone,

$$H_{\text{seniority}} = \epsilon_s \boldsymbol{s}^\dagger \boldsymbol{s} + a_0 \boldsymbol{s}^\dagger \boldsymbol{s}^\dagger \boldsymbol{s}\boldsymbol{s}$$

This is the *seniority* scheme [49], known to be useful in classifying many-nucleon states in the jj-coupling scheme (see §7-1). Here, pairs of nucleons with their angular momentum coupled to zero are treated differently from those that are not coupled to $J = 0$. From this property, we see also that IBA-1 has a pairing structure built into the Hamiltonian and can therefore account for many of the observed nuclear properties in which pairing interaction dominates.

On the other hand, if all the terms related to the s-boson operator are ignored, we obtain a system dominated by quadrupole excitations induced by d-bosons,

$$H_{\text{vib limit}} = \epsilon_d \boldsymbol{d}^\dagger \cdot \boldsymbol{d} + \sum_J a_J (\boldsymbol{d}^\dagger \times \boldsymbol{d}^\dagger)_J \cdot (\overline{\boldsymbol{d}} \times \overline{\boldsymbol{d}})_J$$

In this limit, we obtain quadrupole vibrational motion in nuclei similar to that described in §6-1.

If we put all the parameters in Eq. (6-29) to zero except a_1 and a_2, we obtain the SU_3 limit

$$H_{SU_3} = a_1 \boldsymbol{L} \cdot \boldsymbol{L} + a_2 \boldsymbol{Q} \cdot \boldsymbol{Q}$$

This has been used with success in understanding rotation-like structure in ds-shell nuclei from oxygen to potassium. Because of the $\boldsymbol{L}^2$-term, the Hamiltonian gives a spectrum that has an $L(L+1)$ dependence. If the nucleon intrinsic spins are coupled together to $S = 0$ in a nucleus, we have $\boldsymbol{J} = \boldsymbol{L}$, and an $L(L+1)$ spectrum is the same as one with the $J(J+1)$ dependence we have seen earlier in rotational nuclei. The $\boldsymbol{Q} \cdot \boldsymbol{Q}$ term provides a constant in the energy for all the levels in a "band" and can therefore be interpreted as the dependence on the intrinsic structure of the rotating nucleus. In this way, we expect that IBA-1 can explain rotational structure in nuclei as well.

The full model. In practice, IBA-1 is found to be limited by the fact that only excitations of either neutrons or protons can take place. To overcome this restriction, the Hamiltonian given in Eq. (6-29) is expanded to include both neutrons and protons, as well as interactions between them. This gives us

$$H_{\text{IBA}-2} = H_{nn} + H_{pp} + V_{np}$$

where H_{nn} and H_{pp} are, respectively, the neutron- and proton-boson Hamiltonians. The interaction between these two types of bosons is provided by V_{np}. The most general form, known as IBA-2, contains a maximum of 29 parameters, 9 for H_{nn}, 9 for H_{pp}, and 11 for V_{np}. This is too complicated, and a simplified version is found to be adequate for most applications.

The IBA-2 permits a connection to be made with the underlying single-particle basis. All the nucleons in a nucleus can be divided into two groups, those in the inert core and those in the active, or *valence,* space. The core may be taken to be one of the closed shell nuclei (to be discussed in §7-2) and may be treated as the "vacuum" state for the problem. The nucleons in the core are assumed to be inactive except in providing a binding energy to the valence nucleons. Active neutron pairs and proton pairs can be put into the space by boson creation operators acting on the vacuum. The IBA-2 therefore provides a basis to study a wide variety of nuclear structure phenomena, from single-particle to collective degrees of freedom (for more details, see Arima and Iachello [9]).

Interacting boson models belong to a more general type of approach to nuclear structure studies sometimes known as *algebraic* models. We have seen evidence that symmetries play an important role in nuclear structure. For each type of symmetry, there is usually an underlying mathematical group associated with it. Although there are very few exact symmetries, such as angular momentum, there is a large number of approximate, or "broken," symmetries that are of physical interest and can be exploited. One good example of the latter category is isospin, or SU_2 symmetry, the symmetry in interchanging protons and neutrons, or u- and d-quarks. Although isospin invariance in nuclei is broken by Coulomb interaction, it is nevertheless a useful concept, as we have seen earlier on several occasions. One of the aims of group theoretical approaches to nuclear structure problems is to make use of these symmetries to classify nuclear states according to the irreducible representations of the underlying mathematical groups. We have seen some features of such an approach in IBA-1. A few other elementary applications will also be made in the next chapter to classify single-particle states in the nuclear shell model. A general discussion of algebraic models is, however, inappropriate here, in part because of the amount of preparation in group theory required.

Problems

6-1. When two identical phonons, each carrying angular momentum $\boldsymbol{\lambda}$, are coupled together, only states with even J-values ($\boldsymbol{J} = \boldsymbol{\lambda} + \boldsymbol{\lambda}$) are allowed. Show that this is true by counting the number of states for a given total M, the projection of angular momentum on the quantization axis. Use the same method to show that when three quadrupole phonons are coupled together, only states with $J^\pi = 0^+$, 2^+, 3^+, 4^+, 6^+ are allowed.

6-2. Three rotational bands have been identified in ^{25}Mg: a $K^\pi = 5/2^+$ band starts from the ground state ($J^\pi = 5/2^+$) and has three other members, $7/2^+$ at 1.614 MeV, $9/2^+$ at 3.405 MeV, and $11/2^+$ at 5.45 MeV; a $K = 1/2^+$ band with six members, $1/2^+$ at 0.585 MeV, $3/2^+$ at 0.975 MeV, $5/2^+$ at 1.960 MeV, $7/2^+$ at 2.738 MeV, $9/2^+$ at 4.704 MeV, and $11/2^+$ at 5.74 MeV; and a second $K = 1/2^+$ band with four members, $1/2^+$ at 2.562 MeV, $3/2^+$ at 2.801 MeV, $5/2^+$ at 3.905 MeV, and $7/2^+$ at 5.005 MeV. Calculate the moment of inertia and the decoupling parameter, where applicable, for each band.

6-3. The following energy level positions in mega-electron-volts are known for two rotational bands in ^{19}F: $1/2^+$ 0.000, $1/2^-$ 0.110, $5/2^+$ 0.197, $5/2^-$ 1.346, $3/2^-$ 1.459, $3/2^+$ 1.554, $9/2^+$ 2.780, $7/2^-$ 3.999, $9/2^-$ 4.033, $13/2^+$ 4.648, and $7/2^+$ 5.465. Calculate the moment of inertia and the decoupling parameter for each band. Comment on the likelihood of the $11/2^+$ level at 6.5 MeV to be a member of the $1/2^+$-band.

6-4. The ground state of $^{152}_{63}$Eu is known to be 3^- with an electric quadrupole moment of $+3.16 \times 10^2\, e\text{fm}^2$. Find the intrinsic quadrupole moment of the nucleus for the ground state and deduce the value of δ defined in Eq. (6-18), the difference between R_3 and $R_\perp$. What is the shape of this nucleus?

6-5. The following $E2$-transition rates appear in a table of nuclei in terms of natural line width Γ for the $K = 0^+$ band in ^{20}Ne: 2^+ (1.63 MeV) $\rightarrow$ 0^+ (ground) 6.3×10^{-4} eV, 4^+ (4.25 MeV) $\rightarrow$ 2^+ (1.63 MeV) 7.1×10^{-3} eV, 6^+ (8.78 MeV) $\rightarrow$ 4^+ (4.25 MeV) 0.100 eV, and 8^+ (11.95 MeV) $\rightarrow$ 6^+ (8.78 MeV) 1.2×10^{-3} eV. From the information provided,

(a) find the moment of inertia of the band,

(b) find the intrinsic quadrupole moment of the band, and

(c) predict the quadruple moment of the 2^+ member.

Chapter 7

Microscopic Models of Nuclear Structure

The nucleus is a quantum-mechanical many-body system. Powerful and elegant methods have been developed over the years to handle such problems. Several factors, however, contribute to make the nuclear many-body problem somewhat unique. First, the interaction is complicated and still unknown in many aspects. Second, rotational symmetry imposes the condition that each observed state has a definite spin. Thus angular momentum coupling becomes an important issue in any practical calculations. Finally, even in the heaviest nucleus, the nucleon number is not large enough to be treated as an infinite system where many simplifications can be applied.

A nucleus is made up of neutrons and protons. It is therefore natural to adopt a Hamiltonian based on nucleons, interacting with each other through a two-body potential. The eigenfunctions obtained by solving the Schrödinger equation may be used to calculate observables and the results compared with experiments. In principle, such a calculation is possible once the nucleon-nucleon interaction is given. In practice, special techniques are needed and we shall examine a few of the more basic ones.

7-1 Many-Body Basis States

To describe a nucleus using nucleon degrees of freedom, we need to express the wave functions in terms of those for individual nucleons. The first step in a microscopic calculation for the nuclear many-body problem is then to find a suitable set of single-particle wave functions. Antisymmetrized products of such functions form the basis states for our many-body system made of A nucleons.

Mathematically, we can take any complete set of functions as the basis states. However, the Hilbert space is in general infinite in dimension, and truncation of the space to a small finite subset is essential in any practical calculations. The selection of this truncated, or *active,* space depends on the basis states chosen. For this reason, selection of the basis states is an important step in a calculation. As we shall see in the later sections in this chapter, a well-chosen single-particle basis wave function can greatly simplify the problem.

Matrix method to solve the eigenvalue problem. Our calculation is centered around the solution to the many-body eigenvalue problem

$$H\Psi_\alpha(\boldsymbol{r}_1, \boldsymbol{r}_2, \ldots, \boldsymbol{r}_A) = E_\alpha \Psi_\alpha(\boldsymbol{r}_1, \boldsymbol{r}_2, \ldots, \boldsymbol{r}_A) \tag{7-1}$$

where E_α is the energy of the state with wave function $\Psi_\alpha(\boldsymbol{r}_1, \boldsymbol{r}_2, \ldots, \boldsymbol{r}_A)$. The Hamiltonian consists of a sum of the kinetic energy of each nucleon with reduced mass μ_i and the interaction between any two nucleons,

$$H = \sum_{i=1}^{A} \frac{\hbar^2}{2\mu_i} \nabla_i^2 + \sum_{i \neq j} V_{ij} \tag{7-2}$$

To simplify the notation, we shall not make any explicit reference here to the intrinsic spin and other degrees of freedom and we shall use $\boldsymbol{r}_i$ to represent all the independent variables of the system pertaining to nucleon i. From the eigenfunctions obtained, we can find other properties of the system by calculating the matrix elements for the corresponding operators, such as those given in Chapter 4.

For many purposes, it is more convenient to solve Eq. (7-1) using a matrix method. In this approach, we start with a complete set of basis states for the A-particle system, $\{\Phi_k(\boldsymbol{r}_1, \boldsymbol{r}_2, \ldots, \boldsymbol{r}_A)\}$ for $k = 1, 2, \ldots, D$, where D is the number of linearly independent states in the Hilbert space. For mathematical convenience, we shall assume that the basis is an orthogonal and normalized one. Any eigenvector $\Psi_\alpha(\boldsymbol{r}_1, \boldsymbol{r}_2, \ldots, \boldsymbol{r}_A)$ may be expressed as a linear combination of these D basis states,

$$\Psi_\alpha(\boldsymbol{r}_1, \boldsymbol{r}_2, \ldots, \boldsymbol{r}_A) = \sum_{k=1}^{D} C_k^\alpha \, \Phi_k(\boldsymbol{r}_1, \boldsymbol{r}_2, \ldots, \boldsymbol{r}_A) \tag{7-3}$$

Here C_k^α are the expansion coefficients for the αth eigenfunction. In principle, the solution to Eq. (7-1) is independent of the basis states chosen; in practice, the ease of solving the problem depends critically on the choice. As mentioned earlier, this is especially true if we wish to truncate the active space to a manageable one. We shall return to the question of truncation in the next few sections.

Once the basis is fixed, the unknown expansion coefficients C_j^α in Eq. (7-3) may be found by proceeding in the following way. First, we multiply both sides of Eq. (7-1) from the left with $\Phi_j^*(\boldsymbol{r}_1, \boldsymbol{r}_2, \ldots, \boldsymbol{r}_A)$ and integrate over all the independent variables. In terms of Dirac bra-ket notation, the result may be expressed as

$$\langle \Phi_j(\boldsymbol{r}_1, \boldsymbol{r}_2, \ldots, \boldsymbol{r}_A) | H | \Psi_\alpha(\boldsymbol{r}_1, \boldsymbol{r}_2, \ldots, \boldsymbol{r}_A) \rangle = E_\alpha \langle \Phi_j(\boldsymbol{r}_1, \boldsymbol{r}_2, \ldots, \boldsymbol{r}_A) | \Psi_\alpha(\boldsymbol{r}_1, \boldsymbol{r}_2, \ldots, \boldsymbol{r}_A) \rangle$$

Using the expansion given in Eq. (7-3) and the orthonormal property of the basis wave functions, the expression can be reduced to

$$\sum_{k=1}^{D} H_{jk} C_k^\alpha = E_\alpha C_j^\alpha \tag{7-4}$$

where H_{jk} is the matrix element of the Hamiltonian between basis states Φ_j and Φ_k,

$$H_{jk} \equiv \langle \Phi_j(\boldsymbol{r}_1, \boldsymbol{r}_2, \ldots, \boldsymbol{r}_A) | H | \Phi_k(\boldsymbol{r}_1, \boldsymbol{r}_2, \ldots, \boldsymbol{r}_A) \rangle$$

In terms of matrices, Eq. (7-4) may be written as

$$\begin{pmatrix} H_{11} & H_{12} & \cdots & H_{1D} \\ H_{21} & H_{22} & \cdots & H_{2D} \\ \vdots & \vdots & \ddots & \vdots \\ H_{D1} & H_{D2} & \cdots & H_{DD} \end{pmatrix} \begin{pmatrix} C_1^\alpha \\ C_2^\alpha \\ \vdots \\ C_D^\alpha \end{pmatrix} = E_\alpha \begin{pmatrix} C_1^\alpha \\ C_2^\alpha \\ \vdots \\ C_D^\alpha \end{pmatrix} \tag{7-5}$$

The eigenvalues E_α are the roots of the secular equation,

$$\det \begin{vmatrix} H_{11} - E_\alpha & H_{12} & \cdots & H_{1D} \\ H_{21} & H_{22} - E_\alpha & \cdots & H_{2D} \\ \vdots & \vdots & \ddots & \vdots \\ H_{D1} & H_{D2} & \cdots & H_{DD} - E_\alpha \end{vmatrix} = 0$$

Once an eigenvalue E_α is found, the coefficients C_i^α, $i = 1, 2, \ldots, D$, may be obtained, in principle at least, by solving Eq. (7-5) as a set of D algebraic equations. This gives us the eigenvector corresponding to E_α. The complete set of eigenvectors for $\alpha = 1, 2, \ldots, D$ may be viewed as a matrix $\{C_i^\alpha\}$ that transforms the Hamiltonian from the basis representation into a diagonal one. In this way, the eigenvalue problem posted by Eq. (7-1) is solved by diagonalizing the Hamiltonian matrix $\{H_{jk}\}$. Powerful numerical techniques are available to handle eigenvalue problems by matrix diagonalization [153].

Single-particle basis states. In microscopic nuclear structure calculations, the basis states $\{\Phi_k\}$ for many-body wave functions are usually constructed out of products of single-particle wave functions $\phi_i(\boldsymbol{r}_j)$. To ensure proper antisymmetrization among the nucleons, a many-body state is often written in the form of a Slater determinant,

$$\Phi_k(\boldsymbol{r}_1, \boldsymbol{r}_2, \ldots, \boldsymbol{r}_A) = \frac{1}{\sqrt{A!}} \det \begin{vmatrix} \phi_1(\boldsymbol{r}_1) & \phi_1(\boldsymbol{r}_2) & \cdots & \phi_1(\boldsymbol{r}_A) \\ \phi_2(\boldsymbol{r}_1) & \phi_2(\boldsymbol{r}_2) & \cdots & \phi_2(\boldsymbol{r}_A) \\ \vdots & \vdots & \ddots & \vdots \\ \phi_A(\boldsymbol{r}_1) & \phi_A(\boldsymbol{r}_2) & \cdots & \phi_A(\boldsymbol{r}_A) \end{vmatrix} \tag{7-6}$$

where the factor $(A!)^{-1/2}$ is required for normalization. Different sets of single-particle states form different many-body basis states. The choice of single-particle wave functions therefore determines the type of many-body basis states that can be constructed.

The single-particle spectrum is an infinite one. It is bound at the low-energy end by the ground state but extends to infinity at the other end. This is very similar to the energy spectrum of a harmonic oscillator. In fact, we shall see that the harmonic oscillator is often used as the starting point of nuclear single-particle wave functions. If we select a set of states with single-particle energies close to those found in actual nuclei, it is possible to truncate the Hilbert space based on energy considerations. Partly for this reason, it is more convenient to take as basis states the eigenfunctions of a single-particle Hamiltonian,

$$h(\boldsymbol{r}_i)\phi_k(\boldsymbol{r}_i) = \epsilon_k \phi_k(\boldsymbol{r}_i)$$

Here ϵ_k is the single-particle energy. We shall see an example of $h(\boldsymbol{r}_i)$ in Eq. (7-10) of the next section. In terms of such a single-particle Hamiltonian, the many-body

Hamiltonian in Eq. (7-1) may be expressed in the form

$$H = \sum_{i=1}^{A} h(\boldsymbol{r}_i) + \sum_{i \neq j=1}^{A} \widetilde{V}(\boldsymbol{r}_i, \boldsymbol{r}_j) \tag{7-7}$$

where $\widetilde{V}(\boldsymbol{r}_i, \boldsymbol{r}_j)$ is the *residual* two-body interaction, the original nucleon-nucleon interaction V_{ij} in Eq. (7-2) minus contributions already included in $h(\boldsymbol{r}_i)$. We shall not be concerned here with the technical question of transforming the Hamiltonian from the form given by Eq. (7-2) to that in (7-7). One method is given in §7-3, and the formal procedure is given in §7-5. On the other hand, it is clear that if we choose $h(\boldsymbol{r}_i)$ such that a large part of the effect of the two-body interaction in Eq. (7-2) is included, the residual interaction $\widetilde{V}(\boldsymbol{r}_i, \boldsymbol{r}_j)$ will be sufficiently weak that, in some cases, it may even be adequate to ignore it. This gives us various independent particle models. Alternatively, we can make use of the energies ϵ_k to reduce the Hilbert space to a manageable size and solve the eigenvalue problem with the residual interaction in the truncated space. An example is the spherical shell model described in §7-5.

7-2 Magic Number and Single-Particle Energy

The best evidence for single-particle behavior is found in closed shell nuclei, ^{4}He, ^{16}O, ^{40}Ca, ^{90}Zr, and ^{208}Pb. These are nuclei with proton number Z = 2, 8, 20, 40, 82 and neutron number N = 2, 8, 20, 50, 82, 126. They have special features, such as:

- Energies of the first few excited states are higher than those in nearby nuclei, as shown in Fig. 7-1.

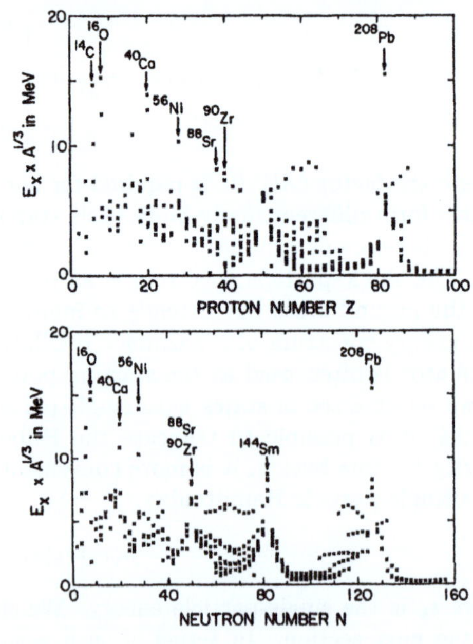

Figure 7-1: Energy of the first excited state of even-even nuclei as a function of proton number (upper) and neutron number (lower). All energies are multiplied by a factor $A^{1/3}$ to account for the general decrease in excitation energy with increasing nucleon number A. Among a total of 389 nuclei, form ^{4}He to ^{256}Fm, there are 372 with spin-parity 2^+ for the first excited state, only 7 with 0^+, 3 with 3^-, 2 with 1^-, and 5 with unknown spin. The highest excitation energy is found in ^{4}He (not shown) with E_x = 20.1 MeV or $E_x \times A^{1/3}$ = 31 MeV. The data are taken from Ref. [95].

- Single-neutron and single-proton removal energies,

$$S_n(Z,N) = E_B(Z,N) - E_B(Z,N\text{-}1) \tag{7-8}$$
$$S_p(Z,N) = E_B(Z,N) - E_B(Z\text{-}1,N) \tag{7-9}$$

are much larger than those in the neighboring even-even nuclei, as shown in Fig. 7-2.

- The intrinsic shape of the ground states is spherical, as can be seen from observations such as electromagnetic transitions.

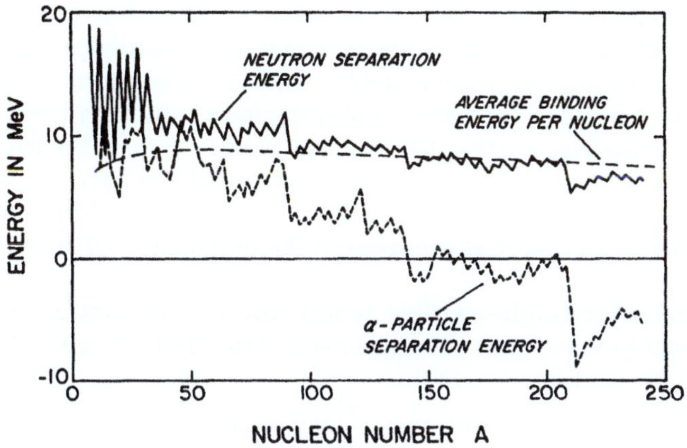

Figure 7-2: Neutron and α-particle separation energies for stable nuclei as a function of nucleon number A. The values are calculated from a table of binding energies.

These properties are sufficiently prominent that the sequence of numbers, 2, 8, 20, 40(50), 82, and 126 are known as magic numbers. One of the early achievements of nuclear physics was in explaining the cause of these magic numbers using an independent particle model, based on a Hamiltonian that is a slight extension of that for a simple, three-dimensional harmonic oscillator.

Independent particle model. We saw in the previous section that an independent particle model is one in which the residual interaction is ignored. In this approximation, the nuclear Hamiltonian is a sum of single-particle terms,

$$H = \sum_{i=1}^{A} h(\boldsymbol{r}_i)$$

Physically, we can think of a nucleon i moving in a potential $v(\boldsymbol{r}_i)$ that is a good representation of the average effect of the two-body interaction the nucleon has with all the other nucleons in the nucleus. If ϵ_k is the kth eigenvalue of this Hamiltonian,

$$h(\boldsymbol{r})\phi_k(\boldsymbol{r}) = \epsilon_k \phi_k(\boldsymbol{r})$$

the many-body Hamiltonian in the independent particle model may be rewritten in terms of the single-particle energies

$$H = \sum_k \epsilon_k n_k$$

Here the summation on k is over all single-particle states and n_k is the number operator measuring the occupancy of single-particle state k.

In this picture, the magic numbers arise because of the fact that the nuclear single-particle spectrum is not a smooth one. Instead, they are grouped into "shells" with relatively large energy gaps between shells. When each group of states is completely filled, the Fermi energy of the nucleus is just below one of these large energy gaps. The ground state of the nucleus is made up by filling all the single-particle states below the Fermi level. To form an excited state in such an independent particle model, a nucleon must be promoted from an occupied single-particle state below to an empty one above. Since the gap is large, it takes more energy to excite the nucleus. For this reason, nuclei satisfying this condition for both neutrons and protons are also called closed shell nuclei. With all the orbits filled, the ground state of the nucleus is tightly bound and spherical in shape, as explained at the beginning of §6-3.

Harmonic oscillator single-particle spectrum. We can construct a simple model to see why energy gaps appear in the single-particle spectrum. A one-body Hamiltonian may be written in the form

$$h(\boldsymbol{r}) = -\frac{\hbar^2}{2\mu}\nabla^2 + v(\boldsymbol{r}) \tag{7-10}$$

where $\boldsymbol{r}$ is the coordinate of the nucleon and μ is its reduced mass. For mathematical convenience, we shall assume for the moment that the potential $v(\boldsymbol{r})$ is a central one that depends only on the magnitude of $\boldsymbol{r}$ but not on its direction. A good approximation of such a potential is given by the harmonic oscillator well,

$$v(r) = \tfrac{1}{2}\mu\omega_0^2 r^2 \tag{7-11}$$

where ω_0 is the frequency. This is a reasonable assumption for the bound nucleons. To provide binding, the potential must have a minimum, and near this minimum it must have a quadratic dependence on the spatial coordinates. Such a form is well represented by that given in Eq. (7-11). Examples of single-particle radial wave functions generated by such a potential are shown in Table 7-1. We expect that the radial dependence may not be realistic near the nuclear surface, especially for single-particle states around the Fermi energy. However, this is not a problem for us here.

For an isotropic, three-dimensional harmonic oscillator potential, each (major) *shell* is characterized by N, the number of oscillator quanta. All states belonging to a given shell are degenerate with energy

$$\epsilon_N = \left(N + \tfrac{3}{2}\right)\hbar\omega_0 \tag{7-12}$$

For each shell, the allowed orbital angular momenta are

$$\ell = N,\ N-2, \ldots,\ 1,\ \text{or } 0$$

Table 7-1: Harmonic oscillator radial wave functions.

$R_{1s}(r) = 2\Big(\frac{\nu^3}{\pi}\Big)^{1/4} e^{-\nu r^2/2}$	$R_{1p}(r) = \sqrt{\frac{2^3}{3}}\Big(\frac{\nu^5}{\pi}\Big)^{1/4} r e^{-\nu r^2/2}$
$R_{1d}(r) = \sqrt{\frac{2^4}{15}}\Big(\frac{\nu^7}{\pi}\Big)^{1/4} r^2 e^{-\nu r^2/2}$	$R_{2s}(r) = \sqrt{\frac{2^3}{3}}\Big(\frac{\nu^3}{\pi}\Big)^{1/4} (\frac{3}{2} - \nu r^2)\, e^{-\nu r^2/2}$
$R_{1f}(r) = \sqrt{\frac{2^5}{105}}\Big(\frac{\nu^9}{\pi}\Big)^{1/4} r^3 e^{-\nu r^2/2}$	$R_{2p}(r) = \sqrt{\frac{2^4}{15}}\Big(\frac{\nu^5}{\pi}\Big)^{1/4} (\frac{5}{2} - \nu r^2)\, r e^{-\nu r^2/2}$
$R_{1g}(r) = \sqrt{\frac{2^6}{945}}\Big(\frac{\nu^{11}}{\pi}\Big)^{1/4} r^4 e^{-\nu r^2/2}$	$R_{2d}(r) = \sqrt{\frac{2^5}{105}}\Big(\frac{\nu^7}{\pi}\Big)^{1/4} (\frac{7}{2} - \nu r^2)\, r^2 e^{-\nu r^2/2}$
$R_{3s}(r) = \sqrt{\frac{2^3}{15}}\Big(\frac{\nu^3}{\pi}\Big)^{1/4} (\frac{15}{4} - 5\nu r^2 + \nu^2 r^4)\, e^{-\nu r^2/2}$	

Note: As approximate single-particle wave functions for a nucleus, the oscillator parameter, $\nu = m\omega_0/\hbar$, may be taken to be $A^{-1/3}$ femtometers squared.

(See, e.g., p. 818 of Ref. [46].) Since each nucleon also has an intrinsic spin $s = \frac{1}{2}$, the number of states, D_N, i.e., the maximum number of neutrons or protons a harmonic oscillator shell can accommodate, is given by

$$D_N = 2 \sum_{\text{allowed } \ell} (2\ell + 1) = 2 \sum_{k=1}^{N+1} k = (N+1)(N+2)$$

where the factor of 2 in front of the summations is to account for the two possible orientations of nucleon intrinsic spin. The total number of states, D_{max}, up to some maximum number of harmonic oscillator quanta, N_{max}, is given by a sum over all N-values to N_{max},

$$\begin{aligned} D_{\text{max}} = \sum_{N=0}^{N_{\text{max}}} D_N &= \frac{1}{3}(N_{\text{max}}+1)(N_{\text{max}}+2)(N_{\text{max}}+3) \\ &\xrightarrow[N_{\text{max}} \gg 1]{} \frac{1}{3}(N_{\text{max}}+2)^3 \end{aligned} \tag{7-13}$$

In arriving at the final result, we have made use of the identity

$$\sum_{k=1}^{n} k^2 = \tfrac{1}{6} n(n+1)(2n+1)$$

From Eq. (7-13) we obtain the values $D_{\text{max}} = 2, 8, 20, 40, 70, 112, 168, \ldots$ for $N_{\text{max}} = 0, 1, 2, \ldots$.

The harmonic oscillator frequency ω_0 is related to the size of the nucleus and, hence, to the nucleon number A. The expectation value of r^2 in a state of $N\hbar\omega_0$ can be obtained from that for the harmonic oscillator potential energy,

$$\left\langle \tfrac{1}{2}\mu\omega_0^2 r^2 \right\rangle_N = \tfrac{1}{2}\left(N + \tfrac{3}{2}\right)\hbar\omega_0$$

where the factor $\frac{1}{2}$ on the right-hand side comes from the fact that, for a particle in a three-dimensional harmonic oscillator well, the average of potential energy is half of the total energy. Using this relation, we obtain the expectation value of r^2 in a state with N harmonic oscillator quanta to be

$$\langle r^2\rangle_N = \frac{\hbar}{\mu\omega_0}\left(N + \tfrac{3}{2}\right) \tag{7-14}$$

The mean-square radius of a nucleus made of A nucleons is given by the average over all occupied harmonic oscillator states for both neutrons and protons,

$$\langle R^2\rangle = \frac{2}{A}\sum_{N=0}^{N_{\rm max}} D_N\langle r^2\rangle_N = \frac{2}{A}\sum_{N=0}^{N_{\rm max}}(N+1)(N+2)(N+\tfrac{3}{2})\frac{\hbar}{\mu\omega_0} \tag{7-15}$$

where the factor 2 in front of the summations arises from the need to consider both neutrons and protons. For simplicity, we shall assume here that neutron and proton numbers are equal to each other. The final result is obtained by substituting the explicit values of D_N given in Eq. (7-13) and $\langle r^2\rangle_N$ in (7-14).

The summation over N in the final form of Eq. (7-15) may be carried out with the help of the mathematical identity

$$\sum_{k=1}^{n} k^3 = \left(\tfrac{1}{2}n(n+1)\right)^2$$

together with those for $\sum k^2$ and $\sum k$ given earlier. The result is

$$\begin{aligned}\sum_{N=0}^{N_{\rm max}}(N+1)(N+2)(N+\tfrac{3}{2}) &= \tfrac{1}{4}(N_{\rm max}+1)(N_{\rm max}+2)^2(N_{\rm max}+3)\\ &\xrightarrow[N_{\rm max}\gg 1]{} \tfrac{1}{4}(N_{\rm max}+2)^4\end{aligned}$$

In the limit of large $N_{\rm max}$, we obtain the result

$$\langle R^2\rangle = \frac{2}{A}\frac{\hbar}{\mu\omega_0}\frac{1}{4}(N_{\rm max}+2)^4$$

which relates the square of the nuclear radius to the value of $N_{\rm max}$.

Alternatively, we can use this relation to express $\hbar\omega_0$, a quantum of harmonic oscillator energy, in terms of $N_{\rm max}$,

$$\hbar\omega_0 = \frac{1}{A}\frac{\hbar^2}{\mu\langle R^2\rangle}\frac{1}{2}(N_{\rm max}+2)^4 \tag{7-16}$$

The number of nucleons A can also be related to $N_{\rm max}$ using Eq. (7-13),

$$A = 2\sum_{N=0}^{N_{\rm max}} D_N = \tfrac{2}{3}(N_{\rm max}+2)^3$$

where the factor of 2 is used to account for the fact that each harmonic oscillator state can take a neutron as well as a proton with a given spin orientation. On inverting the relation, we obtain the expression

$$(N_{\rm max}+2) = (\tfrac{3}{2}A)^{1/3} \tag{7-17}$$

Combining the results of Eqs. (7-16) and (7-17), we obtain

$$\begin{aligned}\hbar\omega_0 &= \frac{1}{A}\frac{\hbar^2}{\mu\langle r^2\rangle}\frac{1}{2}\Big(\frac{3}{2}A\Big)^{4/3} \\ &= \frac{\hbar^2}{\mu\frac{3}{5}(r_0A^{1/3})^2}\frac{3}{4}\Big(\frac{3}{2}A\Big)^{1/3} \\ &= \frac{5}{4}\Big(\frac{3}{2}\Big)^{1/3}\frac{\hbar^2}{\mu r_0^2}A^{-1/3} \approx 41A^{-1/3}\ \text{MeV}\end{aligned} \tag{7-18}$$

where we have adopted a constant-density sphere model to convert $\langle r^2\rangle$ to $\frac{3}{5}(r_0A^{1/3})^2$, as done in Eq. (4-20), and used $r_0 = 1.2$ fm to arrive at the final result, invoked earlier to characterize the energy required to excite a nucleon up one major shell.

Spin-orbit energy. Let us go back to the question of magic numbers. From Eq. (7-13), we find that the first part of the sequence, 2, 8, 20, and 40, is accounted for by, respectively, filling up harmonic oscillator shells with either neutrons or protons up to $N_{\max} = 0, 1, 2, 3$. This gives us an indication that the harmonic oscillator potential is a reasonable starting point for understanding the structure of single-particle states in nuclei. However, deviations are found beyond $N_{\max} = 3$. To correct for this, additional terms must be introduced into the single-particle Hamiltonian beyond what is given by the harmonic oscillator potential of Eq. (7-11).

The departure of the sequence of magic numbers from the values given by $D_{\max}$ in Eq. (7-13) is explained by single-particle spin-orbit energy, suggested by Mayer and Haxel, Jensen, and Suess in 1949 (see Ref. [102]). If the potential that binds a nucleon to the central well has a term that depends on the coupling between $\boldsymbol{s}$, the intrinsic spin of a nucleon, and $\boldsymbol{\ell}$, its orbital angular momentum, the single-particle energies will be a function of the j-value of a state as well. Since $\boldsymbol{j} = \boldsymbol{s} + \boldsymbol{\ell}$, two possible states can be formed from a given ℓ and the energies of the two are different, depending on whether $\boldsymbol{s}$ is parallel to $\boldsymbol{\ell}$ ($j = \ell + \frac{1}{2}$) or antiparallel to $\boldsymbol{\ell}$ ($j = \ell - \frac{1}{2}$). The source of this single-particle spin-orbit term may be traced back to the spin dependence in the nucleon-nucleon interaction. For our purpose here, we shall, for simplicity, take a semi-empirical approach without any concern for the origin.

Let a be the strength of the spin-orbit term. The single-particle Hamiltonian of Eq. (7-10) now takes on the form

$$h(\boldsymbol{r}) = -\frac{\hbar^2}{2\mu}\nabla^2 + \tfrac{1}{2}\mu\omega_0^2 r^2 + a\,\boldsymbol{s}\cdot\boldsymbol{\ell} \tag{7-19}$$

where the parameter a may depend on the nucleon number A and can be determined, for example, by fitting observed single-particle energies. When the spin-orbit term is included, the single-particle energy of Eq. (7-12) becomes

$$\epsilon_{N\ell j} = \big(N + \tfrac{3}{2}\big)\hbar\omega_0 \begin{cases} +\frac{1}{2}a\ell & \text{for } j = \ell + \frac{1}{2} \\ -\frac{1}{2}a(\ell+1) & \text{for } j = \ell - \frac{1}{2}\end{cases}$$

The splitting in energy between the $j_> \equiv \ell + \frac{1}{2}$ single-particle level and the $j_< \equiv \ell - \frac{1}{2}$ level is $a(2\ell+1)/2$. However, the centroid energy of the two groups is not affected.

For $a < 0$, a single-particle state with $j = j_> = \ell + \frac{1}{2}$ is lowered in energy. Since the amount of depression increases with increasing ℓ-value, a $j_>$-state for large ℓ may be pushed down in energy by an amount comparable to $\hbar\omega_0$, the energy gap between two adjacent harmonic oscillator major shells. As a result, the $j_>$-states of the largest ℓ in a shell with N oscillator quanta may be moved closer to the group of states belonging to the $N-1$ shell below. [In practice, as we shall later in Eq. (7-29), one needs also an ℓ^2-dependent term in the single-particle Hamiltonian to lower the centroid energy of states with large ℓ-values so that the $j_<$-states are prevented from moving up to join the states in the harmonic oscillator shell higher up.] Because of spin-orbit splitting, we find that the $j = \frac{9}{2}$ single-particle states for $\ell = 4$ in the $N = 4$ shell are depressed sufficiently in energy that their location is closer to the $N = 3$ group. As a result, the $j = \frac{9}{2}$ states join those of $N = 3$ to form a major shell of 30 single-particle states instead of 20. For this reason, we have 50 instead of 40 as the magic number for neutrons. Similarly, the magic number 82, instead of 70, is obtained if the $j = \frac{11}{2}$ states of the $\ell = 5$ group in the $N = 5$ shell are lowered in energy to join the $N = 4$ group. By the same token, the magic number 126 is formed by summing all the particles in the $N \leq 5$ shells (totaling 112) together with those filling the $j = \frac{13}{2}$ orbit (which accommodates $2j + 1 = 14$ identical nucleons) from the major shell above. Following this line of reasoning, the first magic number beyond the known ones is 184.

A point to be noted here is the absence of a doubly magic (both N and Z magic numbers) nucleus with $Z = 50$. Because of Coulomb repulsion, nuclei beyond ^{40}Ca must have an excess of neutrons over protons to be stable and the amount of neutron excess required increases with Z. For ^{90}Zr ($Z = 40$), we find that the neutron excess is $N - Z = 10$ and for ^{208}Pb ($Z = 82$), the excess increases to $N - Z = 44$. To form a stable nucleus with $Z = 50$, we expect a neutron excess somewhere between 10 and 20. The next higher magic number after 50 is 82. Since $N = 82$ gives too large a neutron number for $Z = 50$, a doubly magic nucleus with $Z = 50$ cannot be constructed. In spite of this, we do find that the element Sn ($Z = 50$) has more stable isotopes than those nearby. Other properties of the stable tin isotopes also support the observation that empirically $Z = 50$ is one of the magic numbers, producing nuclei that are more tightly bound than their neighbors.

Superheavy nuclei. The heaviest closed shell nucleus known is ^{208}Pb with $Z = 82$ and $N = 126$. Calculations indicate that the next stable proton number may be 114 because of the large separation in single-particle energy between two groups of proton orbits, one consisting of $1h_{9/2}$, $1i_{13/2}$, and $2f_{7/2}$ and the other of $3p_{3/2}$ and $2f_{7/2}$. There is a similar separation for the neutron orbits but the energy gap is smaller and no clear indication for a neutron subshell at $N = 114$ is found among empirical evidence. Since $Z = 114$ is not too far from the end of the actinide series at $Z = 103$, there is some possibility that a "superheavy" element with $A = 298$ ($Z = 114$ and $N = 184$) can be made in the laboratory. Alternatively, we may use the known magic number of 126 as the proton number and end up with $A = 310$ as the possible candidate for a superheavy nucleus. Many experimental attempts have been made to find these nuclei and to discover a new group, or "island," of stable nuclei around the next set of magic numbers. As an important step in this direction, the element $Z = 112$ and $N = 165$ was created in the laboratory, as we shall see later in §9-1.

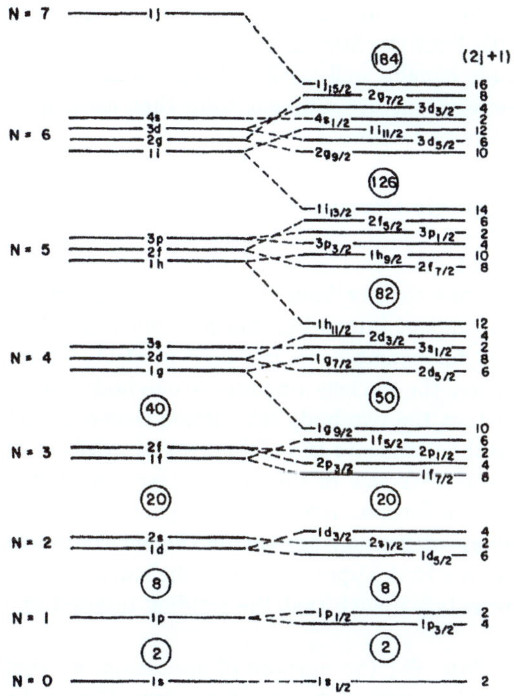

Figure 7-3: Schematic diagram of single-particle energy spectrum for spherical shell model. The lowest three major shells, $1s$, $1p$ and $2s1d$, are the same as those produced by a three-dimensional, isotropic harmonic oscillator well. The higher major shells include also the orbit with the largest j-values lowered in energy from the harmonic oscillator shell above by spin-orbit energy.

Spectroscopic notation. We shall end this section with a description of the standard nomenclature for (spherical) single-particle orbits in nuclear physics. Each orbit is identified by three labels, N, ℓ, and j. The convention is to use spectroscopic notation with a single letter s, p, d, f, g, h, i, j, ... to stand for $\ell = 0$, 1, 2, 3, 4, 5, 6, 7, ... , respectively. The j-value is indicated as a subscript following the letter and the label for major shell is given as a prefix. It is customary to replace N, the number of harmonic oscillator quanta in a major shell, by n, the number of nodes in the (modified) radial wave function. There are at least two conventions to number n, depending whether one counts the node at the origin. We shall follow the one with $n = 1$, 2, ... for the first time (one node), the second time (two nodes), and so on for a particular ℓ-value to appear in the sequence of single-particle orbits arranged in ascending order according to energy. For example, the single-particle orbit with the lowest energy is $\ell = 0$ without a node in the wave function. It is called the $1s_{1/2}$-orbit. A higher $\ell = 0$ orbit found in the $N = 2$ shell is labeled $2s_{1/2}$. The next higher $\ell = 0$ orbit at $N = 4$ is the $3s_{1/2}$, and so on. Figure 7-3 gives a more complete illustration of this way of labeling the single-particle orbits. The alternate convention is basically the same except that instead of starting with 1 for the first occurrence of a particular ℓ-value, it starts with 0. Both conventions are equally well used and some confusion may arise on occasion. Furthermore, both conventions are different from that used in atomic physics and in certain quantum mechanics textbooks.

7-3 Hartree-Fock Single-Particle Hamiltonian

In its most elementary form, the nuclear Hamiltonian, as given earlier in Eq. (7-2), is a sum of two terms, one coming from the kinetic energy of the individual nucleons and the other from the mutual interaction between nucleons. There is no fundamental one-body potential as, for example, in the case of electrons in an atom where the Coulomb attraction from the nucleus provides a one-body interaction. When this is coupled with the fact that the two-body interaction between nucleons is fairly strong, it is not easy in general to truncate the Hilbert space involved down to sizes suitable for practical calculations. We saw in the previous section that one way to solve this difficulty is to make a transformation of the single-particle basis wave functions so that the single-particle energies reflect a large part of the nucleon-nucleon interactions. One of the aims of the Hartree-Fock approach to the nuclear many-body problem is to find a single-particle representation such that the residual interaction is small.

Variational calculation. For the purpose of discussion, we shall start with an arbitrary set of single-particle states as the trial functions. The final result is, in principle, independent of this choice; in practice, it is advantageous to take a set that is convenient from a mathematical point of view, such as harmonic oscillator wave functions.

In the absence of a two-body residual interaction, the ground state is given by the configuration with the lowest A single-particle states occupied and the Fermi energy ϵ_F is determined by the highest occupied single-particle state. The wave function $|\,\Phi_0\,\rangle$ for a system of A nucleons is a Slater determinant constructed out of this set of single-particle wave functions. To simplify the notation, let us use $|\phi_1\phi_2\cdots\phi_A\rangle$ to represent the Slater determinant given by Eq. (7-6). Thus, we have

$$|\Phi_0\rangle = |\phi_1\phi_2\cdots\phi_A\rangle \tag{7-20}$$

If $|\,\Phi_0\,\rangle$ is the true ground state wave function of the system, it must satisfy the variational condition

$$\delta\langle\phi_1\phi_2\cdots\phi_A|H|\phi_1\phi_2\cdots\phi_A\rangle = 0 \tag{7-21}$$

That is, the function $|\,\Phi_0\,\rangle$ is one that produces a minimum in the energy. The aim of a Hartree-Fock calculation is to find a set of single-particle states that, as far as possible, fulfills this condition.

Since variations on the bra $\langle\Phi_0|$ are not independent of those on the ket $|\Phi_0\rangle$, Eq. (7-21) is equivalent to the condition

$$\langle\delta\Phi|H|\Phi_0\rangle = 0 \tag{7-22}$$

In other words, the aim of Eq. (7-21) can be achieved by a variation of either the bra or the ket alone. There are two possible ways to carry out the operation on a many-body wave function. Either we can modify the single-particle wave functions themselves such that Eq. (7-22) is satisfied or, alternatively, we can keep the single-particle basis fixed and alter Φ_0 by adding to it small amounts of Slater determinants made of products of different A single-particle states. As long as there is a complete set of states and all possible variations are applied, these two methods are equivalent to each other.

We shall take the latter approach for our derivation here. With a fixed single-particle basis, each many-particle basis state may be labeled by the single-particle states occupied. For example, the trial ground state wave function Φ_0 in Eq. (7-20) is one with the lowest A single-particle states occupied (and those above empty). A different many-body wave function will have a different set of occupied single-particle states. A linear combination of two or more such many-body basis states then means that some of the single-particle states are partially occupied, and the occupancies of these states take on fractional value between 0 and 1 as a result.

Let us label the basis, for instance, by numbering all the single-particle states with an index $r = 1, 2, \ldots, d$ in ascending order according to single-particle energy ϵ_r. A many-body basis state may be specified in this case by giving the indices of the occupied single-particle states. Thus, the lowest many-body state, as far as the sum of single-particle energies is concerned, may be represented as

$$|\Phi_0\rangle = |1, 2, \ldots, A\rangle$$

Such a scheme of displaying a many-body state is sometimes referred to as the *occupancy representation.*

A variation on $|\Phi_0\rangle$ may be carried out by mixing a small amount of $|\Phi_{kt}\rangle$, made by promoting a particle from single-particle state t below the Fermi energy to single-particle state k above. Such a state may be represented as

$$|\Phi_{kt}\rangle = |1, 2, \ldots, t-1, t+1, \ldots, A, k\rangle$$

Many-body states constructed in this way, by promoting a nucleon from an occupied single-particle state below the Fermi level of the A-nucleon system to an unoccupied state above, are called one-particle one-hole states, or $1p1h$-states for short. Other variations, involving two-particle two-hole ($2p2h$) and more complicated types of excitation, can also be considered, but we shall ignore them here.

An arbitrary variation consisting of all possible $1p1h$-excitations may be written in the form

$$|\delta\Phi\rangle = \sum_{kt} \eta_{kt} |\Phi_{kt}\rangle$$

where η_{kt} denotes the amount of each component in the variation. To ensure that the variations are carried out in small steps, the absolute value of η_{kt} must be kept to be much smaller than unity. If we restrict ourselves to $1p1h$-excitations, Eq. (7-22) may be written as

$$\sum_{kt} \eta_{kt} \langle\Phi_{kt}|H|\Phi_0\rangle = 0$$

Since different $1p1h$-components, represented by different occupied single-particle state k and empty single-particle state t, are independent of each other, it is necessary that each term in the sum vanishes. From this requirement, we obtain the condition

$$\langle\Phi_{kt}|H|\Phi_0\rangle = \langle\Phi_{kt}|\sum_{i=1}^{A} h(\boldsymbol{r}_i) + \sum_{i\neq j=1}^{A} V(\boldsymbol{r}_i,\boldsymbol{r}_j)|\Phi_0\rangle = 0 \tag{7-23}$$

For the one-body part of the Hamiltonian, the only possible nonvanishing contribution comes from cases where the left- and right-hand sides of the matrix element differ by no more than the single-particle state of one nucleon. Similarly, for the two-body part of the Hamiltonian, only matrix elements with occupied single-particle states on the two sides of $V(\boldsymbol{r}_i,\boldsymbol{r}_j)$ differing by no more than two can be nonvanishing. Upon integrating over all the single-particle coordinates other than those acted upon by the operators, the condition expressed in Eq. (7-23) for a one- plus two-body Hamiltonian reduces to

$$\langle k|h(\boldsymbol{r})|t\rangle + \sum_{r}\langle kr|V(\boldsymbol{r}_1,\boldsymbol{r}_2)|tr\rangle = 0 \tag{7-24}$$

where the summation is over all the occupied states.

The first term in Eq. (7-24), $\langle k|h(\boldsymbol{r})|t\rangle$, is the matrix element of the one-body part of the Hamiltonian between single-particle states $|t\rangle$ and $|k\rangle$. Similarly, the matrix element of the two-body part, $\langle kr|V(\boldsymbol{r}_1,\boldsymbol{r}_2)|tr\rangle$, is between two antisymmetrized two-particle states, $|kr\rangle$ and $|tr\rangle$, given by

$$\begin{aligned}
|kr\rangle &\equiv \frac{1}{\sqrt{2}}\begin{vmatrix}\phi_k(\boldsymbol{r}_1) & \phi_k(\boldsymbol{r}_2)\\ \phi_r(\boldsymbol{r}_1) & \phi_r(\boldsymbol{r}_2)\end{vmatrix} = \frac{1}{\sqrt{2}}\big\{|\,\phi_k(\boldsymbol{r}_1)\phi_r(\boldsymbol{r}_2)\,\rangle - |\,\phi_r(\boldsymbol{r}_1)\phi_k(\boldsymbol{r}_2)\,\rangle\big\}\\
|tr\rangle &\equiv \frac{1}{\sqrt{2}}\begin{vmatrix}\phi_t(\boldsymbol{r}_1) & \phi_t(\boldsymbol{r}_2)\\ \phi_r(\boldsymbol{r}_1) & \phi_r(\boldsymbol{r}_2)\end{vmatrix} = \frac{1}{\sqrt{2}}\big\{|\,\phi_t(\boldsymbol{r}_1)\phi_r(\boldsymbol{r}_2)\,\rangle - |\,\phi_r(\boldsymbol{r}_1)\phi_t(\boldsymbol{r}_2)\,\rangle\big\}
\end{aligned} \tag{7-25}$$

Using these results, the two-body matrix element in Eq. (7-24), the second term on the left-hand side, may be expressed explicitly in terms of single-particle wave functions in the following way:

$$\begin{aligned}
&\big\langle kr\big|V(\boldsymbol{r}_1,\boldsymbol{r}_2)\big|tr\big\rangle\\
&\quad= \tfrac{1}{2}\big\langle\phi_k(\boldsymbol{r}_1)\phi_r(\boldsymbol{r}_2) - \phi_r(\boldsymbol{r}_1)\phi_k(\boldsymbol{r}_2)\big|V(\boldsymbol{r}_1,\boldsymbol{r}_2)\big|\phi_t(\boldsymbol{r}_1)\phi_r(\boldsymbol{r}_2) - \phi_r(\boldsymbol{r}_1)\phi_t(\boldsymbol{r}_2)\big\rangle\\
&\quad= \big\langle\phi_k(\boldsymbol{r}_1)\phi_r(\boldsymbol{r}_2)\big|V(\boldsymbol{r}_1,\boldsymbol{r}_2)\big|\phi_t(\boldsymbol{r}_1)\phi_r(\boldsymbol{r}_2)\big\rangle - \big\langle\phi_k(\boldsymbol{r}_1)\phi_r(\boldsymbol{r}_2)\big|V(\boldsymbol{r}_1,\boldsymbol{r}_2)\big|\phi_r(\boldsymbol{r}_1)\phi_t(\boldsymbol{r}_2)\big\rangle
\end{aligned}$$

where we have made use of the symmetry relation

$$\langle\phi_k(\boldsymbol{r}_1)\phi_r(\boldsymbol{r}_2)|V(\boldsymbol{r}_1,\boldsymbol{r}_2)|\phi_t(\boldsymbol{r}_1)\phi_r(\boldsymbol{r}_2)\rangle = \langle\phi_r(\boldsymbol{r}_1)\phi_k(\boldsymbol{r}_2)|V(\boldsymbol{r}_1,\boldsymbol{r}_2)|\phi_r(\boldsymbol{r}_1)\phi_t(\boldsymbol{r}_2)\rangle$$

The derivation from Eq. (7-23) to (7-24) may be carried out in a more elegant way using second-quantized notations. However, we shall not do it here because of the amount of preparation needed to introduce the notations.

Hartree-Fock Hamiltonian. It is perhaps more instructive to write the relation expressed by Eq. (7-24) in an operator form. For this purpose, the left-hand side of the equation may be regarded as the matrix element of a one-body operator, as it operates only on the single-particle state k to its right (or the single-particle state t to its left). However, $V(\boldsymbol{r}_1, \boldsymbol{r}_2)$ is a two-body operator, and we shall see how to "reduce" it to one-body.

Let us distinguish between two sets of single-particle states here by using Greek letters α, β, ... to indicate the original, or trial, single-particle states and the Roman alphabet, r, s, ... , for the Hartree-Fock single-particle states that satisfy Eq. (7-24). A two-body interaction potential may be expressed in the original basis as an operator in the following way:

$$\sum_{ij} V(\boldsymbol{r}_i, \boldsymbol{r}_j) = \sum_{\alpha\beta\gamma\delta} |\alpha\beta\rangle V_{\alpha\beta\gamma\delta} \langle\gamma\delta| \tag{7-26}$$

where

$$V_{\alpha\beta\gamma\delta} \equiv \langle\alpha\beta|V(\boldsymbol{r}_1, \boldsymbol{r}_2)|\gamma\delta\rangle$$

is the matrix element of $V(\boldsymbol{r}_1, \boldsymbol{r}_2)$ between antisymmetrized and normalized two-body wave functions $|\alpha\beta\rangle$ and $|\gamma\delta\rangle$, such as those given in Eq. (7-25). Using this form of the two-body potential, we can write the left-hand side of Eq. (7-24) as the one-body matrix element of the following operator:

$$h_{\text{HF}} = h + \sum_r \sum_{\alpha\beta\gamma\delta} \langle r|\alpha\beta\rangle V_{\alpha\beta\gamma\delta} \langle\gamma\delta|r\rangle \tag{7-27}$$

This is the Hartree-Fock single-particle Hamiltonian operator. The left-hand side of Eq. (7-24) is the matrix element of h_{HF} between single-particle bra $|\, k\,\rangle$ and single-particle ket $|\, t\,\rangle$. The quantity $\langle t|\alpha\beta\rangle$ in Eq. (7-27) is a one-body operator since it is the overlap of a two-body ket with a one-body bra. Except for the implied antisymmetrization in the two-body wave function, we may take the quantity as

$$\langle r|\alpha\beta\rangle \sim \langle r|\alpha\rangle|\,\beta\,\rangle$$

The second term of h_{HF} in Eq. (7-27) may be interpreted as the average one-body potential, or the *mean field*, experienced by a nucleon as the result of (two-body) interactions with each one of the other nucleons in the nucleus.

It is also possible to write Eq. (7-24) as an eigenvalue equation using the operator form given in Eq. (7-27),

$$h|\,t\,\rangle + \sum_r \sum_{\alpha\beta\gamma\delta} \langle r|\alpha\beta\rangle V_{\alpha\beta\gamma\delta} \langle\gamma\delta|rt\rangle = \epsilon_t|\,t\,\rangle \tag{7-28}$$

where ϵ_t is the Hartree-Fock single-particle energy. The solution provides us with a transformation from the set of basis states $|\,\alpha\,\rangle$, $|\,\beta\,\rangle, \ldots$, used as the trial wave functions, to the eigenstates of the Hartree-Fock single-particle Hamiltonian $|\,k\,\rangle$, $|\,t\,\rangle, \ldots$, defined by Eq. (7-28).

The calculation is not as straightforward as it may seem on the surface. To find both $\langle r|\alpha\beta\rangle$ and $\langle\gamma\delta|rt\rangle$ in Eq. (7-28), or the equivalent quantity of $\langle kr|V|tr\rangle$ in Eq. (7-24), we need an *a priori* knowledge of the solution, as these matrix elements are evaluated in the Hartree-Fock basis that forms a part of the end result of the calculation. This means that the calculation must be carried out iteratively starting with an arbitrary set of single-particle wave functions, such as harmonic oscillator wave functions. Using these trial functions, we can evaluate all the necessary matrix elements and solve Eq. (7-28) approximately using the trial functions for $|r\rangle$ and $|t\rangle$. The solutions obtained are only a first approximation, as we have not used the proper wave functions to evaluate the matrix elements to start with. On the other hand, the results represent a better approximation of the "true" Hartree-Fock wave functions than the trial wave functions used as the input. Now we can make improvements by using the first approximation results as the input and carrying out the calculation again. The process is repeated until self-consistency in the solution is achieved; that is, the solutions obtained are essentially identical to the wave functions used to evaluate the matrix elements.

Hartree-Fock calculations have been used extensively to study low-lying states in nuclei, as, for example, that for ^{208}Pb shown in Fig. 7-4. However, since each nuclear wave function is made of a single Slater determinant, it does not correspond to a state with definite spin and isospin. In order to use them to calculate quantities that can be compared with experimental data, states of good J and T must be projected out of the Hartree-Fock wave function. We shall not go into the technical detail of how to carry out the projection or extensions of the topic to projected Hartree-Fock where the variational calculation is carried out after spin and isospin projections. The self-consistent single-particle basis obtained here is, however, important for understanding nuclear ground states as well as the nuclear shell model to be discussed in §7-5.

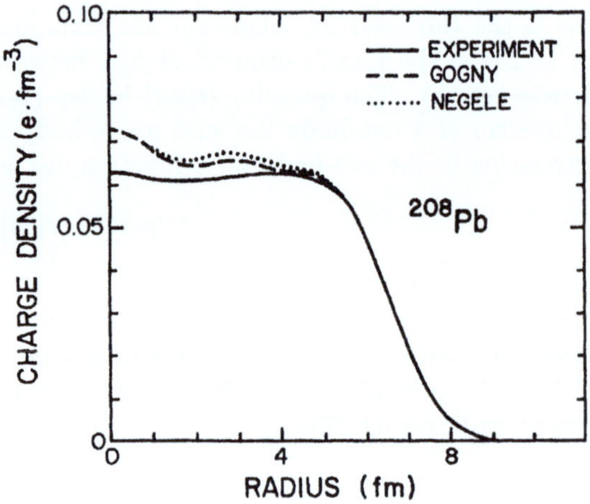

Figure 7-4: Charge density of ^{208}Pb measured by Frois et al. [67] using elastic electron scattering. The experimental data (solid line) in the surface region of the nucleus are well described by the calculated Hartree-Fock results of Gogny [74] and Negele [110]. The main differences are in the interior region.

7-4 Deformed Single-Particle States

So far, we have taken the effective one-body potential arising from the average interaction of a nucleon with all the other nucleons in the nucleus to be spherical. This is a reasonable assumption only if the equilibrium shape of the nucleus itself has the same shape. As we saw earlier in §6-3, for many nuclei, a deformed shape is actually more stable. For these nuclei, it is more appropriate to use a deformed average potential well to generate the single-particle basis states.

Deformed single-particle Hamiltonian. Let us again assume axial symmetry for simplicity. Instead of Eq. (7-19), we can adopt a semi-empirical single-particle Hamiltonian of the following form:

$$h(\boldsymbol{r}) = h_0 + h_\delta + a\,\boldsymbol{\ell}\cdot\boldsymbol{s} + b\,\boldsymbol{\ell}^2 \tag{7-29}$$

Here h_0 is the spherical part, generally taken to be the Hamiltonian of an isotropic three-dimensional harmonic oscillator similar to that given in Eq. (7-10),

$$h_0 = -\frac{\hbar^2}{2\mu}\nabla^2 + \tfrac{1}{2}\mu\omega_0^2 r^2$$

The deformation is produced by h_δ and is often taken to be that due to a quadrupole field,

$$h_\delta = -\tfrac{1}{3}\delta_{\rm osc}\mu\omega_0^2 r^2\sqrt{\frac{16\pi}{5}}\,Y_{20}(\theta,\phi)$$

where $\delta_{\rm osc}$, to be defined later in Eq. (7-32), provides a measure of the departure from a spherical shape. We have already encountered the other two terms in the single-particle Hamiltonian. The spin-orbit term $a\,\boldsymbol{\ell}\cdot\boldsymbol{s}$, given in §7-1, is required to account for the magic numbers, and the term $b\,\boldsymbol{\ell}^2$ is used to give the proper ordering of single-particle states in the spherical limit. Deformed single-particle states, produced as the eigenvectors of the Hamiltonian given in Eq. (7-29), are often referred to as the Nilsson states or Nilsson orbitals.

Labels for deformed single-particle states. With a deformed Hamiltonian, the spin j is no longer a constant of motion and a new set of labels must be found to identify a single-particle state. As we have seen earlier in §6-3 in a discussion of the rotational model, the third component of $\boldsymbol{j}$ remains a good quantum number for axially symmetric nuclei. This gives us Ω, the projection of $\boldsymbol{j}$ on the body-fixed quantization axis (the 3-axis), as one of the labels.

The wave function of a Nilsson state may be expressed as a linear combination of spherical harmonic oscillator states $|\,N\ell j\Omega\,\rangle$,

$$|\,N\Omega\,\rangle = \sum_{\ell j} C_{N\ell j}|\,N\ell j\Omega\,\rangle \tag{7-30}$$

as deformation mixes spherical orbits. The expansion coefficients $C_{N\ell j}$ depend on the value of the deformation parameter $\delta_{\rm osc}$. In the limit of zero deformation, the states become identical with those in the spherical case used in the previous sections. If admixtures between spherical states belonging to different major harmonic oscillator shells are not permitted in constructing the deformed single-particle basis, N, the number of harmonic oscillator quanta remains a constant of motion and may be used as one of the quantities to describe a deformed state.

Labels N and Ω alone cannot uniquely specify a deformed state; two additional quantities, n_3 and λ, are commonly used. The origin of these labels may be summarized in the following way. For large deformations, we can ignore the effects of the $\boldsymbol{\ell}\cdot\boldsymbol{s}$ and

ℓ^2 terms and treat $h_0 + h_\delta$ in a cylindrical coordinate system. The Hamiltonian of Eq. (7-29) can now be written as

$$h = -\frac{\hbar^2}{2\mu}\nabla^2 + \tfrac{1}{2}\mu\{\omega_3^2 x_3^2 + \omega_\perp^2(x_1^2 + x_2^2)\} \tag{7-31}$$

where the oscillator frequency along the symmetric axis has the value,

$$\omega_3 = \omega_0\sqrt{1 - \tfrac{4}{3}\delta_{\rm osc}} = \varpi(1 - \tfrac{2}{3}\delta_{\rm osc})$$

and in the directions perpendicular to it,

$$\omega_\perp = \omega_0\sqrt{1 + \tfrac{2}{3}\delta_{\rm osc}} = \varpi(1 + \tfrac{1}{3}\delta_{\rm osc})$$

In terms of $\omega_\perp$ and ω_3, the deformation parameter $\delta_{\rm osc}$ is given by the relation

$$\delta_{\rm osc} = \frac{\omega_\perp - \omega_3}{\varpi} \tag{7-32}$$

where the average frequency

$$\varpi = \tfrac{1}{3}(\omega_1 + \omega_2 + \omega_3) = \tfrac{1}{3}(2\omega_\perp + \omega_3) \approx \omega_0$$

may be taken to be the same as ω_0, the harmonic oscillator frequency in the spherical case. The parameter $\delta_{\rm osc}$ is very similar but not identical to β of Eq. (6-11). For an axially symmetric nucleus, it can be shown[1] that

$$\beta = \frac{1}{3}\sqrt{\frac{16\pi}{5}}\delta_{\rm osc} + O(\delta_{\rm osc}^2) = 1.057\delta_{\rm osc} + O(\delta_{\rm osc}^2)$$

The difference between $\delta_{\rm osc}$ here, β of Eq. (6-11), and δ of Eq. (6-18) is that $\delta_{\rm osc}$ is given in terms of the harmonic oscillator frequencies for different directions, whereas β and δ are given in terms of the values for various radii. All three ways of parametrizing quadrupole deformation are used in the literature. To first order, $\delta_{\rm osc} \approx \delta \approx 0.945\beta$. More detailed relation between these two quantities are given in Ref. [35].

In the limit of large deformation, the single-particle energy of a deformed state may be expressed in terms of the number of oscillator quantum n_i along each of the three principal (body-fixed) axes,

$$\begin{aligned}\epsilon_{Nn_3\Omega} &= \left(n_3 + \tfrac{1}{2}\right)\hbar\omega_3 + \left(n_2 + \tfrac{1}{2}\right)\hbar\omega_2 + \left(n_1 + \tfrac{1}{2}\right)\hbar\omega_1 \\ &= \left(n_3 + \tfrac{1}{2}\right)\hbar\omega_3 + (n_\perp + 1)\hbar\omega_\perp \end{aligned} \tag{7-33}$$

where, for the axially symmetric case under discussion here, $n_\perp = n_1 + n_2$ is the number of quanta in the direction perpendicular to the symmetry axis. Since the total number of harmonic oscillator quanta is fixed, we have

$$N = n_3 + n_\perp$$

[1] See, e.g., Eq. (2.82) of Ref. [119]

The label n_3 (or $n_\perp$) is a good quantum number in this limit and may be used as a label for Nilsson states even in cases where the deformation is not large.

When the term $a\boldsymbol{\ell}\cdot\boldsymbol{s}$ is included, states with different projections of orbital angular momentum on the symmetry axis are no longer degenerate. As a result, in the limit of large deformation, the projection of the orbital angular momentum along the symmetry axis,

$$\lambda = \pm n_\perp, \pm(n_\perp - 2), \ldots, \pm 1, 0$$

may be used as the fourth label. The set of four labels $[Nn_3\lambda\Omega]$ completely specifies a deformed state within one major shell. Examples of the variation of energy with deformation for the low-lying Nilsson states are shown in Fig. 7-5. More complete results can be found, for example, in Ref. [95].

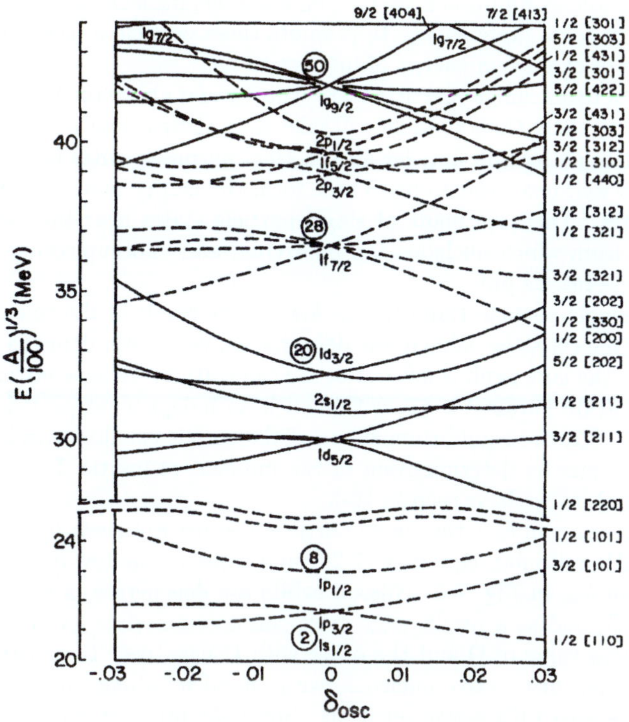

Figure 7-5: Low-lying Nilsson single-particle energies, the eigenvalues of the Hamiltonian given by Eq. (7-29), as a function of deformation parameter δ_{osc}.

Many-body states in the Nilsson scheme. The Nilsson orbitals for a deformed nucleus may be thought of as the equivalent of Hartree-Fock single-particle states in the spherical limit. The major difference, apart from the shape of the potential, is that a Hartree-Fock calculation starts with a nucleon-nucleon potential. In contrast, the Nilsson model, in its most elementary form, adjusts the deformation parameter to fit the nuclear shape. Furthermore, the Hamiltonian given by Eq. (7-29) is a very simple and intuitive one. The fact that it works so well in describing a wide range of nuclear properties has led to a number of interesting developments in nuclear structure. Some of these will be discussed in §9-2.

For an independent particle approximation in the deformed basis, we can ignore the residual interaction. The nuclear Hamiltonian is then a sum of the single-particle Hamiltonian given in Eq. (7-29) for all the active nucleons. The single-particle states may be divided into three groups: the core states, the valence states, and the empty states. To restrict the size of the Hilbert space for a calculation, we shall take some of the single-particle states as permanently filled with nucleons. Since these nucleons are never excited, the only role they play is to provide an average (deformed) potential for the rest of nucleons. In other words, they form an "inert core" for our active nucleons. This is a reasonable assumption if we put, for example, nucleons in the lowest few states as the core, as it takes more energy to promote these particle to the unoccupied states than the range of excitation energy of interest to us.

By the same token, single-particle states that are far above the Fermi energy are not of interest, as any excitations to them will take more energy than we are concerned with. As a result, these states will not, for all practical purposes, enter into our calculations and we may as well leave them out. These form the empty states. The remainder are the active states. They are made of single-particle states near the Fermi level, both occupied ones from which nucleons can be excited and unoccupied one into which the excited nucleons can be put.

Since the single-particle Hamiltonian $h(\boldsymbol{r})$ is the result of the valence nucleons interacting with the core, the value of the deformation parameter depends on the shape of the core. Since the core itself is left out of the calculation, we have no way of determining the deformation parameter $\delta_{\rm osc}$ within such an independent particle approach. As we shall see in §9-2, the equilibrium shape of the core is usually considered separately, in the same manner as determination of the moment of inertia $\mathcal{I}$ is outside a simple rotational model, as we have seen in §6-3.

For a given value of $\delta_{\rm osc}$, the single-particle energies are given by the eigenvalues of the deformed Hamiltonian equation (7-29) and some of the low-lying ones are shown as a function of $\delta_{\rm osc}$ in Fig. 7-5. Nilsson orbits are degenerate in energy with respect to the sign of Ω, and as a result, each orbit can accommodate two identical nucleons, one with positive value of Ω and the other with Ω negative. To construct the ground state configuration of n active nucleons, we proceed by filling up the lowest available single-particle states with active nucleons. Since the order of Nilsson states $\epsilon_{Nn_3\lambda\Omega}$ is different for different deformations, as can be seen in Fig. 7-5, the many-particle states formed depends on the value of $\delta_{\rm osc}$. Let us illustrate this point using a few specific examples.

Examples of Nilsson model calculation. For simplicity we shall examine the ground state spin of a few nuclei at the beginning of the ds-shell. For these, we can take ^{16}O as the inert core and the ds-shell as the active space. In a spherical basis, the $1d_{5/2}$ single-particle energy is the lowest one among the ds-shell orbits, as can be seen in Fig. 7-5 for $\delta_{\rm osc} = 0$. In this case, we expect that the lowest energy configuration of nuclei in the beginning of the ds-shell, from ^{17}O to ^{28}Si, is made up by filling the

$1d_{5/2}$-orbit with nucleons. The even-even nuclei are not of interest here, as the ground states must have $J^\pi = 0^+$ because of pairing interaction. For an odd-mass nucleus, the ground state spin in an independent particle model is given by that of the single-particle state occupied by the unpaired nucleon. Consequently, nuclei ^{19}F, ^{19}Ne, ^{21}Ne, and ^{23}Na are expected to have $J^\pi = \frac{5}{2}^+$ for their ground state spins if they are spherical. Experimentally, they turn out to be different and have, instead, values $\frac{1}{2}^+$, $\frac{1}{2}^+$, $\frac{3}{2}^+$, and $\frac{3}{2}^+$, respectively.

In the deformed basis, the observed values are understood in the following way. The nuclei in the lower half of the ds-shell are known to be deformed and have predominantly prolate spheroidal shapes (axially symmetric with $\delta_{\rm osc} > 0$). From Fig. 7-5, we see that, for positive deformation, the positions of deformed orbitals above the ^{16}O core ($N = Z = 8$) are $\Omega[Nn_3\lambda] = \frac{1}{2}[220]$, $\frac{3}{2}[211]$, $\frac{5}{2}[211]$, in ascending order according to energy. Since each Nilsson orbit can accommodate two identical nucleons, the ground state configurations of the four nuclei in question are those shown in Table 7-2.

Table 7-2: Examples of using Nilsson orbitals.

Nucleus	Proton configuration	Neutron configuration	$K = \sum \Omega_i$
^{19}F	$\frac{1}{2}[220]^1$	$\frac{1}{2}[220]^2$	$\frac{1}{2}$
^{19}Ne	$\frac{1}{2}[220]^2$	$\frac{1}{2}[220]^1$	$\frac{1}{2}$
^{21}Ne	$\frac{1}{2}[220]^2$	$\frac{1}{2}[220]^2\frac{3}{2}[211]^1$	$\frac{3}{2}$
^{23}Ne	$\frac{1}{2}[220]^2\frac{3}{2}[211]^2$	$\frac{1}{2}[220]^2\frac{3}{2}[211]^1$	$\frac{3}{2}$

Note: Occupancies are indicated by the superscripts.

With deformed single-particle states, we do not have a definite spin $\boldsymbol{j}$ for each nucleon. As a result, it is not possible to couple the angular momenta of all the active nucleons to form the nuclear ground state spin $\boldsymbol{J}$. Instead, we shall proceed in the same way as we did in the case of rotational model. Since the projection of $\boldsymbol{j}$ on the body-fixed 3-axis is a constant of motion, the sum

$$K = \sum_i \Omega_i$$

for all the active nucleons is also a constant of motion. Using the fact that K is the projection of $\boldsymbol{J}$ on the 3-axis, the ground state spin of a deformed nucleus must have spin $J = K$. This is the same argument used earlier to deduce spin in the rotational model. From the last column of Table 7-2, we find that $K = \frac{1}{2}$ for ^{19}F and ^{19}Ne and $K = \frac{3}{2}$ for ^{21}Ne and ^{23}Ne. Hence $J = \frac{1}{2}$ for the first two nuclei and $J = \frac{3}{2}$ for the second pair, in agreement with observation. If the deformation were oblate ($\delta_{\rm osc} < 0$), the unpaired nucleon would have been in orbitals $\frac{5}{2}[202]$, $\frac{5}{2}[202]$, $\frac{1}{2}[220]$, and $\frac{1}{2}[220]$ instead, and the ground spins of the nuclei would have been $\frac{5}{2}$, $\frac{5}{2}$, $\frac{1}{2}$, and $\frac{1}{2}$, in contradiction to the observed values.

The residual interaction in the deformed single-particle basis is, in principle, small. However, for any detailed properties of nuclei, we need to go beyond the independent particle model. We shall discuss this topic in the next section for the spherical case. In a deformed basis, the mathematics involved is far more complicated, as we no longer have the simplicity of rotational symmetry, and hence spherical tensors and angular momentum algebra, to help us.

7-5 Spherical Shell Model

We have seen in the previous three sections that independent particle models, using either spherical or deformed basis, are able to account for a number of nuclear properties. For more precise information, it is necessary to include the residual interaction in our study. Since the independent particle states are already good approximations, we can make use of them as the basis for our more detailed studies. The role of the residual interaction may be viewed as introducing configuration mixing among such states so that the wave functions, now made of linear combinations of these "basis" states, give a better description of the actual physical situation.

The many-particle space spanned by different products of the single-particle states is infinite in dimension. To carry out any practical calculations, it is necessary to truncate the Hilbert space to a finite one. Since our choice of single-particle states and, hence, the many-body basis states is based on physical grounds, such as the Hartree-Fock approach discussed in §7-3, we expect that reasonable approximations to the true nuclear wave functions can be achieved in a relatively small part of the complete space. Furthermore, we saw earlier in §7-2 that single-particle orbits are separated into "shells" because of the large energy gaps between groups of orbits. This provides us with a natural way to select the active space. For this reason, the approach is given the name *shell model.* In principle, a shell-model calculation can also be carried out in a deformed basis. However, mathematical convenience makes the approach viable mainly in the spherical limit, and this is what we shall restrict ourselves to.

There are three steps that must be carried out before we can perform the calculations: the choice of a single-particle basis, the selection of an active space, and the derivation of an effective interaction. These three steps are intimately related to each other, as we shall see from the discussion following.

Selection of the shell-model space. In the spherical shell model, each nucleon has an intrinsic spin $\boldsymbol{s}$ and occupies a state of definite orbital angular momentum $\boldsymbol{\ell}$. The many-body basis states formed by putting A nucleons into single-particle states are coupled together to form states with definite total angular momentum $\boldsymbol{J}$ and isospin $\boldsymbol{T}$. There are two ways to carry out the angular momentum coupling. In the LS-coupling scheme, the orbital angular momentum $\boldsymbol{\ell}_i$ and the intrinsic spin $\boldsymbol{s}_i$ of each one of the nucleons are first coupled separately to total orbital angular momentum $\boldsymbol{L}$ and total intrinsic spin $\boldsymbol{S}$:

$$\boldsymbol{L} = \sum_{i=1}^{A} \boldsymbol{\ell}_i \qquad\qquad \boldsymbol{S} = \sum_{i=1}^{A} \boldsymbol{s}_i$$

The total angular momentum, or *spin*, of the state is the vector sum of $\boldsymbol{L}$ and $\boldsymbol{S}$,

$$\boldsymbol{J} = \boldsymbol{L} + \boldsymbol{S}$$

Alternatively, in the jj-coupling scheme, the orbital angular momentum and the intrinsic spin of each nucleon is coupled together first to form the nucleon spin $\boldsymbol{j}_i$,

$$\boldsymbol{j}_i = \boldsymbol{\ell}_i + \boldsymbol{s}_i$$

then the nuclear spin

$$\boldsymbol{J} = \sum_{i=1}^{A} \boldsymbol{j}_i$$

is obtained by summing over the spins $\boldsymbol{j}_i$ of individual nucleons.

In a spherical basis, the Hamiltonian is invariant under a rotation of the coordinate system and J is a good quantum number. Furthermore, isospin T is also a constant of motion if we ignore symmetry-breaking effects due to electromagnetic interaction. For these reasons, Hamiltonian matrix elements between states of different J and T values vanish. If the many-body basis states are grouped together according to their (J, T) values, the Hamiltonian matrix in the complete shell-model space appears in a block-diagonal form; that is, only square blocks of matrix elements along the diagonal corresponding to a given set of (J, T) values are different from zero. The calculation can therefore be carried out separately within the subspace of a specific (J, T) of the full shell-model space. In this way, angular momentum coupling greatly reduces the size of the Hilbert space in which a calculation has to be carried out.

To truncate the Hilbert space, we can follow the same procedure as outlined in the previous section for a deformed basis. The nucleons are divided into two groups, core nucleons and valence nucleons. The single-particle states are separated into three categories, core states, active states, and empty states.

In most nuclear structure investigations, we are primarily interested in a few low-lying states. As a result, only nucleons in single-particle states near the Fermi surface are directly involved. The rest of the nucleons are in low-lying single-particle states and they are seldom excited. For all practical purposes, they can be assumed to form an inert *core*. Their contributions to the Hamiltonian may be separated into two parts. The first is a constant term in energy, made of single-particle energies and mutual interaction between nucleons in the core. Such a constant can usually be absorbed into the definition of the zero point of the energy scale for the A-nucleon system and may be ignored. The only exception occurs in calculations such as those involving the total binding energy of the nucleus. The second is the binding energies of the active nucleons provided by the core. The single-particle energy of an active nucleon consists of the kinetic energy as well as the average interaction energy with all the other nucleons, including those in the inert core. Contributions from the core nucleons cannot be ignored here but can be easily accounted for in the definition of single-particle energies for the active nucleons. The net result is that the effect of the core in a shell-model calculation may be included without explicitly considering the nucleons in it. In this way, the single-particle states occupied by the core nucleons may be left out of the active space.

Similarly, there are single-particle states so high above the Fermi energy that any many-body basis states having nucleons occupying these states will be very high in energy. If our interest is confined to the low-lying region of a nucleus, it is unlikely that there can be any significant contributions from these basis states. Such a set of single-particle states is therefore essentially always empty for our purposes and, as a result, may also be ignored.

The only remaining single-particle states are the few near the Fermi energy and they form the *active,* or *valence,* space from which reasonable approximate wave functions of the nuclear states of interest can be constructed. The aim of the nuclear shell model is to solve the eigenvalue problem in the space constructed out of single-particle states in the active space alone.

Effective Hamiltonian. What is the appropriate Hamiltonian to be used in a shell-model space? We have implicitly assumed a Hartree-Fock single-particle basis for this section. As a result, we have already made a transformation of the nuclear Hamiltonian from its fundamental form given by Eq. (7-2), consisting of a single-particle kinetic energy term and a nucleon-nucleon interaction term, to that of Eq. (7-7), made of single-particle Hartree-Fock energies and residual interaction. Mathematically, we need to make another transformation here from the infinite-dimensional space, specified by all the Hartree-Fock single-particle states, to a finite, truncated shell-model space. Physically, we need to find an *effective* Hamiltonian such that when the active shell-model space is restricted to a manageable size, the effect of the states ignored in the calculation may be accounted for in an efficient manner.

A formal definition of the effective Hamiltonian H_{eff} may be made in the following way. Let P be an operator that projects out a finite shell-model space of dimension d in which we wish to carry out the calculations. If the "true" Hamiltonian is H, the eigenvalue problem in the complete Hilbert space may be written as

$$H\Psi_i = E_i\Psi_i$$

where E_i is an eigenvalue and Ψ_i is the corresponding eigenvector of H. An ideal effective Hamiltonian is one that satisfies the condition

$$H_{\text{eff}}P\Psi_i = E_iP\Psi_i \tag{7-34}$$

In other words, an effective Hamiltonian is one which produces the same eigenvalues and eigenfunctions as those obtained by solving the problem in the complete space using the true Hamiltonian. In general, it is impossible to satisfy this condition for all d eigenvalues in the truncated shell-model space. This is, however, not a problem, as we are interested only in a small number of low-lying states that is much less than d.

The effective Hamiltonian may also be written as a sum of two terms,

$$H_{\text{eff}} = H_0 + V_{\text{eff}} \tag{7-35}$$

where the one-body part, H_0, may be taken, in our case, to be a sum of Hartree-Fock single-particle Hamiltonians,

$$H_0 = \sum_i h(\boldsymbol{r}_i) = \sum_i \epsilon_i \boldsymbol{n}_i$$

Here ϵ_i is the energy and $\boldsymbol{n}_i$ the number operator for single-particle state i.

It is understood that single-particle energy ϵ_i, defined in Eq. (7-28), also includes contributions from the core nucleons. In practice, it is common to replace ϵ_i by the observed energy level positions of single-particle states in the region of interest. The empirical values of ϵ_ι may be found, for example, in nuclei one nucleon away from closed shells where some of the low-lying states are made predominantly by the coupling of one nucleon or one hole to the ground states of the closed shell nuclei. Such states are, in principle, exactly the ones described by Hartree-Fock eigenvectors. If a realistic Hamiltonian is used in the Hartree-Fock calculation, one may expect to obtain essentially the same eigenvalues as the experimental energies. For nuclei away from closed shells, correlations other than $1p1h$ also play an important role, and the Hartree-Fock states may no longer be good approximations of the eigenstates of the complete Hamiltonian. The strength of each single-particle state, in this case, may be shared by several (observed) nuclear states, as can be seen, for example, from the spectroscopic factors of one-nucleon transfer reactions (see §8-2). In such cases, the choice of empirical single-particle energy may not be straightforward.

We shall assume that the effective interaction V_{eff} remains two-body in character, although there is no reason to rule out three-body and higher order terms caused by excitations to basis states outside the shell-model space. Such terms are believed to be small in general and may be ignored.

A formal solution of the effective interaction problem may be obtained in the following way. In addition to the operator P, which projects out the active part of the space from the complete many-body space, we shall also define an operator Q which projects out the rest of the Hilbert space, such that

$$P + Q = 1 \tag{7-36}$$

Being projection operators, they have the properties

$$P^2 = P \qquad\qquad Q^2 = Q$$

To economize on notation, let us write the eigenvalue equation in the complete space as

$$H\Psi = E\Psi \tag{7-37}$$

and

$$H = H_0 + V$$

On applying operator P from the left to both sides of Eq. (7-37), we obtain

$$PH\Psi = PE\Psi \tag{7-38}$$

Using Eq. (7-36) and the fact that E is a number and therefore commutes with operator P, we can express Eq. (7-38) as

$$PH(P+Q)\Psi = EP\Psi$$

or

$$PHP\Psi + PHQ\Psi = EP\Psi \tag{7-39}$$

Similarly, instead of P, we can apply Q from the left to both sides of Eq. (7-37) and obtain an equation similar to Eq. (7-39), except with the roles of P and Q interchanged,

$$QHP\Psi + QHQ\Psi = EQ\Psi \tag{7-40}$$

Equations (7-39) and (7-40) may be regarded as two coupled equations for $P\Psi$ and $Q\Psi$.

We can now proceed to solve these equations by expressing $Q\Psi$ in terms of $P\Psi$. Equation (7-40) may be put into the form

$$EQ\Psi - QHQ\Psi = QHP\Psi$$

This can be rewritten as

$$(E - QH)Q\Psi = QHP\Psi$$

or

$$Q\Psi = \frac{1}{E - QH}QHP\Psi \tag{7-41}$$

We must be careful with the order of operations in such a "formal" solution, as not all the operators commute with each other. Furthermore, the meaning of having operators in the denominator, $(E - QH)^{-1}$, needs to be clarified and we shall do this later.

Substituting the "solution" for $Q\Psi$ given by Eq. (7-41) into (7-39), we obtain the result

$$PH\ P\Psi + PH\frac{1}{E - QH}QH\ P\Psi = EP\Psi$$

or

$$P\Big\{H + H\frac{1}{E - QH}QH\Big\}P\Psi = EP\Psi \tag{7-42}$$

It is useful to recall that the Hamiltonian has the form $H = H_0 + V$, where H_0 is a one-body operator and V is a two-body potential. In terms of H_0 and V, Eq. (7-42) becomes

$$P\Big\{H_0 + V + (H_0 + V)\frac{1}{E - QH}Q(H_0 + V)\Big\}P\Psi = EP\Psi \tag{7-43}$$

We are now in a position to simplify this equation and put it in a form that can be compared with Eq. (7-34).

If all the single-particle states are chosen to be eigenfunctions of $h(\boldsymbol{r})$ and our truncation of the many-body Hilbert space is carried out by restricting the number of active single-particle states, we have the commutation relation

$$PH_0 = H_0P$$

Since P and Q are mutually exclusive, i.e., $PQ = QP = 0$, we can eliminate the last two H_0's on the left-hand side of Eq. (7-43). Among the three H_0's in the expression it is easy to see that the one furthest to the right does not make any contribution to the equation, as $QH_0P\Psi = 0$. The H_0 to its left occurs in the product $P\{H_0(E - QH)^{-1}QV\}P\Psi$. Since everything to the right of this H_0 acts on states in the space projected out by Q, the term is equivalent to $PH_0Q(E - QH)^{-1}QVP\Psi$ and vanishes because $PH_0Q = H_0PQ = 0$. Upon eliminating these two H_0's, Eq. (7-42) may now be written in a form that can be compared with Eqs. (7-34) and (7-35),

$$P\Big\{H_0 + V + V\frac{1}{E - QH}QV\Big\}P\Psi = EP\Psi$$

From the comparison we identify that

$$V_{\text{eff}} = V + V\frac{1}{E - QH}QV \tag{7-44}$$

This is still a formal solution and its usefulness lies mainly in the possibility of making an infinite series expansion for $(E - QH)^{-1}$. This is an advantage, as it permits an order-by-order calculation of the effective potential.

An equivalent way of writing the second term on the right-hand side of Eq. (7-44) is the following:

$$V\frac{1}{E - QH}QV = VQ\frac{1}{E - H_0 - QV}QV$$

In a Hartree-Fock single-particle basis, it is likely that the expectation values of the residual interaction V are smaller than those for H_0 and perhaps smaller than those for $E - H_0$ as well. As a result, the following condition may be true:

$$\frac{1}{E - H_0}QV < 1$$

Under such conditions, we can make use of an infinite series expansion of operator $(E - H_0 + QV)^{-1}$ in powers of $(E - H_0)^{-1}QV$,

$$\begin{aligned} VQ\frac{1}{E - H_0 - QV}QV &= VQ\frac{1}{E - H_0}QV + VQ\frac{1}{E - H_0}QV\frac{1}{E - H_0}QV \\ &\quad + VQ\frac{1}{E - H_0}QV\frac{1}{E - H_0}QV\frac{1}{E - H_0}QV + \cdots \\ &= VQ\sum_{n=1}^{\infty}\Big(\frac{1}{E - H_0}QV\Big)^n \end{aligned}$$

The effective interaction of Eq. (7-44) can now be expressed as

$$V_{\text{eff}} = V + VQ\sum_{n=1}^{\infty}\Big(\frac{1}{E - H_0}QV\Big)^n \tag{7-45}$$

If the series converges, we have a method to evaluate the effective interaction to any order of accuracy desired. Furthermore, since H_0 is diagonal in the basis states we have chosen,

$$H_0|\Psi_i\rangle = \Big(\sum_r \epsilon_r\Big)|\Psi_i\rangle$$

where ϵ_r are the single-particle energies and the summation is over all the occupied single-particle states in $|\,\Psi\,\rangle$ and operator H_0 in the denominator may be replaced by a sum of single-particle energies.

There is, however, no known proof that the series is actually convergent. Furthermore, it is not easy to carry out the calculation in practice beyond the third order or so in a nontrivial P-space. In spite of these difficulties, the effective interaction, obtained by using Eq. (7-45) to roughly second order, has been shown to give shell-model results that are in good agreement with a variety of experimental data, if we start with a realistic potential that fits the free nucleon-nucleon scattering data.

The procedure outlined above to find the effective interaction in a shell-model space is also known as a *renormalization* procedure, as it "renormalizes" an interaction for the complete Hilbert space to one suitable for the truncated space. We have now a complete set of procedures that can generate a effective interaction in a truncated shell-model space starting from one between free nucleons. The steps involved are the following. Using free nucleon-nucleon interactions, we construct an interaction for bound nucleons. This allows us to carry out a Hartree-Fock type of calculation to obtain a physical single-particle basis and a residual interaction in the space. Finally, we use the renormalization procedure outlined above to find an effective interaction to be used in a manageable shell-model space.

The complete process from nucleon-nucleon potential to an effective interaction is very involved and has been proved on many occasions to be useful in nuclear structure investigations. However, for certain practical applications, simplifications are needed and several semi-empirical approaches have been developed to obtain effective shell-model interactions.

Two-body matrix elements. In a finite active space, a two-body operator is completely specified if all the independent two-body matrix elements in the space are given. This may be seen in the following way. In the absence of antisymmetrization and angular momentum coupling, a many-body wave function $|1, 2, \ldots, A >$, expressed as a product of A single-particle wave functions, can always be written as a product of the wave function for $A-2$ particles and that for two particles,

$$\begin{aligned}&|1, 2, \ldots, k, \ldots, t, \ldots, A >_{\text{not antisym.}}\\ &\qquad = |1, 2, \ldots, k-1, k+1, \ldots, t-1, t+1, \ldots, A >_{\text{not antisym}} |k, t >_{\text{not antisym.}}\end{aligned}$$

As we can see from Eq. (7-6), antisymmetrization requires the many-body wave function written as a sum of different products of single-particle states. As a result,

$$\begin{aligned}&|1, 2 \cdots k \cdots t \cdots A >_{\text{antisym.}}\\ &\qquad = \sum_{\substack{(A-2)\\(2)}} C_{(A-2)(2)} |1, 2 \cdots k-1, k+1 \cdots t-1, t+1 \cdots A >_{\text{antisym}} |k, t >_{\text{antisym.}}\end{aligned}$$

The factor $C_{(A-2)(2)}$ is known as a two-particle fractional parentage coefficient and expresses the "fraction" of the antisymmetrized A-body wave function coming from the product of $(A-2)$- and two-body wave functions. For our purpose here, we need not be concerned with the exact values of these expansion coefficients, as we shall not be doing any actual calculations with them. In terms of $C_{(A-2)(2)}$, we can express the many-body matrix element of a two-body operator $\hat{O}(2)$ in the following way:

$$\begin{aligned}&< 1', 2' \cdots A' |\hat{O}(2)| 1, 2 \cdots A >\\ &= \sum_{\substack{(A-2)'\\(2)'}} \sum_{\substack{(A-2)\\(2)}} C_{(A-2)'(2)'} C_{(A-2)(2)}\\ &\quad \times < 1' \cdots k'-1, k'+1 \cdots t'-1, t'+1 \cdots A' | 1 \cdots k-1, k+1 \cdots t-1, t+1 \cdots A >\\ &\quad \times < k', t' |\hat{O}(2)| k, t >\end{aligned}$$

Since the overlap of $|1' \cdots k'-1, k'+1 \cdots t'-1, t'+1 \cdots A'>$ and $|1 \cdots k-1, k+1 \cdots t-1, t+1 \cdots A>$ vanishes unless all $A-2$ single-particles states are the same on both sides, we find the result that the A-body matrix element is zero if more than two single-particle states are different from each other, a result we used earlier in going from Eq. (7-23) to Eq. (7-24). If, on the other hand, no more than two single-particle states are different from each other, the A-body matrix element can be expressed in terms of two-body matrix elements $< k', t'|\hat{O}(2)|k, t >$. Thus, if all the matrix elements in two-particle space are given, we can calculate any Hamiltonian matrix elements in the shell-model space. This is equivalent to saying that the effective interaction V_{eff} is completely specified within the active space. We have already seen an example of this in Eq. (7-26), where we defined the two-body part of the Hamiltonian for a Hartree-Fock calculation in terms of $V_{\alpha\beta\gamma\delta}$.

The two-body matrix elements required here are, however, slightly different from $V_{\alpha\beta\gamma\delta}$, as we wish to work in a subspace with definite spin J and isospin T. For this purpose, it is convenient to have the defining two-body matrix elements for the interaction given also in terms of two-particle states with definite J and T. In this scheme, an antisymmetrized and normalized two-body matrix element may be written as

$$W^{JT}_{rstu} \equiv \langle rsJT|V|tuJT\rangle$$

where $|\,rsJT\,\rangle$ is an antisymmetrized and normalized wave function for two particles, one in single-particle state r and the other in state s, similar to that given in Eq. (7-25). The additional feature here is that the two single-particle wave functions are angular momentum coupled together to final spin J and isospin T. The two-particle wave function $|\,tuJT\,\rangle$ is defined in the same way, except that the single-particle states involved are t and u, instead of r and s. In terms of two-body matrix elements, the effective interaction can be written in the form

$$V_{\text{eff}} = \sum_{\substack{rstu \\ JT}} |\,rsJT\,\rangle W^{JT}_{rstu} \langle tuJT|$$

in analogy with Eq. (7-26). Since the nuclear Hamiltonian is a scalar in spin and isospin, only two-body matrix elements diagonal in J and T are nonvanishing.

Other symmetries of the nuclear Hamiltonian can also help to reduce the number of independent two-body matrix elements required to define an effective interaction in a finite shell-mode space. Because of time-reversal invariance, the matrix elements may be taken to be real and symmetric, i.e.,

$$W^{JT}_{rstu} = W^{JT}_{turs} \tag{7-46}$$

Furthermore, since the wave functions are antisymmetrized, two functions differing only by a permutation of the two single-particle wave functions involved are related to each other by a phase factor,

$$|\,rsJT\,\rangle = (-1)^{j_r+j_s-J-T}|\,srJT\,\rangle$$

This is made of a combination of three separate factors: a minus sign due to the permutation of two fermion states, a factor $(-1)^{1/2+1/2-T}$ due to isospin recoupling of the two single-particle states as given by Eq. (A-11), and a similar factor $(-1)^{j_r+j_s-J}$ for recoupling the spins. Because of these relations, we have the following symmetries among two-body matrix elements:

$$W^{JT}_{rstu} = (-1)^{j_r+j_s-J-T} W^{JT}_{srtu} = (-1)^{j_t+j_u-J-T} W^{JT}_{rsut} = (-1)^{j_r+j_s-j_t-j_u} W^{JT}_{srut} \tag{7-47}$$

Because of Eqs. (7-46) and (7-47), the number of independent two-body matrix elements required to define an effective interaction in a finite shell-model space may be sufficiently small that one may be able to determine them empirically by fitting all the required matrix elements to the available data in the same space.

Semi-empirical effective interaction. Let us use a simple example to illustrate the semi-empirical approach to effective interaction. Some of the low-lying levels in calcium isotopes, ^{41}Ca to ^{48}Ca, may be approximated by a shell-model space made of the $1f_{7/2}$-orbit alone. The inert core here is the ^{40}Ca nucleus, and the 40 nucleons filling the $1s$-, $1p$-, $1d$-, and $2s$-orbits are not to be excited. All the active nucleons in this case are neutrons. Since a $f_{7/2}$-orbit can take a maximum of $2j+1=8$ neutrons, the active space is completely filled when we come to ^{48}Ca.

The binding energy difference between ^{41}Ca and ^{40}Ca provides us with the single-particle energy for the $1f_{7/2}$-orbit,

$$\epsilon_{1f_{7/2}} = -8.36 \text{ MeV}$$

In a similar way, we can calculate the binding energy of the two neutrons in ^{42}Ca with respect to the ^{40}Ca core. The result, -19.84 MeV, is different from twice the value of $\epsilon_{1f_{7/2}} = -8.36$ MeV because of residual interaction between two neutrons. This provides us with one piece of experimental information required to determine the effective interaction. Since the ground state of ^{42}Ca has $J=0$ and $T=1$, we obtain the two-body matrix element for $(J,T)=(0,1)$ from the binding energy of ^{42}Ca with respect to ^{40}Ca after removing contributions due to the single-particle energies of the two neutrons,

$$W^{01}_{1f_{7/2}1f_{7/2}1f_{7/2}1f_{7/2}} = -19.8433 - (2 \times -8.3627) = -3.12 \text{ MeV}$$

Because of antisymmetrization requirements, two neutrons in $f_{7/2}$-orbit can only be coupled to $J=0$, 2, 4, 6. As a result, three additional two-body matrix elements are needed to complete the definition of the effective interaction in this simple shell-model space. These can be found from the energy level positions of the $J=2$, 4, 6 excited states in ^{42}Ca known to be, respectively, at 1.5247, 2.7504, and 3.1893 MeV above the ground state. The corresponding two-body matrix elements, therefore, have values -1.59, -0.366, and 0.0714 MeV.

To simplify the notation, we shall drop the superscript for isospin, as we are dealing with neutrons only, and the subscripts $1f_{7/2}$, as this is the only orbit with which we are concerned in this example. The complete effective Hamiltonian for the $1f_{7/2}$-neutron shell model is given by five pieces of information, one single-particle energy and four two-body matrix elements,

$$\epsilon = -8.36 \text{ MeV} \qquad W^0 = -3.12 \text{ MeV} \qquad W^2 = -1.59 \text{ MeV}$$
$$W^4 = -0.366 \text{ MeV} \qquad W^6 = 0.0714 \text{ MeV}$$

With these five pieces of input obtained from ^{41}Ca and ^{42}Ca, we are now in a position to calculate all the energy levels in the $1f_{7/2}$ shell-model space from ^{43}Ca to ^{48}Ca.

The calculated results for ^{43}Ca to ^{46}Ca are listed in Table 7-3. The binding energies with respect to the ^{40}Ca core are given in the first row and the excitation energies in the remaining rows. In addition, the calculated binding energies for ^{47}Ca and ^{48}Ca are found to be, respectively, 68.58 and 80.29 MeV, compared with the measured values of 63.99 and 73.94 MeV. When we examine the six binding energies in more detail, we find the difference between the calculated and observed values gets progressively further apart as the number of active neutrons increases from three ($A = 43$) to eight ($A = 48$). In fact, it is easy to show that the difference is roughly proportional to the factor $n(n-1)/2$, the number of neutron pairs in the $1f_{7/2}$-shell model space. This means that the effective interaction deduced from the difference between ^{42}Ca and ^{40}Ca turns out to be a little too strong. If we reduce the contribution due to binding energy of ^{42}Ca in the effective Hamiltonian by 0.21 MeV, i.e., increasing each one of the five two-body matrices by 0.21 MeV, the calculated binding energies change to 27.64, 38.58, 46.26, 56.83, 64.20, and 74.44 MeV for ^{43}Ca to ^{48}Ca, in much better overall agreement with the observed values. (On the other hand, if the difference were linearly proportional to the number of active neutrons, the cause would have to be attributed to

Table 7-3: A shell-model calculation in the $1f_{7/2}$-space for the energy levels of calcium isotopes

^{43}Ca			^{44}Ca		
J	Experimental	Calculated	J	Experimental	Calculated
$\frac{7}{2}$	$E_B = 27.78$	$E_B = 28.27$	0	$E_B = 38.91$	$E_B = 39.83$
$\frac{5}{2}$	0.37	0.28	2	1.16	1.52
$\frac{3}{2}$	0.59	1.31	4	2.28	2.51
$\frac{11}{2}$	1.68	1.77	4	3.05	2.75
$\frac{9}{2}$	2.09	2.05	6	3.28	3.19
$\frac{15}{2}$	2.75	3.12	8	5.09	5.30
^{45}Ca			^{46}Ca		
J	Experimental	Calculated	J	Experimental	Calculated
$\frac{7}{2}$	$E_B = 46.32$	$E_B = 48.35$	0	$E_B = 56.72$	$E_B = 59.98$
$\frac{5}{2}$	0.17	0.28	2	1.35	1.52
$\frac{3}{2}$	1.43	1.31	4	2.58	2.75
$\frac{11}{2}$	—	1.77	6	2.97	3.19
$\frac{9}{2}$	—	2.05			
$\frac{15}{2}$	—	3.12			

the single-particle energy instead.) Such overall shifts in the defining matrix elements, in general, do not affect the excitation energies.

The calculated energy level positions of the excited states in ^{43}Ca to ^{46}Ca are also compared with experimental values in Table 7-3. There are obviously more excited states in these nuclei than the ones listed. Since our model space is restricted to the $1f_{7/2}$-orbit alone, only observed levels belonging to this model space may be used in the comparison. In principle, one can identify a $1f_{7/2}$-level by measuring the angular distribution of one-nucleon transfer strengths (see §8-2). In practice, the identification is not always simple, as substantial admixtures from other single-particle states, such as $2p_{3/2}$ and $2p_{1/2}$, are expected. The comparison between calculated and observed values in the table must therefore be viewed with the simplicity of the model space used for this illustrative example in mind. In fact, the agreement is better than we could have expected.

A second nontrivial but still relatively simple example is the $1p$-shell, consisting of nuclei from ^{5}He and ^{5}Li to ^{16}O. The inert core here is the ^{4}He nucleus, with two protons and two neutrons completely filling up the $1s_{1/2}$-orbit. All single-particle states above the $1p$-shell, starting from the ds-shell, are empty. In this space, there are two valence orbits, $1p_{3/2}$ and $1p_{1/2}$, in the jj-coupling scheme. The one-body part of the effective Hamiltonian is therefore defined by two single-particle energies, $\epsilon_{1p_{3/2}}$ and $\epsilon_{1p_{1/2}}$. Because of the symmetries given in Eqs. (7-46) and (7-47), the number of independent two-body matrix elements required to determine the two-body residual interaction for a given J and T is $d_{JT} = n(n+1)/2$, where n is the number of two-particle states with spin-isospin (J, T). The total number of two-body matrix elements in this shell-model space, as can be seen from Table 7-4, is $\sum_{JT} d_{JT} = 15$. The complete $1p$-shell effective interaction therefore comprises a total of 17 parameters, 2 single-particle energies and 15 two-body matrix elements. It is not possible to find 17 energy levels in mass 5 and 6 nuclei to specify the effective Hamiltonian, as we have done earlier for the simpler case of $1f_{7/2}$ shell-model space. On the other hand, since all the $1p$-shell states in nuclei from $A = 5$ to $A = 16$ can be calculated from these 17 parameters, we can use any $1p$-shell data among the available ones in any of these nuclei. In fact, more than 17 pieces of experimental information can be identified in this mass region and a least-squares procedure may be used to deduce the values of the parameters that best fit the data.

Table 7-4: Allowed two-particle states in $1p$-shell.

	$T=0$			$T=1$		
	$J=1$	$J=2$	$J=3$	$J=0$	$J=1$	$J=2$
$1p_{3/2}^2$	1	—	1	1	—	1
$1p_{1/2}^2$	1	—	—	1	—	—
$1p_{3/2}1p_{1/2}$	1	1	—	—	1	1
Total	3	1	1	2	1	2
d_{JT}	6	1	1	3	1	3

This was done by Cohen and Kurath [45]. The calculation serves two useful purposes. The first is to demonstrate that the idea behind a semi-empirical effective Hamiltonian is a sound one. Once the 17 parameters are determined from fitting a given number of pieces of data, the effective Hamiltonian obtained may be used to calculate the shell-model values corresponding to the data used. Since a fitting procedure was used, the calculated results are not necessarily identical to the observed ones used as input. The quality of the agreement serves as a measure for the validity of such an approach. Normally the success of a least-squares fit to some functional form is given by the value of the χ^2 for the overall fit and the standard deviation for each independent parameter obtained. A small χ^2, among other things, indicates that the functional form used to make the fit is reasonable. Here we are dealing with a highly nonlinear least-squares fitting procedure involving matrix calculations. It is therefore not easy to give a figure of merit, analogous to the role of χ^2-values, for the functional form used. The fact that the calculated results agree well with the original input is a good indication of the power of the effective Hamiltonian approach. A second use of the 17 parameters obtained from the fit is that we now have an effective $1p$-shell Hamiltonian that can be employed for investigating other nuclear properties in the same space. This has been used extensively with success.

A similar project for the ds-shell composed of nuclei from ^{17}O to ^{40}Ca has also been carried out [147]. Here, the valence orbits are $1d_{5/2}$, $1d_{3/2}$, and $2s_{1/2}$. The effective Hamiltonian is given by 3 single-particle energies and 63 two-body matrix elements (see Problem 7-6). The calculated results represent some of the best description of the low-lying states in nuclei from mass 18 to 40.

Examples of shell-model results. It is useful here to give some other examples on what a microscopic shell-model calculation can produce. Instead of introducing new physical phenomena, we shall make use of observations that are already familiar. In §6-1 we have seen that some of the excited states in certain nuclei can be understood as collective vibrations of nucleons. There, the observed properties were described in terms of harmonic vibrations of the collective coordinates $\alpha_{\lambda\mu}(t)$, the shape parameters. One of the nuclei exhibiting such properties is ^{62}Ni. The ground state spin-parity of ^{62}Ni is 0^+, typical of an even-even nucleus, and the first excited state is 2^+ at 1.17 MeV. A triplet of states, 0^+, 2^+, and 4^+, is observed at slightly less than twice this energy at 2.05 to 2.34 MeV. These three groups of states are interpreted as quadrupole vibrations built upon a spherical nucleus with the ground state as the zero-phonon state, the first excited state as the one-phonon state, and the triplet of 0^+, 2^+, and 4^+ states as the two-phonon states. The $E2$-transition rates, given earlier in Table 6-1, also confirm this interpretation. Here, we shall take a microscopic approach and treat ^{62}Ni using the shell model. The valence nucleons are taken to be the six neutrons outside a ^{56}Ni core and the shell-model space consists of single-particle orbits $1f_{5/2}$, $2p_{3/2}$, and $2p_{1/2}$. The three single-particle energies are taken from ^{57}Ni. The low-lying energy level positions calculated with an effective interaction, obtained using a renormalization procedure, are compared with the observed values in Fig. 7-6. It is seen that the energy level positions obtained indeed display the typical structure of a vibrational nucleus, in agreement with observation.

Using the eigenvectors obtained, we can calculate the electromagnetic properties

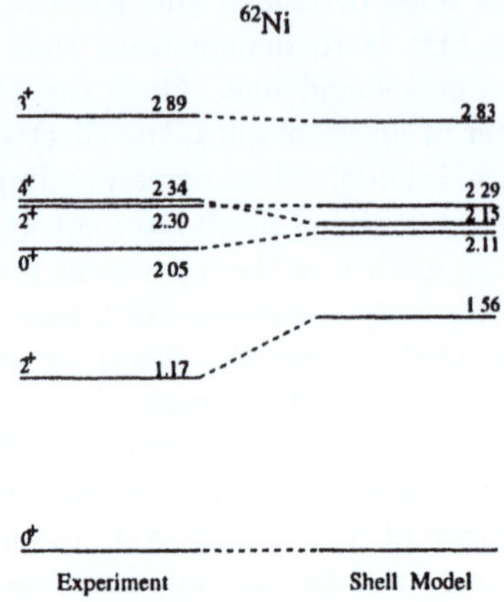

Figure 7-6: Comparison of the observed energy levels of ^{62}Ni with the results of a shell-model calculation in the $1f_{5/2}$-, $2p_{3/2}$-, and $2p_{1/2}$-space using a renormalized effective interaction. The observed quadrupole vibration features of the nucleus is well reproduced by the microscopic calculation.

of the states involved. Here we encounter the question of an effective operator in a truncated shell-model space. For a quadrupole vibrational nucleus, the dominant electromagnetic transitions are $E2$ induced by charged currents. Since electric currents in a nucleus are usually associated with the motion of protons, our calculated results using the *bare* charge of active nucleons in the space, i.e., the free nucleon values used to define electromagnetic operators in §5-3, will be zero, as we have only active neutrons. This is clearly a problem caused by the truncation procedure. In other words, the transition operators must also be *renormalized* before we can expect to obtain any reasonable calculated values in a truncated space. For electromagnetic operators, the usual practice is to give an *effective* charge to neutrons (as well as protons) and adjust its size to fit the observed $E2$-transition rates and quadrupole moments, as we shall see in the next example.

A second example is the low-lying positive-parity states of ^{20}Ne shown in Fig. 7-7. Here we see that the energy level positions display a rotational structure with E_J essentially proportional to $J(J+1)$ up to $J = 8$. For a shell-model calculation, we can take ^{16}O as the inert core and the four valence nucleons, two protons, and two neutrons

Table 7-5: Values of $B(E2; J \to J-2)$ between $K = 0^+$ band members in ^{20}Ne.

		Experimental [4]		Shell model,*	Rotation model,
J_i	J_f	e^2fm^4	W.u.	e^2fm^4	e^2fm^4
2	0	67 ± 4	20.8 ± 1.3	48	57
4	2	72 ± 7	22.5 ± 2.5	58	83
6	4	65 ± 10	20.2 ± 3.0	43	91
8	6	30 ± 4	9.2 ± 1.3	28	96

Note: *An effective charge of $e_p = 1.5e$ and $e_n = 0.5e$ is used.
1 W.u. (Weisskopf unit) = 3.2 e^2fm^4 for ^{20}Ne.

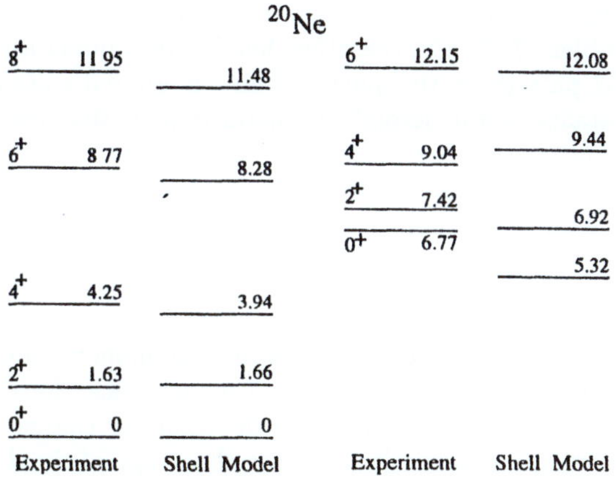

Figure 7-7: Comparison of the observed energy levels for the two lowest observed $K = 0^+$ bands in ^{20}Ne with the results of a shell-model calculation in the $1d_{5/2}$-, $1d_{3/2}$-, and $2s_{1/2}$-space using a renormalized effective interaction.

are in single-particle orbits $1d_{5/2}$, $1d_{3/2}$, and $2s_{1/2}$. The results also give a reasonable description of the energy level positions, as can be seen in the figure. For the calculated $E2$-transition rates listed in Table 7-5, an effective charge of $0.5e$ is used; that is, the charge of an active proton is $1.5e$ and that of an active neutron is $0.5e$. Here we see again that a microscopic interpretation of a collective phenomenon can also be made in terms of a very small number of active nucleons in a highly truncated active space.

Effective operator. The reasons behind the large effective charge for $E2$-transition operators may be traced to the relatively small number of active nucleons used in the calculations. Although the energy level positions are well accounted for, the small number is inadequate to produce the large enhancements seen in the $E2$-transition rates for collective states. Since the deformations here, whether in the equilibrium shape as in the case of ^{20}Ne or in the form of shape vibration as in the case of ^{62}Ni, are predominantly quadrupole in nature, it is not surprising that we find the difference between the effective and real charge for $E2$-transitions to be most pronounced. Since collective motion involves the action of a large number of nucleons, including some of those considered to be a part of the "inert" core in the shell-model calculations, a large effective charge is required. It is also interesting to note that such large enhancements due to core nucleons can be accounted for by essentially an overall factor in the form of an effective charge. The possibility of making the corrections in a simple way lends support to the idea behind renormalizing the operators.

Another demonstration of effective operators can be found in the square of the charge form factor $F^2(q)$, obtained, for example, from electron scattering off nuclei. As we have seen earlier in §4-1, the charge form factor is the Fourier transform of the charge distribution in a nucleus. The measured values for ^{12}C and ^{16}O are shown in Fig. 7-8. For ^{12}C we see that there is only one minimum in $F^2(q)$ at around $q = 1.5$ fm^{-1}. This is exactly what is expected from the Fourier transform of the density of a nucleon in the $1p$-shell (see Problem 7-7). On the other hand, two minima are observed in ^{16}O. A simple shell model puts the active nucleons for the ground state of ^{16}O also in the $1p$-shell. The appearance of the second minimum implies that there is a substantial admixture of configurations having nucleons excited into the ds-shell. If we insist on carrying out the form factor calculation for ^{16}O using only active nucleons in the $1p$-shell space, a correction factor, for example, in the form

$$f(r) = 1 - e^{-\alpha r^2}$$

must be introduced. This gives the observed second minimum without having to invoke configurations involving the ds-shell into our shell-model space. Such a correction factor may be regarded as a renormalization of the operator for charge form factor. The effective operator produced as a result simulates the shape of the ds-shell form factor for valence nucleons in the $1p$-shell.

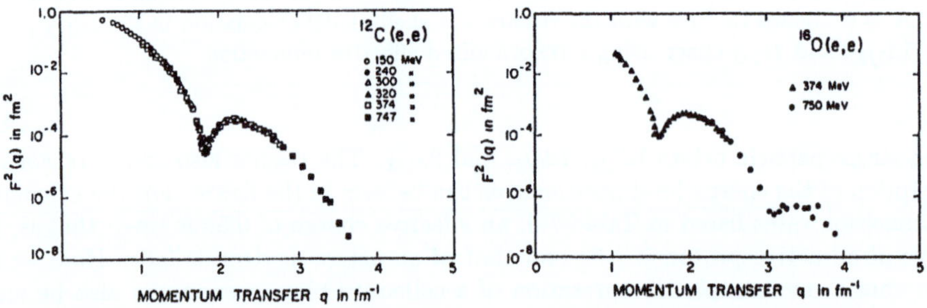

Figure 7-8: The square of charge form factors for ^{12}C and ^{16}O obtained from elastic electron scattering. A single minimum in $F^2(q)$ for ^{12}C is expected on the ground that the active nucleons are in the $1p$-shell. The second minimum in ^{16}O shows the presence of active particles in higher single-particle orbits. (Plotted using data from Refs. [127, 117].)

Not every operator requires a large renormalization as we have used above for electric quadrupole transitions. For example, Gamow-Teller transitions throughout the ds-shell have been found to be given by the bare operator without noticeable modifications [39]. This may be related to the fact that β-decay is not a collective phenomenon like, for instance, $E2$-transitions.

Besides effects due to truncation of the shell-model space, renormalization of the excitation operators from their bare nucleon values may also be required because of mesonic and other degrees of freedom in nuclei. When a nucleon is embedded inside a nucleus, we expect such processes as the exchange of virtual mesons to be different from the situation when the nucleon is a free, isolated particle. In recent years one of the interesting developments in the nuclear shell model has been in the direction of obtaining such renormalization effects from field-theoretical approaches. In this way, a better and more fundamental understanding of the behavior of nucleons inside a nucleus may be reached.

Because of its direct connection with the individual nucleon degrees of freedom in a finite nucleus, the shell model can also be used to "simulate" data for testing other models. This is similar to numerical simulations used in many other fields to supplement data and to help us to probe aspects of nature that are difficult to examine experimentally. Such numerical "experiments" do not substitute actual observations involving real nuclei. However, they provide a convenient avenue to test our models and our understanding of the physical situation and are useful as a tool to further our knowledge of nuclei.

7-6 Other Models

In the previous chapter we have seen that certain nuclear properties can be understood from a macroscopic point of view in terms of the collective degrees of freedom. Alternatively, one can start from the individual nucleon degrees of freedom and try to understand observed phenomena from a microscopic point of view, as we have been doing in this chapter. Calculations starting from the individual nucleon degrees of freedom are attractive, as we can make connections with the interaction between nucleons. This, in turn, allows us to make contact with the fundamental strong interaction between nucleons. Unfortunately, the nuclear many-body problem, similar to many-body problems in other branches of physics, is not simple to solve. For this reason, several techniques, in addition to the ones described above, have been developed so that we may be able to examine certain specific aspects of some problems in a more convenient way. It is perhaps useful to mention some of these very briefly here, even though both the scope of this book and the background knowledge required preclude any detailed discussions.

If we are concerned only with a limited range of behavior in a few special states, it seems superfluous to invoke the nuclear shell model and solve the complete eigenvalue problem. As we have seen earlier in the discussion on Hartree-Fock techniques, $1p1h$-(one-particle one-hole) excitations in a many-body system can be handled with relative ease. This is an advantage, as $1p1h$-excitations constitute the dominant components in a variety of processes. For example, nuclear states that have large $1p1h$-components are strongly excited by electromagnetic processes, such as inelastic electron scattering, and by strong interaction probes in the form of intermediate-energy nucleon scattering, such as the ones described later in §8-4. One of the primary concerns in such studies is to establish the correct correlations between different $1p1h$-components so as to be able to produce, for example, the observed strong enhancement in strengths. One of the difficulties here is to have the proper ground state wave function upon which we can build the excitations. A simple independent particle model is often inadequate, as correlations resulting from residual two-body interaction are largely absent. The random-phase approximation (RPA) solves this problem by concentrating only on certain types of important correlations and is thus able to account for the strong $1p1h$-excitations observed in many nuclei with a relatively simple calculation. A more detailed discussion of RPA can be found in Fetter and Walecka [62].

The idea behind independent particle approximation can also be generalized by including other types of correlation. Since one-body terms are so much easier to handle than two-body residual interactions, it is preferable to incorporate, as much as possible, the effect of nucleon-nucleon interaction into the mean field experienced by individual nucleons. There is a large variety of mean-field theories. One of the attractive features here is that the approach can also be extended easily to a relativistic one (see, e.g., Celenza and Shakin [43]). Although we have limited ourselves to nonrelativistic quantum mechanics here, many aspects of nuclear structure and nuclear scattering, especially those involving intermediate- and high-energy probes, require a relativistic treatment. For this reason, mean-field theories are becoming an increasingly more important tool in nuclear studies.

Besides vibrations and rotations, nuclei also display clustering behavior. The simplest example is the decay of the ground state of ^{8}Be into two α-particles. One explanation of this phenomenon is that nucleons prefer to form α-particle clusters in nuclei. Since the binding energy per nucleon of an α-particle is approaching the maximum value that can be attained inside a nucleus, there is relatively little force of attraction left between different α-particle "clusters." In the case of ^{8}Be, binding energy actually favors the formation of two separate α-particles. For this reason, the ground state of ^{8}Be is unstable toward α-particle emission, even though it is stable against β-decay and nucleon emission. Another example is the observation of nuclear "molecular" states, such as in the separation of an excited ^{24}Mg nucleus into two ^{12}C clusters [38]. For a shell model to split a group of nucleons into two or more separate clusters, a single-particle basis, far larger than anything one can contemplate in practice, is required. Special techniques such as the generator coordinate method [148] and the two-centered shell model have been developed for studying such phenomena.

We have seen from shell-model studies that the presence of energy gaps in a single-particle level spectrum is important in understanding nuclear structure. The most naive collective models, however, ignore this feature, as only smooth variations with nucleon number and other macroscopic properties of nuclei are incorporated into the picture. Improvements to the collective models can be achieved if local variations due to shell closures can be included. Such "shell" corrections are essential since, as deformation grows, the energies of individual single-particle states are modified in such a way that the energy gaps observed for spherical nuclei disappear and new ones at different neutron and proton numbers appear. For example, such shell corrections have been found to be important in improving the vibrational model description of some of the bulk properties in nuclei [108]. We shall see an example of such "shell" corrections in §9-2 for the case of superdeformed nuclei.

For illustrative purposes, we have separated nuclear properties into collective and single-particle behaviors. In practice, both types of phenomena are present in the same nucleus. Furthermore, they can couple with each other to form states with both types of behavior coexisting with each other. What we have mainly done so far is to examine the two extremes separately so as to illustrate the underlying physics. Specific states are identified as being either single particle or collective depending on which one of these two extremes dominates the property of the state. In fact, states that can be identified in such a simple manner constitute only a minority among the multitude of known ones. For the bulk of states, all physical principles are at play. A thorough understanding of nuclear structure will require us to examine these states as well; however, we shall ignore them here.

Problems

7-1. If $^{25}_{12}$Mg is a spherical nucleus, what are the most likely spin, parity, and isospin of its ground state? If, instead, it is a deformed nucleus with prolate deformation, what are the most likely spin and parity?

7-2. The nuclei $^5_2$He, $^5_3$Li, $^{17}_8$O, $^{17}_9$F, $^{41}_{20}$Ca, $^{41}_{21}$Sc, $^{209}_{82}$Pb, and $^{209}_{83}$Bi may be considered to be made of a neutron or a proton outside a closed shell core. Use an independent particle model to deduce their ground state spin and parity. From this information calculate the ground state magnetic dipole moment of each nucleus. Do the same for the one-hole nuclei of $^3_1$H, $^3_2$He, $^{15}_7$N, $^{15}_8$O, $^{39}_{19}$K, $^{39}_{20}$Ca, $^{207}_{81}$Tl, and $^{207}_{82}$Pb.

7-3. The energies of ^{17}O with respect to the ^{16}O core are −4.15 MeV for the $5/2^+$ state, −3.28 MeV for the $1/2^+$ state, and +0.93 MeV for the $3/2^+$ state. Assuming these values are the single-particle energies of the *ds*-orbits, use an independent particle model to find the relative energies of the lowest $1/2^+$, $3/2^+$, and $5/2^+$ states in ^{39}Ca with respect to the ^{40}Ca core.

7-4. The *ds*-shell single-particle energies with respect to ^{16}O core are $\epsilon_{1d5/2} = -4.15$ MeV, $\epsilon_{2s1/2} = -3.28$ MeV, and $\epsilon_{1d3/2} = +0.93$ MeV. A particular effective interaction has the following set of two-body matrix elements for $(J,T) = (0,1)$:

$$\begin{aligned}
\langle 1d_{5/2}, 1d_{5/2}; J=0, T=1|V|1d_{5/2}, 1d_{5/2}; J=0, T=1\rangle &= -2.0094 \text{ MeV}\\
\langle 1d_{5/2}, 1d_{5/2}; J=0, T=1|V|1d_{3/2}, 1d_{3/2}; J=0, T=1\rangle &= -3.8935 \text{ MeV}\\
\langle 1d_{5/2}, 1d_{5/2}; J=0, T=1|V|2s_{1/2}, 2s_{1/2}; J=0, T=1\rangle &= -1.3225 \text{ MeV}\\
\langle 1d_{3/2}, 1d_{3/2}; J=0, T=1|V|1d_{3/2}, 1d_{3/2}; J=0, T=1\rangle &= -0.8119 \text{ MeV}\\
\langle 1d_{3/2}, 1d_{3/2}; J=0, T=1|V|2s_{1/2}, 2s_{1/2}; J=0, T=1\rangle &= -0.8385 \text{ MeV}\\
\langle 2s_{1/2}, 2s_{1/2}; J=0, T=1|V|2s_{1/2}, 2s_{1/2}; J=0, T=1\rangle &= -2.3068 \text{ MeV}
\end{aligned}$$

(a) Calculate the ground state binding energy of ^{18}O with respect to ^{16}O and compare the result obtained from a table of binding energies. What are the excitation energies of the two other 0^+ states in this space?

(b) Obtain the ground state wave function of ^{18}O and use it to calculate the relative probability for finding a neutron in the $1d_{5/2}$, $2s_{1/2}$, and $1d_{3/2}$ single-particle states in ^{18}O. The results are essentially the spectroscopic factors for one-neutron pickup reactions.

7-5. If the wave function of the lowest 1^+ state in ^{18}O is

$$|J^\pi = 1^+,\ T=1\rangle \ = \ |1d_{3/2}2s_{1/2}; J^\pi = 1^+,\ T=1\rangle$$

find the magnetic dipole moment of this state.

7-6. Use the same method as outlined in Problem 6-1 to show that when two nucleons are in an orbit with spin j, the allowed J-values for two-particle states are 0, 2, 4, ... , $2j-1$ for $T=1$ and $J = 1, 3, 5, \ldots, 2j$ for $T=0$. Construct a table similar to Table 7-4 to give the number of states of each J and T for two nucleons in the three *ds*-shell orbits, $1d_{5/2}$, $1d_{3/2}$, and $2s_{1/2}$. Use this information to show that the total number of two-body matrix elements required to define a two-body potential in the *ds*-shell is 63.

7-7. The explicit forms of the radial wave function for a spherical harmonic oscillator potential well are given in Table 7-1. Use these to show that the form factor, the Fourier transform of the radial density $\rho_{n\ell}(r) = |R_{n\ell}(r)|^2$, is positive for the 1s-orbit in the region $q = 0$ to infinity [i.e., no node in $F^2(q)$] and changes sign once at $q^2 = 6\nu$ for the 1p-orbit [one node in $F^2(q)$]. For the 1d-orbit, the sign changes twice, i.e., there are two nodes in $F^2(q)$.

Chapter 8

Nuclear Reactions

A large fraction of our knowledge on the properties of nuclei is derived from nuclear reactions. When an incoming particle is scattered off a target nucleus, the outcome depends on a combination of three factors: the reaction mechanism, interaction between the projectile and the target, and the internal structure of the nuclei involved. Different probes complement each other in what we can learn from an investigation. Furthermore, it is often possible to select the bombarding energy and the reaction in such a way that we can focus on particular aspects of the problem, as we shall see in a few examples in this chapter.

Nuclear reaction is a large subject by itself. We can give here only an overview of some of the more important topics. In order to highlight the basic points, it may be necessary to sacrifice some of the details in our discussion. For some of the more established topics, such as Coulomb scattering, excellent review articles are available. These will be referred to at the appropriate places. For some of the fast developing topics, only the current literature can provide us with the latest information.

8-1 Coulomb Excitation

When a projectile carrying a charge ze approaches a target consisting of Z protons, the strength of the Coulomb field between them may be characterized by the Sommerfeld number of Eq. (4-64),

$$\eta = \left[\frac{1}{4\pi\epsilon_0}\right]\frac{zZe^2}{\hbar v} = \alpha z Z\frac{c}{v}$$

where v is the velocity at which one particle approaches the other when they are still separated by large distances and α is the fine structure constant. Classically, the distance of closest approach R_s is given by the condition

$$\left[\frac{1}{4\pi\epsilon_0}\right]\frac{zZe^2}{R_s} = \tfrac{1}{2}\mu v^2 \tag{8-1}$$

where μ is the reduced mass of the projectile. In terms of η, we have the result

$$\tfrac{1}{2}R_s = \frac{\alpha\hbar c z Z}{\mu v^2} = \eta\frac{\hbar}{\mu v}$$

From this expression, we see that the Sommerfeld number may be viewed as the ratio between $\frac{1}{2}R_s$ and the reduced de Broglie wavelength,

$$\frac{\lambda}{2\pi} = \frac{1}{2\pi}\frac{h}{p} = \frac{\hbar}{\mu v}$$

A small η means that the Coulomb field is weak compared with the available kinetic energy in the scattering. Under such conditions, the wave function of the incident particle is not greatly modified by the Coulomb field and Born approximation applies. In Coulomb excitation we are, however, primarily interested in the opposite limit, with $\eta \gg 1$. In this case, the two particles are never close enough to each other for nuclear force to play a role, and excitation of the target or the projectile nucleus is accomplished through Coulomb interaction. Such a process is known as *Coulomb excitation.* Because of its intrinsic interest, the subject is also treated in a number of standard quantum mechanics texts, such as Merzbacher [103] and Messiah [104].

There are several reasons why Coulomb excitation is of interest in nuclear physics. First, the reaction mechanism is well known and may be regarded essentially as the inverse of electromagnetic decay discussed in §5-3. Second, experimentally, very strong Coulomb fields can be created by bombarding nuclei with a beam of heavy ions. Finally, when this advantage is coupled with the precision that can be achieved in charged particle experiments, we have a powerful tool for investigating certain properties of nuclei.

Multipole expansion. We shall follow the approach used in §5-3 and treat electromagnetic interaction between the projectile and the target as a perturbation to the nuclear Hamiltonian. The contribution of Coulomb excitation may be written as

$$H'(t) = \left[\frac{1}{4\pi\epsilon_0}\right]\frac{zZe^2}{|\boldsymbol{r}_p(t) - \boldsymbol{r}|} - \text{monopole term}$$

where $\boldsymbol{r}_p(t)$ is the location of the projectile at time t and $\boldsymbol{r}$ that of the target nucleus. It is necessary to take away the contributions from the "monopole term" here, as it can only deflect the projectile without exciting any of the internal degrees of freedom. For simplicity, we shall not be concerned with this term from now on.

In the region $r_p > r$, the perturbation $H'(t)$ may be expanded in terms of spherical harmonics using the relation

$$\frac{1}{|\boldsymbol{r}_p(t) - \boldsymbol{r}|} = \sum_{\lambda=0}^{\infty}\sum_{\mu=-\lambda}^{\lambda}\frac{4\pi}{2\lambda+1}\frac{r^\lambda}{r_p^{\lambda+1}(t)}Y_{\lambda\mu}(\theta,\phi)Y^*_{\lambda\mu}(\theta_p,\phi_p) \tag{8-2}$$

as we have done earlier in §4-6. In the above expression, quantities pertaining to the target state have the form $r^\lambda Y_{\lambda\mu}(\theta,\phi)$, the same as $\boldsymbol{O}_{\lambda\mu}(E\lambda)$, the operator for $E\lambda$ excitation given in Eq. (5-25). We can take advantage of this similarity to make a connection between Coulomb excitation and electromagnetic decay.

Furthermore, the reaction is reminiscent of electron scattering off nuclei discussed in §4-1. There we found that, since the scattering is due primarily to electromagnetic interaction, the cross section is proportional to that for point-particle scattering. Nuclear effects enter as form factors modifying the purely point-particle results. In this spirit, the Coulomb excitation cross section, from an initial state i to a final state f, may be expressed as

$$\left(\frac{d\sigma}{d\Omega}\right)_{fi} = \frac{1}{2J_i+1}\sum_{M_f M_i}|P_{M_f M_i}|^2\left(\frac{d\sigma}{d\Omega}\right)_{\text{point}} \tag{8-3}$$

where J_i is the initial spin of the target nucleus and M_i and M_f are, respectively, the projections of the initial and final spins on the quantization axis. The square root of the constant of proportionality is given by the expression

$$P_{M_f M_i} = \frac{1}{i\hbar}\int_{-\infty}^{+\infty}\langle J_f M_f \xi|H'(t)|J_i M_i \zeta\rangle\, e^{i(E_f-E_i)t/\hbar}\,dt$$

The integral may be written as a product of two parts, a nuclear transition matrix element and an integral independent of the nuclear states involved.

On substituting the expansion of Eq. (8-2) into $H'(t)$, we reduce $P_{M_f M_i}$ to a sum over products in the following form:

$$P_{M_f M_i} = \frac{4\pi z e}{i\hbar}\sum_{\lambda\mu}\frac{1}{2\lambda+1}\langle J_f M_f \xi|\boldsymbol{O}_{\lambda\mu}(E\lambda)|J_i M_i \zeta\rangle\, S_{\lambda\mu}(E\lambda)$$

The matrix element $\langle J_f M_f \xi|\boldsymbol{O}_{\lambda\mu}(E\lambda)|J_i M_i \zeta\rangle$ gives the dependence of the cross section on nuclear wave functions and may be related to the reduced transition probability $B(E\lambda; J_i\zeta \rightarrow J_f\xi)$ for $E\lambda$-transition given in Eq. (5-28). The integral over time is contained in the factor $S_{\lambda\mu}(E\lambda)$ and may be expressed as

$$S_{\lambda\mu}(E\lambda) = \int_{-\infty}^{+\infty} e^{i(E_f-E_i)t/\hbar}\frac{1}{r_p^{\lambda+1}(t)}Y_{\lambda\mu}(\theta_p,\phi_p)\,dt$$

Because of the spherical harmonics in the integrand, the integral is a function of the scattering angles; however, it is independent of the nuclear wave functions involved. The derivation of $S_{\lambda\mu}(E\lambda)$ is basically quite simple, even though the actual steps are complicated by angular momentum couplings. The final form, given by Alder et al. [6, 7] may be expressed in the following manner:

$$S_{\lambda\mu}(E\lambda) = \frac{1}{va^\lambda}Y_{\lambda\mu}(\tfrac{1}{2}\pi,0)\mathcal{F}_{\lambda\mu}(\theta,\varrho) \tag{8-4}$$

where ϱ, the "adiabaticity parameter," is related to the energy required to excite the target nucleus from an initial state at energy E_i to a final state at E_f,

$$\varrho = \eta\frac{E_f - E_i}{2E_i} \tag{8-5}$$

The quantity $\mathcal{F}_{\lambda\mu}(\theta,\varrho)$ is an integral having the form

$$\mathcal{F}_{\lambda\mu}(\theta,\varrho) = \int_{-\infty}^{+\infty} e^{i\varrho(\varepsilon\sinh x + x)}\frac{(\cosh x + \varepsilon + i\sqrt{\varepsilon^2-1}\sinh x)^\mu}{(\varepsilon\cosh x + 1)^{\lambda+\mu}}\,dx$$

where $\varepsilon = (\sin\tfrac{1}{2}\theta)^{-1}$.

The spherical harmonics in Eq. (8-4), with $\theta = \frac{1}{2}\pi$ and $\phi = 0$, may be written explicitly in terms of λ and μ,

$$Y_{\lambda\mu}(\tfrac{1}{2}\pi, 0) = \begin{cases} (-1)^{(\lambda+\mu)/2}\sqrt{\dfrac{2\lambda+1}{4\pi}}\dfrac{\sqrt{(\lambda-\mu)!(\lambda+\mu)!}}{(\lambda-\mu)!!(\lambda+\mu)!!} & \text{for } \lambda+\mu = \text{even} \\ 0 & \text{otherwise} \end{cases}$$

The differential scattering cross section for Coulomb excitation in Eq. (8-3) now reduces to

$$\left(\frac{d\sigma}{d\Omega}\right)_{fi} = \sum_{\lambda=1}^{\infty}\left(\frac{2ze}{a^\lambda \hbar v}\right)^2 B(E\lambda; J_i\zeta \rightarrow J_f\xi)\,\frac{df(E\lambda, \varrho)}{d\Omega} \tag{8-6}$$

Here $a = \frac{1}{2}R_s$, half the distance of closest approach given in Eq. (8-1). The angular dependence is contained in the differential

$$\frac{df(E\lambda, \varrho)}{d\Omega} = \frac{4\pi^2}{(2\lambda+1)^3}\sum_{\mu}\left|Y_{\lambda\mu}(\tfrac{1}{2}\pi, 0)\mathcal{F}_{\lambda\mu}(\theta, \varrho)\right|^2\left(\frac{d\sigma}{d\Omega}\right)_{\text{point}}$$

The integral

$$f(E\lambda, \varrho) = \int \frac{df(E\lambda, \varrho)}{d\Omega}\,d\Omega$$

is the total excitation cross section in the classical limit. Explicit values for different multipole order λ are shown in Fig. 8-1 for the limiting case of $\eta \rightarrow \infty$.

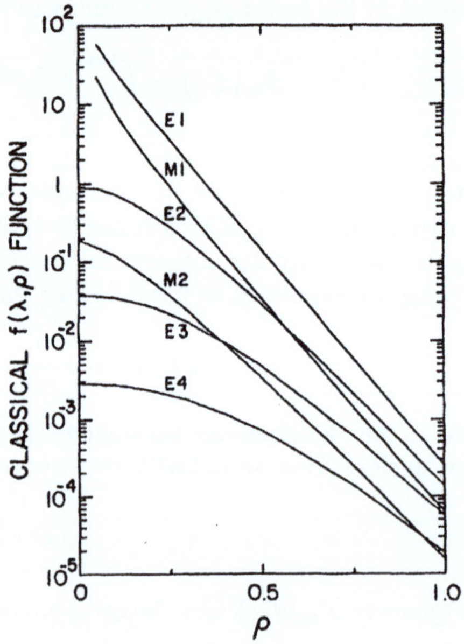

Figure 8-1: Classical Coulomb excitation function $f(\lambda, \varrho)$ for different electric ($E\lambda$) and magnetic ($M\lambda$) multipole transitions. The function $f(\lambda, \varrho)$, with ϱ defined by Eq. (8-5), gives the total excitation cross section in the classical limit of $\eta \rightarrow \infty$, corresponding to the situation of large Z and very low-energy incident particles. (Plotted using values given in Ref. [6].)

So far we have discussed only electric multipole excitations. Magnetic multipole excitations are also present in a Coulomb excitation. The form of the differential cross section for the magnetic case is similar to that for electric multipoles given by

Eq. (8-6), except that $B(E\lambda)$ is replaced by the corresponding $B(M\lambda)$, the reduced transition probability for magnetic multipole λ. Differences in the angular distribution between $M\lambda$- and $E\lambda$-transitions are contained in the differences between $df(M\lambda)/d\Omega$ and $df(E\lambda)/d\Omega$. These are also given in Alder et al. [6, 7], and the values for low-order magnetic multipoles are plotted in Fig. 8-1.

Multiple scattering. Coulomb excitation is useful in creating excited states in the target or projectile nucleus. Because of the intense electromagnetic fields accompanying heavy-ion scattering, states up to very high spin can be excited, as shown later in Fig. 9-2. However, the probability drops rapidly with multipolarity order. An examination of Fig. 8-1 shows that, for example, the probability of $E4$-excitation is reduced by about two orders of magnitude compared with $E2$-processes. Furthermore, the strengths for higher multipole transitions are weaker in general, as the nuclear matrix elements involved are smaller in size. This is evident also from, for example, the values of single-particle estimates given in §5-4. As a result, multiple low-order excitations may become competitive with a single higher multipolarity transition in exciting high-spin states.

Consider a hypothetical even-even nucleus with low-lying level scheme shown in Fig. 8-2 as an example. On the left of the diagram, we have first-order processes promoting the nucleus from the 0^+ ground state to 2^+ and 4^+ excited states. Since $E2$ Coulomb excitations are so much stronger than $E4$-excitations, a succession of two $E2$-processes may be comparable or even stronger in strength than a single $E4$-transition to reach the 4^+ excited state. In fact, because of large reductions in the sizes of the integral $f(\lambda, \varrho)$ with increasing λ, multiple processes through successive low-multipolarity Coulomb excitations may become the preferred path for a nucleus to reach final states of relatively high spins.

There is another class of second-order process shown on the right of Fig. 8-2. In this case, the first excitation brings the nucleus to, for example, a particular magnetic

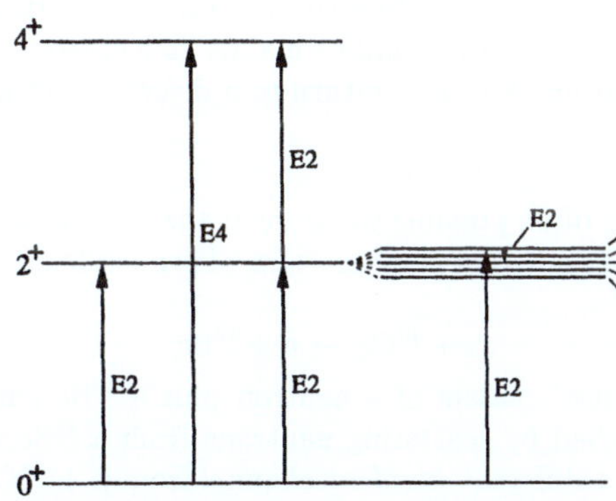

Figure 8-2: Low-lying energy level spectrum of a hypothetical even-even nucleus showing first-order and second-order Coulomb excitation processes.

substate M of a $J^\pi = 2^+$ level. Instead of proceeding to a higher state, the second "scattering" takes the nucleus to a state having a different M-value for the same 2^+ level. This is known as *reorientation* effect. Such a process is sensitive to the expectation value of $O(E2)$ of the level involved, the 2^+-state in the example here. The matrix element, in this case, is related to the quadrupole moment of the excited state. As a result, the quadrupole moment of an excited state may be deduced from the cross section of Coulomb excitation. The value obtained depends somewhat on the nuclear model used; however, this is not a serious problem in general, as reliability of the calculation may be checked against several other properties of the nucleus at the same time. Besides electric quadrupole moments, magnetic dipole and electric hexadecapole (2^4) moments can also be deduced through second-order Coulomb excitation processes. In this way, the static moments of excited states are determined for a large number of nuclei.

8-2 Compound Nucleus Formation

We have seen in the previous two chapters that single-particle and collective degrees of freedom form the two extreme points of view on nuclear structure. A parallel situation exists in nuclear reaction studies in terms of the two limiting situations of direct reaction and compound nucleus formation. In the former case, one assumes that only one nucleon, or a cluster of nucleons, in the projectile interacts with one of the nucleons, or a cluster, in the target without exciting the internal degrees of freedom in any of the clusters or the rest of the nucleus. The basis for taking such a *direct reaction* point of view is the short time, of the order of 10^{-22} s, it takes for the projectile and the target to interact once with each other. Since this is comparable to the transit time for an incident particle with kinetic energy greater than the order of 1 MeV per nucleon to travel over a distance on the order of the nuclear diameter, the probability for interacting more than once is small. On the other hand, if the incident energy is much lower, the projectile and the target may "fuse" together for a long time, for example, of the order of 10^{-14} s. In this case, a *compound nucleus* is formed as the intermediate state. In this section we shall be mainly concerned with reactions involving compound nucleus formation. We shall return to a discussion of direct reactions in the next section.

Reaction channel. It is often possible to arrive at the same final state of a reaction, or exit channel, starting from different combinations of projectile and target. For example, consider the reaction

$$p + {}^{40}\text{Ca} \rightarrow n + {}^{48}\text{Sc}$$

The final state in this case consists of a neutron plus a ^{48}Sc nucleus. The same exit channel can also be reached by scattering neutrons from a ^{48}Sc target. The n + ^{48}Sc system here constitutes a different *incident channel* from p + ^{48}Ca.

Since we are dealing with microscopic objects, time-reversal invariance holds. Thus the reaction $p + {}^{48}\text{Ca} \rightarrow n + {}^{48}\text{Sc}$ may also take place with time order going in the opposite direction, $n + {}^{48}\text{Sc} \rightarrow p + {}^{48}\text{Ca}$, and the roles of incident and exit channels are reversed. On occasion it may be more convenient to speak of a *reaction channel* without specifying whether it is an incident channel or an exit channel.

For a reaction involving two particles, we need three sets of quantum numbers to label the channel in an unambiguous way. For instance, in the exit channel we need a set of labels to specify the wave function of the emerging particle, a second set to identify the wave function of the residual nucleus, and a third set to describe the relative motion between these two particles, as illustrated in §B-4. Returning again to the ^{48}Ca$(p,n)^{48}$Sc example above, the reaction can also leave the ^{48}Sc nucleus in an excited state. Since the wave function for an excited state of ^{48}Sc is different from that for the ground state, we have a different exit channel, distinguishable from the ground state channel by the wave function of the residual nucleus. Furthermore, the wave function for the relative motion of the two particles may be decomposed in terms of partial waves, each with a definite orbital angular momentum ℓ. In principle, each partial wave forms a different reaction channel. On the other hand, the orbital angular momentum between two particles is not usually observed in a reaction, and we may sometimes wish to refer to a reaction channel as the sum of all the partial waves instead.

At low incident energies, where the kinetic energy is less than 1 MeV per nucleon, the de Broglie wavelength is longer than the dimension of a typical nucleus. Under such conditions, the scattering cannot be very sensitive to the details in the structure of nuclei involved. Once the two nuclei in the incident channel come into contact, their nucleons have the time to interact with each other by coming into contact many times. As a result, the identities of the two original nuclei are lost. For a short time, these two nuclei form a single entity, a *compound nucleus*. After formation, memory of the entrance channel is no longer retained because of the numerous intervening interactions. Subsequent evolution of the system is determined primarily by the amount of excitation energy available in the system. At low energies, the lifetimes for such systems are relatively long, as the number of open exit channels is small. As a result, the width Γ of a compound nuclear state is narrow. At the same time, the density of states at such low energies is small so that D, the mean spacing between neighboring states, is large. With $D \gg \Gamma$, we find that isolated resonances dominate the reaction cross section.

Scattering cross section. One of the important features of a compound nucleus reaction is the absence of any dependence between formation and decay of the system. Let σ_α be the cross section for forming a compound nucleus $\mathcal{N}$ through an incident channel α. The decay of $\mathcal{N}$ to a particular exit channel β with final state consisting of particles b and B is characterized by transition probability $\mathcal{W}_\beta$ or partial width $\Gamma_\beta = \hbar\mathcal{W}_\beta$. There may be several such channels open, for example,

		decay product	exit channel
$\mathcal{N}$	$\longrightarrow$	$a+A$	α
	$\longrightarrow$	$b+B$	β
	$\longrightarrow$	$c+C$	γ
	$\longrightarrow$	$\cdots$	$\cdots$

The total width of the decay is given by the sum of all the partial widths,

$$\Gamma = \Gamma_\alpha + \Gamma_\beta + \Gamma_\gamma + \cdots \tag{8-7}$$

and Γ_β/Γ is the probability for $\mathcal{N}$ decaying through channel β. The reaction cross section from an entrance channel α to an exit channel β is given by the product of the probability to form the compound nucleus $\mathcal{N}$ and that for $\mathcal{N}$ to decay through β,

$$\sigma_{\beta\alpha} = \sigma_\alpha \frac{\Gamma_\beta}{\Gamma} \tag{8-8}$$

To make further progress, we shall make use of some of the results given in §B-4.

Let us assume that in each reaction channel there is a radius R_c, the *channel radius*, outside which there is no interaction between the scattered particle and the residual nucleus (ignoring the long-range Coulomb interaction here for simplicity). Thus, in the outside region ($r > R_c$), the particles may be considered to be free and the wave functions are given by plane waves (or Coulomb wave functions in the more general case). In the inside region ($r < R_c$), the situation is complicated because of interaction between nucleons in the two components, and there is little hope of obtaining a reasonable solution. At the boundary $r = R_c$, the logarithmic derivative of the modified radial wave function u_c of each channel,

$$\rho_c = \left(\frac{r}{u_c}\frac{du_c}{dr}\right)_{r=R_c} \tag{8-9}$$

must be continuous from the inside to the outside region. In general, ρ_c is complex.

In a scattering problem, we are primarily interested in the asymptotic behavior of the system. The only information of the wave function for $r < R_c$ we need is contained in the value of its logarithmic derivative at the boundary. In other words, as far as the outside region is concerned, the information of the inside region is completely contained in a set of logarithmic derivatives, and the values of these derivatives may be used as the "boundary conditions" to fix the asymptotic wave function of interest to the scattering problem. In the absence of better knowledge, we can take the logarithmic derivatives as parameters characterizing the inside region. This method of treating scattering is akin to that used in solving electrostatic problems, where we exclude regions containing sources and replace them by appropriate boundary conditions. In this way, the problem is reduced to a more manageable one.

To simplify the discussion further, we shall assume that only s-wave scattering is different from zero and, as a result, a single logarithmic derivative ρ_0 is adequate to specify the problem completely. In this case, the reduced radial wave function of Eq. (B-10) has the asymptotic form

$$u_0(r) \sim \{e^{-ikr} - \eta_0 e^{ikr}\}$$

The quantity $\eta_0 = \exp\{2i\delta_0\}$ is the inelasticity parameter and δ_0, the complex phase shift for $\ell = 0$. By taking the logarithmic derivative of $u_0(r)$ and equating it to ρ_0 at $r = R_c$, we can relate η_0 to ρ_0,

$$\eta_0 = \frac{\rho_0 + ikR_c}{\rho_0 - ikR_c} e^{-2ikR_c}$$

Using Eq. (B-19), we obtain the elastic scattering cross section

$$\sigma^{\rm el} = \frac{\pi}{k^2}|1 - \eta_0|^2 = \frac{\pi}{k^2}\left| e^{2ikR_c} - 1 - \frac{2ikR_c}{\rho_0 - ikR_c}\right|^2 \tag{8-10}$$

From Eq. (B-40), we arrive at the expression for the reaction cross section

$$\sigma^{\rm re} = \frac{\pi}{k^2}(1 - |\eta_0|^2) = \frac{\pi}{k^2}\frac{-4kR_c\Im\rho_0}{(\Re\rho_0)^2 + (\Im\rho_0 - kR_c)^2} \tag{8-11}$$

If ρ_0 is real, corresponding to the case of scattering from a real potential, $\sigma^{\rm re}$ vanishes, as expected. Furthermore, since the reaction cross section cannot be negative, the absolute value of η_0 must be less than or equal to unity. This, in turn, implies that the imaginary part of ρ_0 must be negative.

Breit-Wigner formula for isolated resonances. For a reaction of the type described above, the cross section has a resonance structure similar to that of an alternating current electric circuit. The maximum of $\sigma^{\rm re}$ occurs at $\Re\rho_0 = 0$, where $\Re\rho_0$ is the real part of ρ_0. Let E_c represent the energy where this takes place. The real part of ρ_0 may be expanded as a power series in E around E_c,

$$\Re\rho_0 = a(E - E_c) + \cdots$$

where a is a parameter characterizing the leading-order term of the real part of ρ_0. Similarly, the leading order of the imaginary part of ρ_0 may be expressed in terms of a (positive) parameter b,

$$\Im\rho_0 = -b + \cdots$$

The two cross sections in Eqs. (8-10) and (8-11) near resonance energy E_c may now be written as

$$\begin{aligned}\sigma^{\rm el} &= \frac{\pi}{k^2}\left|e^{2ikR_c} - 1 - \frac{2ikR_c}{a(E - E_c) - i(b + kR_c)}\right|^2\\ \sigma^{\rm re} &= \frac{\pi}{k^2}\frac{4kR_cb}{\{a(E - E_c)\}^2 + (b + kR_c)^2}\end{aligned}$$

in terms of parameters a and b.

We now make the following identifications:

$$\Gamma = 2\frac{b + kR_c}{a} \qquad \Gamma_\alpha = \frac{2kR_c}{a} \qquad \Gamma^{\rm re} = 2\frac{b}{a}$$

where Γ is the total width and Γ_α is the partial width for the entrance channel. The total reaction width is then

$$\Gamma^{\rm re} = \sum_{i\neq\alpha}\Gamma_i$$

and the total width is $\Gamma = \Gamma_\alpha + \Gamma^{\rm re}$. This allows us, in turn, to rewrite Eqs. (8-10) and (8-11) in the following way:

$$\sigma^{\rm el} = \frac{\pi}{k^2}\left|e^{2ikR_c} - 1 - \frac{i\Gamma_\alpha}{(E - E_c) - i\frac{1}{2}\Gamma}\right|^2 \tag{8-12}$$

$$\sigma^{\rm re} = \frac{\pi}{k^2}\frac{\Gamma^{\rm re}\Gamma_\alpha}{(E - E_c)^2 + (\frac{1}{2}\Gamma)^2} \tag{8-13}$$

The cross sections are expressed here in terms of the relevant widths instead of parameters a and b as we have done above.

The elastic channel has two parts, a nonresonating part with amplitude proportional to $1-\exp\{2ikR_c\}$ and a resonating part containing an energy-dependent factor $(E-E_c)$ in the denominator. The contribution of the nonresonating part corresponds to a smooth background in the cross section and is usually called *shape-elastic* or *potential* scattering. At $E \approx E_c$, the elastic cross section is dominated by the resonating part,

$$\sigma^{\text{cel}} \approx \frac{\pi}{k^2} \frac{\Gamma_\alpha^2}{(E-E_c)^2 + (\frac{1}{2}\Gamma)^2}$$

This is called *compound elastic* scattering cross section, as it differs from shape-elastic scattering by the fact that a compound nucleus is formed before the system returns to the entrance channel.

We can now go back to Eq. (8-8) and calculate σ_α, the cross section for forming the compound nucleus through entrance channel α. Since shape-elastic scattering does not involve the formation of a compound nucleus, we can ignore it here. The cross section for compound nucleus formation is, then, a sum of compound-elastic and reaction contributions,

$$\sigma_\alpha = \sigma^{\text{cel}} + \sigma^{\text{re}} = \frac{\pi}{k^2} \frac{\Gamma^{\text{re}}\Gamma_\alpha + \Gamma_\alpha^2}{(E-E_c)^2 + (\frac{1}{2}\Gamma)^2} = \frac{\pi}{k^2} \frac{\Gamma\Gamma_\alpha}{(E-E_c)^2 + (\frac{1}{2}\Gamma)^2} \tag{8-14}$$

where we have made use of the fact that $\Gamma = \Gamma_\alpha + \Gamma^{\text{re}}$. The cross section for the reaction from entrance channel α to exit channel β given in Eq. (8-7) may now be written as

$$\sigma_{\beta\alpha} = \frac{\pi}{k^2} \frac{\Gamma_\beta\Gamma_\alpha}{(E-E_c)^2 + (\frac{1}{2}\Gamma)^2}$$

This is known as the Breit-Wigner one-level formula.

Overlapping resonances. So far the discussion has been confined to the low-energy region where the density of states is small. The idea of compound nucleus formation applies also at higher energies where individual level width is comparable to or greater than the average level spacing ($\Gamma \gtrsim D$). Since resonances are now overlapping each other, it is more meaningful to examine the average values of the various cross sections that enter into the scattering.

Assuming that the cross section to form a compound nucleus for a particular state is still given by the Breit-Wigner form of Eq. (8-14), we can define an average in a small energy interval W in the following way:

$$\overline{\sigma}_\alpha = \frac{1}{W} \int_{E-W/2}^{E+W/2} \sum_i \frac{\pi}{k^2} \frac{\Gamma^i\Gamma_\alpha^i}{(E'-E_i)^2 + (\frac{1}{2}\Gamma^i)^2} dE' = \frac{\pi}{k^2}\frac{2\pi}{W} \sum_i \Gamma_\alpha^i$$

where Γ^i is the total width of the ith resonance and Γ_α^i is the partial width for decaying into channel α. The summation is over all the resonances in the energy interval W centered at E. The interval must be small enough that the underlying conditions for the resonances are not too different from each other and yet are large enough so that

$W \gg \Gamma^\imath$. Our discussion is still limited to s-wave scattering for simplicity. We shall not attempt a more general discussion here, as it will also involve angular momentum coupling factors.

Using the fact that the number of levels in the interval is given by W/D, where D is the average level spacing, we can define a mean width,

$$\overline{\Gamma}_\alpha = \frac{D}{W}\sum_\imath \Gamma^\imath_\alpha$$

The quantity $\overline{\Gamma}_\alpha/D$ is known as the (s-wave) *strength function* and the average compound nucleus formation cross section may be expressed in terms of it,

$$\overline{\sigma}_\alpha = \frac{\pi}{k^2} 2\pi \frac{\overline{\Gamma}_\alpha}{D} \tag{8-15}$$

This quantity may be related to ρ_0, the logarithmic derivative of the wave function at $r = R_c$ given by Eq. (8-9). Since the density of states is now high, the probability for the compound nucleus to decay through the entrance channel is small. We can say that the nucleus appears to be "black" to the incident channel.

In the limit of a completely absorptive nucleus, the wave function of the interior region may be approximated by an incoming term, $u_0(r) \sim \exp\{i\kappa r\}$, alone. As a result, ρ_0 is purely imaginary and may be written as

$$\rho_0 = -i\kappa R_c$$

where κ is the wave number for $r < R_c$. On substituting this value into Eqs. (8-10) and (8-11), we obtain the average value of the compound nucleus formation cross section for channel α in the energy region

$$\overline{\sigma}_\alpha = \frac{\pi}{k^2}\frac{4\kappa k}{(\kappa + k)^2} \approx \frac{4\pi}{\kappa k} \tag{8-16}$$

since $k \ll \kappa$ for low-energy scattering from an attractive potential well. Comparing this expression with Eq. (8-15), we obtain the result,

$$\frac{\overline{\Gamma}_\alpha}{D} = \frac{2k}{\pi\kappa}$$

Furthermore, no resonance can be expected from Eq. (8-16).

In practice, resonances are observed at high energies. These are due primarily to coupling of a large number of small resonances, for example, to a state in the vicinity that is strongly excited due to some special features in the nuclear structure. Such a strongly excited state is often called a *doorway* state. The decay of a compound nucleus in the high-energy region depends on the number of accessible final states and is therefore dominated by the density of final states and other statistical considerations. This is the subject of Hauser-Feshbach theory, for which we shall refer the reader to the original literature [82].

8-3 Direct Reaction

Stripping and pickup reactions. A good example of a direct reaction is a (d,p) process in which a deuteron, with more than a few mega-electron-volts of kinetic energy, incidents on a target nucleus. In the exit channel, a proton is observed. Since the deuteron is a loosely bound system of a proton and a neutron, we can envisage that the neutron is captured into one of the single-particle orbits of the target nucleus without disturbing the rest of the nucleons, and the proton continues on to become the scattered particle. The process may also be viewed as one in which a neutron is "stripped" from the projectile. For this reason, the reaction is known as a one-neutron *stripping* reaction. States in the final nucleus that are strongly excited by such a reaction are those formed predominantly by a nucleon coupled to the ground state of the target nucleus. Other stripping reactions, such as (t,p), transfer two nucleons from the projectile to the target. Even more complicated reactions, such as those involving the transfer of a cluster of nucleons, are also possible. To qualify as a direct reaction, both the target nucleus and the internal structure of the cluster transferred must be undisturbed by the reaction. The residual nucleus is simply the coupled product of the cluster and the ground state of the target nucleus. This condition is generally difficult to meet for transfer reactions involving large numbers of nucleons.

The complement of a stripping reaction is a *pickup* reaction. In this case, one or more nucleons are taken away from the target nucleus without changing the structure among the rest of the nucleons. A good example is the reaction ^{40}Ca(^{3}He,^{4}He) ^{39}Ca. The states in the residual nucleus, ^{39}Ca here, strongly excited by a pickup reaction are the one-hole states, i.e., those formed by removing one of the particles in ^{40}Ca and leaving the remaining 39 nucleons unchanged in their relative motion.

Born approximation. The scattering cross section in a direct reaction, stripping as well as pickup, may be obtained using first Born approximation. The reaction mechanism is relatively straightforward here because of the simple relation between initial and final nuclear states underlying the direction reaction assumption. The Schrödinger equation for the process may be written in the form of a standard second-order differential equation,

$$(\nabla^2 + k^2)\psi(\boldsymbol{r}) = \frac{2\mu}{\hbar^2}V(\boldsymbol{r})\,\psi(\boldsymbol{r}) \tag{8-17}$$

where $k^2 = 2\mu E/\hbar^2$, with E as the kinetic energy in the center-of-mass system. A formal solution of Eq. (8-17) for the outgoing wave function may be expressed in terms of a Green's function $G(\boldsymbol{r},\boldsymbol{r}')$, as done in §B-6,

$$\psi(\boldsymbol{r}) = e^{i\boldsymbol{k}_i\cdot\boldsymbol{r}} + \frac{2\mu}{\hbar^2}\int G(\boldsymbol{r},\boldsymbol{r}')V(\boldsymbol{r}')\,\psi(\boldsymbol{r}')\,d^3r' \tag{8-18}$$

where we have chosen $\boldsymbol{k}_i$ to be along the direction of the incident particle and the function $\exp\{i\boldsymbol{k}_i\cdot\boldsymbol{r}\}$ is the solution of the homogeneous part of Eq. (8-17), i.e., for $V(\boldsymbol{r}) = 0$.

We shall take the Green's function here to have the explicit form

$$G(\boldsymbol{r},\boldsymbol{r}') = -\frac{1}{4\pi}\frac{e^{ik|\boldsymbol{r}-\boldsymbol{r}'|}}{|\boldsymbol{r}-\boldsymbol{r}'|} \tag{8-19}$$

It satisfies the equation

$$(\nabla^2 + k^2)G(\boldsymbol{r},\boldsymbol{r}') = \delta(\boldsymbol{r}-\boldsymbol{r}')$$

More generally, we can include in the defining equation for $G(\boldsymbol{r},\boldsymbol{r}')$ a part of $V(\boldsymbol{r})$, for example, the part representing the average effect of the target nucleons on the incident particle. This is similar in spirit to the mean-field approach used in nuclear structure investigations to obtain the single-particle states for shell-model calculations in §7-3. To keep the discussion simple here, we shall take the Green's function to have the elementary form given by Eq. (8-19).

Using Eq. (8-19), the formal solution for the scattering wave function of Eq. (8-18) may be written as

$$\psi(\boldsymbol{r}) = e^{i\boldsymbol{k}_i\cdot\boldsymbol{r}} - \frac{\mu}{2\pi\hbar^2}\int \frac{e^{ik|\boldsymbol{r}-\boldsymbol{r}'|}}{|\boldsymbol{r}-\boldsymbol{r}'|}V(\boldsymbol{r}')\psi(\boldsymbol{r}')\,d^3r'$$

We shall take the range of potential $V(\boldsymbol{r}')$ to be short. The effect of a (long-range) Coulomb interaction may be included as a part of the Green's function or the optical model potential to be discussed in the next section. In this limit, we may approximate the argument of the exponential function in the asymptotic region by the first two terms in the expansion

$$\begin{aligned} ik|\boldsymbol{r}-\boldsymbol{r}'| &= k\sqrt{r^2 - 2\boldsymbol{r}\cdot\boldsymbol{r}' + r'^2} \\ &= kr - \boldsymbol{k}_f\cdot\boldsymbol{r}' + O(r'^2) \\ &\approx kr - \boldsymbol{k}_f\cdot\boldsymbol{r}' \end{aligned}$$

where $\boldsymbol{k}_f = k\boldsymbol{r}/r$ is taken along the direction of the emerging particle. The formal solution of the scattering equation now becomes

$$\psi(\boldsymbol{r}) \approx e^{i\boldsymbol{k}_i\cdot\boldsymbol{r}} - \frac{\mu}{2\pi\hbar^2}\frac{e^{ikr}}{r}\int e^{-i\boldsymbol{k}_f\cdot\boldsymbol{r}'}V(\boldsymbol{r}')\,\psi(\boldsymbol{r}')\,d^3r' \tag{8-20}$$

where we have taken $|\boldsymbol{r}-\boldsymbol{r}'| \simeq r$, correct in the asymptotic region where the scattered particle is observed.

Comparing Eq. (8-20) with the asymptotic form of the scattering wave function given in Eq. (B-5), we obtain the scattering amplitude

$$f(\theta) = -\frac{\mu}{2\pi\hbar^2}\int e^{-i\boldsymbol{k}_f\cdot\boldsymbol{r}'}V(\boldsymbol{r}')\,\psi(\boldsymbol{r}')\,d^3r' \tag{8-21}$$

This is only a formal, or integral equation, solution for the scattering amplitude, as the expression involves an unknown function $\psi(\boldsymbol{r}')$, the solution to the scattering equation. Equation (8-21) is useful in that it provides us with a starting point to expand scattering cross section in terms of a Born series. In the (first) Born approximation, the unknown function $\psi(\boldsymbol{r}')$ in Eq. (8-21) is replaced by its first term in Eq. (8-18), and we obtain an approximate form of the scattering amplitude,

$$f(\theta) \approx -\frac{\mu}{2\pi\hbar^2}\int e^{-i\boldsymbol{k}_f\cdot\boldsymbol{r}'}V(\boldsymbol{r}')\,e^{i\boldsymbol{k}_i\cdot\boldsymbol{r}'}d^3r' \tag{8-22}$$

This result may be used to find the differential scattering cross section for stripping and pickup reactions.

The expression in Eq. (8-22) can be simplified further by expressing the results in terms of the momentum transfer vector,

$$\boldsymbol{q} = \boldsymbol{k}_i - \boldsymbol{k}_f$$

and by expanding the plane wave in terms of spherical harmonics as in Eq. (B-10),

$$e^{i\boldsymbol{q}\cdot\boldsymbol{r}'} = \sum_\ell i^\ell\sqrt{4\pi(2\ell+1)}\,j_\ell(qr')\,Y_{\ell 0}(\theta') \tag{8-23}$$

where $j_\ell(qr')$ is the spherical Bessel function of order ℓ. The angle θ' is between vectors $\boldsymbol{q}$ and $\boldsymbol{r}'$ and is one of the variables of integration in Eq. (8-22). The scattering angle θ, on the other hand, is between vectors $\boldsymbol{k}_f$ and $\boldsymbol{k}_i$.

Angular distribution. The discussions in the previous paragraph ignore the internal structure of particles participating in the scattering. Since in stripping and pickup reactions we are dealing with a change in the nuclei involved, the wave functions of both the initial and final nuclei must enter into the expression for the scattering amplitude. Let us take the asymptotic forms of the initial and final wave functions of the scattering system to be

$$\begin{aligned}\Psi_i &\rightarrow e^{i\boldsymbol{k}_i\cdot\boldsymbol{r}}\Phi_i\\ \Psi_f &\rightarrow e^{i\boldsymbol{k}_f\cdot\boldsymbol{r}}\Phi_f\end{aligned}$$

where Φ_i is the product of the internal wave functions of the incident particle and the target nucleus and Φ_f that of the wave functions of the scattered particle and the residual nucleus. To simplify later discussions, plane waves $\exp\{i\boldsymbol{k}_i\cdot\boldsymbol{r}\}$ and $\exp\{i\boldsymbol{k}_f\cdot\boldsymbol{r}\}$ are used to describe the motions of the particles in, respectively, the incident and exit channel. Physically, this means we are assuming that the particles do not interact with each other (except in the small region where the reaction takes place). As we shall see in the next section, it is more reasonable to consider also effects, such as Coulomb interaction, by using an optical model.

As a concrete example, let us take the case of $^{40}\text{Ca}(d,p)^{41}\text{Ca}$. For this reaction, we have

$$\begin{aligned}\Phi_i &= \{\phi(d)\times\phi(^{40}\text{Ca})\}\\ \Phi_f &= \{\phi(p)\times\phi(^{41}\text{Ca})\}\end{aligned}$$

where $\phi(d)$, $\phi(p)$, and $\phi(^{40}\text{Ca})$ and $\phi(^{41}\text{Ca})$ are, respectively, the wave functions describing the internal structure of the deuteron, proton, and ground states of ^{40}Ca and ^{41}Ca.

The multiplication symbols here imply that the wave functions are coupled together to some definite values in angular momentum and isospin.

In the spirit of direct reaction, the deuteron wave function may be taken as the (weakly) coupled state of a proton and a neutron,

$$\phi(d) = \{\phi(p) \times \phi(n)\}$$

To simplify the argument and avoid complications due to angular momentum recoupling, we shall treat the proton purely as a spectator in the entire scattering process. If the neutron is captured into a single-particle state of the target nucleus with orbital angular momentum ℓ_t, the wave function of the residual nucleus may be expressed as

$$\phi_{\ell_t}(^{41}\mathrm{Ca}) \sim \{\phi(n)\phi(^{40}\mathrm{Ca})Y_{\ell_t m_t}(\theta', \phi')\}$$

where spherical harmonics $Y_{\ell_t m_t}(\theta', \phi')$ is the orbital angular momentum wave function of the single-particle state in which the neutron is captured. Using these wave functions, the scattering amplitude for $^{40}\mathrm{Ca}(d,p)^{41}\mathrm{Ca}$ may be written as

$$\begin{aligned} f(\theta) \approx\ & -\frac{\mu}{2\pi\hbar^2}\int e^{-i\boldsymbol{q}\cdot\boldsymbol{r}'}\Big\langle\big\{\phi(p) \times (\phi(n)\phi(^{40}\mathrm{Ca})Y_{\ell_t m_t}(\theta', \phi'))\big\}\Big| \\ & \times V(\boldsymbol{r}')\Big|\big\{\phi(^{40}\mathrm{Ca}) \times (\phi(p) \times \phi(n))\big\}\Big\rangle\, d^3r' \end{aligned} \tag{8-24}$$

The role of the potential $V(\boldsymbol{r}')$ here is to strip the neutron from the deuteron and put it into the residual nucleus. For our purposes, it may be approximated by a delta function at the nuclear surface,

$$V(\boldsymbol{r}') = V_0\,\delta(r' - R) \tag{8-25}$$

to simplify the derivation. Here R is the radius of the residual nucleus. The meaning of this approximation is that the neutron is stripped off the incident deuteron and captured by the $^{40}\mathrm{Ca}$ on contact. The strength V_0 represents the probability for such a process to take place and may be treated as a parameter related to the absolute magnitude of the scattering cross section.

Once we integrate Eq. (8-24) over the coordinates of both nucleons and $^{40}\mathrm{Ca}$, no nuclear wave functions are left in the expression. The exponential factor may be expanded in terms of spherical harmonics using Eq. (8-23), and the first Born approximation scattering amplitude of Eq. (8-22) reduces to

$$\begin{aligned} f(\theta) &\approx -\frac{2\mu}{\hbar^2}V_0\sum_\ell i^\ell\sqrt{\frac{2\ell+1}{4\pi}}\,j_\ell(qR)\int Y_{\ell 0}(\theta')Y^*_{\ell_t m_t}(\theta', \phi')\sin\theta'\,d\theta'\,d\phi' \\ &= -\frac{2\mu}{\hbar^2}V_0\sum_\ell i^\ell\sqrt{\frac{2\ell+1}{4\pi}}\,j_\ell(qR)\,\delta_{\ell\ell_t}\delta_{m_t 0} \\ &= -\frac{2\mu}{\hbar^2}V_0 i^{\ell_t}\sqrt{\frac{2\ell_t+1}{4\pi}}\,j_{\ell_t}(qR) \end{aligned} \tag{8-26}$$

In integrating over the angular variables, we have made use of the orthonormal condition of spherical harmonics given in Eq. (B-18). The only angular dependence remaining at the end in Eq. (8-24) is contained in the argument of the spherical Bessel function, as we shall see in the next paragraph. Since we have used a plane wave here to approximate the solution to the scattering equation, it is known as the plane wave Born approximation (PWBA). A more rigorous derivation can be found in standard references on direct reactions such as Tobocman [137] and Satchler [123].

From Eqs. (8-26) and (B-7), we find that the differential cross section for direct reaction is given by

$$\frac{d\sigma}{d\Omega} \sim |j_{\ell_t}(qR)|^2$$

The momentum transfer depends on the scattering angle θ as shown in Eq. (4-13),

$$q = \sqrt{k_i^2 + k_f^2 - 2k_i k_f \cos\theta} \approx 2k \sin\left(\frac{\theta}{2}\right) \tag{8-27}$$

where we have taken $k \approx k_i \approx k_f$, valid if the incident energy is sufficiently high. The angular distribution is characterized by the angular momentum transferred and given by the factor $|j_{\ell_t}(2kR\sin(\frac{1}{2}\theta))|^2$, as shown in Fig. 8-3. For example, since $j_0(\rho) \sim \sin\rho/\rho$, we see that, for an $\ell_t = 0$ transfer, the angular distribution peaks at 0°. For higher ℓ_t-value transfers, there is no longer a maximum at 0°, as can be seen, for example, from $j_1(\rho) \sim \sin\rho/\rho^2 - \cos\rho/\rho$ for $\ell_t = 1$. As the value of ℓ_t is increased, the first maximum in the angular distribution shifts to successively larger angles, as the first peak of $|j_{\ell_t}(\rho)|^2$ appears at successively larger values of ρ with increasing ℓ_t. This is a feature observed in direct reactions, as can be seen, for example, in the ^{20}Ne(d,n) ^{21}Ne reaction shown in Fig. 8-4.

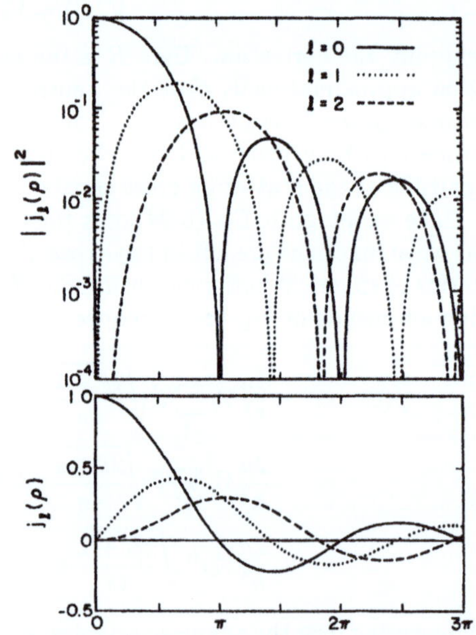

Figure 8-3: Spherical Bessel functions $j_\ell(\rho)$ and characteristic angular distribution of stripping reaction given by $j_\ell^2(\rho)$. The plots are made as functions of $\rho = qR$, with q being the momentum transfer. The relation with scattering angle θ is given by Eq. (8-27).

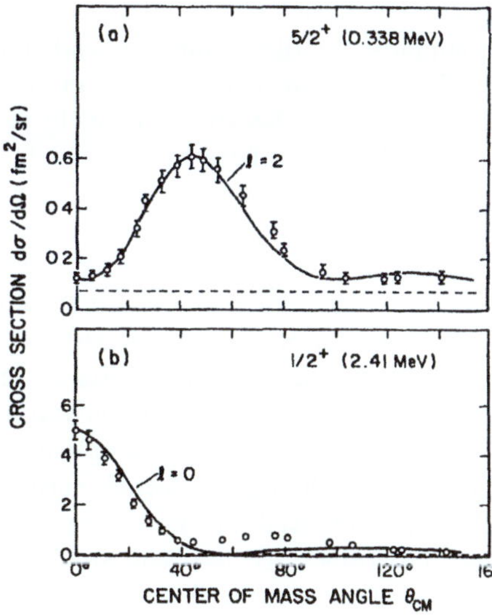

Figure 8-4: Angular distribution of neutrons observed in the reaction ^{20}Ne$(d,n)^{21}$Ne leading to (a) the 0.338-MeV, $J^\pi = 5/2^+$ level in ^{21}Ne showing a typical $\ell = 2$ one-nucleon transfer and (b) the 2.41-MeV, $J^\pi = 1/2^+$ level, an $\ell = 0$ transfer. Solid lines are calculated results using DWBA for direct reaction and dashed lines are the results of a Hauser-Feshbach calculation for compound nucleus reaction [40].

Although the PWBA method correctly gives the essential features in the angular distribution of direction reactions, it lacks predictive power. This is, in part, due to *distortion* of the incident and scattered waves as a result of the average, or "optical," potential experienced by the incoming and scattered particles, as we shall see in the next section. Furthermore, there does not seem to be an easy way to derive the interaction strength V_0 of Eq. (8-25), and as a result, the magnitude of the angular distributions cannot be deduced from PWBA. A more accurate picture of the scattering is given by the distorted wave Born approximation (DWBA) where, instead of plane waves, more realistic wave functions are used for the relative motion between the projectile and the target nucleus and between the scattered particle and the residual nucleus.

8-4 The Optical Model

Besides compound nucleus formation and direct reaction, we may also be interested in the average result of a reaction at some fixed bombarding energy. For such purposes, it is possible to invoke the analogy of an optical wave passing through a "cloudy" crystal ball. In a nuclear reaction, the scattered wave may be divided into two categories: elastic scattering, in which only the direction of the wave propagation is changed, and inelastic scattering, in which the particles are scattered into an exit channel different from the incident one. The former may be compared with refraction of optical waves and the latter with absorption due to the fact that the crystal ball is cloudy.

The aim of the optical model is to find a potential that describes smooth variations of the scattering cross section as a function of incident energy E and target nucleon number A. The scattering itself may be quite complicated; however, if we are only interested in the averaged properties, away from resonances and states strongly excited by direct reactions, it is possible to simplify the situation by a large extent. The basic idea is very similar to the mean-field approach we have seen in the previous chapter for nuclear structure studies.

To simplify the discussion, we shall for the most part consider only elastic scattering. There are two main sources of contribution to the cross section. The first is potential scattering, described earlier in Eq. (8-12). The second comes from multiple scattering with intermediate states involving the excited states of nuclei participating in the scattering. Not all such scattering returns the system back to the incident channel and, as a result, some of the incident flux is lost. Rather than trying to calculate the cross section for each one of the inelastic channels exactly, we shall attempt to represent their average contributions by making the potential complex. The same idea can also be extended to scattering between hadrons in general, but we shall not do it here. Our primary concern will be with nucleon-nucleus scattering, and we shall return later for a brief discussion on applying the optical model for pion-nucleus scattering.

There are three aspects of an optical model potential we shall touch upon in this section. First, we shall give a formal derivation so as to make a connection between the optical model potential and averaging over contributions involving a large number of intermediate states. Second, semi-empirical forms of the optical model potential have been used over the years with great success for low-energy (<200-MeV) scattering. We shall give an example here to provide some feeling of the form of the optical model potential used in practice and its dependence on incident energy and other variables. Third, we wish to make some contact to scattering at the nucleon-nucleon level by giving a "microscopic" foundation to the optical model potential.

Formal derivation of the optical model potential. Consider the case of a free nucleon scattering off a nucleus made of A nucleons. Let $\boldsymbol{r}_0$ represent the coordinate of the projectile and $\boldsymbol{r}_i$, for $i = 1, \ldots, A$, those of the A nucleons in the target. To keep the notation simple, we shall suppress any explicit reference to spin and other degrees of freedom. Our aim is to solve the Schrödinger equation,

$$H(\boldsymbol{r}_0; \boldsymbol{r}_1, \boldsymbol{r}_2, \ldots, \boldsymbol{r}_A)\Psi(\boldsymbol{r}_0; \boldsymbol{r}_1, \boldsymbol{r}_2, \ldots, \boldsymbol{r}_A) = E\Psi(\boldsymbol{r}_0; \boldsymbol{r}_1, \boldsymbol{r}_2, \ldots, \boldsymbol{r}_A) \tag{8-28}$$

with boundary conditions appropriate for scattering. As usual, it is impossible to solve exactly the many-body problem, and we shall seek an approximate solution adequate to understand the average results in a scattering.

For the time being we shall ignore the necessary antisymmetrization between the projectile and the nucleons in the target. The Hamiltonian for the complete system, consisting of the projectile and the target nucleus, may be separated into three parts,

$$H(\boldsymbol{r}_0; \boldsymbol{r}_1, \boldsymbol{r}_2, \ldots, \boldsymbol{r}_A) = T_0 + \sum_{i=1}^{A} V(\boldsymbol{r}_{0i}) + H_A(\boldsymbol{r}_1, \boldsymbol{r}_2, \ldots, \boldsymbol{r}_A) \tag{8-29}$$

where $\boldsymbol{r}_{0i} = \boldsymbol{r}_0 - \boldsymbol{r}_i$, for $i \neq 0$, is the relative coordinate between the projectile and the ith nucleon in the target. The operator T_0 describes the kinetic energy of the projectile and $H_A(\boldsymbol{r}_1, \boldsymbol{r}_2, \ldots, \boldsymbol{r}_A)$ is the Hamiltonian operating only among the nucleons in the target. The interaction between the projectile and the target nucleons is provided by the potential $V(\boldsymbol{r}_{0i})$.

We shall assume that the A-body eigenvalue problem for the target nucleus has already been solved and that a complete set of solutions $\{\Phi_i\}$ is available for the

Schrödinger equation:

$$H_A(\boldsymbol{r}_1, \boldsymbol{r}_2, \ldots, \boldsymbol{r}_A)\Phi_i(\boldsymbol{r}_1, \boldsymbol{r}_2, \ldots, \boldsymbol{r}_A) = \epsilon_i\Phi_i(\boldsymbol{r}_1, \boldsymbol{r}_2, \ldots, \boldsymbol{r}_A) \tag{8-30}$$

Furthermore, we shall take that Φ_i is normalized to unity and is a part of an orthogonal set of eigenfunctions. The general solution for the complete system, including both the projectile particle and the target nucleus, may be expressed as a linear combination of the products of $\chi_i(\boldsymbol{r}_0)$ and $\Phi_j(\boldsymbol{r}_1, \boldsymbol{r}_2, \ldots, \boldsymbol{r}_A)$,

$$\Psi(\boldsymbol{r}_0; \boldsymbol{r}_1, \boldsymbol{r}_2, \ldots, \boldsymbol{r}_A) = \sum_{ij} \chi_i(\boldsymbol{r}_0)\Phi_j(\boldsymbol{r}_1, \boldsymbol{r}_2, \ldots, \boldsymbol{r}_A) \tag{8-31}$$

where $\chi_i(\boldsymbol{r}_0)$ is the wave function of the projectile. If our primary interest is in elastic scattering, the only part of Ψ that is of interest to us here is $\chi_0\Phi_0$, where both the target nucleus and the projectile are in their respective lowest energy states. Our problem here is to obtain χ_0 (as Ψ_0 is assumed to be already known).

We shall first construct an equation for χ_0 using the method of projection operators. The approach is very similar to that used earlier in §7-5 to find a renormalized Hamiltonian in nuclear structure calculations when the active space is reduced to a small subset of the complete shell-model space. Let P be a projection operator for the ground state of the target. We may write P as

$$P = |\Phi_0\rangle\langle\Phi_0| \tag{8-32}$$

with the understanding that any integration to be carried out is over the coordinates of the target nucleons only. When P acts on the wave function of Eq. (8-31), we obtain the result

$$P\Psi = \chi_0\Phi_0$$

We may also define an operator Q that projects out the rest of the states,

$$Q = 1 - P$$

It is easy to see that

$$P^2\Psi = P\Psi \qquad Q^2\Psi = Q\Psi \qquad PQ\Psi = QP\Psi = 0 \tag{8-33}$$

as P and Q are operators projecting out different parts of the complete space.

Since $P + Q = 1$, the Schrödinger equation (8-28) may be written as

$$(E - H)(P + Q)\Psi = 0 \tag{8-34}$$

On multiplying from the left by P and making use of the relations given in Eq. (8-33), we obtain $P\Psi$ in terms of $Q\Psi$,

$$(E - PHP)\,P\Psi = (PHQ)\,Q\Psi \tag{8-35}$$

Similarly, on multiplying Eq. (8-34) from the left by Q, we obtain

$$(E - QHQ)\,Q\Psi = (QHP)\,P\Psi$$

This result may be used to express $Q\Psi$ formally in terms of $P\Psi$,

$$Q\Psi = \frac{1}{E - QHQ} QHP\,P\Psi$$

When this "solution" for $Q\Psi$ is substituted into the right-hand side of Eq. (8-35), we obtain an expression for $P\Psi$,

$$\Big\{E - PHP - PHQ\frac{1}{E - QHQ}QHP\Big\}P\Psi = 0$$

On multiplying this equation from the left by $\langle\Phi_0|$ and integrating over the coordinates of the target nucleons with the help of the explicit form of P given in Eq. (8-32), we arrive at a relation for χ_0,

$$\Big\{E - \langle\Phi_0|H|\Phi_0\rangle - \langle\Phi_0|HQ\frac{1}{E - QHQ}QH|\Phi_0\rangle\Big\}\chi_0 = 0 \tag{8-36}$$

This is the equation we must solve to obtain χ_0.

The zero point of the energy scale is still arbitrary at this point, and we may set it at the ground state of the target nucleus. In other words, we can choose ϵ_0 in Eq. (8-30) to be 0 to simplify the form of Eq. (8-36). With this definition, we have the result

$$H_A\Phi_0 = \epsilon_0\Phi_0 = 0$$

Equation (8-36) can now be written as

$$\Big\{E - T_0 - \langle\Phi_0|V|\Phi_0\rangle - \langle\Phi_0|VQ\frac{1}{E - QHQ}QV|\Phi_0\rangle\Big\}\chi_0 = 0 \tag{8-37}$$

where $V \equiv \sum_{i=1}^{A} V(\boldsymbol{r}_{0i})$ is the potential acting between the projectile and target nucleons. In arriving at the result, we have made use of Eq. (8-29) and the fact that T_0 operates only on the projectile coordinates and therefore cannot act on Φ_0. This gives us the relation

$$QT_0|\Phi_0\rangle = T_0Q|\Phi_0\rangle = 0$$

The operator $(E - QHQ)^{-1}$ in Eq. (8-37) is meaningful only in the sense of an infinite series expansion of the form

$$\frac{1}{E - QHQ} = \frac{1}{E}\Big\{1 + \frac{1}{E}QHQ + \frac{1}{E}QHQ\frac{1}{E}QHQ + \cdots\Big\}$$

The physical meaning of each term in this expansion may be interpreted in the following way. Each time the Hamiltonian acts between a pair of nucleons, there is an interaction, or "scattering," between these two particular nucleons. The product QHQ implies that the interaction takes place with the target nucleus in one of its excited states. Higher power terms, such as $(E^{-1}QHQ)^n$, represent multiple interactions of order n. The last term of Eq. (8-37), therefore, contains multiple scattering to all orders weighted by energy factor E^{-1} to the appropriate powers.

We can put Eq. (8-36) into the familiar form of an eigenvalue equation,

$$(E - T_0 - \mathcal{V}(\boldsymbol{r}_0))\chi_0 = 0 \tag{8-38}$$

where the equivalent potential is given by

$$\mathcal{V}(\boldsymbol{r}_0) = \langle\Phi_0|V|\Phi_0\rangle + \langle\Phi_0|VQ\frac{1}{E - QHQ}QV|\Phi_0\rangle \tag{8-39}$$

Since we have not yet made any approximation in arriving at Eq. (8-38), we do not have any better chance of solving it than the original equation given by Eq. (8-28). The aim of an optical model is to replace the equivalent potential $\mathcal{V}(\boldsymbol{r}_0)$ by an optical model potential $U_{\rm opt}$, such that the equation

$$(E - T_0 - U_{\rm opt})\chi_0 = 0$$

can be solved.

In general $\mathcal{V}(\boldsymbol{r}_0)$ is *nonlocal*; that is, the potential acting at one point of space may depend on the value of the wave function at a different point. The actual eigenvalue equation takes on the form

$$(E - T_0)\chi_0(\boldsymbol{r}_0) = \mathcal{V}(\boldsymbol{r}_0)\chi_0(\boldsymbol{r}_0) + \int f(\boldsymbol{r}_0, \boldsymbol{r}'_0)\chi_0(\boldsymbol{r}'_0)d\boldsymbol{r}'_0$$

where $f(\boldsymbol{r}_0, \boldsymbol{r}'_0)$ is a function of both $\boldsymbol{r}_0$ and $\boldsymbol{r}'_0$. This greatly complicates the problem and, in practice, one often approximates the potential with a local one. Furthermore, the derivation here may have given the impression that all the scattering into the Q-space eventually returns the target to the ground state. This is certainly not true in general. To represent the loss of flux from the incident channel by scattering that ends up in other exit channels, the optical model potential is usually complex.

Phenomenological optical model potential. The origin of the optical model potential is the average interaction between nucleons in the projectile with those in the target nucleus. It is, in principle, possible to derive the potential from nucleon-nucleon interaction. Before carrying out such a calculation, it is useful to take a more phenomenological approach to the problem, in part, to anticipate the type of results we can expect to obtain.

Based on the fact that the range of nuclear force is short, we expect the radial dependence of an optical model potential $U_{\rm opt}(\boldsymbol{r})$ to follow closely the density distribution in a nucleus. For this reason, a two-parameter Fermi form given earlier in Eq. (8-21),

$$f(r, r_0, a) = \frac{1}{1 + \exp\{(r - r_0A^{1/3})/a\}} \tag{8-40}$$

is often used. In optical model studies this is known as the Woods-Saxon form. The potential is complex in general,

$$U_{\rm vol}(\boldsymbol{r}) = -\{V_0 f(r, r_v, a_v) + iW_0 f(r, r_w, a_w)\} \tag{8-41}$$

In a semi-empirical approach, V_0 and W_0, the depths of the real and imaginary parts of the potential well, are taken as free parameters to be determined by fitting experimental data. If we assume the radial dependence of both parts, $f(r, r_v, a_v)$ and $f(r, r_w, a_w)$, to follow the form given by Eq. (8-40), we may take the radii r_v and r_w and the surface diffuseness a_v and a_w also as adjustable parameters.

The potential given by Eq. (8-41) is only the *volume term*, in the sense that it depends on the distribution of matter in the whole nucleus. In addition, optical model potentials are known to have a spin dependence. When a nucleon is scattered from a nucleus, the result is sensitive to the relative orientation of the nucleon spin before and after the scattering. One way to measure such a dependence is through the *analyzing power* parameter A_y.

Let us define the transverse polarization of the incident and scattered nucleons as positive (or up) if they are oriented in the same direction as the unit vector

$$\hat{\boldsymbol{n}} = \frac{\boldsymbol{k} \times \boldsymbol{k}'}{|\boldsymbol{k}|\,|\boldsymbol{k}'|}$$

given in Eq. (B-6) and negative if they are aligned opposite to $\hat{\boldsymbol{n}}$. Here, $\boldsymbol{k}$ and $\boldsymbol{k}'$ are the wave vectors of, respectively, the incident and scattered particles. Writing the cross section for scattering from positive initial polarization to both positive and negative final polarization as σ_+ and from initial negative polarization to both positive and negative final polarization as σ_-, the analyzing power is given by the ratio of their difference to their sum,

$$A_y = \frac{\sigma_+ - \sigma_-}{\sigma_+ + \sigma_-} \tag{8-42}$$

For elastic scattering, the same information can also be obtained by starting from an unpolarized incident beam and measuring the difference between the cross sections leading to positive and negative polarization for the scattered particle. The result, normalized in the same way as Eq. (8-42), is called *polarization*.

The fact that A_y is nonzero in general in nucleon-nucleus scattering is strong evidence for the presence of spin dependence in the optical model potential. A *spin-orbit* term may be used to represent such an effect,

$$U_{\text{s.o.}}(\boldsymbol{r}) = \boldsymbol{\sigma} \cdot \boldsymbol{\ell} \left(\frac{\hbar}{m_\pi c}\right)^2 \frac{1}{r} \left\{ V_s \frac{d}{dr} f(r, r_{sv}, a_{sv}) + i W_s \frac{d}{dr} f(r, r_{sw}, a_{sw}) \right\} \tag{8-43}$$

Again there are six parameters, V_s, r_{sv}, a_{sv}, W_s, r_{sw}, and a_{sw}, to be adjusted to fit scattering data. Note that the square of the pion Compton wavelength, $(\hbar/m_\pi c^2)^2$, is roughly 2 and the approximate numerical value is often used in its place.

The reason for using derivatives of the volume density distributions as the radial dependence comes from analogy with the Thomas spin-orbit potential for the force felt by atomic electrons in the Coulomb field of a nucleus. For an electron, the spin-orbit term originates from the interaction of its intrinsic magnetic dipole moment $\boldsymbol{\mu}_e$ with the magnetic field $\boldsymbol{B}(\boldsymbol{r})$ it feels because of its own orbital motion around the nucleus. The value of $\boldsymbol{B}(\boldsymbol{r})$ may be found by a Lorentz transformation of the electrostatic field $\boldsymbol{E}(\boldsymbol{r})$ provided by the nucleus, stationary in the laboratory, into a frame of reference at rest with the orbiting electron. The result is

$$\boldsymbol{B}(\boldsymbol{r}) = -\frac{1}{c[c]} \boldsymbol{v} \times \boldsymbol{E}(\boldsymbol{r}) = -\frac{\hbar}{m_e c r [c]} \frac{dV}{dr} \boldsymbol{\ell}$$

where the factor inside the square brackets is needed to convert the expression from cgs to SI units. The orbital angular momentum of the electron is given by

$$\boldsymbol{\ell}\hbar = \boldsymbol{r} \times \boldsymbol{p} = m_e(\boldsymbol{r} \times \boldsymbol{v})$$

To relate to an electrostatic potential $V(r)$, we have made use of the relation

$$\boldsymbol{E}(\boldsymbol{r}) = -\frac{dV}{dr}\frac{\boldsymbol{r}}{r}$$

The spin-orbit interaction energy for an atomic electron is then given by

$$W(\boldsymbol{r}) = -\boldsymbol{\mu}_e \cdot \boldsymbol{B}(\boldsymbol{r}) = \frac{e\hbar^2}{m_e^2 c^2 r}\frac{dV}{dr}\,\boldsymbol{\sigma}\cdot\boldsymbol{\ell}$$

where we have made use of the fact that the dipole moment $\boldsymbol{\mu}_e$ is given by $(e\hbar[c]/m_e c)\,\boldsymbol{\sigma}$. Equation (8-43) has the same form. The use of pion mass m_π instead of that of the electron may be regarded as a convention for the definition of the spin-obit potential well depths V_s and W_s. For the same reason, the electron charge is inappropriate and is absorbed into the definition of the well depths.

For charged particle scattering, a Coulomb term may be included in the optical model potential. The form is usually obtained by approximating the target nucleus as a uniformly charged sphere,

$$U_{\rm C}(r) = \begin{cases} \left[\dfrac{1}{4\pi\epsilon_0}\right]\dfrac{zZe^2}{2R_c}\left(3 - \dfrac{r^2}{R_c^2}\right) & \text{for } r < R_c \\ \left[\dfrac{1}{4\pi\epsilon_0}\right]\dfrac{zZe^2}{r} & \text{for } r \geq R_c \end{cases}$$

where R_c is the Coulomb radius. The quantities z and Z are, respectively, the charge numbers of the projectile and the target nucleus. It may be tempting to treat R_c as a free parameter also. In practice, the scattering results are not sensitive to the details of the Coulomb potential, and it is quite adequate to use the value $R_c = 1.2A^{1/3}$ fm.

The complete phenomenological optical model potential is the sum of volume, spin-orbit, and Coulomb terms:

$$U_{\rm opt}(\boldsymbol{r}) = U_{\rm vol}(\boldsymbol{r}) + U_{\rm s.o.}(\boldsymbol{r}) + U_{\rm C}(\boldsymbol{r})$$

The total number of adjustable parameters is 12 if we do not include R_c. In a typical scattering experiment, the angular distribution of the differential scattering cross section, as well as analyzing power and other quantities, where possible, is observed. The number of independent pieces of data is usually greater than 12 and there is no difficulty to obtain a complete set of the parameter values by fitting calculated optical model results to the measured quantities. A large amount of information has been accumulated in this way, and we have now a fairly clear picture of the energy and mass dependence of these parameters for proton scattering off nuclei up to a laboratory energy of 200 MeV. One of the sets obtained by fitting proton scattering data on a variety of nuclei from $A = 40$ to $A = 208$, and laboratory proton energy from 80 to 180 MeV [126], is given in Table 8-1 as illustration.

There are, however, several problems associated with the phenomenological approach. The first is that, although we have a good picture for proton scattering, the knowledge does not extend to other projectiles. For example, even the neutron optical model potential is not as well known, as there are far less experimental data available for

Table 8-1: Proton-nucleus scattering optical model potential parameters for $40 \le A \le 208$ and proton laboratory energy $80 \le T_p \le 180$ MeV [126].

$V_0 =$	$105.5(1 - 0.1625 \ln T_p) + 16.5(N - Z)/A$
	$r_v = \begin{cases} 1.125 + 0.0010 T_p & \text{for } T_p \le 130 \text{ MeV} \\ & \text{(except } T_p \le 180 \text{ MeV for Ca)} \\ 1.255 & \text{for } T_p > 130 \text{ MeV (except Ca)} \end{cases}$
	$a_v = 0.675 + 3.1 \times 10^{-4} T_p$
$W_0 =$	$6.6 + 0.0273(T_p - 80) + 3.87 \times 10^{-6}(T_p - 80)^3$
	$r_w = 1.65 - 0.0024 T_p$
	$a_w = 0.32 + 0.0025 T_p$
$V_{s.o.} =$	$19.0(1 - 0.166 \ln T_p) - 3.75(N - Z)/A$
	$r_{sv} = 0.920 + 0.0305 A^{1/3}$
	$a_{sv} = \begin{cases} 0.768 - 0.0012 T_p & \text{for } T_p \le 140 \text{ MeV} \\ 0.60 & \text{for } T_p > 140 \text{ MeV} \end{cases}$
$W_{s\,o} =$	$7.5(1 - 0.248 \ln T_p)$
	$r_{sw} = 0.877 + 0.0360 A^{1/3}$
	$a_{sw} = 0.62$

Note: V_0, W_0, $V_{s\,o}$, $W_{s.o.}$ and T_p are in mega-electron-volts; r_v, a_v, r_w, a_w, r_{sv}, a_{sv}, r_{sw} and a_{sw} are in femtometers.

neutron-nucleus scattering. Because of its phenomenological nature, the approach does not lend itself easily to extrapolation to regions where experimental data are scarce. The second is that the forms of radial dependence used in Eqs. (8-41) and (8-43) are found to be inadequate as we move to higher bombarding energies. One remedy is to use more complicated expressions involving additional parameters; however, on aesthetic grounds alone, this is not desirable. Finally, the parametrization is not unique. The 12 parameters are interdependent in a complicated way, and there are often other sets of values that can also provide an equally good description of the data.

Microscopic optical model potential. An optical model potential for nucleon-nucleus scattering represents the average interaction between the incident nucleon and nucleons in the target nucleus. It is, therefore, a function of the nucleon-nucleon interaction. A microscopic model of the potential may be constructed by convoluting the fundamental nucleon-nucleon interaction with the nuclear density. Such a *folding* model has been known to be quite successful in describing nucleon-nucleus scattering data if an appropriate nucleon-nucleon interaction is used as the starting point. For simplicity we shall restrict ourselves to the case that the incident particle is a nucleon and ignore any internal structure it may have. The first term of Eq. (8-39) suggests that we may be able to approximate the nucleon-nucleus optical model potential by

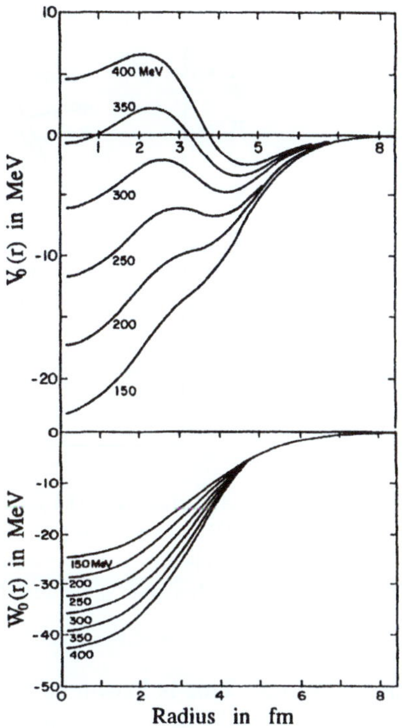

Figure 8-5: Radial shapes of the volume term of a proton-nucleus optical model potential at different bombarding energies. The upper curves are the real part and the lower one the imaginary part of the potential. The results are calculated using a folding procedure with a Paris potential as the interaction between nucleons. (Adapted from Ref. [139].)

the integral

$$U_{\text{opt}}(\boldsymbol{r}_0) \approx \langle \Phi_0(\boldsymbol{r}_1, \boldsymbol{r}_2, \ldots, \boldsymbol{r}_A)| \sum_{i=1}^{A} V(\boldsymbol{r}_{0i}) |\Phi_0(\boldsymbol{r}_1, \boldsymbol{r}_2, \ldots, \boldsymbol{r}_A)\rangle \tag{8-44}$$

where the integration is taken only over the target nucleons.

One must be careful here with antisymmetrization between the incident nucleon and the one in the target nucleus with which it interacts. Let us consider the simplest case in which the incident nucleon undergoes only one interaction with one of the nucleons in the target. When a nucleon emerges from the scattering, there is no way to identify whether the observed particle is the same one as the incident nucleon or the one in the target nucleus with which it interacted. Both possibilities must be included. For this reason, the matrix element on the right-hand side of Eq. (8-44) should be a sum of two terms,

$$\langle \Phi_0 | V | \Phi_0 \rangle = \langle \Phi_0 | t_D | \Phi_0 \rangle + \langle \Phi_0 | t_E | \Phi_0 \rangle \tag{8-45}$$

where t_D is the operator for the direct part of the reaction in which the scattered nucleon is the same one as the incident particle and t_E is the operator for the exchange part in which the incident nucleon is absorbed by the target nucleus and the scattered particle is one of the nucleons originally in the target. We shall see later that the difference between the contributions from these two terms is important in understanding the radial shape of the optical model potential at high energies.

For the nucleon-nucleon potential $V(\boldsymbol{r}_{0i})$, it is tempting to take a naive approach and replace it by a free nucleon-nucleon interaction obtained, for example, from nucleon-

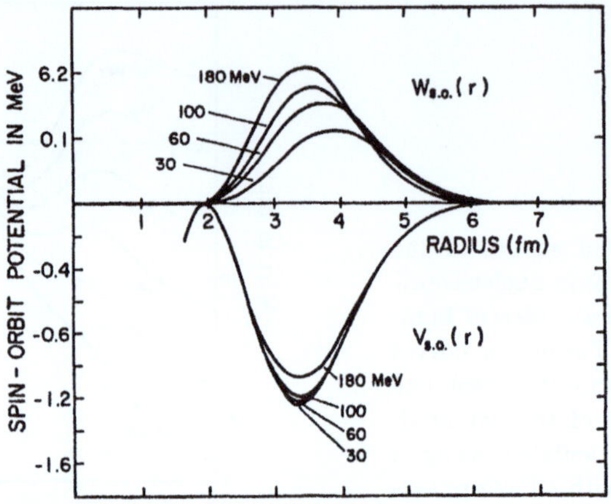

Figure 8-6: Radial shapes of the real and imaginary parts of the spin-orbit term of an optical model potential for proton-nucleus scattering at different bombarding energies. The calculation is based on a folding procedure using a Paris potential for the interaction between nucleons. (Adapted from Ref. [24].)

nucleon scattering. This is known as *impulse approximation* (IA) and, in practice, is found to be too crude to fit experimental data on nucleon-nucleus scattering. Just as with nuclear shell-model calculations, an effective nucleon-nucleon interaction is required here, as one of the two interacting nucleons is embedded in the nucleus. The requirements on the effective interaction are somewhat less stringent than in the corresponding shell-model case. It is usually possible to approximate the nuclear medium as an infinite nuclear matter to simplify the calculations. To take care of the fact that a real nucleus has a large surface region, with density varying from very small to saturation value in infinite nuclear matter, a *density-dependent* effective potential is often used. In other words, the operators t_D and t_E in Eq. (8-45) are made to be functions of the nuclear density ρ. The effective potential in different regions is calculated using nuclear matter of appropriate densities. Furthermore, the free nucleon-nucleon interaction itself is energy dependent and, as a result, both t_D and t_E are functions of the bombarding energy as well.

In terms of single-particle wave functions $\phi_i(\boldsymbol{r}_i)$, the target nucleus density may be expressed as the following operator:

$$\rho(\boldsymbol{r}) = |\Phi_0\rangle\langle\Phi_0| \approx \sum_{i=1}^{A} \phi_i^*(\boldsymbol{r})\phi_i(\boldsymbol{r}) \tag{8-46}$$

Using this, the optical model potential may be related to an integral over the function

$$f(\boldsymbol{r}_0, \boldsymbol{r}) = \sum_{i=1}^{A} \phi_i^*(\boldsymbol{r}) t_D(\boldsymbol{r}_0, \boldsymbol{r}; \rho, E)\phi_i(\boldsymbol{r}) + \sum_{i=1}^{A} \phi_i^*(\boldsymbol{r}) t_E(\boldsymbol{r}, \boldsymbol{r}_0; \rho, E)\phi_i(\boldsymbol{r}_0) \tag{8-47}$$

constructed by convoluting, or "folding," the nucleon-nucleon interaction between the incident and the target nucleons with nuclear density. In order to emphasize the de-

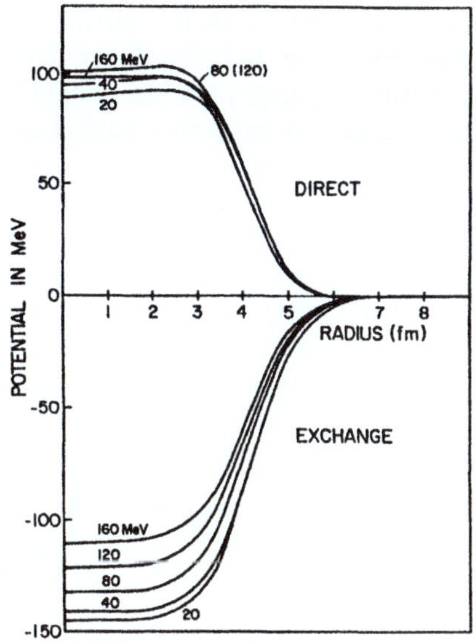

Figure 8-7: Real part of the volume term of a proton-nucleus optical model potential at different bombarding energies. The upper curves are the direct part of a microscopic potential given by Eq. (8-48) and the lower curves are the exchange part given by Eq. (8-49). The nucleon-nucleon interaction comes from a Paris potential. (Taken from Ref. [140].)

pendence on nuclear density and bombarding energy, we have put ρ and E explicitly into the arguments of t_D and t_E.

The direct term of such an optical model potential is relatively easy to evaluate, as the integral involving the first term of Eq. (8-47) may be expressed as

$$U^D_{\rm opt}(\boldsymbol{r}_0, E) = \int \rho(\boldsymbol{r}) t_D(\boldsymbol{r}_0, \boldsymbol{r}, \rho, E)\, d^3r \tag{8-48}$$

For $\rho(\boldsymbol{r})$ we can use the approximate form given in Eq. (8-46). However, the same transformation cannot be carried out for the exchange term, as the two single-particle wave functions have different arguments, $\boldsymbol{r}$ and $\boldsymbol{r}_0$. As a result, the contribution of the exchange term is nonlocal in general. A "local momentum" approximation is usually used to reduce the exchange term to the following simpler form:

$$U^E_{\rm opt}(\boldsymbol{r}_0, E) \approx \int \rho(\boldsymbol{r}_0, \boldsymbol{r}) t_E(\boldsymbol{r}, \boldsymbol{r}_0, \rho, E) j_0(k|\boldsymbol{r}_0 - \boldsymbol{r}|) d^3r \tag{8-49}$$

where

$$\rho(\boldsymbol{r}_0, \boldsymbol{r}) = \sum_{i=1}^{A} \phi_i^*(\boldsymbol{r}) \phi_i(\boldsymbol{r}_0)$$

in analogy with Eq. (8-46). Here, $j_0(\xi)$ is the spherical Bessel function of order zero.

At laboratory energy below 200 MeV, the folding potential produces results very similar to those derived from phenomenological approaches. At higher energies, however, the Woods-Saxon radial shape used in the semi-empirical approach is found to be inadequate. From folding potential calculations we find that, as the bombarding energy is increased, the radial shape of the volume term in the optical model potential changes to a "wine-bottle" shape, as shown in Fig. 8-5. The shape for the spin-orbit potential, shown in Fig. 8-6, however, retains essentially the same form as given by Eq. (8-43). The shape changes in the volume term come from the differences in the energy dependence of the direct and exchange parts of the folding potential. As illustrated in Fig. 8-7, the direct part is repulsive and the exchange part is attractive. Since the exchange part has a slightly sharper energy dependence than the direct part, the cancellation between repulsive and attractive parts, when we sum the two terms to produce $U_{\rm opt}$, produces an energy dependence that cannot be represented by a simple Woods-Saxon form. In Fig. 8-8 the results for the elastic scattering of 362-MeV protons off a ^{40}Ca target are given as an illustrative example to show that a microscopic optical model potential is quite capable of describing intermediate-energy proton-nucleus scattering to very large momentum transfers.

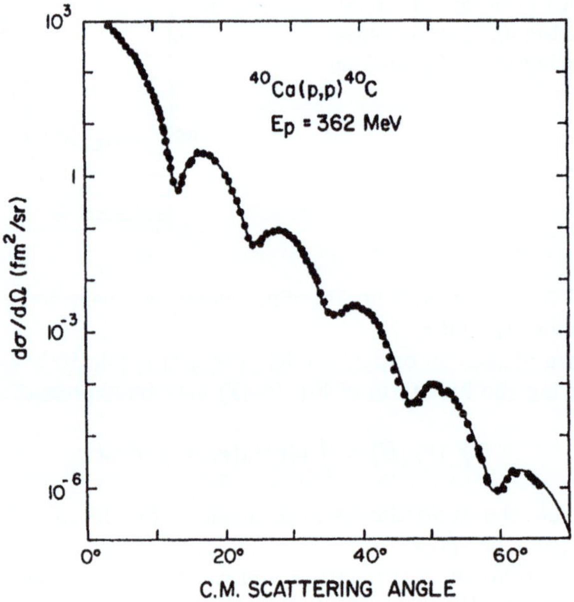

Figure 8-8: Differential cross section for elastic scattering of 362-MeV protons off ^{40}Ca. The continuous curve is the calculated result using a microscopic optical model potential. The diffraction-like pattern is typical in scattering at small angles found in many different types of processes [66].

Besides elastic scattering, an optical model potential is also useful in understanding the cross section for other types of reactions. For example, in cases where the scattering is dominated by direct reaction, contributions from potential scattering and multipole scattering to the same final state are regarded as "background" and the effect may be represented by an optical model potential. The reduction in the incident flux because of other open reaction channels is taken care of by the imaginary part of the potential. From a slightly different prospective, we can view the optical model potential as an average potential that "distorts" both the incident and scattered waves from their plane wave states we saw in the previous section. The contributions of a direct reaction to specific states may be regarded as terms in addition to scattering due to the optical model potential and favors only specific final states. This is the essence of DWBA for direct reaction. The scattering is separated into two parts, a "background" given by the optical model potential and a direct reaction contribution because of special features in the final state involved. The calculated results obtained this way, as we shall see in the next section, have been found to give a fairly good description of the observed cross sections.

8-5 Intermediate-Energy Nucleon Scattering

In §8-3 we saw that direct reaction is a good way to investigate certain aspects of nuclear structure as well as interaction between free and bound nucleons. The main reason is that the reaction mechanism is relatively simple. This is especially true in nucleon-nucleus scattering when the incident energy is in the intermediate range.

Intermediate-energy nucleons are usually taken to mean those with laboratory kinetic energy in the range of 100 to 1000 MeV. At much lower energies, the transit time of a nucleon through a nucleus is sufficiently long that multiple scattering may take place frequently enough to complicate the reaction. At much higher energies, good resolution is difficult to achieve and the increased production rates of pions and other secondary particles make the condition unfavorable for studying nucleon-nucleus interaction.

We shall again restrict ourselves to reactions involving two-body final states to simplify the analysis. Our main emphasis is on proton inelastic scattering, commonly referred to as (p,p') reactions, and charge exchange reactions induced by nucleons, namely, (p,n) and (n,p) reactions. Furthermore, we shall ignore elastic proton scattering here, as some of the primary interests are already covered in the previous section in the discussion of optical models. Besides scattering cross sections, observables related to changes in nucleon spin orientation can also be measured, as we have seen in §3-7; however, for simplicity, we shall not discuss them here. Very interesting data can also be obtained by scattering antiprotons from nuclear targets. The information helps us to understand the connection between nucleon-nucleus reaction and the internal degrees of freedom of nucleons. An example of the differential cross section for scattering off ^{12}C is shown in Fig. 8-9. Unfortunately, a meaningful discussion of the topic of antiproton-nucleus scattering requires additional preparations than what we wish to do here.

Scattering amplitude. We saw earlier in Eq. (8-21) that scattering amplitude in (first) Born approximation may be expressed in terms of the matrix elements of the nucleon-nucleon interaction potential between initial and final states of the nucleon-nucleus system,

$$f(\theta) = -\frac{\mu}{2\pi\hbar^2}\langle\chi_{k_f}(\boldsymbol{r}_0)\Phi_f(\boldsymbol{r}_1,\boldsymbol{r}_2,\ldots,\boldsymbol{r}_A)|\sum_{i=1}^{A}V(\boldsymbol{r}_{0i})|\chi_{k_i}(\boldsymbol{r}_0)\Phi_i(\boldsymbol{r}_1,\boldsymbol{r}_2,\ldots,\boldsymbol{r}_A)\rangle \tag{8-50}$$

where μ is the reduced mass of the scattering nucleon and $\chi_{k_i}(\boldsymbol{r}_0)$ and $\chi_{k_f}(\boldsymbol{r}_0)$ are, respectively, the wave functions of the incident and scattered nucleons in the Born

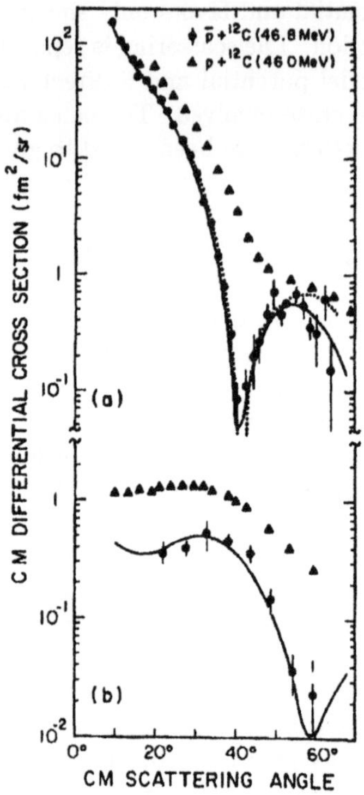

Figure 8-9: Differential cross section for proton (triangles) and antiproton (circles) scattering off ^{12}C at 46 MeV incident energy. Figure (*a*) is for elastic scattering and (*b*) is for inelastic scattering leading to the 2^+ state at 4.44 MeV excitation. The solid curves are obtained from a coupled channel calculation and the dotted curve is a theoretical one for elastic scattering. (Taken from Ref. [72].)

approximation. The wave functions $\Phi_i(\boldsymbol{r}_1, \boldsymbol{r}_2, \ldots, \boldsymbol{r}_A)$ and $\Phi_f(\boldsymbol{r}_1, \boldsymbol{r}_2, \ldots, \boldsymbol{r}_A)$ describe the initial and final nuclear states.

There are three distinctive parts that enter in a calculation of the scattering amplitude $f(\theta)$ here. First, we need an optical model potential with which we can solve for functions $\chi_{k_i}(\boldsymbol{r}_0)$ and $\chi_{k_f}(\boldsymbol{r}_0)$. In this way, the effect of anything other than those due to direct reaction may be accounted for on the average. This is the spirit of the distorted wave approach mentioned at the end of the previous section. Second, we need a potential $V(\boldsymbol{r}_{0\imath})$ that supplies the interaction between a free nucleon and a nucleon embedded in a nucleus. It is this potential that induces the direct reaction over and above the "background" produced by the optical model potential. Third, we must have both the initial and final nuclear wave functions Φ_i and Φ_f, in particular the relationship between them. All three parts are related to the fundamental nucleon-nucleon interaction.

Let us start with the purely nuclear structure problem of relating the initial and final states of the target nucleus. For simplicity, we shall restrict ourselves to targets made of even-even nuclei where many of the studies have been carried out. For such targets, the spin and parity of the ground state of the initial nucleus are 0^+. The angular momentum transferred to the nucleus as a result of the scattering, in this case, is given by the spin and parity of the final nuclear state. Our basic assumption in Eq. (8-50) for the scattering amplitude is that the incident nucleon interacts with only one of the nucleons in the target nucleus. In the case of a (p, p') reaction, the process may be thought of as one in which the incident proton excites the target nucleus by promoting one of the nucleons to a higher single-particle state. For a (p,n) reaction, the incident proton is captured and one of the neutrons in the target is ejected in the process. In either case, we find that states strongly excited by the these reactions are those made predominantly of one-particle one-hole ($1p1h$) excitations built upon the ground state of the target nucleus.

The relation between initial and final nuclear wave functions may be expressed in terms of a "transition density." In the form of an operator, the $1p1h$-transition density of interest here may be written as

$$\rho_{\rm tr}(1p1h) = \sum_{ph} a_{ph}|\phi_p\rangle\langle\phi_h|$$

where $\phi_h(\boldsymbol{r})$ is one of the occupied single-particle states in the target nucleus and $\phi_p(\boldsymbol{r})$ one that is empty before the scattering. In principle, we should also couple $|\phi_p\rangle\langle\phi_h|$ to some definite spin and isospin so that the operator $\rho_{\rm tr}(1p1h)$ is a spherical tensor of definite ranks. However, we shall dispense with this complication in the following discussion to simplify the argument.

If a state $|\,(1p1h)\,J^\pi\,\rangle$ is made up entirely of a linear combination of $1p1h$-excitations built upon the ground state, we can impose the normalization condition

$$\left|\langle(1p1h)\,J^\pi|\rho_{\rm tr}(1p1h)|\text{ground state}\rangle\right|^2 = 1$$

on the transition density operator. In this way, the state $|\,(1p1h)\,J^\pi\,\rangle$ may be expressed in terms of the transition density operator acting on the ground state wave function,

$$|(1p1h)\,J^\pi\rangle = \rho_{\rm tr}(1p1h)|\text{ground state}\rangle = \sum_{ph} a_{ph}|\phi_p\rangle\langle\phi_h|\text{ground state}\rangle$$

As we have seen in §7-5, such a $1p1h$-state is an eigenstate of the Hamiltonian in the limit that the two-body residual interaction can be ignored.

In general, an eigenvector of the nuclear Hamiltonian contains other components as well. The wave function of the final nuclear state, in general, has the form

$$\begin{aligned}|\Phi_f(\boldsymbol{r}_1,\boldsymbol{r}_2,\ldots,\boldsymbol{r}_A)\rangle &= \sum_{ph} a_{ph} \sum_{i=1}^{A} |\phi_p(\boldsymbol{r}_i)\rangle\langle\phi_h(\boldsymbol{r}_i)|\Phi_i(\boldsymbol{r}_1,\boldsymbol{r}_2,\ldots,\boldsymbol{r}_A)\rangle \\ &\quad + \text{other components} \end{aligned} \tag{8-51}$$

where Φ_i and Φ_f are, respectively, the initial and final nuclear wave functions. From this, we obtain the expansion coefficients a_{ph} for the $1p1h$-transition operator

$$a_{ph} = \sum_{i=1}^{A}\langle\Phi_f(\boldsymbol{r}_1,\boldsymbol{r}_2,\ldots,\boldsymbol{r}_A)|\phi_p(\boldsymbol{r}_i)\rangle\langle\phi_h(\boldsymbol{r}_i)|\Phi_i(\boldsymbol{r}_1,\boldsymbol{r}_2,\ldots,\boldsymbol{r}_A)\rangle$$

The transition density is a quantity between two specific nuclear states, and as such, it is independent of the probe and the reaction mechanism. For this reason, the same transition density enters into all other $1p1h$-excitation processes between the same pair of nuclear states. This gives us the opportunity to check the quantity obtained in a

(p,p') reaction against, for example, electromagnetic transitions and inelastic electron scattering. For (p,n) and (n,p) reactions, the transition densities are related to β-decay rates and cross sections of charge exchange reactions induced by other probes such as pions and light ions.

So far we have considered the scattered nucleon in a (p,p') reaction to be one and the same as the incident nucleon and distinguishable from those in the target. As we have already seen in the previous section, this is only the direct part of the scattering amplitude which, with the help of Eq. (8-51), may be expressed as

$$f_D(\theta) = -\frac{\mu}{2\pi\hbar^2}\sum_{ph} a_{ph}\langle\chi_{k_f}(\boldsymbol{r}_0)\phi_p(\boldsymbol{r})|V(\boldsymbol{r}_0,\boldsymbol{r})|\chi_{k_i}(\boldsymbol{r}_0)\phi_h(\boldsymbol{r})\rangle \tag{8-52}$$

The result is obtained after integrating over the coordinates of all other nucleons in the target not involved in this particular scattering. Components other than those related to one-particle one-hole excitation of the target ground state disappear from the expression, as they do not contribute to the direct reaction amplitude in the limit that only $1p1h$-excitations are allowed. Their importance comes mainly in terms of their total weight in Φ_f and, consequently, the fraction of $1p1h$-components present in the state and the overall size of the scattering amplitude $f_D(\theta)$.

To ensure proper antisymmetrization, we must also include an exchange part to the scattering amplitude. In analogy to Eq. (8-52), this may be written as

$$f_E(\theta) = -\frac{\mu}{2\pi\hbar^2}\sum_{ph} a_{ph}\langle\chi_{k_f}(\boldsymbol{r}_0)\phi_p(\boldsymbol{r})|V(\boldsymbol{r}_0,\boldsymbol{r})|\chi_{k_i}(\boldsymbol{r})\phi_h(\boldsymbol{r}_0)\rangle \tag{8-53}$$

Both $f_D(\theta)$ and $f_E(\theta)$ are two-body matrix elements involving either the incident nucleon or the scattered nucleon and one of the nucleons in the nucleus. The reason we can reduce the amplitude to such a simple form comes from the direct reaction assumption that only a single interaction takes place between the incident nucleon and one of the nucleons in the target, with the rest of the nucleons acting merely as "spectators" in the reaction.

Nucleon-nucleus interaction potential. What is an appropriate interaction potential $V(\boldsymbol{r}_0,\boldsymbol{r})$ to use in Eqs. (8-52) and (8-53) for the scattering amplitudes? The simplest approach is to apply an impulse approximation and equate the interaction with one occurring between free nucleons. As we have seen earlier in optical model potentials, this turns out to be too crude an assumption because of the influence of the nuclear medium on the target nucleons. For a semi-empirical approach, we can take a phenomenological one-boson exchange potential consisting of a sum of Yukawa forms, each with a different range. The strength of each term in such a potential may be taken as an adjustable parameter to reflect the fact that we do not have a complete knowledge of the interaction between free and bound nucleons. An example of such a potential is the Michigan three-Yukawa (M3Y) potential [29]. With a sum of only three Yukawa terms, fairly good descriptions have been provided for the observed differential cross sections in many (p,p') reactions.

A more realistic approach is to use an effective interaction based on free nucleon-nucleon scattering with corrections for the influence of the nuclear medium. An example is the one developed by Franey and Love [65]. Alternatively one can take a nuclear matter approach and develop a density-dependent potential, as described earlier for optical model potential studies. In both cases, good descriptions of the observed results up to very large momentum transfers have been obtained for both differential scattering cross sections and spin observables.

Let us recapture what is happening when an intermediate-energy nucleon is scattered off a nucleus. Before the incident nucleon is within the range to interact with one of the nucleons in the nucleus, it is in the field of the nuclear optical model potential. The wave function of the incident nucleon is modified by the average potential. Once the projectile is in the range of the nuclear force of the target, the interaction promotes one of the target nucleons to a different single-particle state. When the scattered nucleon leaves the region, it travels again through the field of an optical model potential. The three steps of a calculation—optical model potential, nucleon-nucleon interaction, and nuclear wave functions—are distinct parts of the problem and may be treated quite independently of each other. On the other hand, all three are the results of interaction between nucleons and can be calculated from the same nucleon-nucleon interaction potential. It is therefore possible to solve the problem in a self-consistent manner and obtain all three parts from a given nucleon-nucleon potential. This is an interesting development, as there are only rare occasions in many-body problems that such an approach can be carried out in practice. Partly because of this possibility, a large amount of work, both experimental and theoretical, has been carried out in intermediate-energy nucleon-nucleus scattering.

Relativistic and other effects. In addition to the above interests, intermediate-energy nucleon-nucleus scattering may also be used to understand the underlying reaction mechanism. For example, above we have implicitly assumed a nonrelativistic Schrödinger approach. However, the kinetic energy of the incident nucleon here is a large fraction of its rest mass energy, and as a result, relativistic effects may be important. Besides simple kinematic effects that require Lorentz invariance in the place of Galilean invariance, we may also need to replace the Schrödinger equation with a Dirac equation. The main difference here may be viewed in the following way. In the Schrödinger approach, the nucleon, being a spin-$\frac{1}{2}$ particle, is described by a two-component wave function to account for the fact that the intrinsic spin of a nucleon can either point up (projection along the quantization axis $+\frac{1}{2}$) or down (projection $-\frac{1}{2}$). In a relativistic quantum-mechanical treatment, a four-component wave function is required to describe a spin-$\frac{1}{2}$ particle, with the upper two components describing the two possible directions of the nucleon spin and the lower two components accounting for the two possible directions of the antinucleon spin. A fully relativistic treatment of the nucleon-nucleus scattering, therefore, differs from the Schrödinger approach by the presence of the two lower components.

At low energies, the influence of the lower components on the behavior of the nucleon is very small and may be replaced by spin-dependent terms in the potential, as we have done earlier. At higher energies, such a simple substitution may not be adequate and the Dirac equation may have to be solved for the scattering. There are indications that, for certain observables in intermediate-energy nucleon-nucleus scattering, particularly those related to changes in the polarization direction between incident and scattered nucleons, a relativistic treatment is needed (see, e.g., Celenza and Shakin [43] and Danos, Gillet, and Cauvin [47] and references therein).

One possible interest in charge exchange reactions is to relate strong and weak interaction processes, as mentioned earlier in §5-6. Intermediate-energy (p, n) and (n, p) reactions are ideal here, as the reaction mechanism is sufficiently simple and the nuclear matrix elements involved are the same as those in nuclear β-decay. Apart from kinematic factors, the only distinction between nuclear β-decay and charge exchange reactions induced by intermediate-energy nucleons is expected to be the difference in their "coupling" constants. If this is true, the ratio between these two processes should be independent of the target nucleus used, and this indeed is found to be the case. As a result, sum rule and giant resonance studies have been extended into charge exchange processes, as we have seen earlier in §6-2.

Alternatively, intermediate-energy nucleon-nucleus scattering can be viewed as a good way to obtain information on the interaction between free and bound nucleons. This is made possible by the fact that two of three ingredients in a reaction calculation may be checked by other means. For example, we have seen that the optical model potential is the same one as that entering into elastic scattering. From the success in describing elastic scattering, we can establish the validity of an optical model potential before using it in either (p, p') or charge exchange reactions. We have also seen that the nuclear structure question involved in the scattering process is identical to those occurring in other reactions. By comparing the transition density with, for example, intermediate-energy inelastic electron scattering, we have a fairly reliable way to find out whether the nuclear structure information is correct. The net result is that the interaction $V(\boldsymbol{r}, \boldsymbol{r}_0)$ in Eqs. (8-52) and (8-53) becomes the least well known part of the three and may therefore become the primary focus of a study. Furthermore, different transitions are sensitive to different parts of the interaction potential. By carefully selecting the initial and final states, it is possible to emphasize a particular aspect of $V(\boldsymbol{r}, \boldsymbol{r}_0)$ for examination.

Finally, if we are confident of all three points above, we can start to ask the finer and more detailed question of whether there are any exotic effects related to, for example, the internal degrees of freedom of nucleons. The energy involved here is certainly high enough that, for example, intermediate states involving the excitation of a nucleon into a Δ-particle can take place, particularly in view of the strong P_{33}-resonance in the pion-nucleon channel. Since a Δ is a distinguishable particle from a nucleon, it does not suffer from the effect of the Pauli exclusion principle due to the presence of other nucleons. Instead of particle-hole excitations, we can imagine Δ-hole excitations to take place. There is already some evidence that such nonnucleonic degrees of freedom may be present in the observed data.

8-6 Meson-Nucleus Reactions

Interaction of mesons with nucleons and nuclei provides us with one way to understand the exchange of virtual mesons between nucleons. We have seen earlier in §3-6 and §3-9 that such exchanges are responsible for the long- and intermediate-range parts of the nuclear potential. Meson scattering is also of interest from the point of view that they are bosons. Since bosons can be absorbed and created in the reaction, we expect to learn something new about scattering that cannot be achieved with baryons and leptons. It is an integral part of hadron-nucleus scattering studies and forms an essential element of our understanding of hadrons.

Experimentally, intense sources of pions are available from "meson" factories, such as LAMPF (Los Alamos Meson Physics Facility), SIN (Swiss Institute for Nuclear Research), and TRIUMF (Tri-University Meson Facility) (see, e.g., Ref. [109]). There are two features that are special to pion scattering. The first is the strong P_{33}-resonance that produces a Δ-particle from a pion and a nucleon. As we have seen in §2-6, the strength of this resonance at pion laboratory energy of 195 MeV is so overwhelming that pion-nucleus reactions at energies below a few hundred mega-electron-volts are dominated by the formation of Δ-particles. The second is that pions have three charge states, π^+, π^0, and π^-. As a result, single-charge exchange as well as double-charge exchange reactions are possible. The study of pion-nucleus reactions can be carried out in a variety of ways, including pion absorption, elastic and inelastic scattering, as well as charge exchange reactions.

Pion absorption. There are two different types of pion absorption studies that can be made, stopped pion and fast pion. In order to enhance the probability, the pion to be absorbed must be slowed down sufficiently such that it is essentially at rest with respect to the nucleus. One way to "stop" a π^- is to capture it first in an atomic orbit to form a π-*mesic* atom. It may happen that the π^- is initially occupying one of the higher "electronic" orbits of the atom. If this is true, the negative pion will eventually cascade down to a low-lying orbit through atomic electromagnetic decay. Since the pion mass is far larger than that of an electron ($m_\pi \approx 300m_e$), the Bohr radius is much smaller. As a result, the wave function of a low-lying π^- in an atom has a significant overlap with that of the nucleus, as we have seen earlier in the analogous situation of muonic atoms in §4-5. However, being a hadron, a pion behaves quite differently from a muon, particularly in the nuclear medium. Because of strong interaction, a pion is readily absorbed by the nucleus once it is close enough for the short-range force to be effective.

When a π^- is absorbed, all its rest mass energy of $\sim$140 MeV is transferred to the nucleus in the form of excitation energy. Since this is about 16 times the average binding energy of a nucleon in a nucleus, it is difficult for a single nucleon to take up the full amount and conserve momentum at the same time. It is therefore likely that a cluster of nucleons, such as an α-particle cluster, is involved. Alternatively, the internal degrees of freedom within a nucleon may be excited. Even though the peak of the lowest energy resonance, the Δ-channel, is still far away, the small possibility remains an important consideration.

In contrast to stopped pions, the absorption of "fast" pions may be defined as a reaction involving an incident pion and no scattered pion,

$$\pi + A \rightarrow A^*$$

Since the incident particle carries both energy and momentum, it is again impossible for a single nucleon in the nucleus to absorb the pion and conserve both energy and momentum at the same time. From the relatively large cross sections observed in the reaction

$$\pi^+ + d \rightarrow p + p$$

we can conclude that the two-nucleon process is important in the absorption of fast pions by nuclei. This idea is also corroborated by the relatively large cross sections observed for (π^+, pp) reactions on nuclei in general. Since it involves two nucleons, fast-pion absorption is sensitive to two-particle correlations in nuclei and, as a result, is one of the ways to make such studies. A concise review of the subject can be found, e.g., in Ashery and Schiffer [11].

The inverse of pion absorption is pion production. When a nucleus is bombarded by electrons, protons, or other particles, pions are produced if sufficient energy is available. The reaction usually results in final states with three or more particles. Furthermore, many other exit channels are also open at these energies, and as a result, both the measurement and the analysis are complicated. For this reason, we shall not be concerned with such reactions here. An example of the pion production cross section in nucleon-nucleon scattering was shown earlier in Fig. 3-4.

Pion scattering. Pion scattering studies may be divided into three categories: elastic and inelastic scattering, single-charge exchange (SCX) reactions, and double-charge exchange (DCX) reactions. Alternatively, because of the strong P_{33}-resonance, measurements are often divided into three groups, depending on whether the energy is below the resonance, on the resonance, or above the resonance. We have already seen in §4-5 that there are strong on-resonance enhancements of the $\pi^+ + p$ and $\pi^- + n$ cross sections over those for $\pi^- + p$ and $\pi^+ + n$ scattering. This made it possible to use pion scattering to distinguish between neutron and proton density distributions in a nucleus. We shall be mainly interested here in the other aspects of pion-nucleus scattering.

At energies no higher than 50 MeV, far below the P_{33}-resonance, the average interaction of pions with nuclei may be represented by an optical model potential. There are several different possible ways to construct such an average potential for pion-nucleus scattering. An example is that given by Stricker, Carr, and McManus [133]. It makes use of the fact that, at such low energies, the scattering is dominated by s- and p-partial waves.

Let us examine first the amplitude for pion-nucleus scattering in the limit that only s- and p-waves are contributing. Since pions are isospin $t = 1$, pseudoscalar ($J^\pi = 0^-$) particles, pion-nucleon scattering amplitude may be expressed in terms of the isospin operator $\boldsymbol{t}$ for the pion and $\boldsymbol{\tau}$ for the nucleon and approximated as

$$f_{\pi N} = b_0 + b_1 \boldsymbol{t}\cdot\boldsymbol{\tau} + (c_0 + c_1 \boldsymbol{t}\cdot\boldsymbol{\tau})\boldsymbol{k}\cdot\boldsymbol{k}' \tag{8-54}$$

where $\boldsymbol{k}$ and $\boldsymbol{k}'$ are, respectively, the initial and final pion wave number vectors. The coefficients b_0 and b_1 are related to s-wave scattering from a nucleon followed by absorption on a neighboring nucleon, and the coefficients c_0 and c_1 are related to the corresponding p-wave process. These coefficients are complex in general, as pions can be absorbed by nucleons. Their values may be found, for example, by fitting calculated results to experimental data on π-mesic atoms.

The pion-nucleus optical model potential that generates a scattering amplitude of the form of Eq. (8-55) may be expressed as the operator

$$U_{\rm opt}(\boldsymbol{r}) = -\frac{2\pi}{\mu}\Big\{(b(r)+B(r)) - \nabla\cdot\big\{L(r)(c(r)+C(r))\big\}\nabla + \frac{p_1-1}{2}\nabla^2 c(r) + \frac{p_2-1}{2}\nabla^2 C(r)\Big\} \tag{8-55}$$

where μ is the reduced mass of a pion. The kinematic factors

$$p_1 = 1 + \frac{\hbar\omega}{M_N c^2} \qquad\qquad p_2 = 1 + \frac{\hbar\omega}{2M_N c^2}$$

come from transformation between frames of reference attached to the center of mass of the pion-nucleon system and the pion-nucleus system. They are functions of the total pion energy $\hbar\omega$ and nucleon mass M_N. The other factors,

$$b(r) = p_1\{\bar{b}_0\rho(r) - e_\pi b_1\delta\rho(r)\} \qquad\qquad B(r) = p_2 B_0\rho^2(r)$$

$$c(r) = p_1^{-1}\{c_0\rho(r) - e_\pi c_1\delta\rho(r)\} \qquad\qquad C(r) = p_2^{-1}C_0\rho^2(r)$$

$$L(r) = \{1 + \frac{4\pi}{3}\lambda[c(r)+C(r)]\}^{-1} \qquad\qquad \delta\rho(r) = \rho_n(r) - \rho_p(r)$$

may also depend on e_π, the charge of the pion. The neutron, proton, and nucleon densities in the nucleus, $\rho_n(r)$, $\rho_p(r)$, and $\rho(r)$, are normalized, respectively, to N, Z,

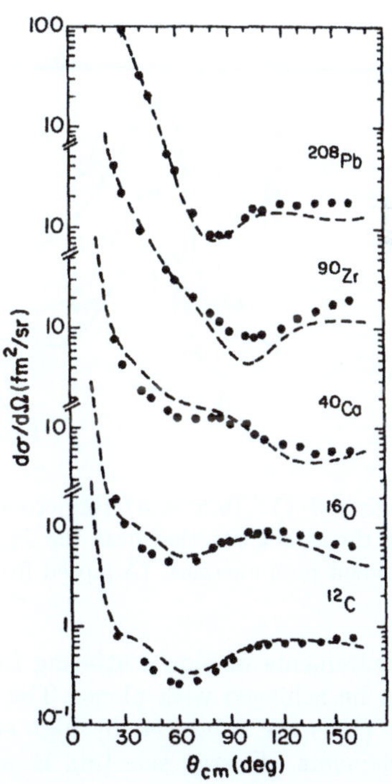

Figure 8-10: Elastic scattering of low-energy pions from nuclei. The continuous curves are calculated results using the values of b_0, b_1, c_0, c_1, and λ given in Eq. (8-56) (adapted from Ref. [133]).

and A. These factors express various first-order correlations between nucleons in a nucleus. Second-order correlations in s-waves are included in Eq. (8-55) through the factor

$$\bar{b}_0 = b_0 - \frac{3k_F}{2\pi}(b_0^2 + 2b_1^2)$$

where k_F, the Fermi momentum of a nucleon in a nucleus, is taken to be 1.4 fm^{-1}. A typical set of parameters for 50-MeV incident pions has the values

$$\begin{array}{llll} \lambda=1.4 & & & \\ b_0=-0.057+0.006i & \text{fm} & c_0=0.75\ +0.03i & \text{fm}^3 \\ b_1=-0.134-0.002i & \text{fm} & c_1=0.428+0.014i & \text{fm}^3 \\ B_0=-0.02\ +0.25i & \text{fm}^4 & C_0=0.36\ +1.2i & \text{fm}^6 \end{array} \qquad (8\text{-}56)$$

As can be seen from examples shown in Fig. 8-10, such an optical model potential gives a good description of experimental data on elastic scattering of both π^+ and π^- off a variety of nuclei at low energies.

At pion energies far above 200 MeV, the influence of P_{33}-resonance diminishes and the nucleus becomes much less absorptive to pions, as can be seen from the examples shown in Fig. 8-11. In the energy range 300 to 800 MeV, pion-nucleon scattering is dominated by many overlapping resonances. In this case, we expect that the cross section for pion-nucleus scattering at comparable energies may be understood by convoluting these resonances, using nucleon Fermi motion inside the nucleus as a smoothing function. However, not enough data are available yet for a more detailed discussion.

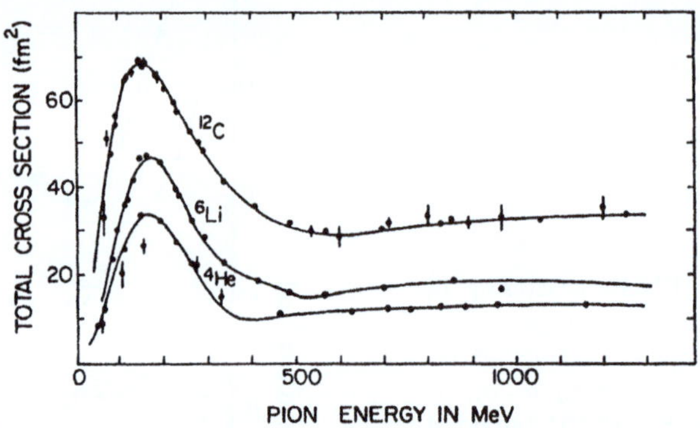

Figure 8-11: Total scattering cross section of pions off ^{4}He, ^{6}Li, and ^{12}C showing the strong reaction near the P_{33}-resonance and smooth variations at higher incident pion energies. (Adapted from Ref. [23].)

Measurements of pion scattering from nuclei are limited by the energy resolution that can be achieved with pions. The problem is caused partly by the fact that the incident pions are produced by high-energy protons striking a thick target made of heavy elements. Energy selection is accomplished by passing the broad spectrum of particles produced through electromagnetic fields. Both the limited initial flux and the short lifetimes of pions put stringent limitations on what can be achieved. The same difficulties are also present in the detection equipment, as the pion energies we have here are still relatively low for some of the more efficient detection techniques to work well. As a result, measurements of pion-nucleus scattering are usually carried out on light nuclei where the low-lying nuclear levels are well separated in energy. Both elastic and inelastic data are available, and they have been useful in complementing the information obtained with other probes.

Charge exchange reactions. Pion scattering involving the exchange of one unit of charge, (π^+, π^0) and (π^-, π^0), are among the most extensively studied π-nucleus reactions. Examples of (π^+, π^0) scattering off ^{14}C and ^{60}Ni at different angles are shown in Fig. 8-12 as examples. Except around the P_{33}-resonance, the processes are similar to (p, n) and (n, p) reactions and their results are often compared. At energies above the resonance, (π^-, π^0) reactions have some advantage over competing (n, p) reactions, as intense intermediate-energy neutron beams with well-defined energies are difficult to obtain. On the other hand, the particle emerging from an SCX reaction is π^0, a neutral particle that is usually detected by the γ-rays produced in its decay through the reaction $\pi^0 \rightarrow \gamma + \gamma$. This puts some constraint on the types of SCX measurements

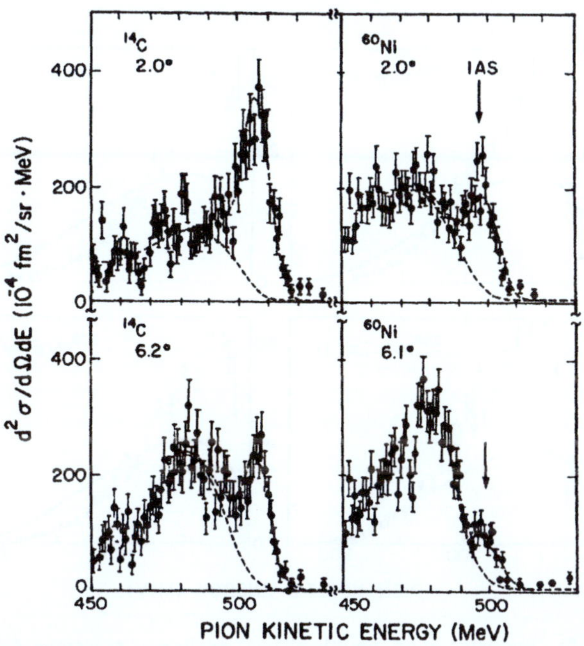

Figure 8-12: Energies of π^0 observed in (π^+, π^0) single-charge exchange reactions induced by 500-MeV pions on ^{14}C and ^{60}Ni at scattering angles indicated. The continuous curves are polynomial fits to the background (taken from Ref. [16]).

that can be carried out.

There are two different double-charge exchange reactions that can be studied with pions, (π^+, π^-) and (π^-, π^+). They are interesting for two reasons. First, these processes must involve at least two nucleons and are therefore useful for investigating nucleon correlations inside a nucleus. Second, the nuclear matrix element that enters into the scattering cross section is related to double β-decay, a process important for understanding the nature of weak interaction itself, as described in §5-6. For all practical purposes, DCX reactions are unique to pions. A nucleon, being an isospin-$\frac{1}{2}$ particle, can only induce SCX reactions. The only way to induce DCX with conventional nuclear probes is to use heavy ions. Here we have the complication that the probe itself can be excited by the reaction as well.

Because of the small cross section, DCX studies tend to be concentrated on strong transitions leading to isobaric analogue states in light nuclei separated by a pair of neutrons or protons. The main interest has been centered around effects involving the internal degrees of freedom of the nucleons. The results seem to indicate that an important role may be played by processes involving intermediate states with nucleons excited to become Δ-particles. The prospects of using DCX to relate strong and weak interactions and to understand nucleon correlations are quite promising. As illustration, examples of inclusive pion double-charge exchange reaction cross sections are shown in Fig. 8-13.

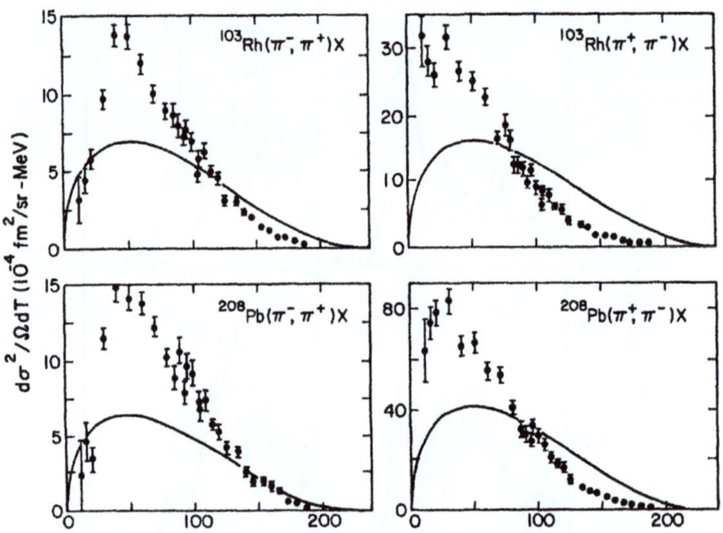

Figure 8-13: Inclusive pion double-charge exchange reactions (π^+, π^-) and (π^-, π^+) on ^{103}Rb and ^{208}Pb at incident pion energy 240 MeV. The emerging pions are detected at $\theta_{\text{lab.}} = 130°$. The smooth curves are classical estimates of Hufner and Thies [86] based on Boltzmann equations (taken from Gram [77]).

Kaons and other mesons. In addition to pions, kaons have been available for scattering off nuclei. Kaons are "strange" mesons involving either an s-quark (K^- and $\overline{K}^0$, strangeness $\mathcal{S} = -1$) or an $\bar{s}$-quark (K^+ and K^0, $\mathcal{S} = +1$). Conservation of strangeness requires that, when a K^--meson is absorbed by a nucleus, one of the nucleons changes into a "strange" baryon such as a Λ-particle ($m_\Lambda c^2 = 1115.6$ MeV) or a Σ-particle ($m_{\Sigma^0} c^2 = 1192$ MeV). The nucleus A becomes a *hypernucleus* ${}_{\rm Y}A^*$ in the process. Because of the large kaon mass ($\sim$500 MeV/c^2), the nucleus is left in a highly excited state, much more so than the case of pion absorption. On the other hand, since there is no light baryon with $\mathcal{S} = +1$, a K^+-meson cannot be absorbed by a nucleus. Many new and different avenues of study are opened up when we use a hadronic probe with nonzero strangeness. In terms of new insights into nuclear structure problems, this is similar to what studies of nuclei far away from the valley of stability can provide us.

Besides pions, interaction between nucleons is also mediated by other mesons such as ρ and ω. For this reason, reactions of these mesons with nucleons and nuclei are of interest. The difficulty is an experimental one; there does not seem to be any easy way to produce intense beams of mesons other than pions and kaons. Some of the information on the interactions of ρ and ω with nucleons, and baryons in general, must be obtained from their production rates in the decay of heavier particles.

Problems

8-1. Show that in the scattering of particle a, with mass M_a, off target nucleus b, with mass M_b, the momentum transfer q from a to b has the same form in both the laboratory and center-of-mass coordinates.

8-2. Use a table of binding energies to calculate the Q-value for the ${}^{120}\mathrm{Sn}(d,p){}^{121}\mathrm{Sn}$ reaction leading to the ground state of ${}^{121}_{50}\mathrm{Sn}$.

8-3. For a Yukawa potential $V(r) = V_0 e^{-r/r_0}/r$, with range $r_0 = \hbar/mc$ given in terms of the mass m of the boson exchanged, find the angular distribution for elastic scattering in first Born approximation due to the potential. Show that in the limiting case of a zero-mass boson, the result is identical to Rutherford scattering.

8-4. The angular distribution of an $\ell = 2$ transfer, ${}^{20}\mathrm{Ne}(d,n){}^{21}\mathrm{Na}$ reaction leading to the $J^\pi = 5/2^+$ state at 2.14 MeV in ^{21}Na peaks at 36° for deuterons with center-of-mass energy 6.0 MeV. Use a plane wave Born approximation to deduce the radius of ^{21}Na. Compare the result with that given by $R = 1.2A^{1/3}$ fm.

8-5. Show that, for direct reactions in a plane wave Born approximation, only $\ell = 0$ transfers have maxima in the differential scattering cross section at scattering angle $\theta = 0°$. For $\ell > 0$ transfers, the forward direction is a minimum and the first maximum in the differential cross section occurs at increasingly larger angles with increasing ℓ-values of the transfer.

8-6. Calculate the radius of the lowest orbit of π^- in a π-mesic atom with Z protons in the nucleus. Assuming a two-parameter Fermi form Eq. (4-22) for the distribution of nucleons in the nucleus, find the overlap between the wave function of the π^-

and the nucleus consisting of A nucleons. Take $c = 5.0$, $z = 0.5$ fm, $Z = 50$, and $A = 120$.

8-7. Find the angular distribution of neutrons emerging from a (n, n) reaction on ^{208}Pb. If the incident energy is sufficiently low, multiple scattering may take place. Assume that each neutron suffers two elastic collisions with the nucleons in the target nucleus before leaving and that the angular distribution of each scattering is given by $|j_0(qR)|^2$, where R is the radius of the target nucleus. Take the incident neutron energy to be 10 MeV in the center of mass and ignore any energy dependence in the scattering cross section.

8-8. Show that for low-energy, hard-sphere scattering, the cross section is equal to $4\pi R^2$, where R is the radius of the potential well.

Chapter 9

Nuclei under Extreme Conditions

For the most part, we have been looking at stable nuclei and their low-lying excited states. This is, to a large extent, dictated by the availability of data. Until quite recently, most measurements have been confined to nuclei that can be easily made into targets and, as a result, the studies are restricted essentially to those in the valley of stability. With heavy-ion accelerators and radioactive beams, a large number of new species can now be examined in the laboratory. When this is combined with improved detection techniques that allow many aspects of the reaction to be measured at the same time, we enter into a new era of nuclear physics.

9-1 Overview of Heavy-Ion Reactions

The term *heavy ion* is generally used to mean nuclei heavier than ^{16}O. For nuclei with $A > 16$, the internal structure becomes sufficiently complex that, when two heavy ions scatter off each other, many reaction channels are open. If, in addition, one of the ions involved is an unstable one, as it is possible in radioactive beam experiments, a number of "exotic" studies can be made. In this section, we shall look at some of the general features of reactions involving complex nuclei before going on in the later sections to examine a few of the more exciting phenomena in detail.

Nuclei far away from the valley of stability. When two heavy ions are fused together into a single entity, the result is usually a neutron-deficient system. As we have seen in Chapter 1, nuclei must increase their neutron excess with increasing nucleon number to stay stable. For example, below $Z = 20$, the ratio of neutron to proton numbers for stable nuclei is $N/Z \approx 1$. For medium heavy nuclei such as zirconium ($Z = 40$), $N/Z \approx 1.3$, and for lead ($Z = 82$), $N/Z \approx 1.5$.

The composite system formed by fusing two heavy ions takes on the average N/Z ratio of the two. The only exception is that a few nucleons may be discarded during initial stages of the reaction. As a result, the neutron excess is much smaller than the value appropriate for the combined system. For example, when two $^{90}_{40}Zr$ nuclei are joined together, the composite system is $^{180}_{80}Hg$. Since the lightest stable mercury isotope is ^{196}Hg, it implies that the composite system is "deficient" by roughly 16 neutrons. The fact that ^{180}Hg is unstable toward β^+-decay with a half-life of 2.9 s for the ground state is another indication that we have a neutron-deficient nucleus at the edge of the valley of stability. The possibility of making such proton-rich species gives us a new window to study nuclear physics. Furthermore, a number of different projectile-target combinations can be used to reach the same final nucleus, providing us with a chance here to examine such nuclei from different angles and to see if their properties can be understood from what we have learned from stable nuclei.

In practice, direct fusion of two heavy ions is not the normal way to create neutron-deficient nuclei. The reason comes from the "extra" energy required to penetrate the Coulomb barrier. As a result, the composite system is usually in a highly excited state, with a large amount of excess energy as well as angular momentum. Many decay channels are open under such circumstances, and the probability of decaying into other nuclei by particle emission or fission becomes high. To enhance the formation of the desired compound nucleus, it is preferable to select a reaction that can carry away some of the energy and angular momentum by emitting a few nucleons and γ-rays. In this way, neutron-deficient nuclei are made up to the point where proton emission begins to dominate the decay mode.

To make neutron-rich nuclei on the other side of the valley of stability, radioactive beams may be used. When a beam of high-energy particles strikes a thick target, a large variety of nuclei is produced. Since the environment is highly nonequilibrium, many of the species created are unstable and far away from the valley of stability. With suitable projectile and target combinations, it is possible to enhance the production of selected unstable particles. For example, nuclei such as ^{8}He ($\tau_{1/2}$ = 122 ms), ^{22}O ($\tau_{1/2}$ = 2.3 s), and ^{132}Sn ($\tau_{1/2}$ = 40 s) are produced in abundance in this way, and they have far more neutrons compared with their stable counterparts of ^{4}He, ^{16}O, and ^{120}Sn. As we shall see in the next chapter, neutron-rich nuclei are important also in the synthesis of elements beyond $A \sim 56$ in supernova explosions. Indeed, the condition to make radioactive beams is very similar to the environment under which certain heavy nuclei are created in an exploding star. For this reason, astrophysics interests have often been one of the motivations for constructing these facilities, such as the Isotope Separator On Line Facility for Production of Radioactive Ion-Beams (ISOLDE) at the European Laboratory for Particle Physics (CERN); the Unstable Beam Facility at the Institute for Nuclear Study (INS), University of Tokyo; the Exotic Beam Facility at Argonne National Laboratory (ANL); the Isotope Separation Accelerator (ISAC) Facility at TRIUMF, Canada's national meson research facility; and the Radioactive Ion Beams at Louvain-la-Neuve, Belgium.[1]

In addition to neutron- and proton-rich nuclei, heavy-ion beams can also be used to create superheavy nuclei. We have seen earlier that nuclei beyond ^{208}Pb are unstable because of the Coulomb repulsion between the large numbers of protons. In fact, the only ones we find outside the laboratory are those with lifetimes comparable to or longer than the age of the solar system or decay products of such long-lived nuclei.

In general, lifetimes of nuclei beyond ^{208}Pb decrease with increasing nucleon number. However, as we have seen in §7-2, it is quite possible that, because of shell structure, a region of relative stable nuclei may be found beyond the heaviest one we are

[1] For more information on ISOLDE, see http://www.cern.ch/ISOLDE; on INS,http://npsun1.ins.u-tokyo.ac.jp; on the Exotic Beam Facility, http://www.phy.anl.gov/div/origins/yellow-book; on ISAC, http://www.triumf.ca/isac/lothar/isac.html, and on Louvain-la-Neuve, http://www.cyc.ucl.ac.be/CYC/rib/rib-en.html.

aware of so far. Since it is unlikely that any such nuclei are stable (otherwise they would have been discovered), the only way we can detect their presence is through their decay scheme to lighter ones. As a result, it is necessary to make first the lighter nuclei below the superheavy ones before we can make the identification. Furthermore, from a practical point of view, the only way to build up to the superheavy "island" of relative stability is go one step at time. A recent addition to such a list is the element $(A, Z) = (277, 112)$, created at GSI (Gesellschaft für Schwerionenforschung, Darmstadt, Germany) using a beam of $^{70}_{30}$Zn on a $^{208}_{82}$Pb target [84]. The identification of the element is through a sequence of α-particle decays, as shown in Fig. 9-1. With radioactive beam facilities, additional means to create such particles become available as well.

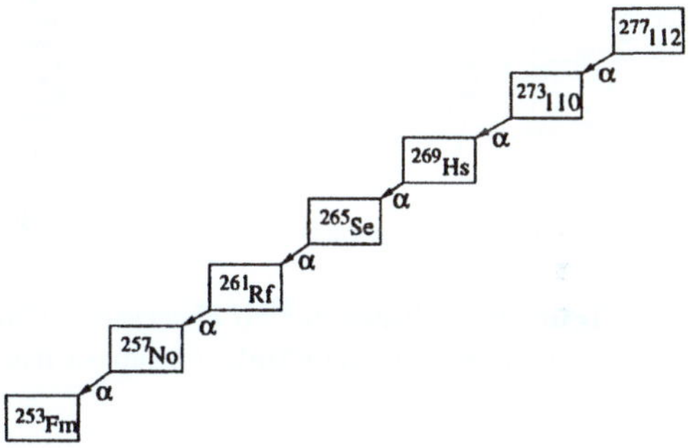

Figure 9-1: Identification of element 112 by its α-particle decay chain. Two such chains, each with a different set of α-particle energies, have been reported, one ending at $^{257}_{102}$No and the other at $^{253}_{100}$Fm [84].

Coulomb effects. Another interesting aspect of heavy-ion reactions is the strong Coulomb field created in the process. First, we saw in §8-1 that the strength of Coulomb excitation is proportional to the product of charges carried by the projectile and the target. With heavy ions, we are able to increase the charge of the projectile by a large factor over that of light ions. For example, if $^{238}_{92}$U is used, the Sommerfeld number η of Eq. (4-64) is increased by a factor of 92 over that for a proton. As a result, low-energy heavy ions are the projectiles of choice for Coulomb excitation reactions. An example is given in Fig. 9-2, where many high-spin states in ^{238}U were first found by Coulomb excitation using a beam of ^{209}Bi at 5 to 6 MeV per nucleon.

Second is the interest at slightly higher energies. If the projectile has sufficient energy to tunnel through the Coulomb barrier of the target nucleus, a compound nucleus of charge $Z = Z_1 + Z_2$ is formed. If the sum of the two proton numbers is greater than 137, the inverse of the fine structure constant, a very interesting situation in quantum electrodynamics may develop as a result. To see this, let us return for the moment to the simple problem of hydrogen-like atoms with a single electron outside a nucleus having Z protons. Nonrelativistically, the energy levels are given by the solution to the Schrödinger equation as

$$E_n = -\left[\left(\frac{1}{4\pi\epsilon_0}\right)^2\right]\frac{m_e e^4}{2\hbar^2}\frac{Z^2}{n^2} = -\frac{\alpha^2 m_e c^2}{2}\frac{Z^2}{n^2}$$

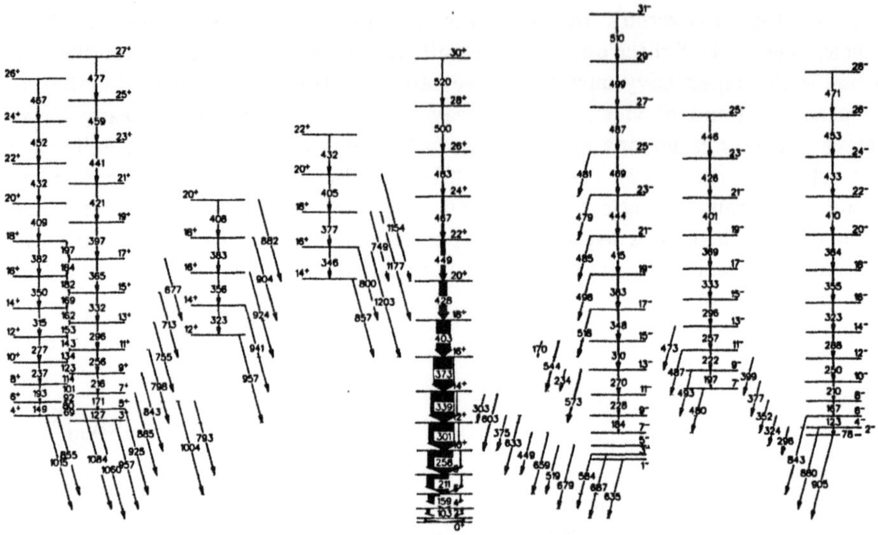

Figure 9-2: States in ^{238}U observed in Coulomb excitation using a beam of ^{209}Bi ions at 1130 and 1330 MeV. (Adapted from Ref. [142].)

where n is the principal quantum number that labels the atomic energy levels. The lowest state is 1s ($n = 1$, $\ell = 0$). For a hydrogen atom, we have $Z = 1$ and the ground state has energy

$$E_{1s} = -\frac{\alpha^2 m_e c^2}{2} = -13.6 \text{ eV}$$

The factor $\alpha^2 m_e c^2/2 = R_y$ is known as the Rydberg energy, the ionization energy for a hydrogen atom in the ground state.

More generally, we can solve the Dirac equation for a hydrogen-like atom assuming, for simplicity, a point nucleus of charge $+Ze$. In this case, we obtain instead the total energy for the ground state of the system,

$$E_{1s} = m_e c^2 \sqrt{1 - (Z\alpha)^2}, \tag{9-1}$$

where α is the fine structure constant. For $Z = 1$, this also yields a value of 13.6 eV for the ionization energy of a hydrogen atom, as expected. For $Z > 1$, the expression is valid up to some critical value $Z_{\text{cr}} = 1/\alpha \approx 137$. The limitation is usually not a problem, as all the known nuclei have Z values much less than Z_{cr}. Even among the man-made elements, the highest Z-value known so far is 112, as we saw earlier. However, in heavy-ion collisions, it is possible for the compound nucleus, formed by fusing two heavy nuclei, to have a Z-value far in excess of 137. In this case, a *supercritical* field is created as a result.

Some corrections to the result $Z_{\text{cr}} \approx 137$ are necessary, as nuclei are not point charges. The exact value depends somewhat on the charge distribution inside a nucleus.

For a uniform charged sphere of radius reasonable for nuclei, a result of $Z_{cr} < 200$ is obtained. This higher value can also be exceeded in heavy-ion collisions.

The physical meaning of a supercritical field may be seen from the following arguments. As the charge number of a nucleus is increased, the eigenvalues of atomic levels decrease from those given roughly by Eq. (9-1) until the critical value is reached. When this happens, the 1s-level becomes degenerate with the negative-energy continuum filled with equal numbers of electrons and positrons. The charge-neutral vacuum is no longer the state of minimum energy. To lower the energy, positrons are released and the remaining vacuum becomes a charged one. This phenomenon is referred to as the "spontaneous" decay of the neutral vacuum. Experimentally, the presence of a supercritical field may be identified by the appearance of narrow positron peaks when two heavy nuclei collide with each other. However, in spite of initial hopeful signs, no positive identification is known to date.

High spin and large deformation. We have seen in §1-3 that, when two heavy ions approach each other, it is possible for the combined system to acquire angular momentum in excess of $100\hbar$. The experimental arrangement of interest to us here is one that the relative kinetic energy between the two particles is still quite low but sufficient for one to tunnel through the Coulomb barrier of the other. Since the impact parameter here is comparable to the sum of their radii, we have essentially a *grazing collision*, as shown schematically in Fig. 9-3. In such cases, only a small number of nucleons in the projectile and the target are in close proximity of each other for nuclear interaction to take place between them, and consequently, it takes some time to transform the relative kinetic energy in the system into internal excitation. At the mean time, the energy appears in the form of rotational energy, as if the projectile and target nuclei are revolving around each other. The composite nucleus may be said to be still rather "cool" in the sense that most of the nucleons remain in their lowest

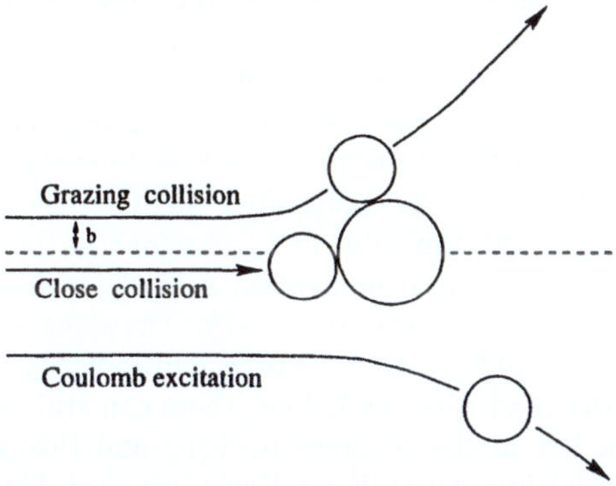

Figure 9-3: Schematic diagram showing the different possibilities in a heavy-ion collision. At low energies, Coulomb excitation dominates. At slightly higher energies, the ions come into contact with each other and we have grazing collisions. At even higher energies, head-on collisions become possible.

single-particle states.

The time available for two heavy ions to overlap each other and coalesce in a heavy-ion scattering under such conditions is very short. To gain some perspective, we can make an order-of-magnitude estimate of the amount it takes for the two nuclei to go around each other once. For simplicity, let us assume that the relative speed between the two ions is 1/15 of the speed of light, corresponding to about 5 MeV/nucleon in energy. If the centers of two ions are 10 fm apart, the time to go around each other once at speed $c/15$ is on the order of 10^{-21} to 10^{-20} s. Compared with the interaction time of the order of 10^{-22} to 10^{-20} s, we see that the two systems do not have enough opportunity to reach an equilibrium with respect to each other. This is quite different from the case of of a normal compound nucleus formation and, consequently, the two ions maintain more or less their original shapes. The resulting intrinsic shape of the composite system is highly deformed, as shown schematically in Fig. 9-4. For this reason, it is not surprising that some of the largest deformations are found in heavy-ion collisions at relatively low energies.

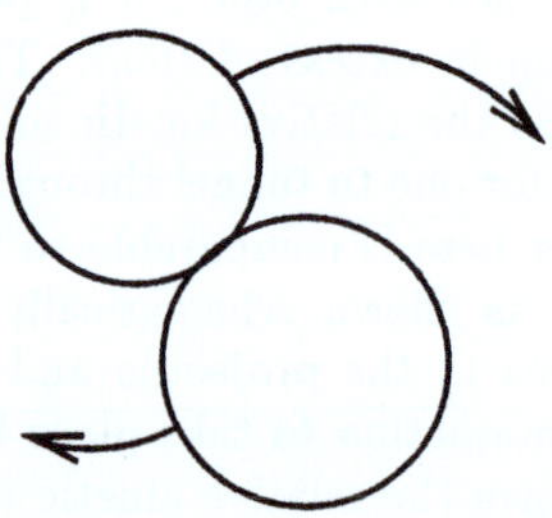

Figure 9-4: Schematic diagram showing a grazing collision between two heavy ions. Without adequate time to equilibrate, the two ions retain more or less their original shapes and revolve around each other to conserve the large amount of angular momentum in the composite system.

The most likely channel for the composite system to decay is fission. If for some reason that fission is inhibited, a part of the excess energy may be discarded by nucleon evaporation and γ-ray emission. The angular momentum carried away by emitting a nucleon is, however, quite small on the average. The amount may be estimated using an approximation that the maximum angular momentum ℓ_m carried away by a nucleon is given by $\hbar kR$, where k is the wave number of the nucleon and R is the radius of the composite system. For a neutron, the average kinetic energy is around 2 MeV, as it is difficult for an individual nucleon to acquire much more energy in the collision of two heavy ions, regardless of whether the composite system is fully equilibrated or not. For nucleon number A around 150, the value of ℓ_m obtained in this way is around $2\hbar$ (see Problem 9-2). The average angular momentum actually carried away by a nucleon is lower than this value and is more likely to be $\sim 1\hbar$. The same is also true for γ-rays, as the probability decreases rapidly with increasing multipolarity.

At the end of nucleon and γ-ray emissions, there can still be a substantial amount of angular momentum left in the remnant nucleus, and this appears in the form of nuclear spin. If the excitation energy is relatively low, such high-spin states are likely to be members of a yrast band, a rotational band consisting of the lowest member in energy of each spin. Once the nucleus is in one of these states, all subsequent decays proceed predominantly through γ-ray emission from one yrast level to the next one just below, as we have seen earlier in §6-3. The process of populating a yrast band is shown schematically in Fig. 9-5.

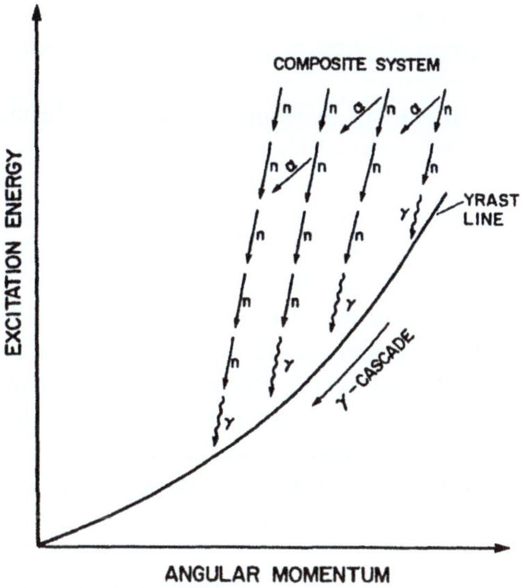

Figure 9-5: Schematic diagram showing the formation of yrast levels. The composite system created in a heavy-ion collision contains large quantities of excess energy and angular momentum. The amounts carried away by neutron, α-particle, and γ-ray emissions are very limited and the nucleus is likely to be left in a high-spin state, often a member of the yrast band. Thereafter, the nucleus decays through γ-ray cascade from one member of the band to the next.

The maximum angular momentum that can be attained by a nucleus depends on several different considerations. For light nuclei, the limiting factor is the highest spin to which the valence nucleons can be coupled together. For example, in the $1p$-shell, the maximum allowed J is 5 for six particles coupled together to $T = 0$. In the *ds*-shell, the maximum value is $J = 14$ for 12 active nucleons. In heavier nuclei, both the number of active orbits and the average spin of single-particle orbits are larger. For such nuclei, the maximum value is more likely to be limited by other considerations, such as stability against fission.

High-spin nuclear states usually lie quite high in excitation energy, in part due to the amount of energy associated with rotation. Since the density of states in such regions is high, the lifetimes of most states are short because of the large number of open decay channels. As a result, it is usually impossible to resolve individual levels. The yrast levels are, however, the exception, as their decays are dominated by γ-ray transitions within the band and, as a result, their lifetimes are long compared with other levels in the vicinity. The narrow widths of these γ-rays stand out against the background made of the decays of short-lived levels in the same region.

The highest J-values are observed in nuclei with "superdeformed" bands. These are formed when two heavy ions are fused together into a highly deformed shape, with the ratio between polar and equatorial axes as large as 2. Such a configuration is usually not the lowest one in energy and is therefore not commonly found in ground state bands. The high-spin members can be "followed" by observing the γ-ray cascade from one member to the next until the lowest energy members of the band merge into those made predominantly of ground state configurations.

When a nucleus is in a state with very high J-values, it is almost classical in the sense of the correspondence principle. Further progress in experimental techniques may lead to identifications of even higher spins and thus allows us to trace the development of a nucleus from a purely quantum-mechanical state to a classical one. This may also be of interest to the study of the transition from quantum-mechanical to classical description of physical phenomena in general.

Deep inelastic scattering. If the collision energy is slightly above the Coulomb barrier between a pair of heavy ions, many reaction channels are open and compete with each other. In such cases, it is usually impossible to investigate each type of final state individually and the inclusive cross sections are studied instead. The reaction is often referred to as *deep-inelastic* collision, similar to the situation of high-energy electron scattering discussed earlier in §4-4. The cross section is large here, especially for the heavier nuclei ($\sigma \sim 10$ to 20 fm^2). The reaction is often accompanied by the transfer of as many as 20 nucleons from one nucleus to the other. At the same time, up to 100 MeV of kinetic energy and $50\hbar$ of angular momentum are shifted from relative motion to excitation energies in the final nuclei.

Since many nucleons are involved here, it is possible to adopt a statistical mechanics approach to study the collective degrees of freedom associated with the process. A good starting point for such a macroscopic view is the master equation

$$\frac{d}{dt}P_n(t) = \sum_m [W_{nm}P_m(t) - W_{mn}P_n(t)] \tag{9-2}$$

where $P_n(t)$ is the probability that, at time t, the system is in a group of closely related states n and W_{nm} is the transition probability per unit time from the group of states m to the group n. The meaning of Eq. (9-2) is simply that the probability of finding a group of states is given by the difference between the sum of those for transferring into the group and those for leaving the group. This is the usual approach used in studying transport phenomena. For the transport description to be valid for the collective degrees of freedom of interest here, the time scales involved in the system must satisfy the condition

$$t_{\text{equ}} \ll t_{\text{coll}} \ll t_{\text{Poincaré}}$$

where t_{equ} is the time it takes for the noncollective degrees of freedom to reach equilibrium with respect to each other, t_{coll} is the time required for the collective degrees of freedom to reach equilibrium, and $t_{\text{Poincaré}}$ is the Poincaré recurrence time, the time for the system to return to its original point in phase space. The condition imposed on the relation between different time scales is necessary here, as we are using the master equation to deal only with the collective degrees of freedom for a system of two heavy ions. This requires that the noncollective degrees of freedom of the system, such as single-particle excitations, have reached an equilibrium already and do not participate in the transport process. The extent to which these conditions are met and the degree of success for a transport theory description of deep-inelastic collisions can be seen in the agreement between the calculated results and experimental observation of the collision of ^{136}Xe on ^{209}Bi at 1130 MeV laboratory energy shown in Fig. 9-6.

As an alternative to statistical mechanics methods, we can take a microscopic view and treat individual nucleons as the basic components. The entire deep-inelastic collision process is, in this case, governed by the time-dependent Schrödinger equation (B-1) and described by a wave function $\Psi(\boldsymbol{r}, t)$. As the two heavy ions approach each other, they collide and evolve into the final state. A general solution to Eq. (B-1) is, however,

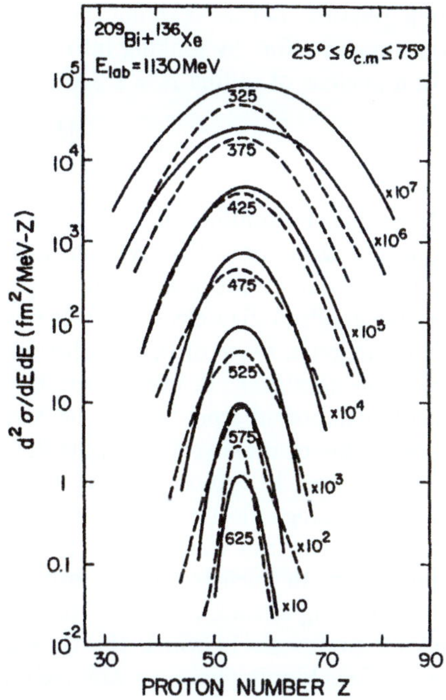

Figure 9-6: Double-differential cross section for the reaction ^{136}Xe on ^{209}Bi at E_{lab}=1130 MeV. Each curve is the sum over a final energy bin of 50 MeV wide centered around the value indicated. Solid lines are the measured values of $d^2\sigma/dE\,dZ$ integrated over center-of-mass angles $25° \leq \theta \leq 75°$ and dashed lines are the calculated results using a transport theory. (Taken from Ref. [92].)

impossible for the complicated case of two heavy ions scattering off each other. One way to simplify the calculations somewhat is to use a time-dependent Hartree-Fock approach and take the time-dependent differential equation (B-1) as a difference equation that gives the changes in the wave function, $\Delta\Psi(\boldsymbol{r},t)$, in the time interval Δt,

$$\Delta\Psi(\boldsymbol{r},t) = \frac{1}{i\hbar} H\,\Psi(\boldsymbol{r},t)\Delta t$$

At a given time t, we have a system of $A_1 + A_2$ nucleons whose motions are given by the Hamiltonian H. The difference between the wave function at time t and $t + \Delta t$,

$$\Delta\Psi(\boldsymbol{r},t) = \Psi(\boldsymbol{r},t+\Delta t) - \Psi(\boldsymbol{r},t)$$

is the result of the action of the Hamiltonian on the system in the small time interval Δt. By solving the difference equation, we obtain the changes in the wave function $\Delta\Psi(\boldsymbol{r},t)$ and, thus, the wave function $\Psi(\boldsymbol{r},t+\Delta t)$ describing the system at time $t+\Delta t$. In this way, the time evolution of the system may be traced out in small steps. The Hamiltonian equation is, however, still a very complicated one to solve in view of the large number of nucleons involved. A further simplification is to invoke the Hartree-Fock approximation, which gives a fairly realistic description of the nuclear physics, as we have seen for the time-independent case in §7-3.

Ultra-relativistic collisions. At the other end of the energy scale we have heavy-ion collisions at ultra-relativistic energies. This is a topic of interest to both nuclear and particle physics. Let us examine the reasons for putting the enormous investment into this particular effort by looking into the energy densities involved.

In a nucleus of radius $R = 1.2A^{1/3}$, the energy density is given by

$$\rho_A \approx \frac{M_A c^2}{(4\pi/3)R^3} \approx 130 \text{ MeV/fm}^3$$

For three quarks confined in a nucleon of radius $r \approx 1$ fm, the value is even higher, roughly 250 MeV/fm^3. If heavy ions are made to collide with each other at center-of-mass kinetic energy far larger than their total rest mass energy, for example, in excess of several hundred giga-electron-volts per nucleon, the energy density is even higher.

We can get a sense of the implications of such large energy densities from the following considerations. In nuclear scattering, we seldom have to be concerned with the chemical compound from which the target is made (except for, of course, practical considerations of making and supporting the target in an actual experimental setup). This is because, in typical nuclear scattering experiments, the energy involved is on the order of mega-electron-volts. Chemical binding energies are, on the other hand, measured in electron-volts. For this reason, we do not need to include any chemical considerations in nuclear physics, except in special cases such as neutrino mass measurements using tritium β-decay. The same is true in the analogous situation of particle physics experiments in which the typical energy involved is GeV or higher. In these cases, nuclear binding energies are irrelevant for all practical considerations. (The only possible exception is in high-energy lepton scattering discussed in §4-4.)

At ultra-relativistic energies, we expect the "bags" that confine quarks inside hadrons to become irrelevant, in the same way as chemical binding in nuclear experiments and nuclear binding in particle physics experiments. When two heavy ions are in a situation that the quarks inside their nucleons are essentially "free," we have a chance of creating a state of matter in a finite volume that is quite different from nuclear matter. It is expected that the region is dominated by quarks, antiquarks, and gluons. For this reason, it is generally referred to as a *quark-gluon plasma*, or QGP for short. In addition to interests in quantum chromodynamics, this state of matter is likely to be similar to that which existed at the beginning of the "big bang," postulated in cosmology as the event that gave birth to our universe. The study is, therefore, a question of central importance in cosmology as well. Since we are now entering into a region with little or no observational data, many new and unexpected things can happen and they, in turn, will lead us into new areas of physics we have not yet thought about.

9-2 High-Spin States in Nuclei

One of the modern interests in nuclear physics is high-spin states and the associated large deformation found in medium and heavy nuclei. We have seen earlier that pairing force is a dominate feature in nuclear interaction. As a result, two identical nucleons have a strong tendency to couple to angular momentum zero. Because of this, all even-even nuclei have $J^\pi = 0^+$ for the ground state. To build up any large spin values requires many nucleons to break away from this trend and align their spins more or less in the same direction. Such an alignment, in turn, implies large deformation, in contrast to nearly spherical shapes found by putting particles pairwise in $+m$ and $-m$ magnetic substates.

We saw earlier in §6-3 that collective states, under appropriate conditions, form rotational bands. In fact, it is usually through such band structures that rotational features are identified. In general, the spin of a state cannot be measured directly. Instead, it is deduced from such observations as the multipolarities of transitions to states of known spin. A simple example is one that decays by electromagnetic radiation to a state with $J^\pi = 0^+$. If the transition is known to be $E2$ from, for example, the angular distribution of the γ-rays emitted, we can identify the spin-parity of the state to be 2^+ using angular momentum conservation. (For decays to final states other than 0^+, the procedure is more complicated and usually requires transitions to more than one state of known spin and parity.) The high-spin values we are interested in here are usually deduced from a sequence of strong decays from one member of a band to the next one lower in energy. In the simple case of an even-even nucleus with the last member of the band known to be 0^+ and all the decays being $E2$, the sequence of decays identifies the members of the band to be 0^+, 2^+, 4^+, ..., as we have seen earlier in §6-3 for $K = 0$ bands. If n such (correlated) $E2$-decays are observed in a nucleus, the highest member of the band must have $J = 2n$. Many examples have been observed in the laboratory and the largest spin value known now exceeds $60\hbar$.

Because of pairing, spherical shape and small deformation dominate low-lying states. The kind of large deformation we are interested in here is usually found slightly above ground state configurations. However, the excitation energies involved, other than rotational, cannot be too large, or else the system is unstable toward fission and it will not be possible to observe the high-spin states. The question why large deformation occurs at such low excitation energies is an interesting one. However, before we can address this problem, let us examine first some of the evidence for assuming that alignment is an important feature underlying large deformation.

Band crossing and backbending. One of the interesting observations made on high-spin states is the presence of small but sudden changes in the moment of inertia of certain nuclei. On a plot of E_J as a function of $J(J+1)$, the changes are usually too insignificant to be noticed. However, if the moment of inertia $\mathcal{I}$ is plotted against the square of the frequency of rotation, local variations are amplified. A group of these may appear in the form of a Z-shaped curve, as shown in Fig. 9-7, and hence the name "backbending." To make such a plot, we need the local values of the rotational frequency ω and moment of inertia. The former is not a quantity that can be measured directly but may be inferred by making an analogy with classical rotational frequency through, for example, the relation

$$\hbar\omega = \frac{dE}{d\sqrt{J(J+1)}}$$

for a $K = 0$ band. The value may be approximated by the difference between the energies of two adjacent members, E_J and E_{J-2},

$$\hbar\omega \approx \frac{\Delta E}{\Delta\sqrt{J(J+1)}}\bigg|_{J-2}^{J} \xrightarrow[J\to\infty]{} \frac{1}{2}(E_J - E_{J-2}) \qquad (9\text{-}3)$$

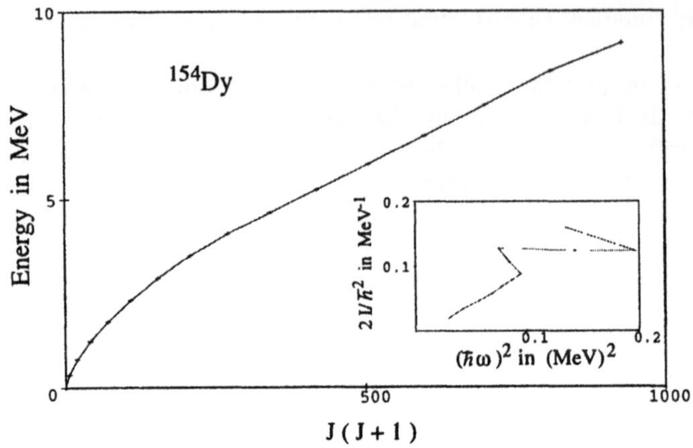

Figure 9-7: Backbending in ^{154}Dy. When the energy of each level in the band is plotted against $J(J+1)$, a typical rotational spectrum is found, with a slow variation of the momentum of inertia due to centrifugal stretching. However, when the relation is examined in more detail in the insert by plotting $2\mathcal{I}/\hbar^2$ against $(\hbar\omega)^2$, a sudden change is observed around $J = 20$, indicating a shift in the intrinsic structure of the nucleus. (Plotted using data from Ref. [14].)

Similarly, the local value of the moment of inertia may be found from the relation

$$\frac{2\mathcal{I}}{\hbar^2} = \frac{4J-2}{E_J - E_{J-2}}$$

and calculated from the energy difference between two adjacent members.

A simple explanation of the phenomenon may be made using the idea of band crossing. Consider a nucleus having two rotational bands, A and B, each with a slightly different moment of inertia and band head position. It may happen that, below some energy E_c, members of band A lie below the corresponding ones of band B, and the other way around above E_c. On a plot of E_J versus $J(J+1)$, the two curves cross each other, a relatively common occurrence in medium and heavy nuclei. If the structure of the two bands are quite different from each other, there will not be any coupling between them and each one appears as a separate entity. This is also true even when the structure of the two bands are similar if members of the two bands with the same J-value are well separated in energy, as the coupling between them is expected to be weak. The exception occurs around E_c, where the "unperturbed" positions of the states in the two bands with the same J^π are almost degenerate in energy. In this case, even a very small interaction can cause mixing between them, as we saw earlier in the analogous situation of isospin mixing in §4-8. The result is strong mixing in a few adjacent states and appears as a local variation of $\mathcal{I}$. Since the moments of inertia are very similar between the two bands, there will not be any strong indications of the change, for example, in a plot of E_J versus $J(J+1)$, as one moves across E_c. One way to amplify the sudden variation is to use a more sensitive plot, such as $2\mathcal{I}/\hbar^2$ versus $(\hbar\omega)^2$, shown as an example in the insert of Fig. 9-7.

From the point of view of rotational alignment, it may be more instructive to view the same explanation from the following perspective. Consider again a low-lying rotational band in an even-even nucleus for simplicity. The intrinsic state is formed mostly by zero-coupled pairs of particles filling up the lowest available (deformed) single-particle states. As we move up in angular momentum, the rotational energy increases. At some point, it may be more advantageous instead to increase the spin value by breaking a zero-coupled pair and aligning the angular momenta of the two (identical) particles to the maximum possible value of $(2j-1)$ (as the value $2j$ is forbidden by the Pauli principle). This is more likely to take place in medium and heavy nuclei where the particles are occupying single-particle states with large j-values. In such cases, the energy required to break a pair may be less than the increase in the rotational energy at large J-values. This picture is confirmed by other observations, such as changes in the single-particle occupancy in the region of backbending. Since the structure is modified by breaking a zero-coupled pair, we may view the situation as one that the rotational band is now taking on a different intrinsic shape. From a band-crossing point of view, this new intrinsic state is one that actually belongs to another rotational band that crosses the present one. We shall see soon that alignment of single-particle angular momenta is an important ingredient in forming high-spin states.

Superdeformation. In §6-3 we saw that rotational bands are associated with nuclei having nonspherical intrinsic shapes. To reach the kind of high spin values of interest here, deformation much larger than the usual ones found in ground state rotational bands is needed. In recent years, highly deformed shapes with the ratio of semi-major to semi-minor axes around 2:1 have been observed in a number of nuclei, starting from $^{132}_{58}$Ce [111] and $^{152}_{64}$Gd [113] in the 1980s. These are the superdeformed bands where spin values beyond $60\hbar$ have been identified. From the measured lifetimes of and transition rates between members of the bands, one can infer that the value of β, the deformation parameter defined in Eq. (6-11), reaches values around 0.6. If we take a first-order approximation that δ_{osc} of Eq. (7-32) is roughly equal to β, we find that the frequency ratio of $\omega_\perp$ to ω_3 is 2:1. Since the ratio of the root-mean-square radius in each direction is proportional to that for the inverse of the oscillator frequencies,

$$< r_1^2 >^{1/2} : < r_2^2 >^{1/2} : < r_3^2 >^{1/2} = \frac{1}{\omega_1} : \frac{1}{\omega_2} : \frac{1}{\omega_3}$$

we obtain the ratio $< r_3^2 >^{1/2} : < r_\perp^2 >^{1/2} = 2 : 1$ for an axially symmetric nucleus. Even larger deformations with axis ratio of $3 : 1$, the case of *hyperdeformation*, have been speculated to exist. However, at the time of writing, there is no known definitive experimental confirmation for such exotic states.

In the ground state region, the deformations are much smaller, typically with $\beta <$ 0.3. The presence of large deformation at relatively low excitation energies raises two questions. The first is that, since they are so different, how can such large deformations be excited, or *populated*, in an experiment. The second question is concerned why such "unusual" states can be observed at all in a region with high level-densities. In other words, why are they not admixed with other configurations nearby and buried as small components among the multitude of others at roughly the same excitation energies? Instead, members of the superdeformed bands have relatively long half-lives and the

γ-rays emitted in transitions from one member to another stand out as sharp lines over and above the background from a large variety of sources.

Heavy-ion reaction has been the method of choice to excite superdeformed bands and high-spin states. The bombarding energy is usually kept as low as possible so that only a minimum amount of excess energy is put into the system. At the same time, the incident energy must be high enough for the ions to penetrate the Coulomb barrier with sufficient probability for the results to be observed. Except for small amounts carried away by nucleons and γ-rays emitted in the reaction, the bulk of the angular momentum in the collision remains with the compound nucleus formed. For example, ^{132}Ce is produced by the reaction ^{100}Mo(^{36}S,4n)^{132}Ce using a 155-MeV ^{36}S beam [122]. Similarly, ^{152}Dy is populated by the reaction ^{120}Sn(^{36}S,4n)^{152}Dy with a 170-MeV ^{36}S beam [124]. The known superdeformed bands in these nuclei are shown schematically in Fig. 9-8 using data from the Evaluated Nuclear Structure Data File (ENSDF) [79] and Ref. [91].

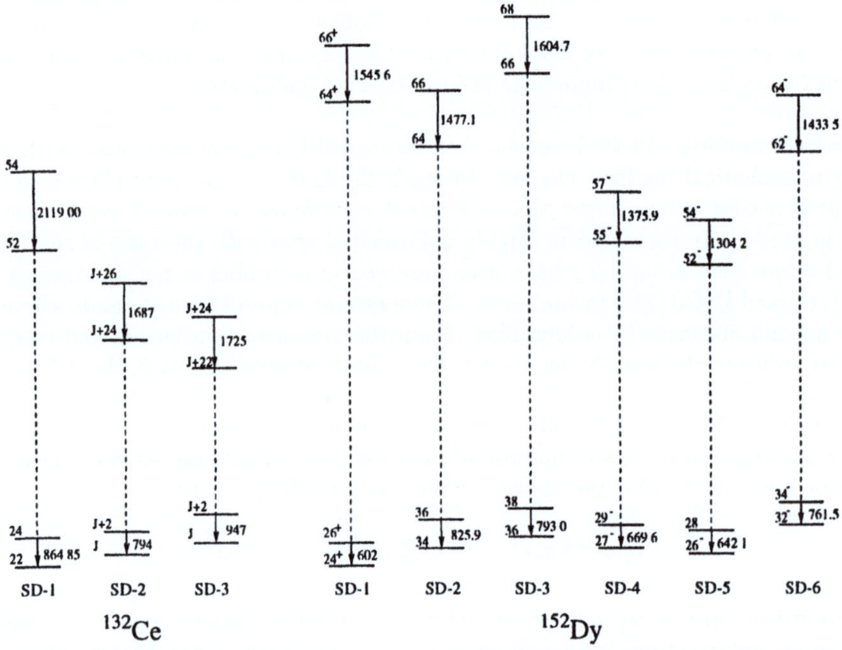

Figure 9-8: Schematic diagram showing the superdeformed bands in ^{132}Ce and ^{152}Dy. The spin values of bands SD-2 and SD-3 are only observed relative to an unknown band head marked as J in each case. (Plotted using data from Refs. [79, 91].)

The deformation of a rotational band is not a quantity that can be observed directly. The usual way to deduce its value is to make use of the intrinsic quadrupole moment Q_0 and moment of inertia $\mathcal{I}$. For a constant-density ellipsoidal nucleus, these two quantities are related to the deformation parameter β of Eq. (6-11) by the relations

$$\begin{aligned} \mathcal{I} &\approx \tfrac{2}{5}AM_N R_0^2(1+0.31\beta)+O(\beta^2) \\ Q_0 &\approx \frac{3}{\sqrt{5\pi}}ZR_0^2\beta(1+0.36\beta)+O(\beta^3) \end{aligned}$$

(See Problem 9-5.) If the spins of the states are known, the (transition) quadrupole moment may be deduced by measuring the lifetimes of the members. For example, Eqs. (6-21) and (6-22) can be used for the case of a $K=0$ band to find the reduced transition rates $B(E2; J \to J-2)$ and their values are related to the intrinsic quadrupole moment Q_0 through Eq. (6-21). Alternatively, we can use Table 6-1 and Eq. (6-21) to give the result

$$\mathcal{W} = 1.83\times 10^8 E_\gamma^5 e^2 Q_0^2 \frac{J(J-1)}{(2J+1)(2J-1)}$$

where E_γ is the transition energy in mega-electron-volts. This gives us the quadrupole moment Q_0 in units of femtometer squared directly in terms of the observed transition rate $\mathcal{W}$ measured in units of inverse seconds.

Dynamic and kinematic moments of inertia. The more likely situation in superdeformation studies is that only the transition energy E_γ is known. In this case, the nature of the band may be deduced using the dynamic moment of inertia

$$\mathcal{I}^{(2)} = \hbar\frac{dJ_x}{d\omega} = \hbar^2\left(\frac{d^2E}{dJ_x^2}\right)^{-1} \tag{9-4}$$

where J_x is the projection of total angular momentum on the rotation axis[2] in units of $\hbar$ and is given by

$$J_x = \sqrt{J(J+1)-K^2}$$

Similar to what we did earlier in Eq. (9-3), we can find the value of $\mathcal{I}^{(2)}$ using the observed γ-ray energies emitted in the decay from one member of the band to the one just below. For a J to $J-2$ transition, we can approximate dJ_x as

$$\Delta J_x = \sqrt{J(J+1)-K^2} - \sqrt{(J-2)(J-1)-K^2} \xrightarrow[J\to\infty]{} 2$$

For $d\omega$ in Eq. (9-4), we can use the approximation

$$\begin{aligned} \Delta\omega &= \frac{1}{\hbar}\Delta(\hbar\omega) \\ &\xrightarrow[J\to\infty]{} \frac{1}{2\hbar}(E_J - E_{J-2}) - \frac{1}{2\hbar}(E_{J-2} - E_{J-4}) \\ &= \frac{1}{2\hbar}\{E_\gamma(J\to J-2) - E_\gamma(J-2\to J-4)\} \equiv \frac{1}{2\hbar}\Delta E_\gamma \end{aligned}$$

The net result is that

$$\mathcal{I}^{(2)} \xrightarrow[J\to\infty]{} \hbar^2\frac{4}{\Delta E_\gamma} \tag{9-5}$$

[2]Since deformation parameter $\gamma = 0°$ here, the rotation is around the smallest axis, taken to be the x-axis here.

independent of the actual value of the spins involved.

A related quantity, the kinematical moment of inertia,

$$\mathcal{I}^{(1)} = \frac{\hbar^2}{2}\left(\frac{dE}{d(J_x^2)}\right)^{-1} = \hbar\frac{J_x}{\omega} \tag{9-6}$$

is also used on occasion to characterize rotational bands. At high spin values, we have $J_x \approx J$, and $\mathcal{I}^{(1)}$ reduces to be the same as the classical expression for the moment of inertia (the *static* moment of inertia) given in Eq. (6-13). By differentiating Eq. (9-6) with respect to ω, we arrive at a relation between $\mathcal{I}^{(1)}$ and $\mathcal{I}^{(2)}$,

$$\mathcal{I}^{(2)} = \mathcal{I}^{(1)} + \omega\frac{d\mathcal{I}^{(1)}}{d\omega}$$

For a rigid rotor, all three moments of inertia, $\mathcal{I}$, $\mathcal{I}^{(1)}$, and $\mathcal{I}^{(2)}$, are independent of the rotation frequency ω and the second term on the right-hand side vanishes. This leads us to the conclusion that, for a rigid rotor,

$$\mathcal{I}^{(2)} = \mathcal{I}^{(1)} = \mathcal{I} = \tfrac{2}{5}MR_0^2(1 + \tfrac{1}{3}\delta) \tag{9-7}$$

where we have made use of the value of the moment of inertia for a rigid body given by Eq. (6-26).

The combination of Eqs. (9-5) and (9-7) provides us with a way to estimate the deformation of a band and to see if it is superdeformed without knowing the spin values. Let us use ^{84}Zr as an example. In the reactions ^{58}Ni(^{29}Si,$2pn$)^{84}Zr at bombarding energy of 128 MeV and ^{58}Ni(^{32}S,$\alpha 2p$)^{84}Zr at 135 MeV, a group of nine transitions, with $E_\gamma = 2716$, 2599, 2435, 2272, 2114, 1959, 1808, 1663, 1526 keV, was found [89]. From the intensities of and coincidences among the γ-rays, one can identify that they come from $E2$-decays, cascading down from one member of a rotational band to the next one just below. However, without transitions observed going to states with known spin and parity, the absolute values of the spin and parity of the band members cannot be established. Using Eq. (9-5) to calculate the values of $\mathcal{I}^{(2)}$ and Eq. (9-3) for $\hbar\omega$, the values of $\mathcal{I}^{(2)}$ are plotted as a function of $\hbar\omega$ in Fig. 9-9. We see that the results are fairly constant, around 25 $\hbar^2$/MeV or slightly above. In contrast, the values for a normal band (band 3) in the same nucleus, shown as crosses, decrease rapidly with increasing value of $\hbar\omega$. To have a feeling of the size of deformation, the value for a rigid rotor with $\delta \sim 0.6$ is also shown in the plot for comparison. A value of $\beta = 0.53$ for the superdeformed band is established subsequently by measuring the lifetimes of the states. The β-value is also consistent with a cranked Woods-Saxon calculation to be discussed below.

Shape coexistence in nuclei. One of the unique features of superdeformed bands is that the decays are going mainly to states within the band. In contrast, normal deformed bands have much more "side feedings," or interband decays. The obvious explanation is that superdeformed bands have very different structure from nearby states, especially in view of the fact that they exist in a region of high level densities.

At the same time, we saw above that the spin values involved are high and the moments of inertia are large for the superdeformed bands. The combination of these

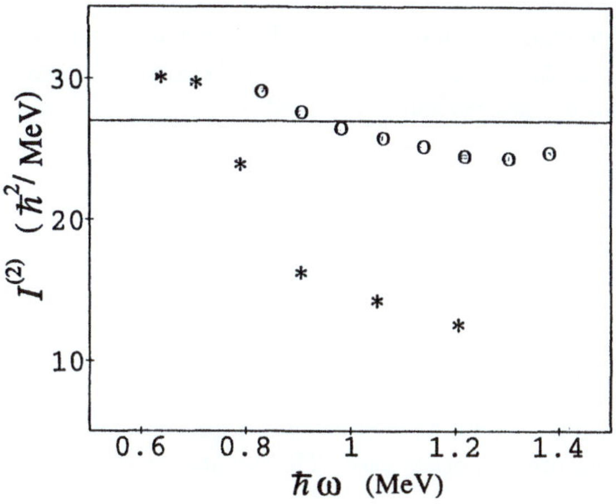

Figure 9-9: Comparison of the values of $\mathcal{I}^{(2)}$ deduced for a superdeformed band (circles) and those of a normal band (asterisks) in ^{84}Zr. The value for a rigid rotor with $\delta \sim 0.6$ is shown as a straight line for comparison. (Plotted using data from Ref. [89].)

two factors means that most of the excitations are in the form of rotational energy. The intrinsic state underlying the band must be fairly low in energy relative to the ground state configuration. The interesting question here is how can such very different structures coexist in the low-lying regions of nuclei. In other words, what are the reasons for quite different intrinsic shapes to have very similar energies. Earlier, we saw that, primarily because of pairing, the lowest configurations in nuclei tend to be spherical in shape or nearly so. Here, in superdeformed bands, we find that highly elongated shapes can also be quite low in energy.

In principle, one can perform a microscopic calculation similarly to what was described in §7-5 and see if there is a group of states at relatively low excitation energies with large quadrupole moments and strong $E2$-transition rates among themselves. It is obvious that many active particles are needed to build up the high spin values of interest here. Furthermore, to reproduce the large deformation, an enormously large number of (spherical) single-particle basis states is required. Such a calculation is out of the question in practice, and a significant part of the theoretical effort in superdeformation studies is devoted to finding suitable alternatives in understanding the new phenomena.

It is actually fairly simple to demonstrate why we can have highly deformed configurations that are relatively stable. We have seen that one of the important reasons for the existence of closed shell nuclei ^{4}He, ^{16}O, ^{90}Zr, and ^{208}Pb is the large energy gaps in the spherical single-particle spectrum shown in Fig. 7-3. In fact, nuclei with either neutrons or protons filling up all the single-particle orbits up to one of these large energy gaps are also relatively stable. Based on this observation, we find that one of the necessary conditions for stable deformed nuclei is large energy gaps in deformed single-particle spectra.

For this purpose, we shall consider only axially symmetric nuclei and, to further simplify the calculation, make use of the harmonic oscillator single-particle Hamiltonian in Cartesian coordinates we saw earlier in Eq. (7-31),

$$h = -\frac{\hbar^2}{2\mu}\nabla^2 + \tfrac{1}{2}\mu\{\omega_3^2 x_3^2 + \omega_\perp^2(x_1^2 + x_2^2)\}$$

The eigenvalues are given by Eq. (7-33). For our purpose here, it is more instructive to express the result in terms of the deformation parameter $\delta_{\rm osc}$ of Eq. (7-32)

$$\epsilon_{Nn_3\Omega} = \hbar\omega_0\left\{N + \tfrac{3}{2} - \tfrac{1}{3}\delta_{\rm osc}(2n_3 - n_\perp)\right\}$$

where $N = n_3 + n_\perp$, with n_3 the number of oscillator quanta along the symmetric axis and $n_\perp$ the number along the two other axes. The results are shown in Fig. 9-10. Since

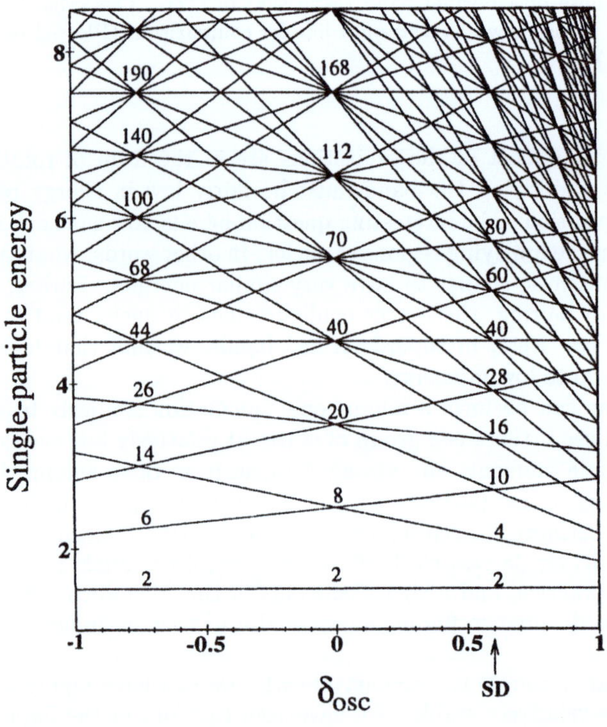

Figure 9-10: Single-particle energies of an axially symmetric harmonic oscillator Hamiltonian in units of $\hbar\omega_0$, showing the existence of shell structure at large deformations that are different from the spherical case. The starting point of superdeformation, with frequency ratio $\omega_\perp : \omega_3 = 2 : 1$, is marked as "SD."

we have not included terms depending on $\boldsymbol{\ell}^2$ and $\boldsymbol{\ell} \cdot \boldsymbol{s}$, the orbits are degenerate with respect to Ω as well as orbital angular momentum. We see that, at large deformations, beyond those shown in the Nilsson diagram of Fig. 7-5, there are new shell structures in the form of large energy gaps in the single-particle spectrum. This happens for both prolate and oblate deformation. The degeneracies, corresponding to magic numbers in the case of the spherical shell model, are marked for the special cases of $\delta_{\text{osc}} = +0.6$, corresponding to $\omega_\perp : \omega_3 = 2 : 1$, and $\delta_{\text{osc}} = -0.75$, for $\omega_\perp : \omega_3 = 1 : 2$. The fact that shell structure in this simple case happens to match with those for superdeformation is purely coincidental. In any actual applications, a far more realistic single-particle Hamiltonian must be used and several correction factors applied before one can make comparisons with observations.

Nilsson-Strutinsky approach. Since a complete microscopic understanding of superdeformation is impossible, some practical and reliable alternatives must be found to analyze the physics behind the experimental observations. Most of the work is based on the method of Strutinsky [134] and Myers and Swiatecki [108] (referred to hereafter simply as the Strutinsky method, following general practice in the literature). We are interested here, among others, in the question of why large deformation can exist at low excitation energies in certain nuclei. For this purpose, we shall calculate the equilibrium shapes of nuclei for different deformations and see if there are minima in the energy surface far away from the spherical limit. Our main concern here shall primarily be with the basic approach, leaving the detailed methods to the literature (see, e.g., Refs. [1, 146, 56]).

From a computational point of view, the simplest way to solve a nuclear many-body problem is to take a macroscopic approach as, for example, we have done for binding energies using semi-empirical mass formulas in §4-9. The major shortcoming in following such a line of investigation is the failure to account for local departures from smooth trends. In the case of binding energies, the Weizacker mass formula, for example, is able to provide a good description of the general trend but failed to account for the sharp increases near closed shells. For superdeformation studies, it is essential to make the "shell corrections," as we are involved with phenomena that depends on the detailed nuclear structure in deformed single-particle orbits.

Among the methods in use, the Strutinsky approach is perhaps best developed among several similar ones. In broad outline, the calculation is divided into two parts, a macroscopic part, involving the bulk parameters of the system, to account for the smooth variations, and a microscopic part, involving individual nucleon degrees of freedom, to make corrections for local variations, such as those coming from shell closure. In terms of energy, such a division may be expressed as

$$E = E_{\text{macro}} + \delta E_{\text{micro}} \tag{9-8}$$

The macroscopic part, E_{macro}, may be obtained from, for example, a liquid drop model. Let us use E_{micro} to represent the contributions from the microscopic part. If we average E_{micro} over local variations, the result is a part of the smooth trend and must have already been accounted for by E_{macro}. For this reason, the second term on the right-hand side of Eq. (9-8) is

$$\delta E_{\text{micro}} = E_{\text{micro}} - \langle E_{\text{micro}} \rangle$$

By taking out the smoothed value $\langle E_{\rm micro}\rangle$ in the microscopic result, we avoid any double counting.

Macroscopic calculation. In §4-9, we saw that a crude liquid drop model in the form of a Weizacker semi-empirical calculation can achieve an overall accuracy of a few percent using a set of parameters obtained from fitting known binding energies. For the special case of the variation in energy as a function of deformation we are interested in here, the model can be further simplified in the following ways. Since nuclear matter is fairly stiff, we expect the volume term in Eq. (4-56) to be unchanged by deformation and it may be ignored. Its contributions may be incorporated into the definition of the zero point of our energy scale. Similarly, symmetry and pairing contributions may be left out of the calculations if we are not interested in the variations with neutron and proton numbers. Analogous to what we did in §4-11 for fission, the major contributions to deformation come from surface and Coulomb terms. On the one hand, deformation increases the surface area compared with a sphere of the same volume. On the other hand, the Coulomb energy is decreased when the protons are pushed further away from each other as a result of the shape change. The macroscopic energy for a deformed nucleus with N neutrons and Z protons is then

$$E_{\rm macro}(N,Z,{\rm def.}) = E_{\rm surface}(N,Z,{\rm def.}) + E_{\rm Coulomb}(N,Z,{\rm def.})$$

The shape of the nucleus can still be parametrized using Eq. (6-1),

$$R(\theta,\phi) = R_0\Big\{1+\sum_{\lambda\mu}\alpha_{\lambda\mu}Y_{\lambda\mu}(\theta,\phi)\Big\}$$

where, for simplicity, we have taken out dependences on time t. For all practical purposes, only the lowest few multipoles, $\lambda = 2$, 4, are necessary and this is used in most of the known calculations.

We have seen on many earlier occasions that the nuclear density does not drop off sharply at the nuclear surface, as implicitly implied in Eq. (6-1). To account for a diffused surface region, Dudek [54] uses the following formula for the surface energy term:

$$E_{\rm surface}(N,Z,{\rm def.}) = -\frac{c_s A^{2/3}}{8\pi^2 R_0^2 a^3}\int d^3r\int d^3r'\left(\frac{|\boldsymbol{r}-\boldsymbol{r}'|}{a}-2\right)\frac{e^{-|\boldsymbol{r}-\boldsymbol{r}'|/a}}{|\boldsymbol{r}-\boldsymbol{r}'|}$$

where a characterizes the diffuseness range. When $a \to 0$, we recover the sharp-drop-off, constant-density form. The quantity

$$c_s = a_s\left\{1-\kappa_s\left(\frac{N-Z}{N+Z}\right)^2\right\}$$

is the effective surface energy parameter with a_s and κ_s as adjustable parameters. As before, the volume remains to be

$$V = \frac{4\pi}{3}R_0^3$$

with average radius $R_0 = r_0 A^{1/3}$. Similarly, the Coulomb energy may be written as

$$E_{\text{Coulomb}}(N, Z, \text{def.}) = \frac{1}{2}\left(\frac{3Ze}{4\pi R_0^3}\right)^2 \left\{\int d^3r \int d^3r' \frac{1}{|\boldsymbol{r} - \boldsymbol{r}'|} - \int d^3r \int d^3r' \left(1 + \frac{|\boldsymbol{r} - \boldsymbol{r}'|}{2a_{\text{den}}}\right) \frac{e^{-|\boldsymbol{r}-\boldsymbol{r}'|/a_{\text{den}}}}{|\boldsymbol{r} - \boldsymbol{r}'|}\right\}$$

where a_{den} has the value $0.99/\sqrt{2}$.

Microscopic contributions. The contributions from the microscopic part is taken to be a sum of proton and neutron single-particle energies. We have seen earlier in §6-1 that, for example, the energy of a state can be reasonably represented by single-particle contributions alone by considering each nucleon to be moving in the average field generated by the interaction with all the other particles. If the potential $\overline{V}$ in the single-particle Hamiltonian

$$\left\{-\frac{\hbar^2}{2m}\nabla^2 + \overline{V}\right\}\psi_i = \epsilon_i \psi_i$$

is well chosen, any residual interaction remaining between particles is expected to be small and may be ignored. Such a mean-field approach has been shown to be quite adequate for a variety of nuclear phenomena and the calculations involved are far simpler than a fully microscopic one.

In the case of deformed nuclei, the potential must also reflect the fact that the mean field is deformed. One way to do this is to use a Woods-Saxon potential together with a Coulomb term for protons,

$$\overline{V} = V_{\text{WS}} + V_{\text{WS}}^{\text{s.o}} + \frac{1+\tau_3}{2} V_{\text{Coul}}$$

Here, τ_3 is the projection of the isospin operator on the symmetry axis. To account for deformation, the Woods-Saxon potential differs slightly in form from that of Eq. (8-40) for optical model potentials. The radial dependences used for the volume and spin-orbit terms are, respectively,

$$\begin{aligned} V_{\text{WS}}(\boldsymbol{r}, \text{def.}) &= V_0 \frac{1 + \kappa\{(N-Z)/(N+Z)\}}{1 + \exp\{d_\Sigma(\boldsymbol{r}, \text{def.})/a\}} \\ V_{\text{WS}}^{\text{s.o}}(\boldsymbol{r}, \text{def.}) &= -\lambda \left(\frac{\hbar}{2mc}\right)^2 (\nabla V_{\text{WS}}(\boldsymbol{r}, \text{def.}) \times \boldsymbol{p}) \cdot \boldsymbol{s}\frac{2}{\hbar^2} \end{aligned}$$

where $d_\Sigma(\boldsymbol{r}, \text{def.})$ represents the perpendicular distance of a point located at $\boldsymbol{r}$ to the nuclear surface given by Eq. (6-1), $\boldsymbol{p} = i\hbar\nabla$ is the linear momentum operator, and $\boldsymbol{s}$ is the spin operator. The other four quantities V_0, κ, a, and λ are adjustable parameters. For $58 \leq Z \leq 74$, Werner and Dudek [145] adopt the values

$$V_0 = -49.6 \text{ MeV} \qquad a = 0.7 \text{ fm}$$

$$\kappa = \begin{cases} +0.86 & \text{for protons} \\ -0.86 & \text{for neutrons} \end{cases} \qquad \lambda = \begin{cases} 36 & \text{for protons} \\ 35 & \text{for neutrons} \end{cases}$$

The form of the Coulomb term is taken as

$$V_{\text{Coul}}(\boldsymbol{r}, \text{def.}) = Ze \int_{\Sigma} \frac{d^3r'}{|\boldsymbol{r} - \boldsymbol{r}'|}$$

where the subscript Σ reminds us that the integral is over the (deformed) nuclear volume (for protons).

Shell corrections. The single-particle energy ϵ_i found by solving the eigenvalue equation

$$\left\{-\frac{\hbar^2}{2m}\nabla^2 + V_{\text{WS}} + V_{\text{WS}}^{\text{s.o.}} + \frac{1+\tau_3}{2} V_{\text{Coul}}\right\}\psi_i = \epsilon_i \psi_i \tag{9-9}$$

contains some of the contributions already included in the macroscopic term E_{macro}. We must find a way to remove the parts already accounted for before we can calculate the value of δE_{micro} needed in Eq. (9-8).

Since the contributions are single particle in nature, it is convenient to separate them into proton and neutron components. Traditionally, they are indicated, respectively, by using symbols π and ν,

$$\delta E_{\text{micro}} = \delta E_{\text{micro}}(\pi) + \delta E_{\text{micro}}(\nu)$$

According to the prescription of Strutinsky,

$$\begin{aligned}
\delta E_{\text{micro}}(\pi) &= \sum_{i=1}^{Z} \epsilon_i(\pi) - \Big\langle \sum_{i=1}^{Z} \epsilon_i(\pi) \Big\rangle_{\text{shell}} \\
\delta E_{\text{micro}}(\nu) &= \sum_{i=1}^{N} \epsilon_i(\nu) - \Big\langle \sum_{i=1}^{N} \epsilon_i(\nu) \Big\rangle_{\text{shell}}
\end{aligned}$$

The parts already included in E_{macro} are obtained by integrating over the locally averaged, or smoothed, level density $\tilde{g}(e)$ for the single-particle spectrum

$$\Big\langle \sum_{i=1} \epsilon_i \Big\rangle_{\text{shell}} = \int_{-\infty}^{\tilde{\epsilon}_F} e\tilde{g}(e)\, de$$

One way to smooth the single-particle spectrum $\{\epsilon_i\}$, obtained from a realistic mean-field calculation, is to expand the level density in terms of Hermite polynomials $H_m(x)$,

$$\tilde{g}(e) = \frac{1}{\gamma\sqrt{\pi}} \sum_{i=1}^{\infty} e^{-u_i^2} \sum_{m=0}^{p} C_m H_m(u_i)$$

where m is the order of the polynomial, and

$$\begin{aligned}
u_i &= \frac{e - e_i}{\gamma} \\
C_m &= \begin{cases} (-1)^{m/2} \dfrac{1}{2^m (m/2)!} & \text{for } m \text{ even} \\ 0 & \text{for } m \text{ odd} \end{cases}
\end{aligned}$$

The smoothed Fermi energy is given by

$$N = \int_{-\infty}^{\bar{\lambda}(\nu)} \tilde{g}(e,\nu)\,de \qquad\qquad Z = \int_{-\infty}^{\bar{\lambda}(\pi)} \tilde{g}(e,\pi)\,de$$

The values of parameters p and γ are adjusted until the final results are essentially independent of small changes in them. In Ref. [145], they are taken to be $p = 6$ and $\gamma = 1.2\hbar\omega_0$, with $\hbar\omega_0$ given by Eq. (7-18).

Rotation and Routhian. So far in all the discussions, we have stayed in the body-fixed frame of reference and ignored any effect rotation has on the nucleus. At the high spin values we are interested in here, the nucleus is spinning at angular velocities at which the Coriolis force plays an even more important role than what we have seen in §6-3 for $K = \frac{1}{2}$ bands.

Formally, we can take care of rotation by applying a transformation to change both the Hamiltonian and the wave function from the body-fixed intrinsic frame of reference to one that is stationary in the laboratory. Let us write the time-dependent Schrödinger equation in the laboratory system as

$$i\hbar\frac{d\psi}{dt} = h\psi$$

A rotation by angle ωt around axis $\boldsymbol{\omega}$ can be accomplished by the operator

$$T(\omega t) = e^{\imath\boldsymbol{\omega}\cdot\boldsymbol{j}t}$$

In the new system, the time-dependent Schrödinger equation takes on the form

$$i\hbar\frac{d\psi'}{dt} = (h - \hbar\boldsymbol{\omega}\cdot\boldsymbol{j})\psi'$$

For a transformation from a laboratory to a body-fixed intrinsic frame of reference that is rotating around the x'-axis of the intrinsic frame, the result is

$$i\hbar\frac{d\psi'}{dt} = (h - \hbar\omega j_{x'})\psi'$$

In other words, to account for rotation, the Hamiltonian in the intrinsic frame of reference becomes

$$h^\omega = h - \hbar\omega j_{x'}$$

It can be shown that h^ω has the properties of a Hamiltonian and satisfies the following eigenvalue equation:

$$(h - \hbar\omega j_{x'})\psi_i^\omega = \epsilon_i^\omega\psi_i^\omega \tag{9-10}$$

The procedure in arriving at h^ω is very similar to that used by Routh in classical mechanics to handle cyclic variables, such as angular variable [76]. For this reason, ϵ_i^ω is often referred to as the *Routhian*.

The eigenvalue equation in the body-fixed frame of reference, Eq. (9-9), is relatively simple to solve. In the laboratory frame of reference, the nucleus is rotating at some frequency ω and the rotation can, in principle, be quite complicated to deal with.

Instead, we can solve Eq. (9-10) using a suitable value for ω. In this case, nucleons appear as independent particles moving in an average field that is rotating with the frame of reference. The approach belongs to the general one of a cranking model and, for this reason, h^ω is also called a *cranked Hamiltonian.*

The contribution of rotation to the macroscopic term $E_{\text{macro}}(N, Z, \text{def.})$ is quite simple, as the calculation is purely classical. This gives us the result

$$E^\omega_{\text{macro}}(N, Z, \text{def.}) = E^{\omega=0}_{\text{macro}}(N, Z, \text{def.}) + \tfrac{1}{2}\mathcal{I}(N, Z, \text{def.})\,\omega^2$$

where $E^{\omega=0}_{\text{macro}}$ is the macroscopic energy we had earlier in the absence of any rotation. Equation (9-8) now reads as

$$E = E^\omega_{\text{macro}}(N, Z, \text{def.}) + \delta E^\omega_{\text{micro}}(N, Z, \text{def.}) \tag{9-11}$$

with $\delta E^\omega_{\text{micro}}(N, Z, \text{def.})$ calculated using the single-particle Routhian ϵ^ω_i, obtained by solving Eq. (9-10). In fact, instead of energy, it is more convenient in many cases to convert all the energies into the corresponding Routhians and consider the total Routhian instead of E in Eq. (9-11).

If we calculate E, or the equivalent Routhian, for different sets of values of the deformation parameters, we obtain the variation of energy for the rotating nucleus as a function of the parameters. The minimum of such a *potential energy surface* gives us the deformation the nucleus prefers to settle in at a given rotation frequency. In this way, superdeformation may be understood by a relatively straightforward calculation.

9-3 Phase Transition and Quark-Gluon Plasma

At ultra-relativistic energies, our interest in heavy-ion collisions turns to the properties of matter under extremely high energy densities. The only occasion that such conditions have existed in nature is during the short time interval of a fraction of a second after the big bang that gave birth to the universe. For this reason, the topic is of interest to cosmology as well. The same condition is also important to quantum chromodynamics, because of the possibility of freeing quarks from their confined state inside hadrons and transform them into a new phase in which many quarks and gluons are present in a plasma-like state, the *quark-gluon plasma.*

To create such a state of matter, it is necessary to involve energies much higher than anything we have encountered so far in the laboratory. For the Relativistic Heavy-Ion Collider (RHIC) at Brookhaven National Laboratory (BNL), collisions of gold nucleus on gold nucleus can reach an energy of almost 40 TeV in the center of mass. In the case of the Large Hadron Collider (LHC) at CERN, energies of 3 TeV per nucleon are possible. This gives a center-of-mass energy in excess of 10^3 TeV for collisions of lead on lead.

Most of the other high-energy accelerators are designed for electrons and protons. The few that can accelerate more complicated particles are limited to lighter ions and, often, lower energy than what we are interested in here. Even when the energy is sufficient, light-ion reactions are not expected to create a sufficiently large region of quark-gluon plasma that can be observed. For these reasons, most of the experimental data available so far can only give us some rough guidance on what we can expect from RHIC and LHC. We shall briefly review what we have learned from these observations and extrapolate from them what are the initial set of experiments to be carried out at RHIC and LHC once they are in operation. We shall also describe some of the results from lattice gauge calculations to see what are the predications we can make based on our theoretical understanding of the physics of strong interaction.

Transformation of nucleons to quark-gluon plasma. A complete understanding of the strong interaction should include also a knowledge of the conditions under which nucleons are transformed into a state in which quarks are no longer confined. Currently, our knowledge of QCD is not yet quite adequate to carry out this task. To get some idea of the possible conditions for the transition to take place, we can make use of the bag model. There are many versions of the model, and together they have been quite successful in explaining a large number of phenomena in particle and nuclear physics. For our purpose here, we are not concerned with the details that distinguish one bag model from another. It is adequate for us to regard the "bag" as a phenomenological entity that provides an inward pressure of magnitude B to keep, for example, the three quarks confined inside a nucleon.

If the pressure of "quark matter" inside the bag is increased to such an extent that it exceeds B, confinement will no longer be possible and we have a new phase of matter, made of quarks and gluons. This can happen if, somehow, the temperature becomes very high, or the baryon number density becomes very large, or a combination of both. To get some qualitative estimates, let us consider each one of these two possibilities separately using the bag pressure B as the parameter.

If we treat quarks and gluons as massless, noninteracting gas molecules, there are two extreme situations that allow simple estimates to be made on the conditions for a phase transition. The first is the case of zero temperature, and our interest here is to find the critical baryon density required for a transition to take place at this temperature. The pressure in this case comes solely from the effect of the Pauli exclusion principle between the quarks. For a relativistic Fermi gas in volume V, the density of states, the number of states in an interval of momentum dp, is given by

$$dN = g\frac{V}{(2\pi)^3\hbar^3}4\pi p^2\, dp$$

where g is the degeneracy. For simplicity, let us ignore the contributions from anti-quarks and gluons. To simplify the discussion further, we shall restrict the number of quark flavors to the lightest two, on the ground that d- and u-quarks dominate at zero temperature. In this limit, the degeneracy for quarks is

$$g_q = N_{\text{color}} \times N_{\text{spin}} \times N_{\text{flavor}} = 3 \times 2 \times 2 = 12$$

The number of quarks filling up all the states up to the quark Fermi energy ϵ_F is given by the integral

$$\begin{aligned} N_q &= g_q\frac{V}{(2\pi)^3\hbar^3}4\pi\int_0^{\epsilon_F/c} p^2 dp \\ &= g_q\frac{V}{6\pi^2(\hbar c)^3}\epsilon_F^3 \end{aligned}$$

For massless quarks, the energy carried by each one is pc. From this, we obtain the total energy

$$\begin{aligned} E_q &= g_q \frac{V}{(2\pi)^3\hbar^3} 4\pi \int_0^{\epsilon_F/c} p^3 c\, dp \\ &= g_q \frac{V}{8\pi^2(\hbar c)^3} \epsilon_F^4 \end{aligned}$$

The pressure in a relativistic noninteracting gas is given by

$$P = \frac{1}{3}\frac{E}{V}$$

If we equate the pressure from degenerate quarks P_q with the bag pressure B,

$$P_q = \frac{1}{3}\frac{E_q}{V} = g_q \frac{1}{24\pi^2(\hbar c)^3} \epsilon_F^4 = B$$

as the condition for phase transition, we obtain the critical Fermi energy

$$\epsilon_{F,c} = \left(\frac{24\pi^2(\hbar c)^3}{g_q} B \right)^{1/4}$$

The critical baryon density estimated in this way is

$$n_{B,c} = \frac{1}{3} n_{q,c} = \frac{1}{3}\frac{N_{q,c}}{V} = \frac{1}{3}\frac{g_q}{6\pi^2(\hbar c)^3}\epsilon_{F,c}^3 = \frac{4}{3}\left(\frac{g_q}{24\pi^2}\right)^{1/4} \left(\frac{B}{\hbar c}\right)^{3/4}$$

where we have made use of the fact that the baryon number of quarks is $\frac{1}{3}$. By adopting a reasonable value for the bag pressure, $(\hbar^3 c^3 B)^{1/4} = 200$ MeV, we obtain a result of just over 400 MeV for the critical Fermi energy of quarks, $\epsilon_{F,c}$. In terms of baryon density, $n_{B,c} \approx 0.7$ fm^{-3}, about five times higher than the value of 0.14 fm^{-3} for nucleons under normal conditions. Other calculations put the zero-temperature critical baryon density for the quark-plasma phase to be up to 10 times the normal nuclear matter density.

The second extreme condition is high temperature. For simplicity, we shall consider here the situation where quark and antiquark are equal in number, and as a result, the baryon density vanishes. Our noninteracting quarks and antiquarks can be treated as two separate Fermi gases, and the gluons as a Bose gas. The number density, or occupancy probability, for each quantum state is given by the expression

$$n_k = \frac{1}{e^{(\epsilon_k - \mu)/kT} \pm 1}$$

where the plus sign in the denominator is for fermions and the minus sign is for bosons. The chemical potentials μ vanishes for the gluons, as well as for quarks, since we are taking the baryon density to be zero. The total number of particles (quarks, antiquarks, or gluons) and the total energy for each type in volume V are obtained by summing over the contributions from each state k. On replacing the summations by integrations, we obtain the results

$$N = g\frac{V}{(2\pi\hbar)^3}\int_0^\infty \frac{4\pi p^2\,dp}{e^{pc/kT}\pm 1} = g\frac{V}{2\pi^2}\left(\frac{kT}{\hbar c}\right)^3\int_0^\infty \frac{z^2\,dz}{e^z\pm 1}$$

$$E = g\frac{V}{(2\pi\hbar)^3}\int_0^\infty \frac{4\pi p^3 c\,dp}{e^{pc/kT}\pm 1} = g\frac{V}{2\pi^2}\left(\frac{kT}{\hbar c}\right)^3 kT\int_0^\infty \frac{z^3\,dz}{e^z\pm 1}$$

where we have made use of the assumption that all the particles are massless and, hence, $\epsilon = pc$. The integrals over z can be found in, for example, Landau and Lifshitz [94]. For quarks, we have the values

$$N_q = g_q\frac{V}{2\pi^2}\left(\frac{kT}{\hbar c}\right)^3 \times \frac{3}{2}1.202$$

$$E_q = g_q\frac{V}{2\pi^2}\left(\frac{kT}{\hbar c}\right)^3 kT \times \frac{7}{120}\pi^4$$

where the factor 1.202 comes from the Riemann zeta function $\zeta(3) = 1.202\ldots$, obtained when we integrate over z. The same results apply to antiquarks as well. For gluons, we have

$$N_g = g_g\frac{V}{2\pi^2}\left(\frac{kT}{\hbar c}\right)^3 \times 2 \times 1.202$$

$$E_g = g_g\frac{V}{2\pi^2}\left(\frac{kT}{\hbar c}\right)^3 kT \times \frac{\pi^4}{15}$$

with $g_g = 16$ for gluon degeneracy. The total energy is a sum of the contributions from quarks, antiquarks, and gluons:

$$E = E_q + E_{\bar{q}} + E_g$$

If we again equate the pressure

$$P = \frac{1}{3}\frac{E}{V}$$

with the bag pressure $(\hbar^3c^3B)^{1/4} = 200$ MeV, we obtain a temperature $kT \sim 140$ MeV. Other calculations put the value to be as high as 250 MeV.

The region for both $T \neq 0$ and $n_B \neq 0$ is more complicated to estimate than either one of the two extremes we studied above. A reasonable conjecture is that the changes from one to the other should be smooth. Schematically, the phase diagram is represented by that shown in Fig. 9-11.

Lattice gauge calculations. In principle, quantum chromodynamics can provide us with some better guidance on the question of transition to QGP than the simple arguments presented above. The major difficulty here is a practical one in carrying out the calculations. First, the equations generated from QCD considerations are nonlinear, and as a result, cannot be put into a form that can be quantized easily. Second, the

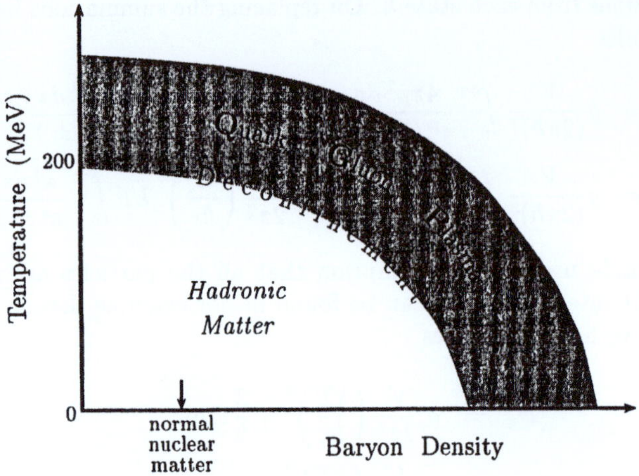

Figure 9-11: Schematic diagram showing the different phases of matter.

interaction involved is strong and we cannot make use of perturbative techniques to find approximate solutions that are reliable. One possible alternative is to take a Feynman path integral approach [63] and formulate the problem numerically on a lattice. Since QCD is a *gauge* theory, a theory that is invariant under a local gauge transformation, we have a lattice gauge calculation. Numerical solution of (continuous) problems by formulating them on discrete lattices is a powerful technique and has been applied successfully to a large variety of problems in, for example, condensed matter physics, fluid mechanics, and engineering. Here, we want to adopt a similar approach, except that the computations are complicated by the fact that gauged fields are involved. In fact, even the power of most modern computers remains to be the main limitation on the types of results one can obtain so far.

In quantum mechanics, the usual way to find the evolution of a state from an initial point in space and time $(\boldsymbol{r}_i, t_i)$ to some final point $(\boldsymbol{r}_f, t_f)$ is to solve the time-dependent Schrödinger equation. The solution may be represented formally as

$$|\boldsymbol{r}_f, t_f> = e^{-iH(t_f-t_i)/\hbar}|\boldsymbol{r}_i, t_i> \tag{9-12}$$

where H is the Hamiltonian operator. In the Feynman path integral approach, we take a different route by evaluating, instead, the classical action

$$S = \int_i^f L\,d\tau$$

Here L is the Lagrangian of the system and the integral is over space and time, as well as over any other independent variables in the system. The value of S usually depends on the path over which the integral is taken from point i to point f. It can be shown that the matrix element of the operator on the right-hand side of Eq. (9-12) is given by the relation

$$< \boldsymbol{r}_f | e^{-iH(t_f - t_i)/\hbar} | \boldsymbol{r}_i > = \frac{1}{Z} \sum_{\text{all paths}} e^{\imath S}$$

where the factor $1/Z$ represents all the necessary constants and normalization factors. The physical meaning of the equation is that time evolution is equivalent to averaging over all possible paths, each one weighted by the action. This is very similar to ensemble averaging in statistical mechanics, and for this reason, the whole approach is sometimes referred to as *statistical field theory.*

In general, the integral for S is not easy to evaluate, especially for the nonlinear QCD Lagrangian we are interested in here. However, if we take a similar approach as used in numerical integration, great simplifications may be achieved. In this scheme, each path is divided into a large number of small "segments." Within each segment, it is possible to replace the integrand by some reasonable average value of L within the small interval, such as that evaluated at two ends of the interval. In this way, the whole space is "transformed" into a lattice of points, and the contribution of each segment to the integral is the product of the average value of the integrand and the size of the segment. The complete integral becomes a sum of the products in all the segments.

Next, we must evaluate the integral once over each one of the infinite number of possible paths. In the case of ensemble averaging in statistical mechanics, the infinite sum is usually carried out analytically by changing it into a derivative of, for example, the partition function. This is not possible here, as the integrand is a complicated nonlinear function. The alternative is to make use of the idea of statistical sampling. In other words, we shall choose the paths randomly and evaluate the action only over the chosen ones. If a sufficiently large sample is taken, our average value should be an accurate representation of the true one. Our task is now turned into a Monte Carlo one on a lattice, and powerful techniques of Monte Carlo calculations may be applied.

Every possible simplification is important here, as the problem is greatly complicated by the fact that we are involved with gauge fields and that quarks are fermions. To maintain gauge invariance, the values of the fields at one lattice point become dependent on those at other points. Even with the most powerful computers in the world at the moment, several further approximations must be made before we can obtain any results. For example, in QCD we have both quarks and gluons, each one described by a field that is a continuous function in space and time. However, in a lattice calculation, only the values of the functions at the lattice points are evaluated. Let us use a to represent the lattice spacing, the distance between two adjacent points on the lattice, and for simplicity, we shall assume that a is the same in all spatial directions and along the time axis by some suitable choice of units. The accuracy of any calculated results clearly depends on the size of a and this, in turn, is given by the number of lattice points taken in the (3+1)-dimension space. Many of the results obtained so far have been limited to fairly modest lattice sizes, such as 12 points in each direction. Even with such limited lattices, a full QCD calculation involving all eight gluons and six flavors of quarks is not possible. Many results are obtained with the gluon fields alone, without the quarks playing any "dynamical" roles.

In spite of the limitations imposed by the available computing power, lattice gauge calculations have provided us with valuable insight into what we can expect from the relation between hadronic matter and QGP. For example, calculations carried out without dynamical quarks show a first-order phase transition at critical temperature around $kT = 260$ MeV. The results with dynamic quarks are less clear, in part, because of the uncertainties due to finite lattice size. Depending on the fermion masses adopted in the calculation, the phase transition can be either first- or second-order. There is also the possibility of hadronic matter crossing over smoothly to QGP without undergoing a phase change.

The interest in lattice gauge calculations is not limited to transition from hadronic matter to quark-gluon plasma. Since it is possibly the only way to carry out the calculations "exactly," it is often quoted as the only fundamental formulation of QCD [93]. Many achievements have been recorded in understanding the low-energy regime, such as hadron spectroscopy, and in determining some of the parameters in the standard model of particle physics, such as the possible determination of the strong coupling constant and certain Cabibbo-Kobayashi-Maskawa (CKM) matrix elements. In many ways, it can also tell us what may lie ahead of the standard model. The strong demand of computational power for such calculations has stimulated novel developments in both hardware and software and has, in many ways, led advances in computing technology.

High-energy nucleon-nucleon scattering. In ultra-relativistic collisions, the reaction is mostly inelastic and there is adequate energy to create a large variety of secondary particles. In the absence of any new physics, the cross section for nucleus-nucleus scattering is expected to be multiples of that for nucleon-nucleon scattering at equivalent energies. To recognize that a region of quark-gluon plasma is created in the process, the signal must appear over and above the complex background coming from ordinary nucleon-nucleon scattering. For this reason, we need to review briefly what is known of nucleon-nucleon scattering at high energies. The subject is a vast one, touching upon many parts of nuclear as well as particle physics. Our interest is limited to the effect it has on the cross section for nucleus-nucleus scattering at ultra-relativistic energies.

In high-energy proton-proton scattering, from which we obtain most of the information on nucleon-nucleon scattering, it is known that the total reaction cross section rises slowly with momentum p. In terms of the Mandelstam variable[3] s, the empirical results for total and elastic cross sections in the range of center-of-mass energy 3 GeV $< \sqrt{s} <$ 100 GeV may be expressed as [83]

$$\begin{aligned}\sigma_{\text{total}} &= (48.0 \pm 0.1) + (0.522 \pm 0.005)(\ln p)^2 + (-4.51 \pm 0.05)\ln p \\ \sigma_{\text{elastic}} &= (11.9 \pm 0.8) + (26.9 \pm 1.7)\, p^{-1.21} + (0.169 \pm 0.021)(\ln p)^2 \\ &\quad +(-1.85 \pm 0.26)\ln p\end{aligned}$$

with p in units of GeV/c. The numerical value is around 40 mb (4.0 fm^2) for σ_{total} in the energy range. The observed result for inelastic cross section at 100 GeV/c is 31.3 ± 1.2 mb. The measured values are shown in Fig. 9-12.

[3]Mandelstam variable $s = (\mathcal{E}/c)^2$, where $\mathcal{E}$ is the total energy of the two colliding particles in the center of mass. In natural units, $c = 1$ and, as a result, $\sqrt{s}$ is often used to represent $\mathcal{E}$ and we shall follow this convention.

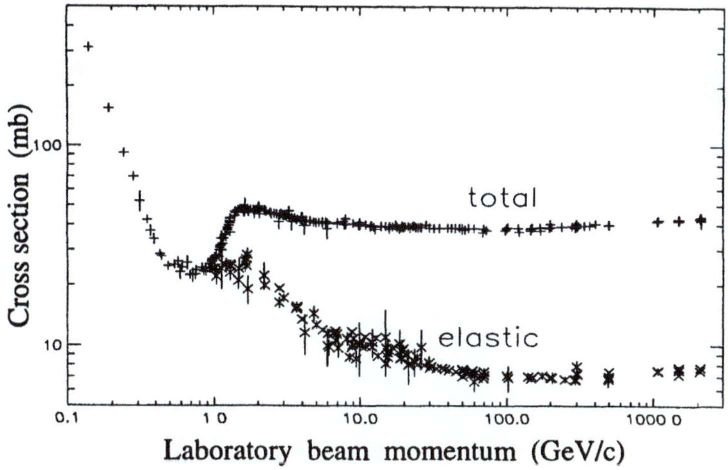

Figure 9-12: Cross sections in high-energy proton-proton scattering in mb (= 10^{-1} fm^2) plotted using data from Ref. [22]. Inelastic scattering, the difference between total and elastic, dominates the reaction and the secondary particles produced are made mainly of mesons, baryons, and antibaryons.

Most of the inelastic cross section goes into production of mesons and baryon-antibaryon pairs. The number of secondary particles produced is given the name *multiplicity*. At high energies, the multiplicity also increases roughly logarithmically with s. Since only charged particles can be measured most readily in an experiment, the best known value is the charge multiplicity $N_{\rm ch}$, the average number of secondary charged particles produced in a pp-collision. The variation of $N_{\rm ch}$ with energy may be represented as [135]

$$N_{\rm ch} = (0.88 \pm 0.10) + (0.44 \pm 0.05)\ln s + (0.118 \pm 0.006)(\ln s)^2 \qquad (9\text{-}13)$$

Around $\sqrt{s} \sim 20$ GeV ($p_{\rm lab} \sim 100$ GeV/c), the value of $N_{\rm ch}$ is around 6. The measured results are shown in Fig. 9-13. Most of the secondary particles are (charged) pions. If we assume equal numbers of π^+, π^0, and π^- are produced on the average, we obtain the total multiplicity to be around 9.

Another quantity of interest in high-energy collisions is the momentum distribution of the secondary particles. This is usually discussed in terms of the *rapidity* variable

$$y = \frac{1}{2}\ln\left(\frac{E + p_L c}{E - p_L c}\right) \qquad (9\text{-}14)$$

where E is the (relativistic) energy and p_L the longitudinal momentum. In the center of mass, $y = 0$ when $p_L = 0$. The range of y is therefore bound by

$$y_{\rm max} = -y_{\rm min} = \frac{1}{2}\ln\left(\frac{s}{p_T^2 + (mc)^2}\right)$$

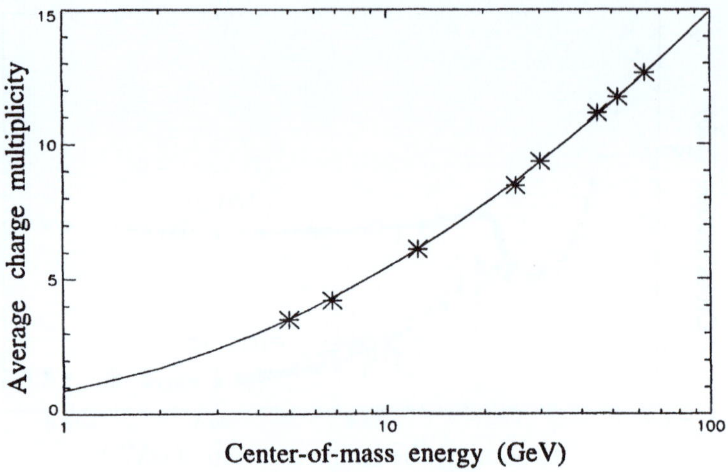

Figure 9-13: Average charge multiplicity as a function of energy. Asterisks represent data from Refs. [135, 107, 25] and the smooth curve is calculated using Eq. (9-13).

where $p_T = \sqrt{p_x^2 + p_y^2}$ is the transverse momentum and $(E/c)^2 = p_T^2 + p_L^2 + (mc)^2$. At high energies, $d\sigma/dy$ is essentially constant except at both extreme values of y.

Since y requires measurements of two quantities, E and p_L, it is often more convenient to make use of the *pseudorapidity*

$$\eta = -\ln\left(\tan\frac{\theta}{2}\right) = \frac{1}{2}\ln\left(\frac{|\boldsymbol{p}| + p_L}{|\boldsymbol{p}| - p_L}\right)$$

that depends only on the scattering angle θ. If the particle energy is high, y and η have approximately the same value.

Nuclear stopping power and transparency. For the past 10 years or so, experiments have been carried out with heavy ions at relativistic energies. The two principal facilities are the Alternating Gradient Synchrotron (AGS) at BNL and Super Proton Synchrotron (SPS) at CERN. Beams from proton to gold are available at AGS up to $29(Z/A)$ GeV/c in energy per nucleon and from proton to lead at SPS up to $400(Z/A)$ GeV/c. The corresponding center-of-mass energy for gold on gold at AGS is $\sqrt{s} \leq 5$ GeV, and for lead on lead at SPS it is $\sqrt{s} \leq 17$ GeV.

We can separate high-energy nucleus-nucleus collisions into two energy regions. The first is the "pure QGP" region where all the nucleons are converted into quarks and gluons. It is estimated that the energy required is $\sqrt{s} \geq 100$ GeV/nucleon. At the AGS and SPS energies, we are still in the second region, the "baryon-rich QGP" region. Here, a substantial fraction of the initial energy in the collision is converted into producing hadronic matter. The process may be discussed in terms of the "stopping power." The opposite way of looking at the same question is "nuclear transparency," the lack of interference when one nucleus passes through another during a collision. One way to have some idea of the stopping power is to consider E_{zd}, the energy in a small cone around the beam axis. The angle is chosen such that particles within the cone are essentially those from the initial projectile without having undergone any energy loss. In the case of complete stopping, the reaction cross section should peak at $E_{zd} = 0$. Observations at SPS energies show that the maxima occur at small values of E_{zd}, consistent with stopping power of around 90% [131].

A large nuclear stopping power implies a baryon-rich environment, as there is not enough energy for nucleons left over from the collision to move away from the central region. This is the situation with experiments at AGS and SPS energies. There are indications for increasing nuclear transparency with higher energies and larger mass numbers of the ions. Thus at RHIC and LHC energies, we expect the remnant nucleons to have adequate energy to move away fast enough for the central region to last for relatively long time scales ($\sim 10^{-24}$ s) as a baryon-free region of quark-gluon plasma.

Signatures of quark-gluon plasma. If a quark-gluon plasma is formed, how can we recognize it? The major difficulty here is that our probe cannot be placed anywhere inside the region. In part, this is because we can at best create a very small volume of this new form of matter with radius measured in femtometers. At the same time, we cannot expect any experimental apparatus to survive the extreme temperature and energy density inside the region where the QGP is present. All "signatures" of QGP must therefore be inferred from the consequences in terms of the particles—baryons, mesons, photons, etc.—we normally encounter in subatomic physics. Based on existing observations, experimental as well as theoretical, the following measurements seem to offer the best possibilities [150, 81, 131].

Strangeness production Since nucleons are made of u- and d-quarks, the strangeness of the initial state in a heavy-ion collision is zero. Heavier quarks, such as strange (s) and charm (c), can also be produced in a reaction if there is sufficient energy. (We shall ignore the even heavier b- and t-quarks, and their antiquarks, in this discussion.) To conserve flavor quantum number, these quarks must be created in the form of quark-antiquark pairs of the same flavor.

Consider first the production of strange quarks. The results are usually discussed in terms of λ, the ratio of $s\bar{s}$ to $u\bar{u}$ and $d\bar{d}$,

$$u\bar{u} : d\bar{d} : s\bar{s} = 1 : 1 : \lambda$$

In pp-collisions, the value of λ is found to be around 0.1 at center-of-mass energy $\sqrt{s} \sim 5$ GeV and rising slowly to about 0.2 at $\sqrt{s} = 50$ GeV. The value obtained from proton-nucleus collision is, within experimental uncertainties, consistent with this trend of values. For silicon on silicon and sulfur on sulfur, the values are almost a factor of 2 larger, around 0.3 at 5 GeV and almost 0.4 at 20 GeV [131]. We have seen earlier in Eq. (9-13) that the average multiplicity in pp-collisions increases with energy. In nucleus-nucleus collision, the increase is even faster, as more energy is available in the center of mass. However, strange particle production is observed to grow faster than nonstrange particles. For example, the ratio of $(K^+ + K^-)/(\pi^+ + \pi^-)$ at 200 GeV/nucleon is found to be a factor 2 larger in nucleus-nucleus collisions than in pp-collisions. As for mesons with two strange quarks, such as ϕ consisting predominantly of $s\bar{s}$, the production in sulfur+uranium is found to be a factor of 2 to 3 higher relative to that in proton+tungsten (and proton-proton) collisions [3].

If strange quarks are in thermal equilibrium with their lighter counterparts, as expected in a quark-gluon plasma, the ratio of strange to nonstrange mesons is given essentially by that of their respective Boltzmann factors $\exp\{-m_q c^2/kT\}$. Here m_q is the mass of the quark involved. Since the temperature of the plasma is comparable to the mass difference between strange and nonstrange quarks, an enhancement of strange meson production is expected compared with nonequilibrium situations, such as in proton-proton and proton-nucleus collisions. However, this is not the only possible explanation for strangeness enhancement. As we shall see soon, we do not yet know of any unique signature for QGP, one for which we can rule out any possible alternative explanations. For this reason, strangeness enhancement remains only one possible way to identify QGP.

Suppression of J/ψ- and ψ'-production In the case of charm meson, J/ψ has the unique advantage that it decays into a pair of leptons, e^+e^- or $\mu^+\mu^-$, with very narrow width, as we have seen in §2-6. This makes it relatively easy to identify the meson, especially in terms of its decay into a muon pair. The background comes mainly from the Drell-Yan process, conversion of a quark-antiquark pair into a lepton pair through the intermediate stage of a virtual photon (see Fig. 9-14 ahead). Since the production of a lepton pair by this mechanism is known to be given by the product of the projectile and target atomic numbers, the ratio of muon pair production from J/ψ-decay to that of the background becomes a measure of the relative production rate of J/ψ. The results show that the cross section for J/ψ-production is lower in nucleus+nucleus collisions than in proton-nucleus collisions [17]. This is opposite to the case of a strange meson where the production is enhanced. Furthermore, the effect is more pronounced for collisions that are more central (i.e., head-on), as measured by the increase in the transverse energy E_T.

The explanation in terms of QGP is that the cross section of J/ψ-production is lowered because of *color screening*. That is, when charm quarks are created in a collision, they must be made as $c\bar{c}$ pairs to conserve color and charm quantum numbers. However, if the production takes place in a QGP medium, the interaction between c and $\bar{c}$ is diminished by the presence of the other quarks, similar to the Debye screening of electric charge in quantum electrodynamics. This makes it more likely for the pair to be split into two separate entities. In addition, there is also a chance for the charm quarks to thermalize with the medium and, thus, decrease further the likelihood of c and $\bar{c}$ to emerge as a single meson. The relation with E_T comes from the inverse temperature dependence of Debye screening length. As temperature rises, the attraction between c and $\bar{c}$ decreases and the chance of dissociation for the $c\bar{c}$ pair increases. Since larger E_T implies a more central collision and more energy (higher temperature) in the collision region, it is not surprising that the observed J/ψ-production rate is decreased.

Again, the above explanation based on the presence of QGP is not the only one. In fact, a $c\bar{c}$ pair can also be separated through interaction with ordinary nucleons as well. Since a J/ψ-particle produced in heavy-ion collisions has to travel through, on the average, a larger region filled with nucleons than in *pp*-collisions, the observed rate is expected to be lower also for this reason. Furthermore, the ratio of ψ' to J/ψ has been observed as well, and its variation with E_T does not seem to follow that expected of a QGP scenario [18].

Two-particle interferometry When two identical particles are located close to each other, there is an interference between them, similar to that for optical waves in a double-slit experiment. This phenomenon may be used to deduce the extent of a source of particles. The principle was first applied in 1956 by Hanbury-Brown and Twiss to measure the angular diameter of a star using the correlation between two photons originating from the source. For this reason, it is known as the Hanbury-Brown-Twiss effect. The correlation exists only for a chaotic source and is absent for a coherent one [150].

The correlation of interest in our case is in the intensity of two identical particles measured in coincidence. By taking two particles at different energy-momentum values, the correlation becomes a Fourier transform of the phase space of the source. As a result, a correlation function of the following form may be constructed [131]:

$$C_2(Q_T, Q_L, Q_0) = 1 + \lambda e^{-(Q_T^2 R_T^2 + Q_L^2 R_L^2 + Q_0^2 \tau^2)/2}$$

where λ measures how chaotic the source is, with $\lambda = 1$ for a complete chaotic one and $\lambda = 0$ for a coherent one. The longitudinal, transverse, and time components of the momentum difference between the two particles, $Q_\mu = p_{1,\mu} - p_{2,\mu}$, are represented respectively by subscripts L, T, and 0. When such a function is fitted to the experimental data, we obtain measures of the longitudinal (R_L) and transverse (R_T) extent as well as the temporal duration (τ) of the source for the two particles.

In the case of heavy-ion reactions, two-particle correlations have been measured for pions and kaons. When two heavy ions collide, a small region of high energy and particle density is created. The particles interact strongly with each other as they are squeezed together. When the compression ceases, the region expands and the interaction intensity decreases. Eventually, the interaction will stop, or *freeze out*. It is likely that most pions and kaons are produced near this stage of development in a heavy-ion reaction. As a result, two-particle correlation "measures" the size of the volume and the time when freeze-out occurs and, thus, provides information on the size and energy achieved in the interaction zone. For example, if there is a phase transition, the latent heat will lead to larger expansion in size.

The results to date from fixed-target measurements indicate that the source size is larger than the radius of the projectile, indicating that the system expands before freeze-out. The time between the onset of expansion and freeze-out is about 5 fm/c along the longitudinal direction and less than 2 fm/c along the transverse direction. However, no long-lived intermediate state can yet be identified from two-particle correlation studies [81].

Thermal radiation In a QGP, large numbers of photons, both real and virtual, are produced through quark-antiquark annihilation,

$$q + \bar{q} \to \gamma + g$$

and quark-gluon scattering,

$$g + q \to \gamma + q \qquad\qquad g + \bar{q} \to \gamma + \bar{q}$$

Reactions such as $g + \bar{q} \to \gamma + \gamma$ are also possible but the probability is much smaller, as they are purely electromagnetic processes. Unlike hadrons, photons interact only weakly with quarks, gluons, and hadrons through electromagnetic interaction. As a result, there is a high probability for photons to emerge from the interaction zone without suffering any collisions. In contrast, hadrons have much shorter mean free path and undergo many interactions before leaving the region. For this reason, they reflect mainly the condition near freeze-out. Photons measured in heavy-ion collisions are therefore expected to be related more directly to the thermodynamic state of the QGP created than hadrons.

Photons are also produced in high-energy collisions from a variety of reasons other than reactions within a QGP. These include processes such as the decay of neutral mesons π^0 and η. For our purpose here, we are interested only in the signal over and above such "background" events. Analyses have indicated that the QGP photons may have a cubic dependence on the temperature and a quadratic dependence on the variation of multiplicity as a function of rapidity. However, experiments so far have not seen anything above the expected background photons. Theoretical investigations suggest that none is to be expected until we reach the conditions in experiments to be carried out at RHIC and LHC.

Dilepton production Similar to photons, leptons also do not interact strongly with the constituents in a QGP. As a result, one can infer the conditions of the plasma by observing the leptons emitted as well.

Lepton pairs $\ell^+\ell^-$ are produced by quark-antiquark annihilation through a virtual photon,

$$q + \bar{q} \rightarrow \text{virtual photon} \rightarrow \ell^+ + \ell^-$$

For all practical purposes here, we can limit the leptons to electrons and muons. The process is very similar to that of Drell-Yan. The main difference is that, in a Drell-Yan process, both the quark and antiquark come from the nucleons participating in the collision process. In contrast, our interest here is in the leptons produced by the collision of a free quark and a free antiquark in QGP. The difference is illustrated by the two diagrams shown in Fig. 9-14.

In a nucleus-nucleus collision, the Drell-Yan cross section scales with the number

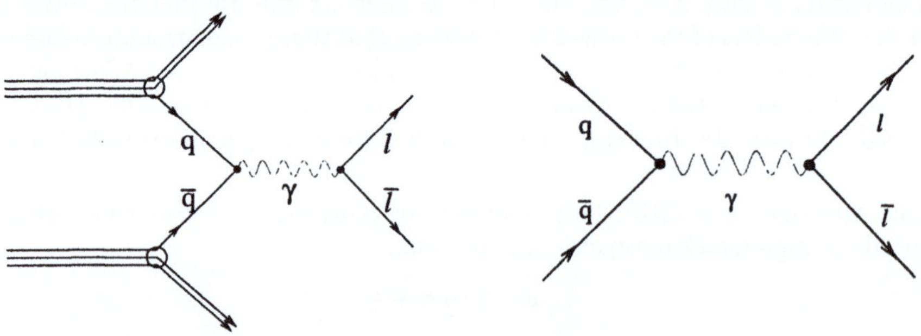

Figure 9-14: Dilepton production from the Drell-Yan process in nucleon-nucleon collision (left) and quark-antiquark annihilation between free quarks (right). In the Drell-Yan process, the quark and antiquark come from the colliding nucleons.

of nucleons involved and forms a dominant part of the lepton pairs produced. There are also other sources of lepton pairs, such as $\pi^+\pi^-$ annihilation and decay of neutral mesons. Our concern is with the lepton pairs produced in a QGP. The cross section depends on the temperature of the plasma and, hence, the initial condition that creates it. Theoretical investigations have shown that the production rate increases with temperature. At temperatures below 300 MeV, the cross section is too small to be seen above those coming from the dominant part by the Drell-Yan process. Once the temperature is above 300 MeV, there is a chance that one can isolate the leptons from QGP. However, one must again wait for experiments to be conducted at RHIC and LHC to reach such high temperatures.

Problems

9-1. Show that

$$E_{1s} = m_e c^2\sqrt{1-(Z\alpha)^2}$$

gives the ionization energy of 13.6 eV for a hydrogen atom.

9-2. If a neutron with 2 MeV of kinetic energy is evaporated from a composite system made of two heavy ions consisting of a total of 150 nucleons, find the maximum angular momentum carried away using the relation $\ell_m = \hbar kR$.

9-3. Two ^{90}Zr nuclei approach each other with kinetic energy 200 MeV in the center of mass. Calculate the total angular momentum in the system assuming that, in the absence of any interaction between them, they will pass each other at a distance of 10 fm between their centers.

9-4. The following γ-ray transitions, with energies given in kilo-electron-volts, were once identified in ^{158}Er: $2^+ \rightarrow 0^+$ 192, $4^+ \rightarrow 2^+$ 335, $6^+ \rightarrow 4^+$ 443, $8^+ \rightarrow 6^+$ 523, $10^+ \rightarrow 8^+$ 579, $12^+ \rightarrow 10^+$ 608, $14^+ \rightarrow 12^+$ 510, $16^+ \rightarrow 14^+$ 473, $18^+ \rightarrow 16^+$ 566, $20^+ \rightarrow 18^+$ 658, $22^+ \rightarrow 20^+$ 738, $24^+ \rightarrow 22^+$ 803, $26^+ \rightarrow 24^+$ 843, $28^+ \rightarrow 26^+$ 855, $30^+ \rightarrow 28^+$ 871, and $32^+ \rightarrow 30^+$ 902. Plot the excitation energy as a function of $J(J+1)$ and calculate the moment of inertia for each state. Use this result to plot $2\mathcal{I}/\hbar^2$ as a function of $\hbar^2\omega^2$ and see if there is any sudden change in the moment of inertia, generally known as "backbending."

9-5. For an axially symmetric nucleus with density given by

$$\rho(r) = \begin{cases} \rho_0 & \text{for } r \le R_0(1+\beta Y_{2,0}(\theta,\phi)) \\ 0 & \text{otherwise} \end{cases}$$

show that the intrinsic (charge) quadrupole moment up to second order in deformation parameter β is given by

$$Q_0 = \frac{3}{\sqrt{5\pi}} Z R_0^2 \beta(1+0.36\beta)$$

and the moment of inertia about the z-axis is given by

$$I = \tfrac{2}{5} M R_0^2 (1+0.31\beta)$$

to first order in β. Here M is the mass of the nucleus and R_0 is the radius of a sphere having the same volume, and ρ_0 may be found from normalization.

9-6. Use Eqs. (6-18) and (6-12) to show that

$$\delta = 0.95\beta + 0.15\beta^2 + O(\beta^3)$$

where δ is the deformation parameter defined by Eq. (6-18).

Chapter 10

Nuclear Astrophysics

On a bright sunny day, one cannot help but be amazed at the enormous amount of energy outpouring from the sun. At a distance of 1.5×10^{11} m, the amount of electromagnetic radiation received by the earth is 1.4 kW/m^2, corresponding to a total output of 10^{26} J/s. The primary source of this energy is fusion of hydrogen into helium. The sun is only one of the billions of stars in our galaxy and our galaxy is, in turn, one of the billions in the universe. Besides burning hydrogen, some stars are converting helium into ^{12}C, and thence to even heavier elements. We shall see that the evolution of a star is intimately related to the thermonuclear reactions taking place inside. Except for hydrogen, some of the helium, and small amounts of deuteron and a few other light nuclei, all other nuclei are made in stars. Understanding stellar evolution, therefore, requires a good knowledge of nuclear physics. At the same time, nuclear physics is incomplete without a clear idea of how all the nuclei are created. We shall begin with an overall picture of nucleosynthesis and the connection between nuclear physics and stellar evolution. Details on some of the topics are given in the remaining sections. The subject of nucleosynthesis is the central theme of several books as well as a large number of review papers.[1] Our aim here is only to examine some of the problems of particular interest to nuclear physics.

10-1 Brief Overview of Stellar Evolution

Most modern views of cosmology are in agreement with the idea that the universe began with an explosion, or "big bang," some 10 to 20 billions years ago. The uncertainties in the models are connected mostly with the very beginning of time, within the first fraction of a second or so. For nuclear astrophysics, there is hardly any need to be concerned with such early times. At the end of approximately the first 3 min [144], about three-fourths of the baryon mass in the universe is in the form of protons and the rest in the form of ^{4}He. Traces of deuteron, ^{3}He, and ^{7}Li are also present but their abundances are down by several orders of magnitude compared with protons (4 to 5 orders for deuteron and ^{3}He and 10 orders for ^{7}Li). The exact amounts of the three less abundant primordial elements are important if we wish to understand the conditions that existed in the first few minutes. However, for stellar evolution and nucleosynthesis we can, for most practical purposes, ignore their presence.

[1]For a recent review, see Ref. [141].

Big-bang nucleosynthesis. The variety of nuclei we observe today in the interstellar space comes from a combination of two different sources. The first is the primordial proton and helium together with minute amounts of other light nuclei. The second is made up of material blown off stars that have ended their lives. We shall start with the first one, as it supplies the bulk of the raw material to form stars.

For proton and helium, the relative abundance is observed to be roughly 3 to 1 in mass. This is essentially determined by the neutron-proton mass difference of 1.29 MeV and the deuteron binding energy of 2.23 MeV. When the universe was at temperatures far above that equivalent in energy to the rest mass of an electron, $kT = m_e c^2$ or $T \sim 6 \times 10^9$ K, electrons and neutrinos were in thermal equilibrium. In such an environment, neutron β^--decay to proton,

$$n \rightarrow p + e^- + \bar{\nu}_e$$

is balanced by electron capture of protons,

$$e^- + p \rightarrow n + \nu_e$$

The abundance of leptons keeps neutrons and protons in thermal equilibrium. Since the chemical potentials of the two species of nucleon are nearly the same, the ratio of neutron to proton numbers is essentially given by

$$\frac{N_n}{N_p} = e^{-Q/kT}$$

Here, Q is the neutron-proton mass difference in terms of rest-mass energy. As the universe expands, the temperature drops and the electron capture rate decreases. When the value falls below $\sim 10^{10}$ K, it becomes impossible to maintain a balance between the two weak interaction processes and we no longer have a thermal equilibrium between neutrons and protons. The weak interaction is said to be "frozen." At the *weak interaction freeze-out* temperature of $T_w \sim 10^{10}$ K, the ratio between the number of neutrons and protons is given by $\exp\{-Q/kT_w\} \sim 0.22$, with 18% of the baryons appearing as neutrons and 82% as protons.

Below the weak interaction freeze-out temperature, free neutrons decay into protons with a half-live of about 10 min. For a neutron to survive much longer time periods, it must be captured by other nucleons to form a bound nucleus. Since most of the nucleons in the universe are in the form of free protons and neutrons at this stage, the most likely candidate to be formed is the deuteron, a bound nucleus made of a proton and a neutron. Unfortunately, the binding energy of a deuteron is very small and this constitutes the major bottleneck in preserving primordial neutrons from β^--decay. Because of the short range of nuclear force, bound nuclei can be made from free neutrons and protons only through random collisions that bring some of them into close contact with each other. The probability of such encounters drops drastically for three or more particles. This leaves us with the deuteron as the only likely bound system that can be made in any significant amount. On the other hand, the small binding energy means that deuterons can also be destroyed easily in random collisions with other particles. The most likely event is with photons, as there is something like 10^9

for each nucleon. For this reason, photodisintegration constitutes an important sink for any deuterons created when the temperature is still sufficiently high.

On further cooling, some deuterons can exist long enough to capture a proton to form ^{3}He. In turn, ^{3}He can capture a neutron and transform it into ^{4}He. We see that once the temperature is sufficiently low for deuterons to last long enough to undergo proton and neutron captures, free neutrons are transformed into bound ones and the total number of neutrons in the universe stays more or less constant until the start of stellar nucleosynthesis at much later stages in the evolution of our universe.

Both ^{5}He and ^{5}B are unstable with half-lives on the order of 10^{-22} s. As a result, further nucleosynthesis beyond ^{4}He by single-nucleon capture is blocked. Together with the fact that ^{4}He has the largest binding energy per particle among all the nuclei involved in this discussion, we find that α-particles become the main reservoir of neutrons at the end of big-bang nucleosynthesis. The exact amount of ^{4}He and, hence, the number of neutrons, available for later nucleosynthesis in stars depends on the condition existing in the short time between weak interaction freeze-out and when deuterons can exist for a sufficiently long time to capture another nucleon. The observed helium in the interstellar medium is between 22% and 28% by weight (with the remainder taken up by protons). This must be close to the amount that existed before stars began to process the primordial material and blow some of the resulting heavier elements into the interstellar medium at the end of their lives. Furthermore, since the universe is observed to be electrically neutral, we can also safely assume that the total number of electrons is the same as the combined number of free and bound protons.

Stellar nucleosynthesis. With all the neutrons locked inside nuclei, the only way for stars to make new species of nuclei is through charged particle reactions. The most likely process is to combine two protons into a deuteron, together with a positron and an electron neutrino, as protons are the most abundant nuclei by far and have the lowest Coulomb barrier. However, the average temperature of the universe now is far too low for two protons to overcome the Coulomb barrier between them and come close enough together for nuclear reactions to take place.

We can make an order-of-magnitude estimate for the kinetic energy required to bring two protons to be close enough for nuclear force to act between them. For simplicity, let us take the range of nuclear interaction to be $r_0 \sim 1$ fm. The Coulomb energy of two protons separated by this distance is

$$E_C = \left[\frac{1}{4\pi\epsilon_0}\right]\frac{e^2}{r_0} = \frac{\alpha\hbar c}{r_0} = 1.4 \text{ MeV}$$

This is equivalent to the thermal energy of a particle at temperature

$$T = \frac{E_C}{k} \approx 2 \times 10^{10} \text{ K}$$

where $k = 8.62 \times 10^{-11}$ MeV/K is the Boltzmann constant. This is three to four orders of magnitude higher in value than the condition existing in the interior of the sun where we know that proton burning is taking place.

There are two reasons why nuclear reaction between protons can start at much lower temperatures. The first is that the fraction of particles with thermal energy E at temperature T is given by a Maxwell-Boltzmann distribution

$$\frac{dN}{N} = \frac{2}{\sqrt{\pi}\, kT}\sqrt{\frac{E}{kT}}\, e^{-E/kT}\, dE$$

The long exponential tail ensures that there is a small fraction of the particles with kinetic energies much higher than kT. The second reason is that, since the collision is a quantum-mechanical process, the probability of tunneling is important for any exothermic process, such as proton burning.

In addition to kinetic energy, a high concentration of particles is also essential to ensure reasonable probability for nuclear reactions to take place. If uniformly distributed, the proton density in the universe is far too low for fusion to take place in any significant way. (Even inside an interstellar "gas," the average density is only about one particle per cubic centimeter, lower than that in a good vacuum chamber on earth.) Both the high temperature and high density required to initiate nuclear reaction actually come as a result of gravitational collapse of primordial matter. However, this cannot happen if the material thrown out from the big bang is distributed uniformly in space. On clear nights, and away from brightly lit cities, we find that stars are concentrated only in certain parts of the sky, the Milky Way. This is only because we are located toward the edge of our galaxy. In fact, ours is a member of a local concentration of galaxies. If we look beyond the *local group*, the distribution of galaxies actually turns out to be fairly uniform. The uniformity is also supported by the isotropic distribution of the 3 K microwave background radiation of photons. If the density is uniform everywhere, the gravitational force felt by a particle will be the same in every direction and there will not be any tendency to coalesce into clusters. In the recent mapping by the Cosmic Background Explorer (COBE),[2] an intrinsic anisotropy of the order of 1 part in 10^5 is observed, and this is found to be adequate for local concentrations of gravity to pull matter together to form planets, stars, and galaxies.

From hydrogen to finite nuclei. Gravitational contraction of matter, dispersed in the universe by the big bang, provides both the concentration of particles and initial energy to start hydrogen burning, the process of converting protons into ^{4}He (and small amounts of heavier nuclei). The reaction produces also two positrons to conserve charge and two neutrinos to conserve lepton number. We shall see in §10-3 that the conversion takes place in several separate steps, as the probability is essentially nil for four protons to be converted directly into an α-particle.

Once hydrogen burning starts, there is a supply of nuclear energy to heat up the star and raise the temperature. Part of the energy produced is radiated into space. We shall not be concerned with the rate of radiation, as it depends on the radial distribution of temperature in a star. One point to note is that, with the higher temperature, there is now a thermal pressure to stop further gravitational contraction. A second point is that, because of higher Coulomb barriers, nucleosynthesis does not proceed beyond ^{4}He until later stages in the evolution of the star. In other words, the star is now in a hydrostatic equilibrium and the condition persists until the hydrogen fuel is exhausted. The amount of time it takes to consume all the available protons depends on the stellar

[2]For more details, see http://www.gsfc.gov/astr/cobe/cobe_home.html.

mass. For our sun, the present status has been going on already for roughly 5 billion years and is expected to remain in more or less the same state for another 5 billion years or so. In more massive stars, the temperature is higher. This leads to much faster rates of nuclear reaction, and the life span of the hydrogen burning stage can be as short as millions of years.

When the hydrogen fuel is used up in a star, production of nuclear energy from fusing protons into ^{4}He stops and the temperature drops. Without adequate thermal pressure to offset gravity, contraction starts again. However, since the outer layer of the star is cooler and less dense, nuclear reactions are slower and some hydrogen remains when the inner core begins to shrink. As a result, the interior of the star goes to a higher temperature beyond that required for hydrogen burning while the outer layer remains essentially unchanged. When the temperature in the stellar core reaches the value to initiate helium burning, nuclear fusion starts again, with helium replacing hydrogen as the fuel. Since the energy released from the interior region is at a higher temperature, it causes the outer layer to expand. As the radius of the star increases, the surface temperature drops, shifting the peak of radiated energy to the longer wavelengths. The result is a red giant.

Helium burning requires a temperature on the order of 10^8 K. Since ^{8}Be is unstable and lives only for $\tau_{1/2} = 6.7 \times 10^{-17}$ s, the conversion takes place mainly through the triple-α reaction

$$^4\text{He} + {}^4\text{He} + {}^4\text{He} \rightarrow {}^{12}\text{C} + \gamma$$

The ^{12}C produced can capture another α-particle to make ^{16}O,

$$^4\text{He} + {}^{12}\text{C} \rightarrow {}^{16}\text{O} + \gamma$$

Further α-particle capture produces even heavier nuclei. However, as we move to heavier and heavier nuclei, the Coulomb barrier increases in height. This calls for higher temperatures that can come only from further gravitational contraction, as shown in Table 10-1. Since this is more likely to take place first at the center, the inner parts of the star go to higher temperatures and densities, and evolve faster through different stages, than those outside. The net result is that the star develops into an onion-like

Table 10-1: Different stages of stellar nucleosynthesis

Fuel	Temperature T (K)	kT (MeV)	Main product
proton	2×10^7	0.002	^{4}He, ^{14}N
^{4}He	2×10^8	0.02	^{12}C,^{16}O
^{12}C	8×10^8	0.07	^{16}O,^{20}Ne,^{24}Mg
^{16}O	2×10^9	0.2	^{20}Ne,^{28}Si,^{32}S
^{20}Ne	1.5×10^9	0.13	^{16}O,^{24}Mg
^{28}Si	3.5×10^9	0.3	$A \approx 56$ nuclei
$A \lesssim 56$	supernova		$A > 56$

structure, with each successive inner layer undergoing a later stage of evolution, as shown later in Fig. 10-6.

Nucleosynthesis by α-particle capture produces mainly $4n$ nuclei. Through collisions, decays, and the presence of small amounts of protons, neutrons, deuterons, and ^{3}He, other nuclear reactions participate also in the process and these produce nuclei outside the $4n$ sequence. Partly for this reason, the relative abundances are lower for species other than the $4n$ nuclei.

The release of fusion energy stops at $A \approx 56$ where the binding energy per nucleon peaks in value, as we saw earlier in Fig. 1-2. This takes place first in the stellar core and most of the nuclei are in the form of ^{56}Fe and ^{56}Ni, the two most stable $A = 56$ isobars. Further evolution of the star depends even more critically on its total mass than any of its early stages. If the value is more than 8 times solar mass, there is enough gravitational energy left in the core at the end of fusion to turn the star into a supernova. The explosion sends shock waves through the outer layers, which are still in the earlier stages of nucleosynthesis. The condition created in this way by a supernova explosion is highly nonequilibrium, making it possible for nuclear reactions to take place that require energy input instead of the purely exothermic ones we have seen so far. The environment reminds one of heavy-ion collisions in which a large number of reaction products are created, including those with $A > 56$.

For smaller stars, the collapse will not be as catastrophic. One possibility is that the star ends its life as a white dwarf, a small star with fairly high temperature but very little energy output. Alternatively, it can turn into a brown dwarf, slowly radiating away the small amount of energy still remaining in the star.

For the purpose of nucleosynthesis, we are interested in the material ejected in a supernova explosion. In addition to protons and helium nuclei left in the surface regions, heavier elements created in the evolution of the star are also sent into the interstellar space. These are mixed with the primordial protons and helium nuclei to form the raw material for future generations of stars. The relative abundance of elements in the solar system, shown in Fig. 10-1, gives clear evidence that the material has already gone through several generations of stellar formation and explosion.

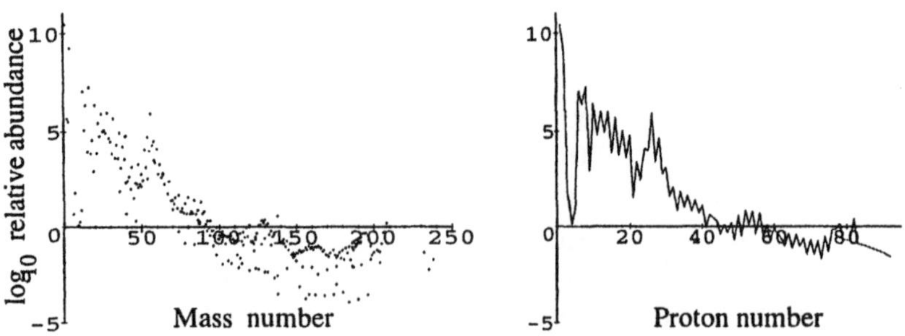

Figure 10-1: Relative abundances of elements in the solar system normalized with respect to that for silicon as 10^6 [42].

10-2 Rate for Nonresonant Reactions

In a region where there are no resonances, the cross section for nuclear reactions is expected to vary smoothly with energy. This is the most likely situation in nucleosynthesis where the kinetic energy involved is very low compared with those encountered in nuclear excitations. Except for the few cases where there happen to be states near the threshold with the same quantum numbers as the reaction channel of interest, the reaction is nonresonant. In this case, the cross section $\sigma(E)$ as a function of energy E may be written as a product of three factors,

$$\sigma(E) = \frac{1}{E} e^{-2\pi Z_1 Z_2 \alpha c/v} S(E) \tag{10-1}$$

where v is the relative velocity between the two nuclei, c is the speed of light, and α the fine structure constant. The inverse energy dependence comes from geometric considerations. As we have seen in the case of nucleon-nucleon scattering in §3-8, the low-energy cross section is given by the square of scattering length a. Similar to the de Broglie wavelength $\lambda = h/p$, the scattering length is inversely proportional to velocity v and, hence, the cross section at low energies is, to a first-order approximation, inversely proportional to E. The exponential factor gives the probability, or penetration factor, for a nucleus with proton number Z_1 to tunnel through the Coulomb barrier of a nucleus with proton number Z_2. The factor $\eta \equiv Z_1 Z_2 \alpha c/v$ is the Sommerfeld number given in Eq. (4-64).

To simplify the notation and to display the energy dependence explicitly, we shall define a quantity b by the relation

$$e^{-2\pi Z_1 Z_2 \alpha c/v} \equiv e^{-b/\sqrt{E}} \tag{10-2}$$

Since we have assumed that there are no resonances involved in the reaction, the nuclear structure cannot have any large variations as a function of energy in the region. Let us represent any effect nuclear structure has on the reaction cross section by $S(E)$. From Eq. (10-1), we can write

$$S(E) = E e^{b/\sqrt{E}} \sigma(E)$$

It is known as the S-factor and we shall see later that it is a convenient quantity in discussing nonresonant reaction rates in nuclear astrophysics.

Unlike laboratory experiments in which the velocities of the projectile and target particles are fixed both in magnitude and in direction, the stellar environment resembles that of a hot gas. Nuclear reactions are initiated through random collisions of the particles involved. Since the rate of a particular reaction depends also on the frequency of collisions between two types of particle, we can define a reaction rate λ as the product of relative velocity v and cross section σ. In a hot gas, the particles have a Maxwellian distribution $p(v)$ in velocity. On averaging over the possible values, we obtain the result

$$\begin{aligned} \lambda &\equiv \int_0^\infty \sigma(E) v p(v)\, dv \\ &= \int_0^\infty \frac{S(E)}{E} e^{-b/\sqrt{E}}\, v \left(\frac{\mu}{2\pi kT}\right)^{3/2} e^{-\mu v^2/(2kT)} 4\pi v^2\, dv \\ &= \sqrt{\frac{8}{\mu\pi}} \left(\frac{1}{kT}\right)^{3/2} \int_0^\infty S(E) \exp - \left\{\frac{E}{kT} + \frac{b}{\sqrt{E}}\right\} dE \end{aligned} \tag{10-3}$$

where the reduced mass μ may be approximated as

$$\mu = \frac{A_1 A_2}{A_1 + A_2}$$

with A_1 and A_2 as the nucleon numbers of the two nuclei involved. For $p(v)$, we have made use of the Maxwell-Boltzmann distribution

$$p(v)\,dv = \left(\frac{\mu}{2\pi kT}\right)^{3/2} e^{-\mu v^2/(2kT)} 4\pi v^2\,dv$$

for a gas at temperature T and with molecular mass μ.

Since $S(E)$ is assumed to be essentially independent of energy, the variation of the integrand in Eq. (10-3) is given by the exponential of

$$f(E) = \frac{E}{kT} + \frac{b}{\sqrt{E}}$$

The maximum of this function occurs at

$$\frac{df}{dE} = \frac{1}{kT} - \frac{1}{2}\frac{b}{E^{3/2}} = 0$$

The value

$$E_{\rm max} = \left(\tfrac{1}{2}bkT\right)^{2/3} \tag{10-4}$$

is therefore the place where the nonresonant reaction rate for charged particles reaches a maximum. Since the energy dependence in the integrand is a product of two exponential terms—$\exp -\{E/kT\}$, which decreases with energy, and $\exp\{-b/\sqrt{E}\}$, which increases with energy—the only effective part of the integral is a small region around $E_{\rm max}$. In fact, the integrand may be approximated by a normal curve [44] centered around $E_{\rm max}$ with a width $\Delta = 4\sqrt{E_{\rm max}kT/3}$ (see also Problem 10-4). Schematically, the dependence is shown in Fig. 10-2.

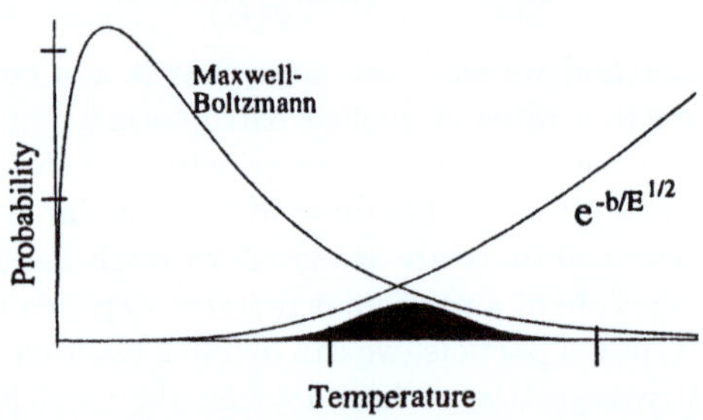

Figure 10-2: Rate of nonresonant nuclear reaction for charged particles in stellar environment. The probability for nuclear reaction is represented by the shaded area (multiplied by a large number to make it visible) resulting from a product of the Maxwell-Boltzmann distribution in the kinetic energy of the particles and the $\exp\{-b/\sqrt{E}\}$ dependence of the Coulomb penetration factor.

10-3 Conversion of Proton into Helium

The sun is classified as a young star with an age somewhere around the middle of its hydrogen burning phase. The radiant energy comes from fusing four protons into a helium nucleus. The actual process, however, takes place in several distinct steps, as direct conversion of four protons into ^{4}He, being a four-body reaction, is extremely unlikely.

The first step is to make a deuteron through the reaction,

$$p + p \rightarrow d + e^{+} + \nu_e \tag{10-5}$$

As we have seen earlier in Table 10-1, the reaction starts around stellar temperature 2×10^7 K. In energy units, it is only 2 keV, far lower than the binding energy of a deuteron. As a result, we do not have to be concerned with photodisintegration as a sink for the deuterons created as, for example, in the case of big-bang nucleosynthesis described in §10-1. Once the density of deuterons is sufficiently high, conversion to ^{3}He through a (p, γ) reaction,

$$p + d \rightarrow {}^3\text{He} + \gamma$$

becomes important. From this point onward, there are several competing routes to change the product into ^{4}He.

PP-chains. One possibility is that two ^{3}He nuclei may collide with each other, initiating the reaction

$${}^3\text{He} + {}^3\text{He} \rightarrow {}^4\text{He} + p + p$$

This is know as the PPI-chain. Since it takes three protons to make each one of the two ^{3}He nuclei required for the production, a total of six protons are involved on the left-hand side of the reaction. On the right-hand side, there are two free protons. The net result is that four protons are converted into a ^{4}He nucleus. In the process two positrons, two neutrinos, and two γ-rays are emitted.

A second possibility is for the ^{3}He to react with one of the ^{4}He nuclei left over from the big bang or produced in the star. This leads to the reaction

$${}^3\text{He} + {}^4\text{He} \rightarrow {}^7\text{Be} + \gamma \tag{10-6}$$

Through electron capture, the ^{7}Be is converted into a ^{7}Li,

$$e^{-} + {}^7\text{Be} \rightarrow {}^7\text{Li} + \nu_e$$

By capturing a proton,

$$p + {}^7\text{Li} \rightarrow {}^4\text{He} + {}^4\text{He}$$

two ^{4}He nuclei are produced. This is known as the PPII-chain. Again, the net result is the conversion of four protons into a ^{4}He, as the starting point is a ^{3}He, made originally from three protons, a ^{4}He, and a proton (as well as an electron).

Instead of electron capture, the ^{7}Be produced in Eq. (10-6) can be changed into a ^{8}B through a (p, γ) reaction,

$$p + {}^7\text{Be} \rightarrow {}^8\text{B} + \gamma$$

Being unstable, ^{8}B decays by β^+-emission into a ^{8}Be with a half-life of 770 ms,

$$^8\text{B} \rightarrow {}^8\text{Be} + e^+ + \nu_e$$

Since ^{8}Be is unstable with respect to α-decay,

$$^8\text{Be} \rightarrow {}^4\text{He} + {}^4\text{He}$$

we end up again with two ^{4}He nuclei. This is the PPIII-chain, and it converts four protons into a ^{4}He through one proton capture, one β^+-decay, and one α-decay. The reactions involved in all three PP-chains are summarized in Table 10-2.

Table 10-2: Proton-proton chains to convert protons into helium

Reaction	Q-value(MeV)
Common to all chains	
$p + p \rightarrow d + e^+ + \nu_e$	1.442
$d + p \rightarrow {}^3\text{He} + \gamma$	5.493
PPI-chain	
$^3\text{He} + {}^3\text{He} \rightarrow {}^4\text{He} + p + p$	12.859
PPII-chain	
$^3\text{He} + {}^4\text{He} \rightarrow {}^7\text{Be} + \gamma$	1.586
$^7\text{Be} + e^- \rightarrow {}^7\text{Li} + \gamma$	0.861
$^7\text{Li} + p \rightarrow {}^4\text{He} + {}^4\text{He}$	17.347
PPIII-chain	
$^7\text{Be} + p \rightarrow {}^8\text{B} + \gamma$	0.135
$^8\text{B} \rightarrow {}^4\text{He} + {}^4\text{He} + e^+ + \nu_e$	18.074

CNO cycle. In addition to the three PP-chains, protons are also converted into ^{4}He through the more elaborate CNO (carbon-nitrogen-oxygen) cycle shown schematically in Fig. 10-3. Even for a young star, there are always some heavier elements present alongside with the dominant components of protons and ^{4}He nuclei. In addition to primordial sources, the interstellar medium is also filled with heavy elements blown off from massive stars that have already gone through their life cycles. Because of the presence of ^{12}C, it is possible to have a different but fairly efficient process for proton burning.

Let us concentrate first on the main part of the CNO cycle represented by the circle in the middle of Fig. 10-3. If we start at the top, from ^{12}C, the cycle may be viewed as a chain of four (p, γ) reactions to capture four protons one after another and convert

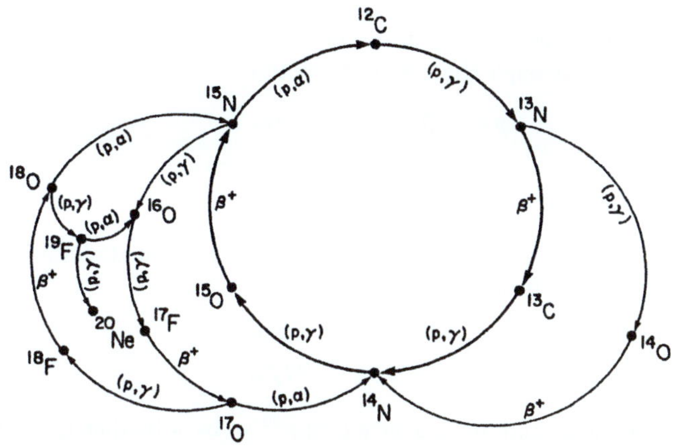

Figure 10-3: Carbon-nitrogen-oxygen (CNO) cycle of nucleosynthesis showing the different reactions involved in converting protons into ^{4}He.

two of them into neutrons by β^+-decays. The six reactions involved are

$$\begin{aligned}
p + {}^{12}\mathrm{C} &\rightarrow {}^{13}\mathrm{N} + \gamma \\
{}^{13}\mathrm{N} &\rightarrow {}^{13}\mathrm{C} + \beta^+ + \nu_e \\
p + {}^{13}\mathrm{C} &\rightarrow {}^{14}\mathrm{N} + \gamma \\
p + {}^{14}\mathrm{N} &\rightarrow {}^{15}\mathrm{O} + \gamma \\
{}^{15}\mathrm{O} &\rightarrow {}^{15}\mathrm{N} + \beta^+ + \nu_e \\
p + {}^{15}\mathrm{N} &\rightarrow {}^{12}\mathrm{C} + \alpha
\end{aligned}$$

The net result is again four protons converted into a ^{4}He together with two positrons and two electron neutrinos, the same final result as the PP-chains. The only exception is that ^{12}C is used as the catalyst here.

There are several side chains to the main CNO cycle that are also of interest. The ^{13}N produced in the (p, γ) reaction on ^{12}C may be converted through another (p, γ) reaction into ^{14}O, which then β^+-decays to ^{14}N,

$$\begin{aligned}
p + {}^{13}\mathrm{N} &\rightarrow {}^{14}\mathrm{O} + \gamma \\
{}^{14}\mathrm{O} &\rightarrow {}^{14}\mathrm{N} + \beta^+ + \nu_e
\end{aligned}$$

The final product returns the process to the main CNO cycle in the form of ^{14}N. Similarly, some of the ^{15}N near the end of the main cycle may be converted back to ^{14}N by the following chain of reactions:

$$\begin{aligned}
p + {}^{15}\mathrm{N} &\rightarrow {}^{16}\mathrm{O} + \gamma \\
p + {}^{16}\mathrm{O} &\rightarrow {}^{17}\mathrm{F} + \gamma \\
{}^{17}\mathrm{F} &\rightarrow {}^{17}\mathrm{O} + \beta^+ + \nu_e \\
p + {}^{17}\mathrm{O} &\rightarrow {}^{14}\mathrm{N} + \alpha
\end{aligned}$$

Again a ^{4}He nucleus is made from four protons by this procedure.

The ^{17}O in the intermediate step above may return some ^{15}N to the main cycle and produce a ^{4}He nucleus through the chain of reaction

$$\begin{aligned} p + {}^{17}\mathrm{O} &\rightarrow {}^{18}\mathrm{F} + \gamma \\ {}^{18}\mathrm{F} &\rightarrow {}^{18}\mathrm{O} + \beta^{+} + \nu_e \\ p + {}^{18}\mathrm{O} &\rightarrow {}^{15}\mathrm{N} + \alpha \end{aligned}$$

The ^{18}O produced in the intermediate stage here may also undergo further proton capture and produce ^{19}F through the process

$$\begin{aligned} p + {}^{18}\mathrm{O} &\rightarrow {}^{19}\mathrm{F} + \gamma \\ p + {}^{19}\mathrm{F} &\rightarrow {}^{16}\mathrm{O} + \alpha \end{aligned}$$

Some of the ^{19}F, in turn, may be converted into ^{20}Ne by a further (p, γ) reaction,

$$p + {}^{19}\mathrm{F} \rightarrow {}^{20}\mathrm{Ne} + \gamma$$

In terms of the amount of energy produced, the PP-chains still constitute the dominant source. However, the CNO cycle is able to generate some of the heavier elements that are of interest from a nucleosynthesis point of view.

10-4 Solar Neutrino Problem

We saw in the previous sections that most of the radiant energy of stars derives from nuclear reactions (and the rest from gravitational contraction). In addition to γ-rays, a comparable number of neutrinos is also emitted in the process. Since most of the reactions take place in the interior of the star, the γ-rays are seldom observed directly. Interactions with the thick outer layers of stellar matter transform the electromagnetic energy into black-body radiation, characterized only by the surface temperature of the star. This is not true for the neutrinos. The small interaction cross section means that most of the neutrinos can emerge from the star without suffering a collision on the way. Their spectrum is therefore a more direct reflection of the conditions existing in the interior of the star. For this reason, there are high hopes for new discoveries to be made through neutrino astronomy: observation of astronomical objects by the neutrinos they emit rather than the traditional means of visible light and other parts of the electromagnetic spectrum.

The major advantage of neutrino astronomy is also its main difficulty. The typical interaction cross section is around 10^{-18} fm^2 (10^{-48} m^2). As a result, most neutrinos can, for example, go through the earth without suffering an interaction (see Problem 10-7). Since the only way we can observe a neutrino is through its rare interactions with matter, huge detectors and highly sensitive apparatus are essential. Fortunately, the earth atmosphere is not a problem here as in optical astronomy, and large setups can be constructed on the surface of the earth rather than high up in mountains and in space stations. In fact, neutrino observatories are usually built deep underground to reduce background from cosmic rays.

Solar neutrino. For an earth-based observatory, the most intense source of stellar neutrinos is the sun, as it is the closest "star." Before examining the expected spectrum, let us summarize first the different types of neutrinos produced in the sun from proton burning. In the previous section, we have seen that neutrinos are produced in the following three reactions in PP-chains:

$$p + p \rightarrow d + e^+ + \nu_e \tag{10-7}$$

$$^8\mathrm{B} \rightarrow {}^8\mathrm{Be}^* + e^+ + \nu_e \tag{10-8}$$

$$e^- + {}^7\mathrm{Be} \rightarrow {}^7\mathrm{Li} + \nu_e \tag{10-9}$$

The neutrino from the $p(p,\gamma)d$ reaction in Eq. (10-7) is the most important one to measure, as one such neutrino must be emitted for each ^{4}He nucleus made, regardless of in which one of the three different chains the actual conversion takes place. Unfortunately, the end-point energy of this reaction is only 0.42 MeV, the lowest one among all the neutrino emission processes involved. The dominant mode of decay for ^{8}B in Eq. (10-8) is to go to the 2^+ first excited state at 3.040 MeV. Since it is broad state, having a width of 1.50 MeV, it complicates slightly the calculation of end-point energy for the neutrino involved.

For electron capture, the energy distribution of the neutrino emitted is in the form of a sharp line, as the final state is two body. However, two discrete lines are produced in the case of ^{7}Be given in Eq. (10-9), one to the ground state of ^{7}Li with $\sim$90% probability and the other to the 0.477-MeV excited state with 10% probability. The end-point energy for the ground state capture is 0.863 MeV, given by the difference between the ^{7}Li binding energy of 39.245 MeV and the ^{7}Be binding energy of 37.600 MeV, less the neutron-proton mass difference of 0.782 MeV. For the excited state, the value is $0.863 - 0.477 = 0.386$ MeV.

In addition to these dominant modes, we must include the following two reactions that are important for the neutrino emitted rather than the fraction of the total energy produced:

$$p + p + e^- \rightarrow d + \nu_e \tag{10-10}$$

$$p + {}^3\mathrm{He} \rightarrow {}^4\mathrm{He} + e^+ + \nu_e \tag{10-11}$$

The first is the *pep* process. It is basically the same as the *pp* process of Eq. (10-7) except that electron capture replaces β^+-decay. Although the rate is greatly diminished because the initial state is three body, the neutrino produced is mono-energetic at 1.442 MeV. The width of the energy distribution is less than 1 keV, arising mainly from the thermal energy distribution of the particles involved in the reaction.

The second addition is the *hep* process of Eq. (10-11). Its importance comes from the fact that it produces the highest energy neutrinos, with an end-point energy of 18.773 MeV. The flux is, however, down by seven orders of magnitude compared with *pp* neutrinos. Another source of high-energy neutrinos comes from ^{8}B decay, given in Eq. (10-8), and it has an end-point energy of 14.9 MeV. As we shall see later, these high-energy neutrinos are the only ones "seen" by the ^{37}Cl detector that has been in operation for more than 30 years.

Neutrinos are also produced from the β^+-decays in the CNO cycle. The dominant ones are

$$^{13}\mathrm{N} \rightarrow {}^{13}\mathrm{C} + \beta^+ + \nu_e$$

$$
\begin{aligned}
{}^{15}\mathrm{O} &\rightarrow {}^{15}\mathrm{N} + \beta^+ + \nu_e \\
{}^{17}\mathrm{F} &\rightarrow {}^{17}\mathrm{O} + \beta^+ + \nu_e
\end{aligned}
$$

The end-point energies are given in Table 10-3 together with those for the other reactions.

Table 10-3: Maximum energy in MeV of neutrino produced in proton burning.

Label	Reaction	E_B(right)	E_B(left)	$E_{\rm max}(\nu_e)$
pp	$p + p \rightarrow d + e^+ + \nu_e$	2.224	0	0.420
^{8}B	$^8\mathrm{B} \rightarrow {}^8\mathrm{Be}^* + e^+ + \nu_e$	56.500	37.738	14.9
hep	$p + {}^3\mathrm{He} \rightarrow {}^4\mathrm{He} + e^+ + \nu_e$	28.296	7.718	18.773
^{13}N	$^{13}\mathrm{N} \rightarrow {}^{13}\mathrm{C} + \beta^+ + \nu_e$	97.108	94.105	1.199
^{15}O	$^{15}\mathrm{O} \rightarrow {}^{15}\mathrm{N} + \beta^+ + \nu_e$	115.491	111.956	1.731
^{17}F	$^{17}\mathrm{F} \rightarrow {}^{17}\mathrm{O} + \beta^+ + \nu_e$	131.763	128.220	1.739

Energy spectrum. In addition to the maximum value, the energy distribution of the neutrinos emitted is also important in what we can expect to detect at an observatory. We have seen earlier that, in a β^+-decay, the probability $\mathcal{W}(p_e)$ for emitting a positron with momentum p_e in an interval dp_e is given by Eq. (5-68),

$$\mathcal{W}(p_e)\,dp_e = CF(Z,E_e)\,p_e^2(E_0 - E_e)^2\,dp_e$$

where, for simplicity, we have assumed the rest mass of neutrinos to be zero and all the energy-independent factors, including the sum over nuclear matrix elements, are represented by the factor C. To convert to the probability $P(E_e)$ for a relativistic positron with kinetic energy E_e, we can make use of the relation

$$E_{\rm rel} = E_e + m_e c^2 = \sqrt{(p_e c)^2 + (m_e c^2)^2}$$

This gives us

$$P(E_e)\,dE_e = \frac{C}{c^3}F(Z,E_e)\,(E_e + m_e c^2)\sqrt{E_e^2 + 2E_e m_e c^2}\,(E_0 - E_e)^2\,dE_e$$

Using $E_\nu = E_0 - E_e$, we obtain the energy spectrum of the neutrinos emitted as

$$
\begin{aligned}
P(E_\nu)\,dE_\nu &= \frac{C}{c^3}F(Z,E_0 - E_\nu)\,\{(E_0 - E_\nu) + m_e c^2\} \\
&\quad \times\sqrt{(E_0 - E_\nu)^2 + 2(E_0 - E_\nu)m_e c^2}\;E_\nu^2\,dE_\nu \qquad (10\text{-}12)
\end{aligned}
$$

Let us ignore the Fermi function $F(Z,E_0 - E_\nu)$ for the time being and concentrate on the energy dependence given by the statistical part. If the end-point energy E_0 is much larger than the rest mass energy of the positron, as, for example, in the case of ^{8}B decay, the energy dependence of $P(E_\nu)$ is symmetric around the peak value at $E_\nu = E_0/2$, as shown in (a) of Fig. 10-4. On the other hand, if E_0 is comparable to $m_e c^2$, as, for example, in converting two protons to a deuteron, the function $P(E_\nu)$ is forward peaked, as shown in (b) of Fig. 10-4. In a log-log plot, the forward peaking appears even more pronounced, as we shall see later in Fig. 10-5.

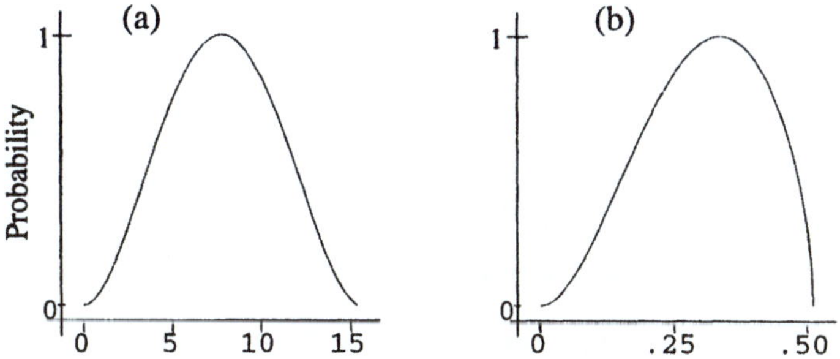

Figure 10-4: Energy spectra of neutrinos. In (a), the end-point energy is much larger than $m_e c^2$ and the shape is almost symmetric. In (b), the end-point energy is comparable to $m_e c^2$ and the spectrum is forward peaked. The vertical scales are normalized arbitrarily to unity at the peak.

Several corrections must be applied before $P(E_\nu)$, shown in Fig. 10-4, can be used as the neutrino spectrum for a specific β^+-decay in proton burning. The first is the Coulomb effect, given by the Fermi function $F(Z, E_0 - E_\nu)$. For β^+-decay, the positrons are given a boost in energy by the nuclear Coulomb field and the spectrum is pushed in a direction such that it appears to be more forward peaked than without the corrections. To conserve energy, the corresponding neutrino spectrum becomes less forward peaked than that given by statistical considerations alone. A second correction arises because the decay takes place in a stellar environment and distribution in the thermal energy of the particles must be taken into account. The main effect here is to smear out the distribution somewhat. In the special case of ^{8}B decay, the dominant final state is a broad one. In this case, the neutrino spectrum is further modified by a convolution of the distribution of the final state with the shape given by Eq. (10-12), as well as the corrections described above. In fact, because of the importance of the high-energy neutrinos emitted, very elaborated calculations have been carried out for the ^{8}B neutrino spectrum [21]. This, as well as spectra from other reactions given above, are summarized in Fig. 10-5.

Neutrino detectors. There are basically two ways to detect neutrinos. The first is by scattering from charged particles. If the energy transfer is sufficiently large, the charged particle recoils with a speed faster than that for light in the same medium. This results in Cerenkov radiation that can be detected using, for example, a scintillation counter. For the low-energy neutrinos coming from hydrogen burning, only scattering from elec-

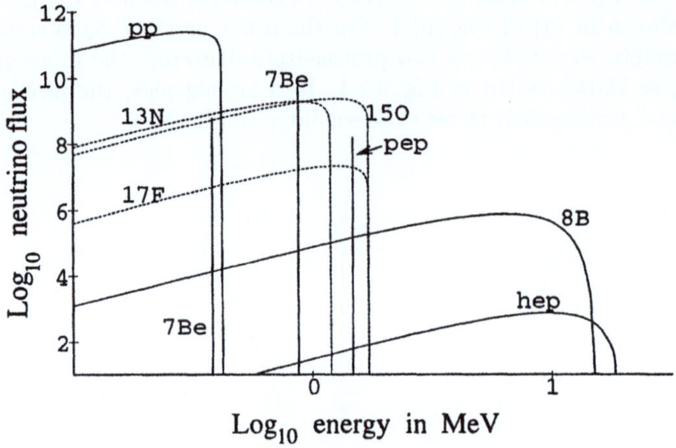

Figure 10-5: Log-log plot of the expected neutrino fluxes from various proton burning reactions in the sun [19]. Solid curves are those from PP-chains and dotted curves from the CNO cycle. Discrete lines come from a *pep* reaction and ^{7}Be from electron capture.

trons has any finite chance of producing such radiations. Furthermore, the medium must be transparent for the radiation to reach the detector. For these reasons, water is used in the Cerenkov detectors at Kamioka. The largest one, the Super Kamiokande, contains 50,000 tons of high-purity water.[3] In principle, any neutrino with energy comparable to the electron rest-mass energy of 0.5 MeV can scatter electrons to produced Cerenkov light. However, the practical limit of the Kamioka detectors is higher, around 7 MeV at the time of writing. As a result, only ^{8}B neutrinos from the sun are seen by the detectors [69].

Instead of ordinary water, the Sudbury Neutrino Observatory (SNO) employs 1000 tons of heavy water.[4] The advantage of using this far more expensive form of water is that, in addition to neutrino-electron scattering, the deuteron in the water is also sensitive to the neutral current reaction,

$$\nu_e + d \rightarrow p + n + \nu_e$$

The neutron liberated in the reaction may be captured by another nucleus through an (n, γ) reaction and a scintillation counter can be used to detect the γ-ray emitted. The minimum neutrino energy required to trigger this reaction is determined by the deuteron binding energy of 2.22 MeV. The advantage of the neutral current reaction is that it is equally sensitive to all three types of neutrinos, ν_e, ν_μ, and ν_τ. In contrast, electron-neutrino scattering favors ν_e (by a factor of 6). Similarly, the charged current reaction

$$\nu_e + d \rightarrow p + p + e^-$$

[3]See http://osksn1.hep.sci.osaka-u.ac.jp/kamioka_Eng/kamioka.html.

[4]For more up-to-date information, see http://snodaq.phy.queensu.ca/SNO/sno.html.

the inverse of that given by Eq. (10-5), is sensitive only to electron neutrinos. The reason that the type of neutrino is of interest here comes from the possibility of neutrino oscillation. We saw earlier in §5-5 that, under weak interaction, the decay products of quarks do not have definite flavor. If this happens also to neutrinos, the neutral lepton emitted in a nuclear β-decay will not be in an eigenstate either, and the observed neutrino flavor may change under suitable conditions, such as those encountered in passing through the outer layers of a star. If this does happen, some of the ν_e emitted from nuclear reactions in the interior of the sun may be transformed into ν_μ or ν_τ and would be missed by detectors sensitive only to ν_e. The neutral current reaction does not have this problem and may therefore be an important source of information on the question of whether there is oscillation in the solar neutrinos.

The second way to detect neutrinos is to use radiochemical methods. Nuclear reactions induced by solar neutrinos are extremely rare because of the small weak interaction cross section. Large detector size alone is insufficient as the signal produced may be lost in the volume. One way to get around the problem is to use reactions that produce radioactive nuclei with half-lives suitable for applying radiochemical techniques. The best known one in this category is the detector at Homestake Gold Mine that has been operating since 1968. It makes use of the reaction

$$\nu_e + {}^{37}\text{Cl} \to {}^{37}\text{Ar} + e^-$$

The chlorine comes in the chemical form CCl_4, an ordinary cleaning fluid. The product, ^{37}Ar, is unstable and decays back to ^{37}Cl by electron capture with a half-live of 35 days. Being a noble gas, the argon can be "flushed" out periodically from the cleaning fluid and counted. The radioactivity of the sample collected in this way gives a measure of the number of ^{37}Ar produced during the period of time and, hence, the number of neutrino reactions that have taken place in the detector. From the known cross section of the reaction, the flux of neutrinos going through the detector may be deduced. As shown in Table 10-4, the minimum energy a neutrino must have before it can be detected in this way is 0.813 MeV. As a result, it is sensitive to neutrinos from ^{8}B, *hep*, the two discrete lines from *pep*, and the higher energy one from the ^{7}Be electron capture reaction. A summary of the parameters for the neutrino-chlorine reactions is given in Table 10-4 together with those for the other reactions described below.

Table 10-4: Threshold energy Q in MeV for various neutrino detectors.

Detector	Reaction	E_B(left)	E_B(right)	Q	$\tau_{1/2}$
Homestake	$\nu_e + {}^{37}\text{Cl} \to {}^{37}\text{Ar} + e^-$	317.100	315.505	0.813	35 days
SAGE GALLEX	$\nu_e + {}^{71}\text{Ga} \to {}^{71}\text{Ge} + e^-$	618.948	617.934	0.232	11.43 days
Molybdenum	$\nu_e + {}^{98}\text{Mo} \to {}^{98}\text{Tc} + e^-$	846.243	843.776	1.685	4.2×10^6 yr
Kamiokande	$\nu + e^- \to \nu + e^-$	—	—	$\gtrsim 5$	0
SNO	$\nu_e + d \to p + n + \nu_e$	2.224	0.0	2.224	0

Two other experiments, Gallium Neutrino Observatory (GNO, more commonly known as GALLEX) and Russian-American Gallium Experiment (SAGE), make use of the reaction

$$\nu_e + {}^{71}\text{Ga} \to {}^{71}\text{Ge} + e^-$$

The threshold energy here is only 0.232 MeV, below the end-point energy of the *pp*-reaction common to all three PP-chains. Since each ^{4}He produced in the PP-chains involves one *pp*-reaction, the *pp*-neutrinos bear almost a direct connection with the energy produced. Such a simple relation has the advantage of making the comparison with measured values much more meaningful. The GALLEX/GNO detector consists of, at the time of writing, 30.3 tons of gallium (12 tons ^{71}Ga) in the form of gallium chloride solution and has been taking data since 1991. A plan is underway to upgrade the detector to 100 tons of gallium. The SAGE detector is made of 57 tons of metallic gallium and has been in operation since 1990. Since gallium is a relatively rare metal, the amounts used in the two detectors represent a significant fraction of the annual production of the metal in the whole world.

Another class of radiochemical "detectors" is represented by the one at Henderson Molybdenum Mine in Colorado. The reaction

$$\nu_e + {}^{98}\text{Mo} \to {}^{98}\text{Tc} + e^-$$

has a neutrino threshold energy of 1.685 MeV and produces ^{98}Tc with a half-live of 4.2×10^6 years. The threshold energy is relatively high. On the other hand, the half-life is so long that extremely lengthy periods of time can be used to accumulate the radioactive product produced in the reaction. In fact, one can take the age of the earth as the period. In this approach, the experiment involves counting the amount of ^{98}Tc in molybdenum ore and comparing the result with expectation. This is very similar in spirit with the chlorine and gallium experiments except that the period is the age of the earth.

The solar neutrino problem. After running the Homestake Mine experiment for more than two solar cycles, it is found that the measured neutrino flux is only about a third of the expected value. Both the gallium and Kamiokande results also support the conclusion that the measured flux is lower than the values expected. The deficiency in the measured solar neutrino flux is therefore "real" in terms of the best available knowledge of the physics involved. For this reason, it has been generally referred to as the *solar neutrino problem* [20].

Because of the extremely small cross section, it is convenient to adopt a new unit, the solar neutrino unit (SNU for short), for any quantitative discussions of the solar neutrino question. One SNU is defined as the flux of neutrinos that produces one reaction a second for every 10^{36} target atoms. The measured value at Homestake Mine is 2.1 ± 0.3 SNU. Some variations with solar cycle are visible in the data, but they are not large enough to be of concern to us if our interest is in the deficiency compared with the expected value.

Since the ν_e + ^{37}Cl reaction is only sensitive to the high-energy neutrinos from ^{8}B decay and the *hep* process, it is more subject to uncertainties in our knowledge of the physics involved. One of the major difficulties is that we do not have a method, independent of neutrinos, to make observations on the conditions existing in the interior of the sun, where the nuclear reactions take place. As a result, models must be constructed on the neutrino spectra we can expect using the best known physical principles and available values for the input parameters. The expected value for the ^{37}Cl measurement, obtained from the "standard solar model," is 7.9 ± 2.4 SNU [19]. The uncertainty represents very conservative estimates of the input parameters that went into the model. The discrepancy of the calculated values from the measured ones is, therefore, statistically significant and may well imply new physics. Partly for this reason, it has stimulated interest in carrying out the newer solar neutrino measurements.

The average measured value of GALLEX at the time of writing is 77 ± 10 SNU. This is only about half of the expected value of 127 SNU from the standard solar model. The measured value of SAGE, 69 ± 13 SNU, is comparable to the GALLEX results. Thus, the gallium measurements also show a deficiency in the flux of solar neutrinos, confirming the existence of the solar neutrino problem.

The original Kamiokande Cerenkov detector was designed to detect proton decay and other high-energy events. It was converted in 1986 to observe low-energy neutrinos by reducing the background. Being a "real-time" detector, it is able to show that the observed neutrinos are coming from the direction of the sun. On the other hand, the present threshold of $\sim$7 MeV makes it sensitive only to the ^{8}B neutrinos. The measured result is about half of that expected from the standard solar model [69]. However, no correlation with solar spot activity was evident from the data.

If we take all the measured results together, it is quite clear that the observed solar neutrino flux is less than the value expected based on the best knowledge we have on the physics involved. This is a very interesting situation, as advances in physics are often made when observations become precise enough to challenge the existing views. In this way, the solar neutrino problem has a chance to become the doorway to new physics we have not yet thought about. No doubt the new detectors of Super Kamiokande and SNO will shed some additional light on the question.

10-5 Helium Burning and Beyond

When the central part of a star runs out of hydrogen fuel, there is a shortage of thermal pressure to maintain the hydrostatic equilibrium and gravitational contraction begins once again. This raises the central temperature of the star until it is high enough for helium burning to start. Since the outer layers of the star are cooler and less dense, nuclear reactions take place at slower rates. As a result, the central part of the star evolves into helium burning while the outer layer continues with the process of hydrogen burning.

Different layers of stellar evolution. The total amount of fusion energy available in helium burning is much less than hydrogen burning, as the binding energy per nucleon of ^{4}He is 7.1 MeV, less than 1.5 MeV away from the peak value around $A \approx 56$. We shall see in a more detailed discussion later that the first step in helium burning is to fuse three ^{4}He into a ^{12}C nucleus. Through an (α, γ) reaction, ^{12}C absorbs an α-particle to form ^{16}O. At the end of this stage, the Coulomb barrier becomes too high for further α-particle capture to take place. The star again repeats the sequence of first exhausting the available fuel and then contracting to even smaller size and higher temperature (and density) until the next group of nuclear fusion reactions can take place. Since the reactions are going on at higher rates in the central part of the star, the evolution takes place on shorter time scales than the outer layers. The net result is that an onion-like structure develops, with successive inner layers of the star undergoing later and later stages of evolution. This is schematically shown in Fig. 10-6.

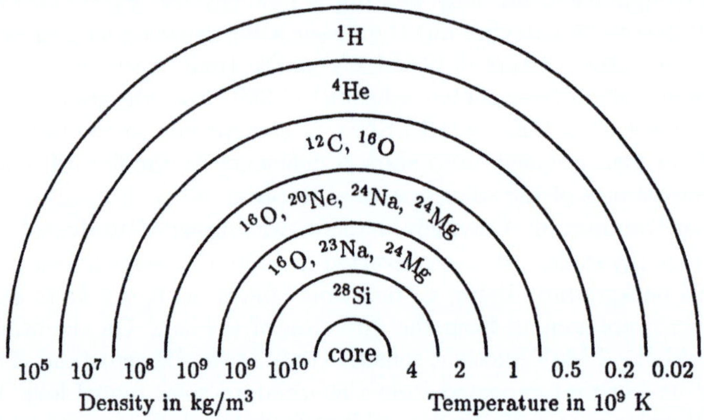

Figure 10-6: Schematic diagram showing the dominant nuclear components, temperature, and density in different layers of a massive star prior to supernova explosion [120].

When the temperature of a layer reaches $kT \sim 1$ MeV, particles as heavy as silicon occupy a significant fraction of the nuclei present. At the same time, γ-rays in thermal equilibrium with the particles are energetic enough to cause photodisintegration of the nuclei present. One consequence of such an environment is that a number of protons and neutrons are knocked out of the nuclei and become the source of other nuclear reactions. In this way, some of the nuclei outside the $4n$ chain are created.

The combination of higher temperature in the stellar interior and smaller binding energy differences in the nuclei involved shortens the duration for each successive stage of nucleosynthesis. In a massive star, 25 solar masses, for example, the hydrogen burning stage takes several million years (compared with the order of 10^{10} years for the sun). The helium burning stage is about an order of magnitude shorter in time. In the next stage, when ^{12}C becomes the dominant fuel, the lifetime is only a few hundred years. Oxygen and silicon combustions at even later stages are estimated to take, respectively, only 6 months and 1 day. At the mean time, the density of the star goes up. At the hydrogen burning stage, a 25-solar-mass star has only five times the density of water. At each successively later stage, the density goes up by more than two orders of magnitude, ending up to be about 30 million times the density of water at the silicon burning stage. This value is, however, still far less than the nuclear matter density of 3×10^{14} that of water (see, e.g., Ref. [64].)

Triple-α process. Let us return first to the end of hydrogen burning. Gravitational contraction starts again, and both stellar temperature and density begin to rise. Conversion of ^{4}He to heavier nuclei becomes possible when the interior of a star reaches temperature $kT \sim 10$ to 20 keV and density $\rho \sim 10^5$ to 10^8 kg/m^3. This is the hydrostatic helium burning stage of a star.

From a nuclear physics perspective, there are several interesting points that merit special attention. The first is the absence of stable $A = 5$ nuclei, as we saw earlier in §10-1. Neither ^{5}He ($\tau_{1/2} = 0.7 \times 10^{-22}$ s) nor ^{5}Li ($\tau_{1/2} = 3.0 \times 10^{-22}$ s) nuclei live long enough for a sufficient number to be built up in a star. As a result, it is impossible to continue the process of proton capture beyond ^{4}He to make the heavier nuclei, as in the hydrogen burning stage. Instead, α-particle capture becomes the important mechanism.

The second is that ^{8}Be is unstable. The ground state decays into two α-particles with $\tau_{1/2} = 6.7 \times 10^{-17}$ s. As soon as ^{8}Be is made through the reaction

$$^4\text{He} + {}^4\text{He} \rightarrow {}^8\text{Be}$$

it decays spontaneously back to two α-particles. However, the half-life is five orders of magnitude higher than those, for example, for the two mass 5 nuclei given in the previous paragraph. As a result, there is always some small amounts of ^{8}Be present at any time. The equilibrium density depends on the rate of creation. At $T \sim 1$ to 2×10^8 K, the value is about one ^{8}Be among 10^9 α-particles. Such a low abundance will not normally be sufficient for further α-particle absorption to form heavier nuclei except for a very fortuitous "accident" in the structure of ^{12}C.

This leads us to the third interesting point, the "triple-α process." The binding energy of ^{12}C is 92.162 MeV and those of ^{8}Be and ^{4}He are, respectively, 56.500 and 28.296 MeV. The reaction

$$^4\text{He} + {}^8\text{Be} \rightarrow {}^{12}\text{C} + \gamma$$

leading to the ground state of ^{12}C, therefore, has a Q-value of 7.366 MeV. Furthermore, detailed nuclear structure calculations show that the ground state wave function of ^{12}C has a relatively small overlap with the product of those for ^{8}Be and ^{4}He. As a result, the reaction cannot be a resonant one and the production rate for this way to create ^{12}C is too small to explain the observed abundance.

Fortunately, there is a 0^+ excited state at 7.66 MeV in ^{12}C. Since the spin and parity of both ^{8}Be and ^{4}He are 0^+, the capture can go through the excited 0^+ state by the strong s-wave channel. The energy difference from the threshold is now less than 300 keV, and the capture can take place almost at the most favorable circumstance for a resonance reaction. Furthermore, the wave function of the 7.66-MeV 0^+ state resembles that of three α-particles arranged in a straight line, rather than in a triangle, as in the case of the ground state. This is shown schematically in Fig. 10-7. The large overlap in the wave functions provides the additional advantage required to make it possible to convert the small concentration of ^{8}Be efficiently into ^{12}C. Since ^{8}Be lasts only for such a short time, the process is essentially one that convert three α-particles to a ^{12}C,

$$^4\text{He} + {}^4\text{He} + {}^4\text{He} \rightarrow {}^{12}\text{C} + \gamma$$

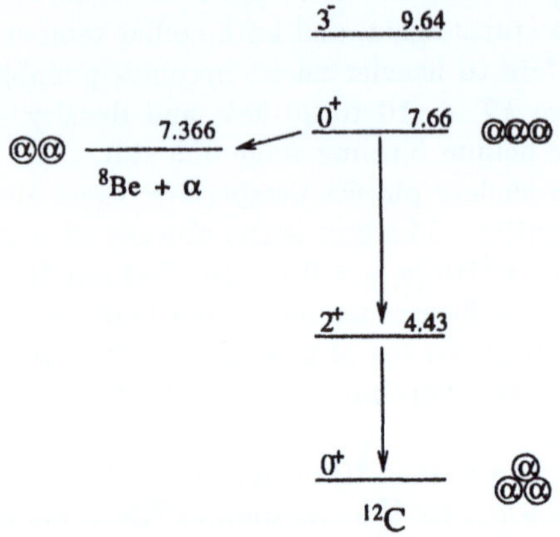

Figure 10-7: Low-lying energy level structure of ^{12}C and the triple-α process. (Plotted using data from Ref. [95].)

and hence the name triple-α process. In the absence of this special 7.66-MeV state, it would have been almost impossible for nucleosynthesis to proceed to heavier elements beyond ^{8}Be in helium burning stars.

From carbon to oxygen. By capturing an α-particle, ^{12}C is converted into ^{16}O. The reaction

$$^4\text{He} + {}^{12}\text{C} \rightarrow {}^{16}\text{O} + \gamma$$

is, by far, the most important one that consumes ^{12}C. Competitive processes such as

$$^{12}\text{C} + {}^{12}\text{C} \rightarrow {}^{24}\text{Mg} + \gamma$$

are far less effective because of the higher Coulomb barriers and unfavorable Q-values [14 MeV in the ^{12}C(^{12}C,γ)^{24}Mg example] [38].

In fact, the triple-α process and ^{12}C$(\alpha,\gamma)^{16}$O are the two most important reactions that control the subsequent production rates and, hence, the relative abundances of heavier elements. Calculations have shown that, in massive stars, the ratio ^{16}O to ^{12}C affects the amount of heavier elements produced in their lifetimes, as well as the properties of the remnant after a supernova explosion [143]. Large values, for example, favor the production of heavier elements, leading to a more massive iron core during the pre-supernova stage, and increase the probability of leaving a black-hole remnant instead of a neutron star. Small values, on the other hand, produce relatively much less heavier elements. The calculated values are plotted in Fig. 10-8 for a few of the more representative elements as illustration.

Although the triple-α reaction cannot be reproduced easily in the laboratory, the cross section may be deduced from other measurements with an uncertainty of about 15%. The primary reason for this is the simple level structure of ^{12}C at the energies involved, as can be seen from looking at Fig. 10-7. This is not true for ^{16}O, and the cross section for the low-energy ^{12}C$(\alpha,\gamma)^{16}$O reaction becomes one of the important measurements in nuclear astrophysics.

The low-lying energy level structure of ^{16}O relevant for the reaction is shown in Fig. 10-9. In the region of excitation energy near the reaction Q-value of 7.162 MeV

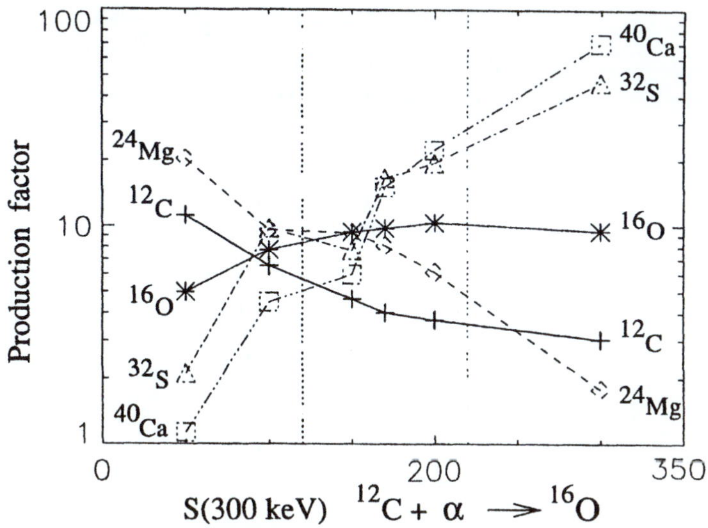

Figure 10-8: Calculated production factors of a few representative elements as a function of the S-factor for the $^{12}C(\alpha,\gamma)\,^{16}O$ reaction. (Values taken from Ref. [143].)

for the $^{12}C(\alpha,\gamma)^{16}O$ reaction, there are no 0^+ states. Consequently, the capture must proceed through channels with orbital angular momentum $\ell > 0$. Since the angular momentum barrier increases almost quadratically with ℓ [proportional to $\ell(\ell+1)/r^2$], the next most likely reaction channels are $\ell = 1$ leading to 1^- states and $\ell = 2$ to 2^+ states. In each of these two channels, there are two states near the threshold, the 2^+ states at 9.85 and 6.92 MeV and the 1^- states at 9.60 and 7.12 MeV. Both the 6.92- and 7.12-MeV states are subthreshold. However, being states with finite widths, the distributions of their reaction strengths extend into regions with positive Q-value. As a result, they make significant contributions to reactions that are extremely low in energy, such as those taking place in the stellar environment of interest to us here.

We can find out the reaction energies involved in helium burning stars from the following consideration. A typical stellar temperature here is $kT \sim 0.015$ MeV ($T \approx 2\times10^8$ K). From Eq. (10-4), the peak of the reaction rate is found to occur at

$$E_{\max} = (\tfrac{1}{2}bkT)^{2/3}$$

Using Eq. (10-2), we obtain the result

$$b = \pi\alpha Z_1 Z_2\sqrt{2\mu c^2} = 0.990 Z_1 Z_2\sqrt{A}$$

in $\text{MeV}^{1/2}$. Here $A = A_1A_2/(A_1+A_2)$ is the reduced mass in atomic mass units (amu), with A_1 and A_2 as the masses of the two nuclei involved. For the $^{12}C(\alpha,\gamma)^{16}O$ reaction,

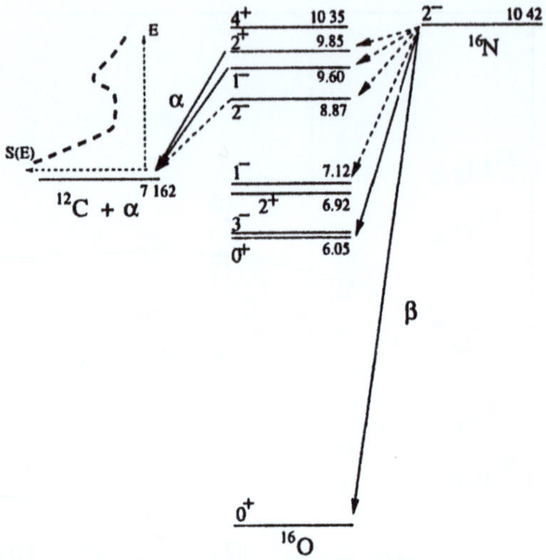

Figure 10-9: Low-lying energy level structure of ^{16}O and the $^{12}C(\alpha,\gamma)^{16}O$ reaction. (Plotted using data from Ref. [136].)

we have $A_1 \approx 4$, $A_2 \approx 12$, $Z_1 = 2$, and $Z_2 = 6$. This gives us

$$E_{\text{max}} = 0.3 \text{ MeV}$$

or 300 keV. In the laboratory, it is impossible to carry out α-particle measurements down to such low energies. First, it is not easy to produce an intense beam when the energy is so low. Second, the Coulomb barrier for the reaction is around 6 MeV and the probability for a 300-keV α-particle to penetrate into a ^{12}C nucleus is so small that one can expect only a few counts a year with the best available accelerator and reasonable target thickness. For this reason, the reaction rate for $^{12}C(\alpha,\gamma)^{16}O$ at energies of astrophysical interest is far less well known than that, for example, for the triple-α process.

Most measurements are carried out at energies on the order of mega-electron-volts, a region where experiments can be reasonably carried out. In these cases, the dominant contributions come from states above the $^{12}C(\alpha,\gamma)^{16}O$ reaction threshold. It is not easy to extrapolate from these measurements the value for the cross section at 300 keV, as the contributions from the subthreshold states, important for astrophysical interests, are insignificant at such high energies. For this reason, the uncertainty was, until recently, almost a factor of 2 for $S(300 \text{ keV})$, the value of the S-factor at E_{max} of 300 keV. As we can see from Fig. 10-8, a more precise value is required to deduce the relative abundances of ^{12}C to ^{16}O in helium burning stars and thence to make predictions of the rate of nucleosynthesis for heavier elements. In fact, the measurement of $S(300 \text{ keV})$ for the $^{12}C(\alpha,\gamma)^{16}O$ has been, on occasion, referred to as one of most important experiments to be carried out in nuclear astrophysics.

Recently, the uncertainty in the $\ell = 1$ part, i.e., contributions from 1^- states, has been reduced greatly using the α-decay rates from low-lying excited states of ^{16}O [15],

$$^{16}\text{O}^* \rightarrow {}^4\text{He} + {}^{12}\text{C}$$

This is the inverse of the $^{12}\text{C}(\alpha,\gamma)^{16}\text{O}$ reaction. The excited states of ^{16}O are obtained from the β^--decay of ^{16}N, a radioactive nucleus ($\tau_{1/2} = 7.13$ s) that can be produced in large quantities from a radioactive beam setup. The equivalent $^{12}\text{C}(\alpha,\gamma)^{16}\text{O}$ cross section is obtained in this way down to ~800 keV in energy, far below the values that can be accessed in direct measurements. As a result, the extrapolation to 300 keV is made more reliable and a value of $S_{\ell=1}(300\text{ keV}) = 80$ keV-barn is obtained. The uncertainty is reduced to about 30% (from a factor of 2 before). Unfortunately, at $E_{\text{max}} = 300$ keV, the 2^+ states are expected to be make comparable contributions. Until the uncertainties in the $\ell = 2$ part of the S-factor are also reduced, the error bar for the complete S-factor remains far too large.

Carbon burning and beyond. When all the available ^{4}He in the central part of a star is used up, the core goes through another stage of gravitational contraction and rise in temperature. When $T \sim 10^9$ K, corresponding to $kT \sim 100$ keV, reactions involving the conversion of any ^{12}C remaining after helium burning become possible, such as

$$\begin{aligned} {}^{12}\text{C} + {}^{12}\text{C} &\rightarrow {}^{24}\text{Mg} + \gamma \\ &\rightarrow {}^{23}\text{Na} + p \\ &\rightarrow {}^{23}\text{Mg} + n \\ &\rightarrow {}^{20}\text{Ne} + \alpha \\ &\rightarrow {}^{16}\text{O} + \alpha + \alpha \end{aligned}$$

The time span for the carbon burning phase is several orders of magnitude shorter than that for helium. Furthermore, only the core can reach the temperature and density required for the reactions to take place. The layer just outside the core still has enough fuel left to continue in the helium burning stage and the layer outside enough to continue hydrogen burning.

Another interesting case in nuclear structure is the absence of significant contributions from the $^{16}\text{O}(\alpha,\gamma)^{20}\text{Ne}$ reaction in stellar nucleosynthesis. The binding energy difference between ^{20}Ne and the sum of those for ^{4}He and ^{16}O is 4.73 MeV. However, the excited states in ^{20}Ne around this energy cannot be formed easily by ^{16}O absorbing an α-particle. This is, to a large extent, because of the unfavorable angular momentum barriers, as can be seen from Fig. 10-10. As a result, the reaction

$$^{16}\text{O} + \alpha \rightarrow {}^{20}\text{Ne} + \gamma$$

does not become important at the end of α-particle capture by ^{12}C to form ^{16}O. On the other hand, because of the relatively low binding energy for α-particles, photodisintegration of ^{20}Ne can be significant when the temperature rises to new heights at the

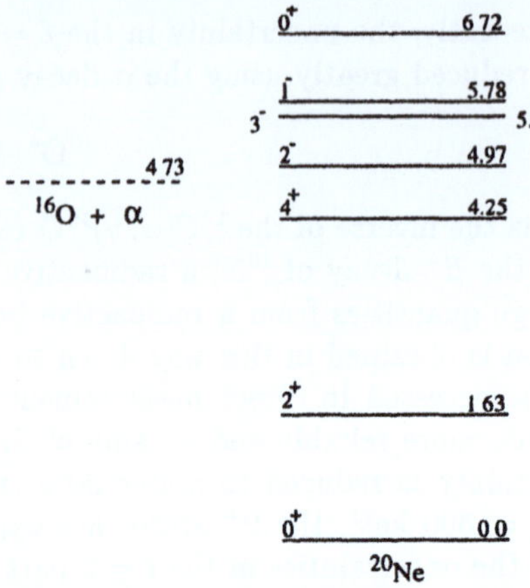

Figure 10-10: Low-lying energy level spectrum of ^{20}Ne and ^{16}O$(\alpha,\gamma)^{20}$Ne reaction. (Plotted using data from Ref. [95].)

end of carbon burning. This produces a short-lived neon burning stage in which the reaction

$$^{20}\text{Ne} + \gamma \rightarrow {}^{4}\text{He} + {}^{16}\text{O}$$

produces α-particles that can be used in the reaction

$$^{4}\text{He} + {}^{20}\text{Ne} \rightarrow \gamma + {}^{24}\text{Mg}$$

to produce ^{24}Mg.

At even higher temperatures, 2 to 3 $\times$ 10^9 K, it is possible to convert ^{16}O into heavier elements, for example,

$$\begin{aligned} ^{16}\text{O} + {}^{16}\text{O} &\rightarrow {}^{32}\text{S} + \gamma \\ &\rightarrow {}^{31}\text{P} + p \\ &\rightarrow {}^{31}\text{S} + n \\ &\rightarrow {}^{28}\text{Si} + \alpha \\ &\rightarrow {}^{24}\text{Mg} + \alpha + \alpha \end{aligned}$$

On the other hand, the reaction ^{12}C + ^{16}O is not considered to be important in nucleosynthesis, as nearly all the ^{12}C is exhausted before the temperature is high enough for the reaction to become significant. (See, e.g., Ref. [44].)

When the temperature is between 3 and 4 $\times$ 10^9 K, conversion of two ^{28}Si to one ^{56}Ni becomes possible. At this stage, γ-rays at the high-energy tail of the Maxwellian distribution are sufficiently energetic for photodisintegration to compete with nuclear fusion. As a result, direct conversion of silicon to nickel is relatively rare. At thermal equilibrium, the radiant energy is proportional to T^4 according to the Stefan-Boltzmann law. In this sea of γ-rays, large numbers of α-particles, protons and neutrons are liberated by photodisintegration, allowing a variety of nuclear reactions to make heavier nuclei up to and just beyond $A \sim 56$. However, temperature and density decrease as we move away from the stellar core. While silicon burning is taking place in the core, the outer layers are still in the various earlier stages of nucleosynthesis.

10-6 Supernova and Synthesis of Heavy Nuclei

Supernovas hold a special place in nucleosynthesis because of the heavy elements they produce. Since binding energy per nucleon decreases beyond $A \sim 56$, it takes energy to create elements that are heavier. The processes we saw in previous sections are ineffective for the present purpose. Through random chance, some heavy elements are produced in the various hydrostatic processes; however, the total amounts are inconsequential. In a supernova explosion, on the other hand, a shock wave is generated when the collapsing core rebounds. On its way through the outer layers of the star, the shock wave provides the ideal condition for endothermic reactions to synthesize elements beyond $A \sim 56$.

Only massive stars end their lives as supernovas. For the core of a star to undergo sudden collapse, there must be sufficient gravitational energy in the system to start with. The remnant of a supernova is often a neutron star or a black hole. This requirement puts the core to be more massive than the Chandrasekhar limit of 1.4 solar mass, as a part of the material is ejected in the ensuing explosion. Furthermore, for the collapse to take place, there cannot be any thermal pressure to act against gravitational force. The core must therefore be totally exhausted of nuclear energy. The combination of these two factors puts the likely candidates to be stars with total mass exceeding 10 times that of our sun. In fact, most model calculations of supernovas treat 20 to 30 solar masses as the typical case. For such massive stars, the core constitutes only about 10% of the total. For smaller stars, such as our sun, the total gravitational energy is inadequate and the more likely end is a white dwarf, a small-size high-temperature object that slowly radiates away any remaining thermal energy.

When all the nucleons in the central part of a massive star are converted into $A \sim 56$ nuclei, principally ^{56}Fe, the temperature reaches around 4×10^9 K. Since all the available nuclear fuel is exhausted, there is no longer the thermal pressure to counterbalance further gravitational contraction and the temperature rises to even higher values. Once the star arrives at this stage of its evolution, the dominant nuclear interactions change into those that consume thermal energy rather than supplement it, as in the earlier stages. The cooling can take place rapidly, on the order of a fraction of a second, depending on the stellar mass. Without the thermal pressure, the inner core of the star implodes with the speed of a free fall. The material is compressed in the process to several times nuclear matter density. Like a stiff spring, the gravitational energy released from the collapse is stored momentarily as potential energy in the compressed nuclear matter. To shed this "extra" energy, the core rebounds and pushes part of the material in the core to the outer layers of the star at supersonic speeds while the remaining material goes into either a neutron star or a black hole. Our main concern here is the explosive nucleosynthesis caused by the ejected material interacting with the nuclei in the outer layers that are still in various earlier stages of hydrostatic burning.

Neutrino cooling. In the previous sections we have been concerned with the production of energy in stars without paying any attention to the mechanism by which energy is radiated into the surrounding. In addition to kinetic energies of the products, γ-rays, positrons, and (electron) neutrinos are produced in copious amounts from various nuclear reactions. Furthermore, since the star as a whole is electrostatically neutral, there are a large number of electrons around, the same number as protons. The positrons that result from conversion of protons into neutrons are quickly annihilated by the surrounding electrons through the reactions

$$e^+ + e^- \to \gamma + \gamma \qquad\qquad e^+ + e^- \to \nu + \bar{\nu}$$

The former is electromagnetic and, therefore, much more likely to occur than the latter weak interaction process by a factor as high as 10^{19}. However, except at the outer layer, the γ-ray pairs produced must penetrate through a thick blank of matter before they can leave the star. Furthermore, the mean free path of γ-rays is relatively short, and as a result, it takes a long time for them to reach the surface. In contrast, the neutrinos suffer at most one interaction on the average before they are outside the star. For this reason, the fraction of energy carried away by neutrinos during the short-lived later stages of the stellar evolution is actually larger than that by γ-rays.

Neutrino cooling is even more important during the final collapse of the core of a massive star. For simplicity, consider a case where the core has 1.5 solar mass (3×10^{30} kg) and it collapses into a neutron star. In addition, we shall assume, again for simplicity, that all the material before the collapse is in the form of ^{56}Ni, with equal number of neutrons and protons. The number of nucleons in the core is then

$$n_{\text{nucleon}} = \frac{3 \times 10^{30}\ \text{kg}}{1.67 \times 10^{-27}\ \text{kg}} = 2 \times 10^{57}$$

To turn into a neutron star, all the protons must be converted into neutrons and the number is half of that for nucleons, or 10^{57}. Most of the conversion is through electron capture, each accompanied by a neutrino. This puts the number of neutrinos emitted to be 10^{57}. A reasonable estimate of the average amount of energy carried away by each neutrino is 15 MeV. This value may be obtained by considering the neutrinos to be a degenerate Fermi gas occupying the same volume as the nucleons. In this limit, the average energy is related to the Fermi energy, in the same way as we have done earlier in arriving at Eq. (4-72) from the Fermi energy of nucleons. The only difference here is that we do not have the isospin degree of freedom as in the case of infinite nuclear matter. Consequently, our density is only half of that of Eq. (4-69) and our Fermi momentum $(1/2)^{1/3}$ is as large as that of Eq. (4-70). The average energy is then $(1/2)^{2/3}$ of $\bar{\epsilon}$ (=23 MeV), or ~15 MeV. For 10^{57} neutrinos, the total amount of energy carried away is

$$E_\nu(\text{total}) \sim 1.5 \times 10^{58}\ \text{MeV} \sim 2 \times 10^{45}\ \text{J}$$

that is, about 10% of the gravitation binding energy of the neutron star.

Core collapse. In addition to neutrino cooling, part of the thermal energy can also be taken away by photodisintegration of nuclei into α-particles and nucleons. At the end of silicon burning, the temperature is over 4×10^9 K, about $\frac{1}{3}$ MeV in energy units. Consequently, photons at the higher end of the Maxwell-Boltzmann distribution have more than adequate energy to dissociate the nuclei present. As example, consider a ^{56}Fe nucleus, the most tight bound member of the $A = 56$ isobar. The energy required to remove an α-particle is only 7.6 MeV. The corresponding amounts for a proton and a neutron are, respectively, 10 and 11 MeV. There is, therefore, fairly high probability for photodisintegration to generate substantial numbers of α-particles and nucleons. The particles produced are important, as we shall see soon, in making some of the nuclei beyond $A \sim 56$ during the rebound after the core collapse. At the same time, the energy consumed in the reactions depletes the thermal energy further.

Without counteraction by thermal pressure, the core contracts further. Since the density is already quite high at the end of silicon burning, in excess of 10^{10} kg/m^3, any further collapse results in electron capture to convert the protons, bound as well as free ones, to neutrons. The neutrinos released in the neutronization process carry energy away from the core, as we have seen earlier. The whole sequence of events from the end of silicon burning to complete collapse takes less than a second, perhaps as short as only milliseconds.

At the end of the collapse, the kinetic energy must have been converted into some form of potential energy. The most likely candidate is in compressing nuclear matter. However, we should not ignore the contributions of the internal energy of matter, through such means as creating mesons and baryons other than nucleons. Unfortunately, we do not have much experimental guidance to make a good estimate of the situation. For a proper treatment, we need an equation of state that can take us beyond ordinary nuclear matter to the high temperatures and densities we are dealing with here. As we have seen in Chapter 9, lattice gauge investigations and ultra-relativistic heavy-ion collision studies may give us some hints on how to do this in the near future.

The observational evidence strongly suggests that the core rebounds after the collapse, with material ejected at supersonic speeds. As far as nucleosynthesis is concerned, we are more interested in the shock waves and ejected material sent through the mantle of the star. Both are fundamental in creating nuclei beyond $A \sim 56$. Furthermore, the rebound is sufficiently energetic that a large fraction of nuclei in the outer layers of the star are blown off into the interstellar space and become a part of the raw material for future star formation and nucleosynthesis. Without these "explosions," nuclei made in one star remain with the star and very little heavy elements become available to make the planets and young stars such as those forming the solar system.

Explosive nucleosynthesis. According to Fowler and Hoyle [64], the mass of a star just before a supernova explosion is 57% in the ^{16}O-rich mantle and 33% in an outer shell consisting of H and ^{4}He. The remaining 10% is in the collapsing core. When the shock wave travels through the different regions, a variety of heavy-ion reactions can take place, such as

$$^{16}\mathrm{O} + {}^{16}\mathrm{O} \rightarrow {}^{28}\mathrm{Si} + {}^{4}\mathrm{He} \qquad\qquad {}^{28}\mathrm{Si} + {}^{28}\mathrm{Si} \rightarrow {}^{56}\mathrm{Ni} + \gamma$$

In the outer layers, the shock wave causes conversion of hydrogen into helium and helium into oxygen. Some elements heavier than $A = 56$ are also produced, but the expected abundances are far less than the observed values shown in Fig. 10-1.

Because of the drastically different environment, nuclear reactions taking place during explosive nucleosynthesis can be quite different from those in hydrostatic burning stages described earlier. Here, we are dealing with higher temperatures and shorter time scales. For example, in explosive hydrogen burning, the (hot) CNO cycle operating at $T \sim 10^8$ to 10^9 K becomes more important than the PP-chains. To appreciate some of the differences, let us again start with the $^{12}\mathrm{C}(p,\gamma)^{13}\mathrm{N}$ reaction at the top of the main circle in Fig. 10-3. Since the time scale is much shorter, it is possible for the ^{13}N created to capture another proton to form ^{14}O before β^+-decay to ^{14}N. Here, it rejoins the main CNO cycle again through one further proton capture to form ^{15}O. The product then β^+-decays to ^{15}N. The cycle is completed through a $^{15}\text{N}(p,\alpha)^{12}\text{C}$ reaction to go back to ^{12}C. Although the nuclei involved are very similar to the corresponding CNO cycle in hydrostatic hydrogen burning, the reaction rates are quite different because of the higher temperature. As a result, the production rates and, hence, the resulting relative abundances of isotopes produced are also different. For example, the observed abundances of ^{12}C, ^{15}N, and ^{17}O are believed to be enhanced by the hot CNO cycle. Similarly, in explosive oxygen burning, ^{32,34}S, ^{35}Cl, ^{36,38}Ar, ^{40}Ca, and ^{46}Ti are produced, and in silicon burning, ^{42}Ca, ^{54}Cr, ^{56,58}Fe, ^{58}Ni, and ^{54}Mn are created relatively more abundantly than otherwise [13].

Creation of heavy elements. For elements beyond iron and nickel, the Coulomb barriers are high because of the large numbers of protons. For proton and α-particle captures, the temperature must rise above 5 to 6 $\times$ 10^9 K to make them sufficiently probable. On the other hand, the average thermal energy is now around 0.5 MeV and photodisintegration is even more important than at the hydrostatic silicon burning stage. The competition between formation of heavy elements by charged particle capture and destruction by photodisintegration produces an equilibrium density too small to be of any importance as far as the final abundances of these elements are concerned.

The bulk of the observed heavy elements are made by capturing neutrons one at a time. The process creates increasingly more neutron-rich nuclei and must be interdispersed with β^--decay to keep the products staying more or less in the valley of stability. Under such circumstances, the observed abundance of an element with mass A is the result of a balance between the production rate to make the element from element $A-1$ and the destruction rate to form the next element $A+1$. The relation between the abundances for different elements may by expressed in terms of the variation in the density of element A,

$$\frac{dN_A}{dt} = N_{A-1}N_n < \sigma_{A-1}v_{A-1} > -N_A N_n < \sigma_A v_A >$$

where N_A and N_{A-1} are, respectively, the number density of elements A and $A-1$, and N_n is the neutron density. The neutron absorption cross sections are given by σ_A and σ_{A-1}. The relative velocities v_A and v_{A-1} control the probabilities for collision between a neutron and the two types of nuclei involved. Both the cross sections and relative velocities are functions of the energy. For our purpose, we are more interested in the results averaged over energy distribution. This is indicated by the angle brackets. The equilibrium density is given by the condition

$$\frac{dN_A}{dt} = 0$$

However, since we are dealing here with a chain of reactions involving neutron captures as well as β^--decays, the relations for different elements are coupled together by the fact that, for example, N_A occurs on the right-hand sides for both dN_A/dt and dN_{A+1}/dt. As a result, the solution must be obtained by solving a set of such equations.

Since neutrons do not carry a net charge, we have the advantage over charged particle reactions in that we do not have the large hindrance factor at low energies due to Coulomb barrier penetration, $\exp\{-b/\sqrt{E}\}$ in Eq. (10-2). The distribution of neutron reaction rates is controlled essentially by the Maxwellian distribution for neutron thermal energy, rather than that shown in Fig. 10-2 for protons and α-particles. At the same time, the density of excited states for heavy nuclei is high, more than 10^5 MeV^{-1}. As a result, we can expect that a neutron can always be captured under the most favorable condition, such as $\ell = 0$, unlike the situation we saw earlier for ^{4}He capture by ^{16}O and ^{20}Ne.

By nuclear physics standards, the neutrons we are interested in here are quite low in energy, from tens of kilo-electron-volts at $T \sim 0.5 \times 10^9$ K to 0.5 MeV at 6×10^9 K. At these energies, the neutron capture cross section is inversely proportional to the velocity. This factor favors reactions at the lower end of the neutron energy distribution. Since the number of reactions per unit time is the product of the cross section σ and the frequency of collision, given by the relative velocity v between the two particles, the rate for

$$(A, Z) + n \rightarrow (A + 1, Z) + \gamma$$

is more or less constant in the energy region. Under such conditions, the half-life for nucleus (A, Z) due to the (n, γ) reaction is inversely proportional to the neutron density N_n and is given by

$$\tau_{(n,\gamma)} = \frac{\ln 2}{N_n < \sigma v >}$$

If we take 0.1 barn (10^{-29} m^2) as a reasonable average value for σ and $v = 10^{-2}c$ (corresponding to $kT \sim 50$ keV), we have

$$\tau_{(n,\gamma)} \sim \frac{10^{23}}{N_n} \text{ s}$$

For neutron density $N_n \sim 10^{11}$ m^{-3}, we obtain a value of $\tau_{(n,\gamma)} \sim 10^5$ yr. Since this is much longer than typical β-decay half-lives, a reaction under such circumstances is known in nucleosynthesis as a *slow process*, or *s*-process for short. If the neutron density is much higher, $N_n \sim 10^{28}$ m^{-3}, the half-life for a nucleus in a chain of (n, γ) reactions is around $\tau_{(n,\gamma)} \sim 10^{-5}$ s, much shorter than typical β-decay times in the mass region of interest. Neutron capture in such cases is called a *rapid process*, or *r*-process for short.

Source of neutron. During the hydrostatic burning stages of a star, free neutrons are not produced in any of the dominant processes. This raises the question of whether there are enough neutrons to provide the density required for the *r*-process and to supply the large numbers needed to build heavy elements all the way from $A = 56$ to beyond $A = 200$.

The first thing to realize here is that the observed abundances of heavy nuclei are down by at least 10^6 compared with that of silicon (and by 10^{10} with respect to hydrogen). Consequently, the total number of free neutrons required is quite modest compared with the total number of nucleons present in the star. Second, neutron density $N_n \sim 10^{28}$ m^{-3}, used earlier as example, is not a large value, at least at the stage immediately after a supernova explosion. The matter density in the oxygen burning shell is 10^{10} kg m^{-3}. This is equivalent to a baryon density on the order of $10^{10}/(1.67 \times 10^{-27}) \sim 10^{37}$ m^{-3}. If we have one free neutron for every 10^9 bound nucleons, equivalent to about 1 for every 10^7 nuclei, we reach the required density for the r-process to take place.

There are several sources for free neutrons. One of these is photodisintegration at the silicon burning stage. The bulk, however, comes (indirectly) from α-capture by carbon, nitrogen and oxygen in the CNO cycle, as shown in Fig. 10-3. For example, α-capture by ^{14}N followed by β^+-decay through the reactions

$$^{14}\mathrm{N} + {}^{4}\mathrm{He} \to {}^{18}\mathrm{F} + \gamma \qquad\qquad {}^{18}\mathrm{F} \to {}^{18}\mathrm{O} + e^+ + \nu_e$$

produces an ^{18}O. Among other possibilities, we have

$$^{18}\mathrm{O} + {}^{4}\mathrm{He} \to {}^{21}\mathrm{Ne} + n$$

that produces a free neutron. Instead of releasing a neutron, we can have a (^{4}He,γ) reaction to create a ^{22}Ne using the ^{18}O produced. A free neutron is released by a (^{4}He,n) reaction on ^{22}Ne,

$$^{22}\mathrm{Ne} + {}^{4}\mathrm{He} \to {}^{25}\mathrm{Mg} + n$$

Again, we can have, instead, the reaction ^{22}Ne(^{4}He,γ)^{26}Mg. Both ^{25}Mg and ^{26}Mg can undergo further (^{4}He,n) reactions,

$$^{25}\mathrm{Mg} + {}^{4}\mathrm{He} \to {}^{28}\mathrm{Si} + n \qquad\qquad {}^{26}\mathrm{Mg} + {}^{4}\mathrm{He} \to {}^{29}\mathrm{Si} + n$$

to free neutrons into the stellar environment. In addition, neutrons are also produced in carbon burning and other processes.

Proton-rich nuclei. Nuclei created by neutron capture tend to be on the neutron-rich side of the valley of stability. Since β^--decay following neutron capture stops at the bottom of the valley, neutron capture cannot be expected to make nuclei on the proton-rich side. For this reason, other reactions must be responsible for their creation. Observed evidence shows that the abundance of proton-rich nuclei are down by two to three orders of magnitude compared with their neutron-rich cousins, suggesting that processes that produce them are less likely than the r- and s-processes.

The possible reactions to form proton-rich nuclei are the (p,γ) reaction we have seen earlier and positron capture

$$(A,Z) + e^+ \to (A, Z+1) + \bar{\nu}_e$$

Since the latter is a weak interaction process, the rate is much lower compared with neutron capture. Another even less probable process is spallation of heavier nuclei by a proton or an α-particle. Finally (γ,n) reactions of the type

$$(A,Z) + \gamma \to (A-1, Z) + n$$

produces nuclei with a higher Z/N ratio than that for the target.

Problems

10-1. What is the gravitational energy released when noninteracting particles with a total mass 1.99×10^{30} kg collapse from an infinite distance of separation to a spherical ball of radius 6.96×10^8 m? Assuming an ideal gas equation of state, what is the temperature of this sphere? Compare the result with the value of 15×10^6 K for the interior of our sun.

10-2. Show that the maximum amount of energy released in the form of electromagnetic radiation from converting four protons to a ^{4}He is given by the binding energy of ^{4}He less twice the sum of the neutron-proton mass difference and the mass of positrons. Ignore any rest mass the neutrino may have.

10-3. Using the fact that three-fourths of the solar mass of 1.99×10^{30} kg consists of protons, calculate the length of time that fusion energy can be generated at the present rate of 1.4 kW/m^2 at a distance of 1.50×10^{11} m in converting four protons to a ^{4}He nucleus.

10-4. Show that the reaction rate near the most effective temperature E_{max}, given by Eq. (10-4), may be approximated by a normal distribution centered around E_{max} with a width $\sigma = \sqrt{2E_{\text{max}}kT/3}$.

10-5. Calculate the energy released in each one of the nuclear reactions in all three PP-chains.

10-6. The inverse β-decay reaction

$$\nu_e + {}^{37}\text{Cl} \rightarrow e^- + {}^{37}\text{Ar}$$

is used to detect neutrinos from the sun. The number of solar neutrinos produced may be estimated from the solar constant (1350 W/m^2). Assume that 10% of the thermonuclear energy is carried away by neutrinos with a mean energy of 1 MeV each and that only about 1% of the neutrino is energetic enough to convert ^{37}Cl to ^{37}Ar. For a detector containing 400 m^3 of tetrachloroethylene (C_2Cl_4), estimate the average number of ^{37}Ar produced in a day if the density of C_2Cl_4 is 1.5 g/cm^3 and about a quarter of the chlorine is ^{37}Cl. The cross section for the reaction may be taken to be 10^{-48} m^2.

10-7. Assuming a cross section of 10^{-48} m^2 for a neutrino to interact with each nucleon, find the difference in the night and daytime detection rate of solar neutrinos for the Super Kamiokande water Cerenkov detector consisting of 50,000 tons of water.

10-8. A neutron star is a compact, dense object made of degenerate neutrons having a density similar to that in the central part of a heavy nucleus.

(a) If the density of nuclear matter is 0.17 nucleons/fm^3 or 2.8×10^{17} kg/m^3, what is the radius of a neutron star having a mass one and a half times that of the sun? (One solar mass = 2.0×10^{30} kg.)

(b) A neutron star is one of the possible remanents of a supernova explosion such as SN 1987a, the one which took place in the Large Magellanic Cloud 160,000 light years away and was first observed on earth on February 24, 1987. When the core of a large star exhausts its nuclear fuel, there is no longer the thermal pressure to counterbalance the gravitational force, and the core of the star collapses. For simplicity, we can consider that all the material in the core of the collapsing star is in the form of ^{56}Ni made of 28 neutrons and 28 protons. Because of the tremendous gravitational force, the protons in ^{56}Ni change into neutrons by capturing atomic electrons through the reaction

$$p + e^- \to n + \nu_e$$

Calculate the number of neutrinos released in converting 1.5 solar mass of ^{56}Ni atoms into neutrons during the gravitational collapse.

(c) If the total cross section for a neutrino to interact with each nucleon is 10^{-48} m^2, how many reactions due to the neutrinos from such a gravitational collapse can one expect in a detector on earth made of 3000 tons of water? Compare this with the number of events (12) observed with such a detector at Kamioka due to supernova SN 1987a.

(d) Assuming that the average energy of each neutrino is 10 MeV in such an event, calculate the total amount of energy carried away by the neutrinos from the gravitational collapse. Compare this value with the rest-mass energy of the sun.

Chapter 11

Nuclear Physics: Present and Future

We have attempted to give in the previous chapters a highly condensed description of the achievements of nuclear physics in its 100 years of history. Our emphases have been on its contributions to the foundation of physics, the properties of nuclei, and the highlights of some of the present research interest. The subject is a rich one and holds tremendous potential for the future. As with any scientific endeavor, we can rely on improvements in our ability to make observations to provide us with new information and further insight. The explosion of technological advances is having the effect of pushing pure science to new heights and nuclear physics is among the chief beneficiaries. In this short chapter, we shall take a somewhat speculative attitude and see if we can foresee some of the possible directions the subject may take in the near future.

Developments in pure science are often controlled by the three-way symbiotic relation of technology, observation, and theoretical understanding. This is especially true in the case of subatomic physics, where experiments often require large accelerators and sophisticated detectors. To make sense of these complicated observations, we construct models that lead us to a better understanding of the part of nature invisible to the naked eye. This knowledge in turn helps us to make progress in our technology both to improve the quality of life itself and to enhance our ability to make better observations. Large facilities, such as the Relativistic Heavy-Ion Collider and neutrino observatories, are expensive to construct and take many years to reach the production phase. From the careful planning that must go into their designs, it is perhaps not hard to foresee what will be the physics to come out initially from these laboratories. We shall describe what are some of these possibilities. There is a second group of interesting problems that technology is ripe for making progress. Examples in this category include computational-based problems, such as lattice QCD, and space-based explorations in astrophysics that are intimately connected with nuclear physics. A third group consists of long-standing fundamental problems that are crying out for solutions. We shall name some of these as well, in part, for their importance to the subject of physics as a whole.

Radioactive beam. Many laboratories around the world are starting to have facilities for producing beams of radioactive nuclei. We have seen that a large part of our knowledge on nuclear physics, and subatomic physics in general, comes from collisions between two particles. With radioactive beams, our horizon is extended to include projectiles made of nuclei that are unstable. Many of these reactions are important by their own right. For example, interactions with unstable nuclei form an integral part of the events going on in stars, such as that we saw in the case of the triple-α process in §10-5. Better knowledge of their reaction rates are essential to improve our understanding of nucleosynthesis and the evolution of stars. For this reason, some of the first experiments planned with radioactive beam facilities often involve nuclear astrophysics interest.

Radiative beams are also sources of unstable nuclei that are of interest. We have already seen an example in the inverse of the $^{12}C(\alpha,\gamma)^{16}O$ reaction in §10-5, using radioactive ^{16}N as the source of excited states of ^{16}O. By observing the α-particle decay, the equivalent reaction cross section was measured to much lower energies, solving one of the long-standing issues in helium burning stars. Similar opportunity exists in a variety of other cases.

Most of our knowledge on nuclear structure is based on observations made on low-lying states of stable nuclei. This may well represent a fairly specialized set of data, those satisfying the conditions for stable nuclei. The true picture of nuclear physics may be obscured by our limited vision, like trying to infer the behavior of all cats, including lions and leopards, from observing the household variety. By checking our existing understanding on unstable nuclei, we can reaffirm the parts that are correct and modify the reminder. In particular, radioactive beams may be possible to extend the number of isotopes observed for a given Z, the number of isotones for a given N, or the number of isobars for a given A. Systematic variations in the structure are particularly simple in these cases, especially if either the neutrons or the protons form a closed shell. The simplicity offered by this new information will certainly lead into expanded interest and improved knowledge in nuclear structure.

Relativistic heavy-ion collision. The central focus of RHIC and LHC is to study the physics of QGP. As we have seen in §9-3, we still do not know of a unique signature for QGP. Many phenomena are possible candidates but all of them have also possible alternative explanations that are based on hadronic matter alone. For this reason, one of the main goals of the first group of experiments at RHIC and LHC are designed specifically to study these events. In the process, we shall also be able to improve our quantitative understanding of QCD, and this by itself is already quite exciting.

For example, we saw in §9-3 that changes in strangeness and charm production may signal the presence of QGP. To establish this, systematic studies must be made using a variety of colliding ions, including cases in which we do not expect anything other than purely hadronic matter at any intermediate steps of the reactions. In relativistic heavy-ion collisions, the amount of energy involved is high, with large numbers and varieties of secondary particles produced. Distributions of these particles as functions of transverse momentum and rapidity (see Eq. 9-14) are indicative of the conditions of the interaction region. Different theoretical models are predicating quite different results. Experimental guidance in this area will definitely help us to understand the physics under such extreme conditions.

The construction of detectors and data acquisition systems for the complicated events expected from these experiments present new challenges to our ingenuity. To leave room for discovering the unexpected, one must record as much information of the collision as possible so that one can analyze the results for phenomena not necessarily expected in the original design. For this reason, detectors are also built to make comprehensive studies of all the particles produced in each collision. Once these facilities are in operation, we can expect a new burst of information that will keep both experimentalists and theorists busy for a long time.

Electron scattering. As we move to phenomena taking place at shorter distances and higher energies, electron scattering stands out in terms of the precision we can achieve. By going to high-momentum transfers, we can probe the nucleus at the fine scales where most of our ignorance lies. For example, it was found that, in the electro-disintegration of deuterons, the idea of one-pion exchange seems to work to far larger momentum transfers than we expect [118]. This implies that something more fundamental than the masses of different mesons being exchanged must be working here. One of the possibilities is the role of *chiral* invariance in nuclei.

The origin of chiral invariance may be traced to QCD. We have seen that u- and d-quarks are far lighter than the others. The corresponding situation at the hadron level, where most of nuclear physics operates, is that pions are much less massive than any of the other particles. In the limit that u- and d-quarks may be treated as massless, helicity (see Eq. 5-51) is conserved and the QCD lagrangian has a special symmetry, generally referred to as $SU_L(2) \times SU_R(2)$ symmetry. In other words, under a rotation in "chiral" space, left-handed and right-handed particles are not mixed with each other (see, e.g., Ref. [58]). At the nuclear physics regime, we can gain a sense of the importance of this approximate symmetry by looking at Eq. (5-53). If pions can be treated as massless, we find that axial-vector current is also conserved. This puts axial-vector current more or less on the same footing as vector currents, as we can see by comparing with Eq. (5-52).

It is often said that chiral symmetry is the connection between strong interaction inside "bags" of quarks with that outside hadrons. It is not appropriate for us here to go into the numerous examples justifying this statement. On the other hand, it is perhaps not difficult to see that nuclear physics can be much simpler if pions can be treated as massless. The importance of exploring the advantage of such a "soft pion" limit cannot be overemphasized if we wish to understand low-energy phenomenology of QCD in nuclei. High-precision electron scattering is certainly one of the most fruitful avenues in this respect.

Another question of interest, where high precision is necessary and can be reached with present-day electron scattering technology, is the difference in charge distribution for a pair of mirror nuclei, in particular, ^{3}H and ^{3}He. Although ^{3}H (tritium) is radioactive, its half-life is sufficiently long (12.33 yr) for a target to be made. Since the $A = 3$ pair constitutes the lightest mirror nuclei, a detailed comparison of their charge distributions can be extremely illuminating. At the same time, it may also give us some information on such fundamental questions as three-body force.

Similar opportunities exist for high-precision studies of the electromagnetic properties for heavier nuclei. We have seen earlier in §4-7 that the question of mesonic current is essential to understand how the electromagnetic operators are modified when nucleons are embedded in nuclei. More generally, the question of "medium" effect—how certain operators are modified when nucleons are bound—is important in understanding nuclear properties. Many aspects of such studies are ideally suited for the high precision that can be reached using electron scattering.

Improvement in the precision of experiments involving particles other than electrons, such as protons and neutrons, are also expected to come with technological progress. This will give us a much better handle on such questions as charge symmetry in nuclear force, parity violation in nuclear interaction, and three-body forces. With the accuracy available to date, many of these fundamental questions cannot be answered in a definitive way.

Kaons and antiparticles. Nuclear physics operates, for the most part, at the level of nucleons and pions. For this reason, strange mesons and antinucleons have not been of any direct concern to us. However, as we saw in §8-4, antiproton scattering off nuclei can help us to understand the force acting between nucleons. Since nucleon-nucleon interaction is only an extension of the strong interaction between quarks, the symmetries of QCD have profound effect on nuclear force and antinucleon scattering is an important source of such information. Antiprotons have been available from the Low-Energy Antiproton Ring (LEAR) of CERN and are produced in many other high-energy accelerator laboratories. Far more antiproton scattering off nuclei can be carried out and the data will be of great interest.

By the same token, interaction of strange mesons, such as kaons, can supplement studies of pions on nuclei. We saw in §2-2 that nucleons and pions are made of u- and d-quarks. Partly for this reason, most of nuclear physics can be understood with these two lightest flavors alone. However, in addition to these "active" quarks, there is also a Dirac "sea" of other quarks together with their antiquarks. These sea quarks are invisible under normal circumstances, in the same way as the electron-positron pairs in the quantum electrodynamics vacuum we saw in §9-1. However, certain observations cannot be understood by the valence quarks alone, such as the relation between nucleon and quark spins. Strange mesons can probe the quark distribution in nucleons from a different angle than those we can observe using mesons, and this will help us to understand the role of strange and perhaps other heavier quarks in nuclei. Kaons are also available from a number of high-energy accelerators. For many years, there have been proposals to build dedicated kaon sources, or "factories," from several laboratories. Once available, it will be a bonus to nuclear as well as particle physics.

Cosmic rays and space observations. A major technological impact on our ability to do science in recent years is the construction of "observatories" outside the earth atmosphere and large arrays of detectors on the surface of earth. Many new phenomena, such as γ-ray bursts and extremely high energy cosmic ray showers, have been recorded and, no doubt, more will be forthcoming in the future. These events complement laboratory studies as well as help us to reach, for example, the high energies that cannot be achieved using available accelerators. A historical example is the first identification of pions in cosmic rays as the "Yukawa" particle. Prior to the discovery, the only particle known in the early part of the 1930s with a mass of $\sim 200m_e$ was the muon. As we know now, the muon cannot be the candidate, as it is not a hadron. Furthermore, it is a fermion and therefore cannot be absorbed and emitted freely as bosons can.

Even though the energies we have achieved in the laboratory are quite impressive, they do not compare with those associated with some of the highest energy particles observed from extra-terrestrial sources. Cosmic rays and γ-bursts have been recorded to reach energies many orders of magnitude higher than what we can hope to achieve in the laboratory in the foreseeable future. Both the production mechanism of these high-energy events and the propagation of these particles in the intergalactic space are of interest.

Most of the observed "new" events are likely to be of interest mainly to cosmology and elementary particle physics. However, the intimate relation of nuclear physics with these subjects makes it imperative for nuclear physicists to take a closer look. This is especially true when these observations are coupled with the possibility offered by neutrino observatories, such as the Sudbury Neutrino Observatory and Super Kamiokande Cerenkov Detector, to study neutrinos coming out of the core of stars without being masked by the thick outer layers of materials.

Computational physics. Advances in computer technology and numerical methods have fundamentally changed the way certain investigations are carried out in physics. Instead of restricting to models with analytical solutions, we can now easily explore new ideas that require extensive calculations and large-scale simulations. We have already seen an example in lattice gauge studies for transition from hadronic matter to quark-gluon plasma in §9-3. In fact, numerical solution using the Feynman path integral approach is likely to be the only way that answers can be obtained for some of the strong interaction problems.

Theoretical investigations in nuclear physics are often computing intensive. This comes in part because of the highly nonlinear nature of the problem, associated with the fact that we are dealing with, for example, eigenvalue problems. In part, it is also because of the nonperturbative nature of the phenomena, involving interactions that are strong. Here, computing is actually an advantage in that the tedium to carry out large calculations can be delegated to machines that are many orders of magnitude faster and more reliable than human being. This applies to numerical work as well algebraic manipulations. In fact, many of the algebraic calculations involved in analytical solutions are often carried on computers as well. Furthermore, if we take good advantage of the visualization tools available on computers, many complicated solutions, both analytical and numerical, can be more readily comprehended than with algebraic symbols alone. In many cases, far more advantage of the visualization capabilities can be used in understanding the results from complicated calculations than what have been done in practice. For example, to be able to "see" a multi-dimensional result is certainly not something that can be achieved easily without computers.

In addition to lattice gauge calculations, several problems in nuclear theory can also benefit from intensive computation. The relativistic shell model is one such example. For the most part in nuclear structure, we are dealing with velocities that are much less than the speed of light. The relativistic effect can nevertheless be important in such cases from the following considerations. In the Schrödinger picture, each spin-$\frac{1}{2}$ fermion has two components, one with spin pointing up and the other with spin down. In the more general Dirac picture, the corresponding wave function has four components, two for the particle and two for the antiparticle. The four-component Dirac equation may be expressed as a set of two coupled equations, each one having only two components. At nonrelativistic energies, the coupling between the "upper" two components and the "lower" two components may be replaced by a spin-orbit term and the two equations decouple from each other in this approximation.

Traditionally, nuclear structure problems are solved by following the time-honored methods used in atomic structure with the Schrödinger approach. However, by going back to the more fundamental Dirac equation, many conceptual difficulties are found to be much easier to handle and better solutions are obtained. We saw one such example in nuclear matter calculations in §4-12 and another one in intermediate-energy nucleon-nucleus scattering in §8-5. Similar successes are found also in a variety of other problems in both nuclear structure and nuclear reaction. The calculations involved in solving four-component equations are more complicated than those in the two-component Schrödinger approach. In addition, we need more experience in handling certain aspects of the Dirac equation in a many-body setting. The results are, however, extremely encouraging and may lead to better understanding of some of the puzzles in nuclear physics.

Another example is the use of sampling, or Monte Carlo, techniques in microscopic calculations. We saw in §7-5 that the Hilbert space in a microscopic calculation can be extremely large. To make progress, drastic truncation of the space as well as renormalization of the interaction has to be carried out. If a larger active space can be used, it can certainly reduce some of the uncertainties introduced by truncation and renormalization. Similar to many other types of problems, one can apply sampling techniques for certain investigations in large spaces and obtain meaningful results by carrying out only a small part of the actual work. The computer is well suited for doing this type of calculation, especially in view of the general trend toward parallel computing by making use of several central processing units at the same time. Furthermore, Monte Carlo techniques are used in a variety of other many-body problems, such as those in condensed matter physics. The advances made there can also be a great help in applying the method to nuclear physics problems.

In §7-5 we saw also that it is possible to start with a realistic nucleon-nucleon scattering potential and modify it so that it is appropriate for bound nucleons. The nuclear wave function obtained may be used to describe the nuclear state involved in an intermediate-energy nucleon-nucleus scattering, as we saw in §8-5. Furthermore, the nucleon-nucleus interaction can also be derived from the same free nucleon-nucleon interaction potential used as the starting point for the nuclear wave function calculation. In each one of the steps, rigorous many-body problem techniques are available and can be applied. The calculation is a rather involved one. On the other hand, both the input nucleon-nucleon potential and the output nucleon-nucleus scattering can be checked directly with independent observations. The comparisons form the tests for many interesting questions, including many-body techniques and our understanding of how nucleons are modified inside the nuclear medium. Far more such large-scale calculations can be performed, especially in view of the higher precision experimental data that can be obtained these days.

Fundamental problems in physics. There are several fundamental questions in physics that involve atomic nuclei. The solar neutrino problem described in §10-4 is a good example. It cuts across several disciplines, particle physics, cosmology, hydrodynamics, just to name a few. In addition to the intrinsic interest in the problem itself, it has the potential of pointing the way to new physics, such as neutrino oscillation. Nuclear physics enters in understanding the processes that produce the particles and in interpreting the measured results. The importance of the problem cannot be understated. This is especially true in view of the great expectation we have in neutrino astronomy. Unless we can understand neutrinos from the sun, the nearest star by many orders of magnitude, we have no hope of extending the observations to any of the other stars.

A second example concerns some of the properties of neutrinos. As we saw in §5-6, double-β-decay is one way to find out whether it is a Majorana or a Dirac particle. The neutrino mass is also an important question in deciding some of the basic properties of the particle. Furthermore, it enters also into the question of "missing" mass in the universe. Answers to these and other questions are interesting by themselves and may also lead to new knowledge of the microscopic world that subatomic physics has adopted as its subject of interest.

A third example is why the QCD effect seems to be totally absent in nuclear phenomena. There are two possible answers. The first is that there is some fundamental symmetry in operation. If this is the case, it will be of interest to find out what this symmetry is and how it functions. Alternatively, and more likely, it is possible that we are not asking the right questions. What are the phenomena in nuclear physics that QCD must be invoked directly? Perhaps, this is one of the questions that will be revealed in relativistic heavy-ion collisions.

Several other basic problems in physics also fit into the category where nuclear physics can be of help in solving them. However, we shall not make the attempt here because of the preparations required to describe them.

Over the century-long history of nuclear physics, we have seen the central emphasis of the subject changing as we improve our knowledge of the subatomic world and our ability to make observations. For example, instead of radioactivity at the beginning of the twentieth century, a large fraction of the present-day effort is in high-energy nuclear physics. Instead of topics that are now in quantum mechanics textbook, we are talking more and more in terms of QCD and cosmology. Such changes form a natural path in the development of our interest in the microscopic regime. As we know more about nuclei, we want to find out more by going into shorter length scales and carrying out more precise measurements. These observations, in turn, are guiding us to new heights in our understanding. Nuclear physics has been, is, and will remain a vibrant part of modern science. It is extremely rich in physics and is of interest to those trying to unravel the mystery of the physical universe.

Appendix A

Parity and Angular Momentum

A-1 Parity Transformation

Parity, or space reflection, transformation is the operation whereby all three coordinate axes in the Cartesian system change sign. That is, if the location of a point in space is given by coordinates (x, y, z) in a particular system, the coordinates of the same point in a system related to the original one by a parity transformation P are $(-x, -y, -z)$,

$$(x, y, z) \xrightarrow[P]{} (-x, -y, -z) \tag{A-1}$$

Such a reflection of the axes changes a right-handed coordinate system to a left-handed one, as illustrated by Fig. 5-3.

In quantum mechanics, the probability of finding a particle at location $\boldsymbol{r}$ is given by the absolute square of its wave function $|\Psi(\boldsymbol{r})|^2$ at the point. Since the probability is an observable, it cannot change its value simply because we have switched from using a right-handed coordinate system to a left-handed one, or vice versa. The wave function itself, however, may change under a parity transformation, subject to the following two conditions. The first is that $|\Psi(\boldsymbol{r})|^2$ must remain invariant, as we saw above. The second is that two successive parity operations must bring the system back to its original state, i.e., $P^2 = 1$. As a result, the wave function $\Psi(\boldsymbol{r})$ can change at most by a sign. States whose wave functions do not change sign under a parity transformation,

$$P\Psi(\boldsymbol{r}) = \Psi(-\boldsymbol{r}) = +\Psi(\boldsymbol{r})$$

are called positive-parity states, and those whose wave functions change sign,

$$P\Psi(\boldsymbol{r}) = \Psi(-\boldsymbol{r}) = -\Psi(\boldsymbol{r})$$

are negative-parity states. A wave function that does not fall into either one of these two categories does not have a definite parity.

In terms of spherical polar coordinates, the radial distance r is not affected by a parity transformation. The only changes are in the angular variables,

$$(r, \theta, \phi) \xrightarrow[P]{} (r, \pi - \theta, \pi + \phi) \tag{A-2}$$

This relation can be shown to be identical as that in Eq. (A-1), for example, by transforming both sides to Cartesian coordinate systems. Because of Eq. (A-2), radial wave functions are not changed by a parity transformation. As a result, the parity of a wave function of a state is given by the angular part alone. For a state $\Psi(\boldsymbol{r})$ with definite orbital angular momentum (ℓ, m), we can decompose the wave function into a product of radial and angular parts,

$$\Psi(\boldsymbol{r}) = R_{n\ell}(r)Y_{\ell m}(\theta, \phi)$$

The angular dependence is described by spherical harmonics $Y_{\ell m}(\theta, \phi)$, the eigenfunctions of orbital angular momentum operators $\boldsymbol{\ell}^2$ and $\boldsymbol{\ell}_0$. The parity of spherical harmonics of order ℓ is $(-1)^\ell$. This can be seen from its explicit form

$$Y_{\ell m}(\theta, \phi) = \frac{(-1)^m}{2^\ell \ell!}\sqrt{\frac{(2\ell+1)}{4\pi}\frac{(\ell-m)!}{(\ell+m)!}}\, e^{\imath m\phi}(1-\eta^2)^{m/2}\Big(\frac{d}{d\eta}\Big)^{\ell+m}(\eta^2-1)^\ell \tag{A-3}$$

where $\eta = \cos\theta$. Since $\cos(\pi - \theta) = -\cos(\theta)$ we have, under a parity transformation,

$$\eta \xrightarrow[P]{} -\eta$$

The transformation of polar angle θ gives a phase factor $(-1)^{\ell+m}$ to $Y_{\ell m}(\theta, \phi)$. The azimuth angle ϕ enters Eq. (A-3) only in the exponential factor $e^{im\phi}$. The transformation from ϕ to $\pi + \phi$ produces a factor $e^{\imath m\pi} = (-1)^m$. The combination of the two gives us the net result,

$$Y_{\ell m}(\theta, \phi) \xrightarrow[P]{} Y_{\ell m}(\pi - \theta, \pi + \phi) = (-1)^\ell Y_{\ell m}(\theta, \phi)$$

For this reason, spherical harmonics of even order have even parity and spherical harmonics of odd order have odd parity.

In additional to parity associated with spatial wave functions, the intrinsic wave function of a particle can also have a definite parity, related to the internal structure of the particle. If the structure is known, such as that for a nucleon from a quark model, the intrinsic parity may be deduced from the wave function. In cases where the internal structure is not known, the intrinsic parity must be determined experimentally using reactions in which the parities of all other particles as well as all the relative angular momenta involved are known.

As an example, we shall see how the intrinsic parity of a pion is determined to be negative. The measurement involves the absorption of π^- by a deuteron. The pion is first captured in the s-state of a deuterium atom, forming a π-mesic atom as a result (see also §8-6). Since the pion is a meson, it can be absorbed by the proton in the deuterium nucleus through the reaction

$$\pi^- + d \to n + n \tag{A-4}$$

Before the reaction, the total angular momentum J of the π-mesic atom is 1, as the intrinsic spin of the pion is 0 (see also §2-7), the spin of the deuteron is 1 (see §3-1), and the orbital angular momentum of the πd-system is 0 (the π^- is in the atomic s-state). Total angular momentum is conserved in the reaction of Eq. (A-4) and, as a result, the final state produced by the reaction must also have $J = 1$.

The two neutrons in the final state, being identical fermions, must be in an antisymmetric state to satisfy the Pauli principle. The symmetry of the wave function of the two-neutron system is determined by L, the relative orbital angular momentum, and S, the sum of the intrinsic spin of the two particles. If the spatial part of the system of two identical fermions is symmetrical (L = even), the total intrinsic spin wave function must be antisymmetrical ($S = 0$). Alternatively, if the spatial part is antisymmetrical (L = odd), the total intrinsic spin wave function must be symmetrical ($S = 1$).

From the fact that $\boldsymbol{J} = \boldsymbol{L} + \boldsymbol{S} = 1$, we find that the possible pairs of (L, S)-values to form $J = 1$ are (0,1), (1,0), (1,1), and (2,1). The combinations (0,1) and (2,1) can be ruled out on the ground that both intrinsic spin and spatial wave functions are symmetric and, therefore, violate the Pauli principle. Similarly, the combination (1,0) is not allowed, as both orbital and intrinsic spin parts are antisymmetric. The only possible combination remaining is $(L, S) = (1, 1)$, which is antisymmetric in the spatial part but symmetric in the intrinsic spin part of the wave function.

The parity of the right-hand side of the reaction given by Eq. (A-4) is therefore $(-1)^L = -1$, independent of the intrinsic parity of neutrons, as there are two involved. Since parity is conserved in the reaction, the left-hand side must also have negative parity. There are three components contributing to the parity of the initial state of the reaction. The parity of the ground state of the deuteron is known to be even ($L = 0$, 2, and both neutron and proton have the same intrinsic parity). The parity of the orbital wave function of the π-mesic atom is positive, as we have seen earlier. As a result, we conclude that the intrinsic parity of π^-, the third component in the initial state, must be negative in order for the parity of the total system to be negative.

For fermions, the intrinsic parity of an antiparticle is opposite to that of its corresponding particle. This can be seen from the structure of the Dirac equation where a particle and an antiparticle are described by a single four-component wave function. Alternatively, it can be determined using such measurements as the polarization of the two photons emitted in the decay of a positronium (e^+e^- system) in the singlet state ($J = 0$). On the other hand, for bosons the parity of both particle and antiparticle must be the same. For more details, see standard particle physics textbooks, such as Perkins [115], and Halzen and Martin [80].

A-2 Spherical Tensor and Rotation Matrix

A quantity $\psi_{JM}(\boldsymbol{r})$ is said to be a *spherical tensor* of angular momentum rank J if it belongs to a group consisting of $2J + 1$ members, each having the same J-value but differing in the M-value, projections along the quantization axis. The possible values of M are $-J$, $-J + 1$, ... , J. Under a rotation of the coordinate axes through Euler angles (α, β, γ) shown in Fig. A-1, the $2J + 1$ members transform among themselves

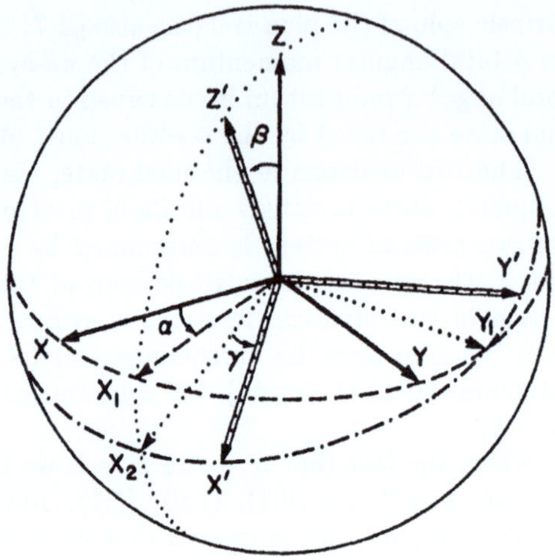

Figure A-1: Rotation of the coordinate axes from (X, Y, Z) to (X', Y', Z') by Euler angles (α, β, γ) in three steps. First, a rotation around the Z-axis through angle α brings (X, Y, Z) to (X_1, Y_1, Z). Second, a rotation around the new Y_1-axis through angle β brings (X_1, Y_1, Z) to (X_2, Y_1, Z_2). In the last step, (X_2, Y_1, Z_2) is brought to (X', Y', Z') by a rotation around Z_2 (same as Z') through angle γ.

according to the relation

$$\psi_{JM'}(\boldsymbol{r}') = \sum_M \psi_{JM}(\boldsymbol{r})\, \mathcal{D}^J_{MM'}(\alpha, \beta, \gamma) \tag{A-5}$$

where coefficients $\mathcal{D}^J_{MM'}(\alpha, \beta, \gamma)$ are the rotation matrices, or $\mathcal{D}$-functions for short. The $2J+1$ components of a spherical tensor ψ_{JM}, for all possible values of M, therefore form an irreducible group under rotation. The set of $2\ell + 1$ spherical harmonics $Y_{\ell m}(\theta, \phi)$, with $m = -\ell, -\ell + 1, \ldots, \ell$, is an example of a spherical tensor of integer rank ℓ. However, spherical tensors are more general quantities and can have half-integer ranks as well. Both wave functions and operators can be spherical tensors, as the requirements can be satisfied by both types of quantities.

There are several possible ways to define the $\mathcal{D}$-function. We shall adopt the convention given by Brink and Satchler [37]. A rotation of the coordinate axes in the way defined in Fig. A-1 can be achieved as three successive infinitesimal rotations represented by the operator

$$\boldsymbol{R}(\alpha, \beta, \gamma) = e^{-i\gamma \boldsymbol{J}_{z'}} e^{-i\beta \boldsymbol{J}_{y_1}} e^{-i\alpha \boldsymbol{J}_z} \tag{A-6}$$

The same transformation is equivalent to a rotation first around the Z-axis by angle γ, followed by a rotation through angle β around the Y-axis, and finally a rotation through angle α around the Z-axis again. That is,

$$\boldsymbol{R}(\alpha, \beta, \gamma) = e^{-i\alpha \boldsymbol{J}_z} e^{-i\beta \boldsymbol{J}_y} e^{-i\gamma \boldsymbol{J}_z} \tag{A-7}$$

The $\mathcal{D}$-function in Eq. (A-5) may be written as the matrix element of $\boldsymbol{R}(\alpha, \beta, \gamma)$,

$$\mathcal{D}^J_{MM'}(\alpha, \beta, \gamma) = \langle \psi_{JM} | \boldsymbol{R}(\alpha, \beta, \gamma) | \psi_{JM'} \rangle \tag{A-8}$$

between components M and M' of a spherical tensor of rank J. The orthogonality relation among the $\mathcal{D}$-functions is given by

$$\sum_{M'}(\mathcal{D}^J_{M'N}(\alpha,\beta,\gamma))^*\mathcal{D}^J_{M'M}(\alpha,\beta,\gamma)=\sum_{M'}\mathcal{D}^J_{MM'}(\alpha,\beta,\gamma)(\mathcal{D}^J_{NM'}(\alpha,\beta,\gamma))^*=\delta_{MN}$$

and

$$\int_0^{2\pi}\int_0^{2\pi}\int_0^{\pi}(\mathcal{D}^{J'}_{M'M}(\alpha,\beta,\gamma))^*\mathcal{D}^J_{N'N}(\alpha,\beta,\gamma)\sin\beta\,d\beta\,d\alpha\,d\gamma=\frac{8\pi^2}{2J+1}\delta_{M'N'}\delta_{MN}\delta_{J'J}$$

where δ_{xy} is the Kronecker delta with value unity if $x=y$ and zero otherwise.

Since ψ_{JM} is an eigenfunction of $\boldsymbol{J}_z$, we can make use of Eq. (A-7) to simplify the $\mathcal{D}$-function,

$$\begin{aligned}\mathcal{D}^J_{MN}(\alpha,\beta,\gamma) &= \langle\psi_{JM}|e^{-i\alpha\boldsymbol{J}_z}e^{-i\beta\boldsymbol{J}_y}e^{-i\gamma\boldsymbol{J}_z}|\psi_{JN}\rangle\\ &= e^{-i(\alpha M+\gamma N)}\langle\psi_{JM}|e^{-i\beta J_y}|\psi_{JN}\rangle\end{aligned}$$

where, in the final form, the matrix element remaining,

$$d^J_{MN}(\beta)\equiv\langle\psi_{JM}|e^{-i\beta\boldsymbol{J}_y}|\psi_{JN}\rangle$$

is called the reduced rotation matrix element. Explicitly, it may written as

$$\begin{aligned}d^J_{MN}(\beta) &= \sum_Q(-1)^Q\frac{\sqrt{(J+M)!(J-M)!(J+N)!(J-N)!}}{(J+M-Q)!(J-N-Q)!Q!(Q+N-M)!}\\ &\quad\times(\cos\beta/2)^{2J+M-N-2Q}(\sin\beta/2)^{2Q+N-M}\end{aligned}$$

The summation is over all possible values of Q that do not lead to negative arguments in the factorials. The phase convention used here is that of Condon and Shortley [37].

Let $\boldsymbol{T}_{kq}$ represent component q of a spherical tensor operator of rank k. The conjugate $(\boldsymbol{T}_{kq})^\dagger$ of $\boldsymbol{T}_{kq}$ is defined by the relation between the Hermitian conjugate of their matrix elements,

$$\langle JM|(\boldsymbol{T}_{kq})^\dagger|J'M'\rangle=\langle J'M'|\boldsymbol{T}_{kq}|JM\rangle^*$$

It is easy to see that $(\boldsymbol{T}_{kq})^\dagger$ is not a proper tensor. If $\boldsymbol{T}_{kq}$ transforms under a rotation according to Eq. (A-5), the transformation of its conjugate is given by

$$(\boldsymbol{T}_{kq'})^\dagger=\Big(\sum_q\boldsymbol{T}_{kq}\mathcal{D}^k_{qq'}(\alpha,\beta,\gamma)\Big)^\dagger=\sum_q(\boldsymbol{T}_{kq})^\dagger(\mathcal{D}^k_{qq'}(\alpha,\beta,\gamma))^*$$

To have the proper transformation, we can make use of the property that

$$(\mathcal{D}^k_{pq}(\alpha,\beta,\gamma))^*=(-1)^{p-q}\mathcal{D}^k_{-p-q}(\alpha,\beta,\gamma)$$

and define an adjoint tensor

$$\overline{\boldsymbol{T}}_{kq}\equiv(-1)^{k-q}(\boldsymbol{T}_{k,-q})^\dagger \tag{A-9}$$

that transforms under a rotation of the coordinate axes in the same way as given by Eq. (A-6) and is, therefore, a proper tensor. It should be pointed out that the phase factor $(-1)^q$ in Eq. (A-9) is unique but the factor $(-1)^k$ is somewhat arbitrary. There are also other conventions found in the literature. The one adopted here has the advantage that it is convenient for tensors of half-integer rank, for example, like those used in Eqs. (2-26) and (2-27).

In problems where spherical symmetry is important, such as those commonly encountered in subatomic physics, spherical tensors are useful for a variety of reasons. Some technical advantages of using spherical tensors are related to the algebra of angular momentum coupling given in the next four sections.

A-3 Angular Momentum Recoupling Coefficients

In general, the product of two spherical tensors is not a spherical tensor. For example, the product of T_{JM}, a spherical tensor of rank J, and $U_{J'M'}$, a spherical tensor of rank J', is a mixture of tensors with ranks $|J-J'|$ to $J+J'$. We can use the product between two ordinary vectors as an illustration. A vector $\boldsymbol{r}$ is a spherical tensor of rank unity and is specified, for example, by giving its projections (r_1, r_2, r_3) on the three axes of a Cartesian coordinate system. The product of $\boldsymbol{r}$ with another vector $\boldsymbol{r}'$, having projections (r'_1, r'_2, r'_3), contains, in general, a total of nine components. We can separate these nine products into three groups. The linear combination

$$S \equiv r_1 r'_1 + r_2 r'_2 + r_3 r'_3 = \boldsymbol{r} \cdot \boldsymbol{r}'$$

is a scalar, as it is invariant under a rotation of the coordinate axes. Three of the quantities transform among themselves like a vector,

$$\boldsymbol{V} \equiv (r_2 r'_3 - r_3 r'_2,\, r_3 r'_1 - r_1 r'_3,\, r_1 r'_2 - r_2 r'_1) = \boldsymbol{r} \times \boldsymbol{r}'$$

as can be seen from the fact it has the standard form of an ordinary vector product between $\boldsymbol{r}$ and $\boldsymbol{r}'$. The remaining five components may be written as

$$T_{ij} \equiv \tfrac{1}{2}(r_i r'_j + r_j r'_i) - \tfrac{1}{3}\delta_{ij} \sum_{k=1}^{3} r_k r'_k$$

and they form a second-rank spherical tensor.

In general, a tensor of definite rank can be *projected* out of a product of two tensors using angular momentum coupling coefficients,

$$(T_{j_1} \times U_{j_2})_{j_3 m_3} \equiv \sum_{pq} \langle j_1 m_1 j_2 m_2 | j_3 m_3 \rangle \, T_{j_1 m_1} U_{j_2 m_2} \tag{A-10}$$

where the Clebsch-Gordan coefficient $\langle j_1 m_1 j_2 m_2 | j_3 m_3 \rangle$ vanishes unless $m_3 = m_1 + m_2$ and $|j_1 - j_2| \leq j_3 \leq j_1 + j_2$. Several different symbols are commonly used in the literature to represent Clebsch-Gordan coefficients,

$$\langle j_1 m_1 j_2 m_2 | j_3 m_3 \rangle \equiv \langle j_1 j_2 m_1 m_2 | j_1 j_2 j_3 m_3 \rangle \equiv C^{j_1 j_2 j_3}_{m_1 m_2 m_3}$$

We shall use the first form in the above expression, as it gives the picture of the overlap of $|\, jm \,\rangle$ with the product of $|\, j_1 m_1 \,\rangle$ and $|\, j_2 m_2 \,\rangle$.

It is often more convenient to express the coupling coefficients between two spherical tensors in terms of Wigner 3j-symbols, related to the Clebsch-Gordan coefficients by a simple factor,

$$\begin{pmatrix} j_1 & j_2 & j_3 \\ m_1 & m_2 & m_3 \end{pmatrix} = \frac{(-1)^{j_1-j_2-m_3}}{\sqrt{2j_3+1}} \langle j_1 m_1 j_2 m_2 | j_3 \, -m_3 \rangle$$

In terms of 3j-symbols, the symmetry in the arguments of the coupling coefficient may be expressed as

$$\begin{aligned} \begin{pmatrix} j_1 & j_2 & j_3 \\ m_1 & m_2 & m_3 \end{pmatrix} &= \begin{pmatrix} j_2 & j_3 & j_1 \\ m_2 & m_3 & m_1 \end{pmatrix} = \begin{pmatrix} j_3 & j_1 & j_2 \\ m_3 & m_1 & m_2 \end{pmatrix} \\ &= (-1)^{j_1+j_2+j_3} \begin{pmatrix} j_1 & j_3 & j_2 \\ m_1 & m_3 & m_2 \end{pmatrix} = (-1)^{j_1+j_2+j_3} \begin{pmatrix} j_1 & j_2 & j_3 \\ -m_1 & -m_2 & -m_3 \end{pmatrix} \end{aligned} \tag{A-11}$$

In other words, 3j-symbols are invariant under an even permutation of the three pairs of arguments, and a phase factor $(-1)^{j_1+j_2+j_3}$ is needed for an odd permutation as well as for the case when all the m-values change sign.

The orthogonality relations between the coefficients are

$$\sum_{m_1 m_2} \begin{pmatrix} j_1 & j_2 & j_3 \\ m_1 & m_2 & m_3 \end{pmatrix} \begin{pmatrix} j_1 & j_2 & j_3' \\ m_1 & m_2 & m_3' \end{pmatrix} = \Delta(j_1 j_2 j_3) \frac{\delta_{j_3 j_3'} \delta_{m_3 m_3'}}{2j_3+1} \tag{A-12}$$

$$\sum_{m_1 m_2 m_3} \begin{pmatrix} j_1 & j_2 & j_3 \\ m_1 & m_2 & m_3 \end{pmatrix} \begin{pmatrix} j_1 & j_2 & j_3 \\ m_1 & m_2 & m_3 \end{pmatrix} = \Delta(j_1 j_2 j_3) \tag{A-13}$$

$$\sum_{j_3 m_3} (2j_3+1) \begin{pmatrix} j_1 & j_2 & j_3 \\ m_1 & m_2 & m_3 \end{pmatrix} \begin{pmatrix} j_1 & j_2 & j_3 \\ m_1' & m_2' & m_3 \end{pmatrix} = \Delta(j_1 j_2 j_3) \, \delta_{m_1 m_1'} \delta_{m_2 m_2'} \tag{A-14}$$

where

$$\Delta(j_1 j_2 j_3) = \begin{cases} 1 & \text{for} \quad |j_1 - j_2| \le j_3 \le j_1 + j_2 \\ 0 & \text{otherwise} \end{cases}$$

In terms of Clebsch-Gordan coefficients, the same relations may be expressed as

$$\begin{aligned} \sum_{m_1 m_2} \langle j_1 m_1 j_2 m_2 | j_3 m_3 \rangle \langle j_1 m_1 j_2 m_2 | j_3' m_3' \rangle &= \Delta(j_1 j_2 j_3) \, \delta_{j_3 j_3'} \delta_{m_3 m_3'} \\ \sum_{m_1 m_2 m_3} \langle j_1 m_1 j_2 m_2 | j_3 m_3 \rangle \langle j_1 m_1 j_2 m_2 | j_3 m_3 \rangle &= (2j_3+1) \Delta(j_1 j_2 j_3) \\ \sum_{j_3 m_3} \langle j_1 m_1 j_2 m_2 | j_3 m_3 \rangle \langle j_1 m_1' j_2 m_2' | j_3 m_3 \rangle &= \Delta(j_1 j_2 j_3) \delta_{m_1 m_1'} \delta_{m_2 m_2'} \end{aligned}$$

The Condon and Shortley phase convention, commonly adopted nowadays, states that in coupling j_1 and j_2 to the maximum possible angular momentum $j_1 + j_2$ and all the projections on the z-axis take on the maximum allowed values, i.e., $m_1 = j_1$ and $m_2 = j_2$, the Clebsch-Gordan coefficient

$$\langle j_1 j_1 j_2 j_2 | j_1{+}j_2 \; j_1{+}j_2 \rangle = +1$$

and

$$\sum_{m_1 m_2} m_1 \langle j_1 m_1 j_2 m_2 | j_3 m_3 \rangle \langle j_1 m_1 j_2 m_2 | (j_3{-}1) \, m_3 \rangle > 0$$

All Clebsch-Gordan coefficients are real in this convention, and the explicit values of a few involving low angular momentum ranks are listed in Table A-1.

Table A-1: Some useful Clebsch-Gordan coefficients.

$\langle \frac{1}{2} m_s \ell m_\ell | jm \rangle$

j	$m_s = +\frac{1}{2}$	$m_s = -\frac{1}{2}$
$\ell + \frac{1}{2}$	$\sqrt{\dfrac{\ell + \frac{1}{2} + m}{2\ell + 1}}$	$\sqrt{\frac{\ell + \frac{1}{2} - m}{2\ell + 1}}$
$\ell - \frac{1}{2}$	$\sqrt{\dfrac{\ell + \frac{1}{2} - m}{2\ell + 1}}$	$-\sqrt{\dfrac{\ell + \frac{1}{2} + m}{2\ell + 1}}$

$\langle 1 m_s \ell m_\ell | jm \rangle$

j	$m_s = +1$	$m_s = 0$	$m_s = -1$
$\ell + 1$	$\sqrt{\dfrac{(\ell+m)(\ell+m+1)}{2(\ell+1)(2\ell+1)}}$	$\sqrt{\dfrac{(\ell-m+1)(\ell+m+1)}{(\ell+1)(2\ell+1)}}$	$\sqrt{\dfrac{(\ell-m)(\ell-m+1)}{2(\ell+1)(2\ell+1)}}$
ℓ	$\sqrt{\dfrac{(\ell+m)(\ell-m+1)}{2\ell(\ell+1)}}$	$\dfrac{-m}{\sqrt{\ell(\ell+1)}}$	$-\sqrt{\dfrac{(\ell-m)(\ell+m+1)}{2\ell(\ell+1)}}$
$\ell - 1$	$\sqrt{\dfrac{(\ell-m)(\ell-m+1)}{2\ell(2\ell+1)}}$	$-\sqrt{\dfrac{(\ell-m)(\ell+m)}{\ell(2\ell+1)}}$	$\sqrt{\dfrac{(\ell+m+1)(\ell+m)}{2\ell(2\ell+1)}}$

$$\begin{pmatrix} j & 0 & j' \\ -m & 0 & m' \end{pmatrix} = \langle jmj' \, {-m'} | 00 \rangle = \frac{(-1)^{j-m}}{\sqrt{2j+1}} \delta_{jj'} \delta mm' \qquad \langle jm00 | j'm' \rangle = \delta_{jj'} \, \delta_{mm'}$$

$$\begin{pmatrix} j & 1 & j \\ -m & 0 & m \end{pmatrix} = (-1)^{j-m} \frac{m}{\sqrt{j(j+1)(2j+1)}}$$

$$\begin{pmatrix} j & 2 & j \\ -m & 0 & m \end{pmatrix} = (-1)^{j-m} \frac{3m^2 - j(j+1)}{\sqrt{(2j-1)j(j+1)(2j+1)(2j+3)}}$$

$$\begin{pmatrix} j_1 & j_2 & j_3 \\ 0 & 0 & 0 \end{pmatrix} = \begin{cases} (-1)^g \sqrt{\dfrac{(2g-2j_1)!(2g-2j_2)!(2g-2j_3)!}{(2g+1)!}} \dfrac{g!}{(g-j_1)!(g-j_2)!(g-j_3)!} & \text{if } 2g = \text{ even} \\ 0 & \text{if } 2g = \text{ odd} \end{cases}$$

where $2g = j_1 + j_2 + j_3$

A-4 Racah Coefficient and 9j-Symbol

When three spherical tensors R_{j_1}, S_{j_2}, and T_{j_3} are coupled together, the final rank J alone is not adequate to specify the product uniquely. In order to distinguish between the different possibilities, an intermediate rank specifying the coupling between two of the three tensors is used. There are two equivalent ways to construct this intermediate coupling. One is to couple the first two tensors R_{j_1} and S_{j_2} together to rank J_{12} and then couple the product to T_{j_3} to obtain the final rank J. The angular momentum structure of the product may be expressed in the form $((R_{j_1} \times S_{j_2})_{J_{12}} \times T_{j_3})_J$. Alternatively, we can couple the last two tensors S_{j_2} and T_{j_3} together first to rank J_{23} and then couple R_{j_1} to the product. This way of coupling may be represented as $(R_{j_1} \times (S_{j_2} \times T_{j_3})_{J_{23}})_J$.

The two forms are not independent of each other and the relation between them is given by

$$\begin{aligned}((R_{j_1} \times S_{j_2})_{J_{12}} \times T_{j_3})_J &= \sum_{J_{23}} \sqrt{(2J_{12}+1)(2J_{23}+1)}\, W(j_1 j_2 J j_3; J_{12} J_{23}) \\ &\quad \times (R_{j_1} \times (S_{j_2} \times T_{j_3})_{J_{23}})_J\end{aligned}$$

where $W(j_1 j_2 J j_3; J_{12} J_{23})$ is the Racah coefficient. It may be expressed as the sum over the products of four Clebsch-Gordan coefficients,

$$\begin{aligned}W(abcd; ef) = \frac{1}{\sqrt{(2e+1)(2f+1)}} &\sum_{\alpha\beta\gamma} \langle a\alpha b\beta | e(\alpha+\beta)\rangle \langle e(\alpha+\beta)\, d(\gamma-\alpha-\beta)|c\gamma\rangle \\ &\times \langle b\beta d(\gamma-\alpha-\beta)|f(\gamma-\alpha)\rangle \langle a\alpha f(\gamma-\alpha)|c\gamma\rangle\end{aligned}$$

However, this is not the way to evaluate a Racah coefficient numerically. It is more convenient to use explicit formulas in terms of its six arguments and these can be found, for example, in Brink and Satchler [37].

A more convenient form of Racah coefficients is the 6j-symbol defined by the relation

$$\begin{Bmatrix} j_1 & j_2 & J_{12} \\ j_3 & J & J_{23} \end{Bmatrix} = (-1)^{j_1+j_2+j_3+J}\, W(j_1 j_2 J j_3; J_{12} J_{23})$$

For example, the symmetry relations of Racah coefficients may be expressed in terms of 6j-symbols in the following manner:

$$\begin{Bmatrix} j_1 & j_2 & j_3 \\ j_4 & j_5 & j_6 \end{Bmatrix} = \begin{Bmatrix} j_2 & j_3 & j_1 \\ j_5 & j_6 & j_4 \end{Bmatrix} = \begin{Bmatrix} j_3 & j_1 & j_2 \\ j_6 & j_4 & j_5 \end{Bmatrix} = \begin{Bmatrix} j_2 & j_1 & j_3 \\ j_5 & j_4 & j_6 \end{Bmatrix} = \begin{Bmatrix} j_4 & j_5 & j_3 \\ j_1 & j_2 & j_6 \end{Bmatrix}$$

The orthogonality relation between two 6j-symbols is given by

$$\sum_j (2j+1) \begin{Bmatrix} j_1 & j_2 & j' \\ j_3 & j_4 & j \end{Bmatrix} \begin{Bmatrix} j_1 & j_2 & j'' \\ j_3 & j_4 & j \end{Bmatrix} = \frac{\delta_{j'j''}}{2j'+1}$$

There are also identities involving products of 3j- and 6j-symbols that can be found in most advanced texts on nuclear structure.

In coupling four spherical tensors together, two intermediate coupling ranks are needed to specify the product uniquely. The different ways of making the intermediate couplings are related to each other through $9j$-symbols,

$$((R_{j_1} \times S_{j_2})_{J_{12}} \times (T_{j_3} \times U_{j_4})_{J_{34}})_J$$
$$= \sum_{J_{13}J_{24}} \sqrt{(2J_{12}+1)(2J_{34}+1)(2J_{13}+1)(2J_{24}+1)}$$
$$\times \begin{Bmatrix} j_1 & j_2 & J_{12} \\ j_3 & j_4 & J_{34} \\ J_{13} & J_{24} & J \end{Bmatrix} ((R_{j_1} \times T_{j_3})_{J_{13}} \times (S_{j_2} \times U_{j_4})_{J_{24}})_J$$

The value of a $9j$-symbol may be expressed as the sum over the products of three $6j$-symbols,

$$\begin{Bmatrix} j_1 & j_2 & J_{12} \\ j_3 & j_4 & J_{34} \\ J_{13} & J_{24} & J \end{Bmatrix} = \sum_{J'} (-1)^{2J'}(2J'+1) \begin{Bmatrix} j_1 & j_3 & J_{13} \\ J_{24} & J & J' \end{Bmatrix} \begin{Bmatrix} j_2 & j_4 & J_{24} \\ j_3 & J' & J_{34} \end{Bmatrix} \begin{Bmatrix} J_{12} & J_{34} & J \\ J' & j_1 & j_2 \end{Bmatrix}$$

The symmetries of $9j$-symbols are

$$\begin{Bmatrix} j_1 & j_2 & j_3 \\ j_4 & j_5 & j_6 \\ j_7 & j_8 & j_9 \end{Bmatrix} = \begin{Bmatrix} j_1 & j_4 & j_7 \\ j_2 & j_5 & j_8 \\ j_3 & j_6 & j_9 \end{Bmatrix} = \begin{Bmatrix} j_7 & j_8 & j_9 \\ j_1 & j_2 & j_3 \\ j_4 & j_5 & j_6 \end{Bmatrix} = \begin{Bmatrix} j_4 & j_5 & j_6 \\ j_7 & j_8 & j_9 \\ j_1 & j_2 & j_3 \end{Bmatrix}$$
$$= (-1)^{j_1+j_2+j_3+j_4+j_5+j_6+j_7+j_8+j_9} \begin{Bmatrix} j_4 & j_5 & j_6 \\ j_1 & j_2 & j_3 \\ j_7 & j_8 & j_9 \end{Bmatrix}$$

and the orthogonality relation is

$$\sum_{J_{13}J_{24}} (2J_{13}+1)(2J_{24}+1) \begin{Bmatrix} j_1 & j_2 & J_{12} \\ j_3 & j_4 & J_{34} \\ J_{13} & J_{24} & J \end{Bmatrix} \begin{Bmatrix} j_1 & j_2 & J'_{12} \\ j_3 & j_4 & J'_{34} \\ J_{13} & J_{24} & J \end{Bmatrix} = \frac{\delta_{J_{12}J'_{12}} \delta_{J_{34}J'_{34}}}{(2J_{12}+1)(2J_{34}+1)}$$

A collection of symmetry and orthogonality relations, as well as relations between $3j$-, $6j$-, and $9j$-symbols, can be found, e.g., in the appendices of Brink and Satchler [37], and Wong [151].

A-5 Wigner-Eckart Theorem

One of the advantages in using tensors of definite spherical ranks is offered by the Wigner-Eckart theorem. The matrix element of an operator of rank k between states with angular momenta J and J' may be separated into two parts, one invariant under a rotation of the coordinate system used and the other expressing the dependence of the matrix element on the coordinate system. Since only projections of tensors on the quantization axis are changed by a rotation of the axes, the invariant part of the matrix element is independent of the projections and, as a result, is only a function of the nature of the operator and the states involved. That is,

$$\langle JM|T_{kq}|J'M'\rangle = (-1)^{J-M}\begin{pmatrix} J & k & J' \\ -M & q & M' \end{pmatrix}\langle J\|T_k\|J'\rangle \qquad \text{(A-15)}$$

where the double-bar matrix element $\langle J\|T_k\|J'\rangle$ represents the invariant part and is generally known as the *reduced matrix element.* The angular momentum dependence of the matrix element is contained in the $3j$-coefficients and is independent of the operator and the states involved, other than their angular momentum ranks. All the physical content of a matrix element is contained in the reduced matrix element. As a result, it may be compared with those of other quantities without being encumbered by dependence on the coordinate system used.

In terms of Clebsch-Gordan coefficients, Eq. (A-15) appears as

$$\langle JM|T_{kq}|J'M'\rangle = (-1)^{2k}\frac{\langle J'M'kq|JM\rangle}{\sqrt{2J+1}}\langle J\|T_k\|J'\rangle$$

Slightly different ways are used by some authors to define the reduced matrix element. In some books, the phase factor and/or the square root in the denominator is absorbed into the definition of the reduced matrix element. Note that the phase factor $(-1)^{2k}$ is essential here, as we deal with operators of half-integer ranks as well.

A-6 Landé Formula

Consider a vector operator $\boldsymbol{V}$. Since it is an operator with spherical tensor rank unity, its matrix element behaves, under a rotation of the coordinate system, in the same way as any other spherical tensor of the same rank, including the angular momentum operator $\boldsymbol{J}$. Using the Wigner-Eckart theorem, the matrix element of component q of $\boldsymbol{V}$ may be expressed in terms of its reduced matrix element as

$$\langle JM|V_{1q}|JM'\rangle = (-1)^{J-M}\begin{pmatrix} J & 1 & J \\ -M & q & M' \end{pmatrix}\langle J\|\boldsymbol{V}\|J\rangle \qquad \text{(A-16)}$$

where q has possible values ± 1 and 0. Similarly, the matrix element of $\boldsymbol{J}$ has the form

$$\langle JM|J_{1q}|JM'\rangle = (-1)^{J-M}\begin{pmatrix} J & 1 & J \\ -M & q & M' \end{pmatrix}\langle J\|\boldsymbol{J}\|J\rangle \qquad \text{(A-17)}$$

Since both reduced matrix elements $\langle J\|\boldsymbol{V}\|J\rangle$ and $\langle J\|\boldsymbol{J}\|J\rangle$ are quantities independent of the coordinate system, they must be multiples of each other, with the ratio

$$\mathcal{R} = \frac{\langle J\|\boldsymbol{V}\|J\rangle}{\langle J\|\boldsymbol{J}\|J\rangle} \qquad \text{(A-18)}$$

independent of M.

Consider, now, the matrix element of the scalar product $\boldsymbol{J}\cdot\boldsymbol{V}$. Since it is a scalar operator, it has nonvanishing matrix elements only along the diagonal, i.e., for $J = J'$ and $M = M'$. In a spherical basis, the scalar product may be expressed as

$$\boldsymbol{J}\cdot\boldsymbol{V} = \sum_q (-1)^q \boldsymbol{J}_{1q}\boldsymbol{V}_{1,-q} \qquad \text{(A-19)}$$

We can check that this is the same as scalar products defined in terms of Cartesian components of the vectors by noting that

$$\begin{aligned} \boldsymbol{J}_{\pm 1} &= \mp\frac{1}{\sqrt{2}}(\boldsymbol{J}_x \pm i\boldsymbol{J}_y) \qquad & \boldsymbol{J}_0 = \boldsymbol{J}_z \\ \boldsymbol{V}_{\pm 1} &= \mp\frac{1}{\sqrt{2}}(\boldsymbol{V}_x \pm i\boldsymbol{V}_y) \qquad & \boldsymbol{V}_0 = \boldsymbol{V}_z \end{aligned}$$

This is slightly different from the definition of angular momentum raising and lowering operators $\boldsymbol{L}_\pm = \boldsymbol{L}_x \pm i\boldsymbol{L}_y$ as the usual convention does not attempt to make them spherical tensor operators.

We can now make an intermediate state expansion of the matrix element of $\boldsymbol{J} \cdot \boldsymbol{V}$,

$$\langle JM|(\boldsymbol{J} \cdot \boldsymbol{V})|JM\rangle = \sum_{M'}\sum_{q}(-1)^q \langle JM|\boldsymbol{J}_q|JM'\rangle\langle JM'|\boldsymbol{V}_{-q}|JM\rangle$$

Since the operator $\boldsymbol{J}$ can change at most the M-value, but not the J-value, of a function on which it acts, a sum over intermediate states of different J-values is not needed. Using the ratio $\mathcal{R}$ defined in Eq. (A-18) and the relations given by Eq. (A-16) and Eq. (A-17), we have the relation

$$\begin{aligned} \langle JM'|\boldsymbol{V}_{-q}|JM\rangle &= (-1)^{J-M'}\begin{pmatrix} J & 1 & J \\ -M' & -q & M \end{pmatrix}\langle J\|\boldsymbol{V}\|J\rangle \\ &= (-1)^{J-M'}\begin{pmatrix} J & 1 & J \\ -M' & -q & M \end{pmatrix}\mathcal{R}\langle J\|\boldsymbol{J}\|J\rangle \\ &= \mathcal{R}\langle JM'|\boldsymbol{J}_{-q}|JM\rangle \end{aligned}$$

With this, we obtain the result

$$\begin{aligned} \langle JM|(\boldsymbol{J} \cdot \boldsymbol{V})|JM\rangle &= \mathcal{R}\sum_{M'}\sum_{q}(-1)^q\langle JM|\boldsymbol{J}_q|JM'\rangle\langle JM'|\boldsymbol{J}_{-q}|JM\rangle \\ &= \mathcal{R}\langle JM|\boldsymbol{J}^2|JM\rangle \\ &= \mathcal{R}J(J+1) \end{aligned}$$

In other words, $\mathcal{R} = \langle JM|(\boldsymbol{J} \cdot \boldsymbol{V})|JM\rangle / J(J+1)$ and

$$\langle JM|V_q|JM'\rangle = \frac{1}{J(J+1)}\langle JM|(\boldsymbol{J} \cdot \boldsymbol{V})|JM\rangle\langle JM|\boldsymbol{J}_q|JM'\rangle \qquad \text{(A-20)}$$

generally known as the Landé formula.

Appendix B

Scattering by a Central Potential

B-1 Scattering Amplitude and Cross Section

The scattering of one particle off another at nonrelativistic energies is described by a time-dependent Schrödinger equation

$$i\hbar\frac{\partial}{\partial t}\Psi(\boldsymbol{r},t) = H\Psi(\boldsymbol{r},t) \tag{B-1}$$

under appropriate boundary conditions. In the center of mass of the two particles, the Hamiltonian has the form

$$H = -\frac{\hbar^2}{2\mu}\nabla^2 + V \tag{B-2}$$

where μ is the reduced mass and V is the potential representing the interaction between the two particles. If H is independent of time t, the time dependence in the wave function may be separated from the rest,

$$\Psi(\boldsymbol{r},t) = \psi(\boldsymbol{r})e^{-iEt/\hbar}$$

Here $\psi(\boldsymbol{r})$ is the eigenfunction of the time-independent Schrödinger equation

$$-\frac{\hbar^2}{2\mu}\nabla^2\psi(\boldsymbol{r}) + (V - E)\psi(\boldsymbol{r}) = 0 \tag{B-3}$$

For simplicity we shall consider $\psi(\boldsymbol{r})$ to be a function of spatial coordinates only and ignore any dependence on other variables, such as spin and isospin.

Incident flux. The usual scattering arrangement involves a collimated beam of projectile particles traveling along the positive z-direction and incident on a target placed at the origin. Except for Coulomb force, interactions between nuclei have short range. For this reason, we shall consider first finite-range potentials and return later to Coulomb interaction in §B-5. Outside the range of the interaction, we can take $V = 0$; both particles are free and their wave functions may be represented by plane waves e^{ikz}, where $k = \sqrt{2\mu E}/\hbar$ is the wave number. (For a Coulomb interaction, Coulomb wave functions must be used instead of plane waves.)

The relation between wave function and intensity of the incident beam is given by the quantum-mechanical probability current density

$$S(r,t) = \frac{\hbar}{2i\mu}\{\psi^*\nabla\psi - \psi\nabla\psi^*\} = \Re\Big\{\psi^*\frac{\hbar}{i\mu}\nabla\psi\Big\}$$

where $\Re$ stands for the real part. For an incident plane wave traveling along the positive z-direction, the number of particles passing through a unit area perpendicular to the z-axis is then

$$S_i = \Re\Big\{e^{-ikz}\frac{\hbar}{i\mu}\frac{d}{dz}e^{ikz}\Big\} = \frac{\hbar k}{\mu} = v \tag{B-4}$$

where v is the velocity of the projectile when it is still outside the interaction region. The value of incident flux S_i depends on the way the plane wave is normalized. Here we have taken it in such a way that $S_i = v$.

Scattered wave. The scattered particle outside the interaction region is described by a spherical wave e^{ikr}/r radiating outward from the center of the interaction region. The particle density in the incident beam is usually sufficiently low that we may ignore any interference between the incident and scattering particles. As a result, the wave function at large r is a linear combination of a plane wave, made of the incident beam and particles not scattered by the potential, and a spherical wave, made of scattered particles. The result may be expressed as

$$\psi(r) \xrightarrow[r\to\infty]{} e^{ikz} + f(\theta,\phi)\frac{e^{ikr}}{r} \tag{B-5}$$

Here, $f(\theta,\phi)$ is the *scattering amplitude* which measures the fraction of incident wave scattered in the direction with polar angle θ and azimuthal angle ϕ. In general, both $\psi(r)$ and $f(\theta,\phi)$ are also functions of the incident wave vector $\boldsymbol{k}$ and scattered wave vector $\boldsymbol{k}'$. However, to simplify the notation, we shall not indicate them unless required in the discussion. Furthermore, the probability for scattering is sufficiently small that the normalization of the incident wave is not affected by particles removed from the incident beam due to scattering.

It is convenient to take the origin of the coordinate system to be at the center of the region where the two particles come into contact with each other. Since the z-axis is chosen to be along the direction the two particles approaching each other outside the interaction zone, the xy-plane is fixed by requiring it to be perpendicular to the z-axis. However, we do not have a natural way to define the orientation of the x- or y-axis in the plane, if all the particles involved have spin $J = 0$, or if the spins of neither the incident nor the target particles are polarized in any given direction and the orientations of the spin of the particles in the final state are not detected. In such cases, the system is invariant under a rotation around the z-axis and the azimuthal angle ϕ cannot be determined uniquely. The wave function of the system must be independent of ϕ and the scattering amplitude becomes a function of the polar angle θ only.

The scattering angle θ is the angle between the incident wave vector $\boldsymbol{k}$ and the scattered wave vector $\boldsymbol{k}'$, as shown in Fig. B-1. For $\theta \neq 0$, $\boldsymbol{k}$ and $\boldsymbol{k}'$ forms a plane, the scattering plane. We may define a unit vector $\boldsymbol{n}$ perpendicular to the scattering plane

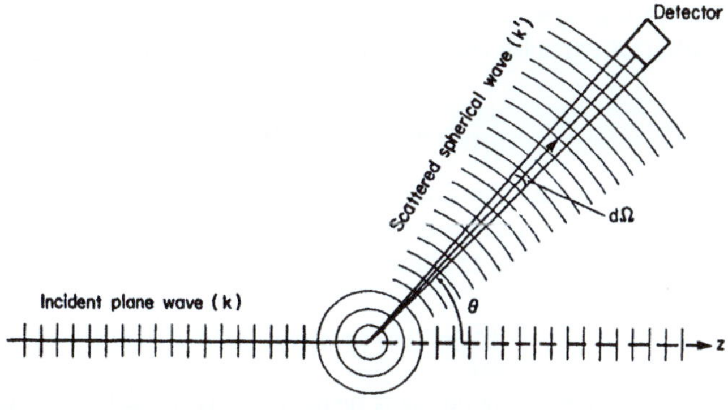

Figure B-1: The scattering plane defined by $\boldsymbol{k}$ and $\boldsymbol{k}'$, respectively, the wave vectors of the projectile and the scattered particle. The scattering angle θ is that between $\boldsymbol{k}$ and $\boldsymbol{k}'$. The scattering is independent of the azimuthal angle ϕ unless the polarization direction of the spin of at least one of the particles is known.

in the following way:

$$\hat{\boldsymbol{n}} = \frac{\boldsymbol{k} \times \boldsymbol{k}'}{|\boldsymbol{k}|\,|\boldsymbol{k}'|} \tag{B-6}$$

The orientation of $\boldsymbol{n}$ depends on the vector $\boldsymbol{k}'$, which, in turn, depends on where the detector is placed. Unless polarization is involved, the choice of the direction of $\boldsymbol{n}$ is arbitrary, usually determined by the convenience of the experimental arrangement. However, if one or both particles involved in the initial state are polarized, or if the spin orientations of one or both of the particles in the final state are detected, spin dependence in the interaction between the two particles may cause a difference in the scattering results that depends on the direction of $\boldsymbol{n}$ relative to that of polarization. Under such conditions, the scattering amplitude is a function of θ as well as ϕ.

Differential cross section. The differential scattering cross section may be expressed in terms of the scattering amplitude $f(\theta)$. The probability current density for the scattered spherical wave is given by the expression

$$S_r = \Re\left\{\left(f(\theta)\frac{e^{ikr}}{r}\right)^* \frac{\hbar}{i\mu}\frac{d}{dr}\left(f(\theta)\frac{e^{ikr}}{r}\right)\right\} = \frac{v}{r^2}|f(\theta)|^2 + O(r^{-3})$$

If the scattered particle is observed by a detector with effective area da placed at distance r from the scattering center, the solid angle subtended by the detector at the origin is

$$d\Omega = \frac{da}{r^2}$$

and the number of particles recorded per unit time is

$$N_r = S_r\,da = S_r r^2\,d\Omega$$

The differential scattering cross section, $d\sigma/d\Omega$, sometimes represented also as $\sigma(\theta)$, is defined as the number of particles scattered into a solid angle $d\Omega$ at angle θ divided by the incident flux,

$$\frac{d\sigma}{d\Omega} = \frac{S_r r^2}{S_i} = |f(\theta)|^2 \tag{B-7}$$

As we have seen in §1-3, it has the dimension of an area and gives a measure of the probability of scattering into a particular direction.

The scattering cross section is the integral of the differential cross section over all solid angles,

$$\sigma = \int \frac{d\sigma}{d\Omega} d\Omega = \int |f(\theta)|^2 2\pi \sin\theta \, d\theta$$

It conveys an idea how much of the incident beam is intercepted by each particle in the target. Since the typical unit of length for nuclei is the femtometer (fm), a convenient unit for scattering cross section is femtometer squared ($= 10^{-30}$ m^2) and that for $d\sigma/d\Omega$ is the femtometer squared per steradian. A derived unit, the barn (1 barn $= 10^{-28}$ m^2), is often used in quoting measured values. Hadronic processes are usually of the order of millibarns (1 mb$= 10^{-31}$ m^2 or 0.1 fm^2), whereas electromagnetic processes are of the order of nanobarns (1 nb$= 10^{-37}$ m^2) and weak interaction processes of the order of femtobarns (1 fb $= 10^{-43}$ m^2), as mentioned in Chapter 1.

B-2 Partial Waves and Phase Shifts

Partial wave expansion. If the interaction potential is a central one, $V = V(r)$, that depends only on the relative distance r, angular momentum is a constant of motion. In this case, it is convenient to decompose the wave function $\psi(\boldsymbol{r})$ into a product of radial and angular parts and write it as a sum over components with definite orbital angular momentum ℓ, or *partial waves*,

$$\psi(r,\theta) = \sum_{\ell=0}^{\infty} a_\ell \, R_\ell(r) \, Y_{\ell 0}(\theta) \tag{B-8}$$

where the coefficients a_ℓ are the amplitudes of each partial wave. Only spherical harmonics $Y_{\ell m}(\theta,\phi)$ with $m = 0$ are involved here, as we are considering systems independent of the azimuthal angle ϕ.

Since $Y_{\ell 0}(\theta)$ is an eigenfunction of the angular part of Eq. (B-3) with eigenvalue $\ell(\ell+1)$, the radial wave function for partial wave ℓ satisfies the equation

$$-\frac{\hbar^2}{2\mu}\left\{\frac{1}{r^2}\frac{d}{dr}r^2\frac{d}{dr} - \frac{\ell(\ell+1)}{r^2}\right\}R_\ell(r) + V(r)R_\ell(r) = ER_\ell(r)$$

In terms of the modified radial wave function

$$u_\ell(r) \equiv rR_\ell(r)$$

the equation may be simplified to

$$\frac{d^2u_\ell(r)}{dr^2} - \left\{\frac{\ell(\ell+1)}{r^2} + \frac{2\mu}{\hbar^2}V(r) - k^2\right\}u_\ell(r) = 0 \tag{B-9}$$

For short-range potentials, $V(r)$ goes to zero as $r \to \infty$. The same is also true for the $\ell(\ell+1)/r^2$ term. In the asymptotic regions, we are left with a simple second-order differential equation of the form

$$\frac{d^2u_\ell(r)}{dr^2} + k^2u_\ell(r) = 0$$

The solution for this equation is the familiar linear combination of $\sin(kr)$ and $\cos(kr)$. That is, at large r, the function $u_\ell(r)$ must take on the form

$$\begin{aligned} u_\ell(r) \xrightarrow[r\to\infty]{} & A_\ell \sin(kr - \tfrac{1}{2}\ell\pi) + B_\ell \cos(kr - \tfrac{1}{2}\ell\pi) \\ &= C_\ell \sin(kr - \tfrac{1}{2}\ell\pi + \delta_\ell) \\ &= C'_\ell\{e^{-i(kr-\frac{1}{2}\ell\pi)} - e^{2i\delta_\ell}e^{i(kr-\frac{1}{2}\ell\pi)}\} \end{aligned} \tag{B-10}$$

where A_ℓ and B_ℓ, or C_ℓ (C'_ℓ) and δ_ℓ, are two constants that must be determined from boundary conditions. The phase factor $\frac{1}{2}\ell\pi$ is included here so that it is more convenient to compare with the asymptotic form of spherical Bessel functions we need to carry out later.

Phase shift. The angle δ_ℓ is known as the *phase shift.* Its physical meaning can be seen by comparing Eq. (B-10) with the partial wave expansion of a plane wave,

$$e^{ikz} = \sum_{\ell=0}^{\infty} \sqrt{4\pi(2\ell+1)}\, i^\ell j_\ell(kr) Y_{\ell 0}(\theta) \tag{B-11}$$

Asymptotically, the spherical Bessel function $j_\ell(kr)$ has the form

$$j_\ell(kr) \xrightarrow[r\to\infty]{} \frac{\sin(kr - \frac{1}{2}\ell\pi)}{kr}$$

and may be compared with that of Eq. (B-10).

In the asymptotic region, a plane wave may be written as

$$\begin{aligned} e^{ikz} \xrightarrow[r\to\infty]{} & \sum_{\ell=0}^{\infty} \sqrt{4\pi(2\ell+1)}\, \frac{i^\ell}{kr} \sin(kr - \tfrac{1}{2}\ell\pi)\, Y_{\ell 0}(\theta) \\ &= \sum_{\ell=0}^{\infty} \sqrt{4\pi(2\ell+1)}\, \frac{i^\ell}{2ikr}\left\{e^{i(kr-\frac{1}{2}\ell\pi)} - e^{-i(kr-\frac{1}{2}\ell\pi)}\right\} Y_{\ell 0}(\theta) \\ &= \sum_{\ell=0}^{\infty} \sqrt{4\pi(2\ell+1)}\left\{\frac{e^{ikr}}{2ikr} - \frac{i^\ell e^{-i(kr-\frac{1}{2}\ell\pi)}}{2ikr}\right\} Y_{\ell 0}(\theta) \end{aligned} \tag{B-12}$$

where we have used the relation $e^{i\ell\pi/2} = i^\ell$ to put the expression into a form convenient for later needs. The difference between Eqs. (B-10) and (B-12) is the phase shift, for example, in the argument of the sine function. Because of interaction induced by potential $V(r)$, the phase of partial wave ℓ in Eq. (B-10) is shifted by a factor δ_ℓ with respect to that of a free particle represented by the plane wave of Eq. (B-12). This is a result we could have anticipated from the beginning. For a real potential, which we have implicitly assumed here, only elastic scattering can take place. Furthermore, if the potential is also a central one, orbital angular momentum ℓ is a good quantum number and the probability current density in each ℓ-partial wave *channel* is conserved. The only thing in the wave function that can change as a result of scattering is the phase angle, and this is represented by the phase shift δ_ℓ. We shall return at the end of this section with an example using a square-well potential as illustration.

In general, elastic as well as inelastic scattering can take place. Such a situation is represented by a complex scattering potential, with the imaginary part representing loss of probability from the incident channel due to such inelastic events as excitation of the target nucleus and projectile particle, absorption of the incident particle by the target, and creation of new particles. In these cases, the phase shifts are also complex in general. We shall return to the case of scattering by a complex potential in §B-4.

Elastic scattering cross section. Using the result of Eq. (B-10), the scattering wave function of Eq. (B-8) in the asymptotic region may be written as

$$\psi(r,\theta) \xrightarrow[r\to\infty]{} \sum_{\ell=0}^{\infty} a'_\ell Y_{\ell 0}(\theta)\frac{1}{r}\sin(kr - \tfrac{1}{2}\ell\pi + \delta_\ell) \tag{B-13}$$

where the unknown coefficients a_ℓ in Eq. (B-8) and C_ℓ in Eq. (B-12) are combined into a single quantity a'_ℓ. Since this is just another asymptotic form of the same wave function as given earlier in Eq. (B-5), we arrive at the equality

$$\begin{aligned} e^{ikz} + f(\theta)\frac{e^{ikr}}{r} &= \sum_{\ell=0}^{\infty} a'_\ell Y_{\ell 0}(\theta)\frac{1}{r}\sin(kr - \tfrac{1}{2}\ell\pi + \delta_\ell) \\ &= \sum_{\ell=0}^{\infty} a'_\ell Y_{\ell 0}(\theta)\Big\{(-i)^\ell e^{i\delta_\ell}\frac{e^{ikr}}{2ikr} - e^{-i\delta_\ell}\frac{e^{-i(kr-\frac{1}{2}\ell\pi)}}{2ikr}\Big\} \end{aligned} \tag{B-14}$$

Using the results of Eqs. (B-12) and (B-13), we can rewrite (B-14) in the following way:

$$\begin{bmatrix} \Big\{\sum_{\ell=0}^{\infty}\sqrt{4\pi(2\ell+1)}\frac{1}{2ik}Y_{\ell 0}(\theta) + f(\theta)\Big\}\frac{e^{ikr}}{r} \\ \\ -\sum_{\ell=0}^{\infty}\Big\{\sqrt{4\pi(2\ell+1)}i^\ell Y_{\ell 0}(\theta)\Big\}\frac{e^{-i(kr-\frac{1}{2}\ell\pi)}}{2ikr} \end{bmatrix} = \begin{bmatrix} \Big\{\sum_{\ell=0}^{\infty} a'_\ell\frac{1}{2ik}Y_{\ell 0}(\theta)(-i)^\ell e^{i\delta_\ell}\Big\}\frac{e^{ikr}}{r} \\ \\ -\sum_{\ell=0}^{\infty}\Big\{a'_\ell Y_{\ell 0}(\theta)e^{-i\delta_\ell}\Big\}\frac{e^{-i(kr-\frac{1}{2}\ell\pi)}}{2ikr} \end{bmatrix} \tag{B-15}$$

The equation is arranged in such a way that terms related to e^{ikr} are on the first line and terms related to e^{-ikr} are on the second line of both sides.

Since the functions e^{ikr} and e^{-ikr} are linearly independent of each other, their coefficients on the two sides of Eq. (B-15) must separately equal each other. From the coefficients for $e^{-i(kr-\ell\pi/2)}$, we obtain the result

$$a'_\ell = \sqrt{4\pi(2\ell+1)}\, i^\ell e^{i\delta_\ell}$$

Substituting this relation back into the coefficients of e^{ikr} in Eq. (B-15), the scattering amplitude may be put in terms of phase shifts as

$$f(\theta) = \frac{\sqrt{4\pi}}{2ik}\sum_{\ell=0}^{\infty}\sqrt{2\ell+1}(e^{2i\delta_\ell} - 1)Y_{\ell 0}(\theta)$$

$$= \frac{\sqrt{4\pi}}{k} \sum_{\ell=0}^{\infty} \sqrt{2\ell+1}\, e^{i\delta_\ell} \sin\delta_\ell Y_{\ell 0}(\theta) \tag{B-16}$$

In terms of the phase shifts, the differential scattering cross section may be written as

$$\frac{d\sigma}{d\Omega} = \frac{4\pi}{k^2} \left| \sum_{\ell=0}^{\infty} \sqrt{2\ell+1}\, e^{i\delta_\ell} \sin\delta_\ell Y_{\ell 0}(\theta) \right|^2 \tag{B-17}$$

by substituting the results of Eq. (B-16) into (B-7).

From the orthogonal condition on spherical harmonics

$$\int_0^{2\pi} \int_0^{\pi} Y^*_{\ell m}(\theta,\phi) Y_{\ell' m'}(\theta,\phi) \sin\theta\, d\theta\, d\phi = \delta_{\ell\ell'}\delta_{mm'} \tag{B-18}$$

we see that the scattering cross section may be reduced to a particularly simple form

$$\begin{aligned} \sigma^{\text{el}} &= \frac{4\pi}{k^2} \sum_{\ell\ell'} \sqrt{(2\ell+1)(2\ell'+1)} e^{i(\delta_\ell - \delta_{\ell'})} \sin\delta_\ell \sin\delta_{\ell'} \int_0^{\pi} Y_{\ell 0}(\theta) Y_{\ell' 0}(\theta) 2\pi \sin\theta\, d\theta \\ &= \frac{4\pi}{k^2} \sum_{\ell=0}^{\infty} (2\ell+1) \sin^2\delta_\ell \\ &= \frac{\pi}{k^2} \sum_{\ell} (2\ell+1) |1 - e^{2i\delta_\ell}|^2 \end{aligned} \tag{B-19}$$

Since we have taken the scattering potential $V(r)$ to be real in this section, only elastic scattering can take place. Later on, when we come to the more general case of a complex scattering potential, inelastic scattering can also take place. The superscript is to remind us that the cross section calculated here is for elastic scattering only.

Relation to scattering potential. A more direct connection between phase shift and scattering potential is provided by the following analysis. By making the substitution $\rho = kr$, Eq. (B-9) may be further simplified to

$$\frac{d^2 u_\ell(\rho)}{d\rho^2} - \left\{ \frac{V(\rho)}{E} + \frac{\ell(\ell+1)}{\rho^2} - 1 \right\} u_\ell(\rho) = 0 \tag{B-20}$$

For a free particle, we have $V(\rho) = 0$ and the corresponding modified radial wave function $f_\ell(\rho)$ for partial wave ℓ satisfies the equation

$$\frac{d^2 f_\ell(\rho)}{d\rho^2} - \left\{ \frac{\ell(\ell+1)}{\rho^2} - 1 \right\} f_\ell(\rho) = 0 \tag{B-21}$$

where $f_\ell(\rho) = \rho\, j_\ell(\rho)$, with $j_\ell(\rho)$ a spherical Bessel function of order ℓ.

The ℓ-dependent term as well as the constant term in Eqs. (B-20) and (B-21) may be eliminated by multiplying Eq. (B-20) with $f_\ell(\rho)$ and subtracting from it Eq. (B-21) multiplied by $u_\ell(\rho)$. The result is

$$\frac{d}{d\rho}\left\{ \frac{df_\ell}{d\rho} u_\ell - f_\ell \frac{du_\ell}{d\rho} \right\} + \frac{V(\rho)}{E} f_\ell(\rho) u_\ell(\rho) = 0 \tag{B-22}$$

When $r \to \infty$, the spherical Bessel function $j_\ell(\rho) \to \rho^{-1}\sin(\rho - \frac{1}{2}\ell\pi)$, as we have seen earlier, and we obtain the results

$$f_\ell(\rho) \to \sin(\rho - \tfrac{1}{2}\ell\pi) \qquad \frac{df_\ell}{d\rho} \to \cos(\rho - \tfrac{1}{2}\ell\pi)$$

and

$$u_\ell(\rho) \to \sin(\rho - \tfrac{1}{2}\ell\pi + \delta_\ell) \qquad \frac{du_\ell}{d\rho} \to \cos(\rho - \tfrac{1}{2}\ell\pi + \delta_\ell)$$

The quantity within the curly brackets in Eq. (B-22) becomes

$$\begin{aligned}\frac{df_\ell}{d\rho}u_\ell - f_\ell\frac{du_\ell}{d\rho} &\longrightarrow \cos(\rho - \tfrac{1}{2}\ell\pi)\sin(\rho - \tfrac{1}{2}\ell\pi + \delta_\ell) - \sin(\rho - \tfrac{1}{2}\ell\pi)\cos(\rho - \tfrac{1}{2}\ell\pi + \delta_\ell) \\ &= \sin\delta_\ell\end{aligned}$$

where the last equality is obtained using standard trigonometric identities. Equation (B-22) now reduces to

$$\frac{d}{d\rho}\sin\delta_\ell = -\frac{V(\rho)}{E}f_\ell(\rho)u_\ell(\rho)$$

or

$$\sin\delta_\ell = -\int_0^\infty \frac{V(\rho)}{E}f_\ell(\rho)u_\ell(\rho)\,d\rho \tag{B-23}$$

This relation determines the phase shift δ_ℓ from a potential $V(\rho)$ up to a multiple of 2π. The general convention to fix this uncertainty is to take $\delta_\ell = 0$ as $E \to 0$. Although Eq. (B-23) expresses δ_ℓ in terms of $V(r)$, the relation is not as direct as it appears on the surface, since $u_\ell(\rho)$ in the integrand depends also on the potential, as can be seen from Eq. (B-20).

Partial wave and bombarding energy. One useful result of partial wave analysis is that, for low bombarding energies, only the phase shifts for $\ell \approx 0$ are substantially different from zero. This can be seen from the following argument. The classical turning radius r_1 is defined as the point where the (repulsive) potential is equal to the incident energy. For partial wave channel ℓ, the effective potential in Eq. (B-9) is

$$\widetilde{V}(r) = V(r) + \frac{\hbar^2}{2\mu}\frac{\ell(\ell+1)}{r^2} \tag{B-24}$$

As a result, we may use the relation

$$E = V(r_1) + \frac{\hbar^2}{2\mu}\frac{\ell(\ell+1)}{r_1^2} \tag{B-25}$$

to determine the classical turning point r_1.

For a short-range potential, the effective potential $\widetilde{V}(r)$ of Eq. (B-24) for large values of r and ℓ is dominated by the repulsive centrifugal barrier term $\ell(\ell+1)/r^2$. (At very small r, the centrifugal term also dominates by virtue of its inverse r^2-dependence; consequently, only in the intermediate range is the nuclear potential important.) As a result, Eqs. (B-20) and (B-21) become the same for large ℓ-values and we obtain

$$\lim_{\ell\to\infty} u_\ell(r) = f_\ell(r)$$

Consequently,

$$\delta_\ell \xrightarrow[\ell\to\infty]{} 0$$

We shall now establish a criterion by which ℓ may be considered as large enough such that phase shifts may be ignored for partial waves of order greater than this value.

Let the range of the potential $V(r)$ be represented by r_0. At low energies, the classical turning radius r_1 is large and we have $r_0 < r_1$. We may therefore ignore the contribution of $V(r_1)$ in the definition of the turning radius. Equation (B-25) can now be approximated by the expression

$$E \approx \frac{\hbar^2}{2\mu}\frac{\ell(\ell+1)}{r_1^2}$$

or

$$(kr_1)^2 \approx \ell(\ell+1)$$

This gives us an approximate value of the turning radius that is independent of $V(r)$. It also implies that the scattering takes place mainly in channels with $\ell \lesssim kr_1$. In other words, for $\ell \gg kr_1$, the phase shifts $\delta_\ell \to 0$.

On the other hand, r_1 is a quantity that depends both on E and ℓ. It is therefore more convenient to use r_0, the range of the potential, instead of r_1 as the condition to determine the highest partial wave that can contribute to the scattering. Since these two quantities are of the same order of magnitude, we obtain the condition

$$\delta_\ell \to 0 \qquad \text{for} \qquad \ell \gg kr_0 \tag{B-26}$$

Classically, no scattering occurs if a point particle approaches a hard sphere with impact parameter b greater than the radius of the sphere r_0. Since $\ell = |\boldsymbol{r}\times\boldsymbol{p}| = \hbar kr$, we arrive at the conclusion that partial waves with $\ell/\hbar > kr_0$ are not scattered. Equation (B-26) is essentially a quantum-mechanical statement of the same criterion.

The range of nuclear potentials is of the order of a femtometer. For nucleon-nucleon collisions at $E = 1$ MeV in the center of mass, $kr_0 \sim 0.2$. Hence only $\ell = 0$, or s-wave, phase shift can be significantly different from zero. This is observed to be true as can be seen, for example, in the values extracted from experimental nucleon-nucleon scattering shown in Fig. 3-3. From the figure, we find that only the s-wave phase shifts are different from zero at low energies and that the sizes of the phase shifts for the higher partial waves, for example p-waves, do not become significant until $E > 10$ MeV. For this reason, nucleon-nucleon collision is often approximated by s-wave scattering for $E < 10$ MeV.

Example of a square-well potential. It is instructive to see the actual relation between phase shifts and scattering potential for a simple case. We shall limit ourselves to s-wave scattering and calculate δ_0 for a square well of radius r_0 and bombarding energy $E = 1$ MeV. For an attractive potential of depth V_0, we have

$$V(r) = \begin{cases} -V_0 & \text{for } r < r_0 \\ 0 & \text{for } r \geq r_0 \end{cases}$$

The radial equation, obtained by solving Eq. (B-9) inside the well, is

$$u_0(r) = \mathcal{A}\sin\kappa r \qquad \text{for} \qquad r < r_0$$

where

$$\kappa = \frac{1}{\hbar}\sqrt{2\mu(E+V_0)}$$

The amplitude $\mathcal{A}$ will be determined later. For a repulsive well, V_0 is a negative quantity. In this case κ becomes purely imaginary for $E < |V_0|$, and instead of a sine function, the radial wave function inside the well is a hyperbolic sine function.

Outside the well, $V(r) = 0$, and the radial wave function is sinusoidal for both attractive and repulsive wells,

$$u_0(r) = \sin(kr + \delta_0) \qquad \text{for} \qquad r > r_0$$

For convenience, we have normalized the wave function to have an amplitude of unity outside the well. The requirement that the logarithmic derivative of the wave function be continuous across the boundary at $r = r_0$ gives us the condition

$$\frac{\sin \kappa r_0}{\kappa \cos \kappa r_0} = \frac{\sin(kr_0 + \delta_0)}{k\cos(kr_0 + \delta_0)}$$

From this result, the s-wave phase shift is found to be

$$\delta_0 = n\pi - kr_0 + \tan^{-1}\left(\frac{k}{\kappa}\tan \kappa r_0\right)$$

where n is to be determined by the condition that $\delta_0 = 0$ at $E = 0$, as we have done for Eq. (B-23). The amplitude of the wave function inside the well is determined by the requirement that $u_0(r)$ itself is continuous across the boundary,

$$\mathcal{A} = \frac{\sin(kr_0 + \delta_0)}{\sin(\kappa r_0)}$$

The results are plotted in Fig. B-2.

For an infinite repulsive potential, the radial wave function cannot penetrate into the well, as shown in Fig. B-2(a), and $u(r) = 0$ for $r \leq r_0$ as a result. Instead of starting at $r = 0$, the nonvanishing part of the wave function is now shifted outward by a distance r_0. The phase shift is then $\delta_0 = -kr_0$. The scattering cross section from Eq. (B-19) becomes

$$\sigma = \frac{4\pi}{k^2}\sin^2\delta_0 = \frac{4\pi}{k^2}\sin^2 kr_0 \approx 4\pi r_0^2$$

a result we expect from comparisons with the scattering of two hard spheres of radius r_0 each. For a finite repulsive well, the radial wave function does not vanish completely inside the well. The amplitude rises exponentially at small r instead of sinusoidally for a free particle, as shown in Fig. B-2(b). The phase shift is still negative, but the magnitude of δ_0 is less than that for an infinite repulsive well.

For an attractive well, the phase shift is positive. If $|V_0|$ is small, the wave function inside the well rises faster near the origin than that of a free particle. As a result, the nodes of the wave function outside the well are shifted closer to the origin, as shown in Fig. B-2(c). As the attractive well becomes deeper, the phase shift grows in magnitude.

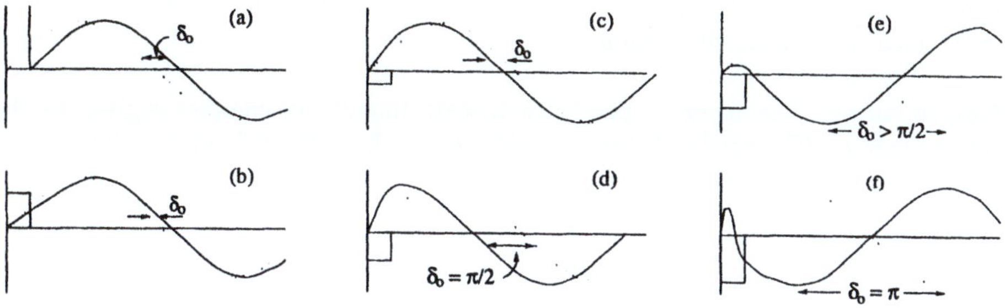

Figure B-2: Radial wave functions for low-energy, s-wave scattering by a square well. For comparison, the corresponding form for a free particle is shown as a dotted curve in each case. The result of an infinite repulsive well is shown in (a) and a finite one in (b). The results for attractive potentials of different depths are shown in (c) to (f). The wave functions inside the well in these cases grow faster near the origin than that for a free particle and the phase shift is positive.

At well depth corresponding to $\delta_0 = \pi/2$, shown in Fig. B-2(d), the scattering cross section becomes $4\pi/k^2$. For $E = 0$, we have the result

$$\sigma = \frac{4\pi}{k^2} \to \infty$$

The meaning of an infinite scattering cross section at zero energy is that the incident particle never emerges from the potential well; i.e., a bound state is formed at $E = 0$. In fact, a bound state appears whenever the phase shift is an odd integer multiple of $\pi/2$. On the other hand, when δ_0 is a multiple of π, the cross section drops to zero and nodes in the wave function appear also inside the well, as can be seen in Fig. B-2(f). In realistic situations, the potential has a more complicated form than a square well; however, the qualitative features discussed above remain true.

B-3 Effective Range Analysis

Scattering length. For low bombarding energies, it is customary to express the scattering results in terms of two parameters: scattering length a and effective range r_e. Since, in general, the cross section must be finite at $E = 0$, we can define a length parameter a by the relation

$$\lim_{k\to 0} \sigma = 4\pi a^2 \tag{B-27}$$

Except for a sign, the *scattering length* is given in terms of the s-wave phase shift by comparing Eq. (B-27) with (B-19),

$$a = \lim_{k\to 0} \Re\left\{-\frac{1}{k}e^{i\delta_0}\sin\delta_0\right\} \tag{B-28}$$

The sign convention adopted here is such that the scattering length is positive if there is a bound state, as for example in the case of isoscalar ($T = 0$) nucleon-nucleon interaction, and $a < 0$ if there is no bound state, as for example in the case of isovector ($T = 1$) nucleon-nucleon interaction.

Effective range. The energy dependence of scattering at low energies is given by the *effective range* r_e. The origin of this parameter comes from the following rationale. For $\ell = 0$, Eq. (B-9) may be written as

$$\frac{d^2u_0(k,r)}{dr^2} - \left\{\frac{2\mu}{\hbar^2}V(r) - k^2\right\}u_0(k,r) = 0 \qquad \text{(B-29)}$$

where we have included the wave number k explicitly in the arguments of the modified radial wave function $u_0(k,r)$ so as to emphasize the energy dependence in the solution. For two different energies, $E_1 = 2\hbar^2k_1^2/2\mu$ and $E_2 = \hbar^2k_2^2/2\mu$, we have two different solutions of Eq. (B-29), $u_0(k_1,r)$ and $u(k_2,r)$, respectively. These functions satisfy the following equations:

$$\begin{aligned} \frac{d^2}{dr^2}u_0(k_1,r) - \left\{\frac{2\mu}{\hbar^2}V(r) - k_1^2\right\}u_0(k_1,r) &= 0 \\ \frac{d^2}{dr^2}u_0(k_2,r) - \left\{\frac{2\mu}{\hbar^2}V(r) - k_2^2\right\}u_0(k_2,r) &= 0 \end{aligned} \qquad \text{(B-30)}$$

By multiplying the first one of Eq. (B-30) with $u_0(k_2,r)$ and the second one with $u_0(k_1,r)$ and integrating the difference over variable r, we obtain the result

$$\int_0^\infty \left\{u_0(k_2,r)\frac{d^2}{dr^2}u_0(k_1,r) - u_0(k_1,r)\frac{d^2}{dr^2}u_0(k_2,r)\right\}dr \\ +(k_1^2 - k_2^2)\int_0^\infty u_0(k_1,r)u_0(k_2,r)\,dr = 0$$

The first integral may be carried out by parts, and we obtain the result

$$\left\{u_0(k_2,r)\frac{d}{dr}u_0(k_1,r) - u_0(k_1,r)\frac{d}{dr}u_0(k_2,r)\right\}\Big|_0^\infty \\ = (k_2^2 - k_1^2)\int_0^\infty u_0(k_1,r)u_0(k_2,r)\,dr \qquad \text{(B-31)}$$

This is true for an arbitrary potential, including $V(r) = 0$.

Consider another function $v_0(k,r)$ satisfying the same equation as Eq. (B-29) except with $V(r) = 0$,

$$\frac{d^2v_0(k,r)}{dr^2} + k^2v_0(k,r) = 0 \qquad \text{(B-32)}$$

Analogous to Eq. (B-31), we have

$$\left\{v_0(k_2,r)\frac{d}{dr}v_0(k_1,r) - v_0(k_1,r)\frac{d}{dr}v_0(k_2,r)\right\}\Big|_0^\infty \\ = (k_2^2 - k_1^2)\int_0^\infty v_0(k_1,r)v_0(k_2,r)\,dr \qquad \text{(B-33)}$$

If the potential has a short range, Eqs. (B-29) and (B-32) are identical to each other in the asymptotic region. As a result, we may require that their solutions have the same form at $r = \infty$,

$$v_0(k,r) \underset{r\to\infty}{=} u_0(k,r) \underset{r\to\infty}{=} \mathcal{A}\sin(kr+\delta_0) \tag{B-34}$$

where the amplitude $\mathcal{A}$ will be determined later. Since the radial wave function $R_0(r)$ itself must be finite at the origin,

$$u_0(k,r) \xrightarrow[r\to 0]{} 0$$

The left-hand side of Eq. (B-31) may be expressed in terms of $v_0(k,r)$ using Eq. (B-34),

$$\begin{aligned}\left\{u_0(k_2,r)\frac{d}{dr}u_0(k_1,r) - u_0(k_1,r)\frac{d}{dr}u_0(k_2,r)\right\}\Bigg|_0^\infty \\ = \lim_{r\to\infty}\left\{v_0(k_2,r)\frac{d}{dr}v_0(k_1,r) - v_0(k_1,r)\frac{d}{dr}v_0(k_2,r)\right\}\end{aligned}$$

Using this, we can subtract Eq. (B-31) from (B-33). The contributions from $r=\infty$ on the left-hand side of the two equations cancel each other and we are left with the result

$$\begin{aligned}v_0(k_1 0)\frac{d}{dr}v_0(k_2,0) - v_0(k_2,0)\frac{d}{dr}v_0(k_1,0) \\ = (k_2^2-k_1^2)\int_0^\infty \{v_0(k_1,r)v_0(k_2,r) - u_0(k_1,r)u_0(k_2,r)\}\,dr\end{aligned} \tag{B-35}$$

However, $v_0(k,r)$ does not vanish at the origin, as can be seen from Eq. (B-34). This may be used to fix the amplitude $\mathcal{A}$ such that $v_0(k,0)=1$. As a result,

$$v_0(k,r) = \frac{\sin(kr+\delta_0)}{\sin\delta_0} \tag{B-36}$$

and Eq. (B-35) simplifies to the form

$$\frac{d}{dr}v_0(k_2,0) - \frac{d}{dr}v_0(k_1,0) = (k_2^2-k_1^2)\int_0^\infty \{v_0(k_1,r)v_0(k_2,r) - u_0(k_1,r)u_0(k_2,r)\}\,dr$$

Alternatively, we obtain

$$\frac{k_2\cot\delta_0(k_2) - k_1\cot\delta_0(k_1)}{k_2^2-k_1^2} = \int_0^\infty \{v_0(k_1,r)v_0(k_2,r) - u_0(k_1,r)u_0(k_2,r)\}\,dr$$

using Eq. (B-36).

If both E_1 and E_2 are close to some value $E = 2\mu k^2/\hbar^2$, the above expression may be written as

$$\frac{d}{d(k^2)}k\cot\delta_0 = \int_0^\infty \{v_0^2(k,r) - u_0^2(k,r)\}\,dr$$

The effective range is defined as twice the integral in the expression at $k=0$,

$$r_e = 2\int_0^\infty \left\{v_0^2(k,r) - u_0^2(k,r)\right\}_{k=0} dr$$

The energy dependence of the s-wave phase shift can now be expressed in the form

$$k\cot\delta_0(k) = (k\cot\delta_0)_{k=0} + \tfrac{1}{2}r_e k^2 + \cdots \tag{B-37}$$

Using the definition of scattering length a in Eq. (B-28), the first term on the right-hand side of Eq. (B-37) can be shown to be equal to $-1/a$. Up to order k^2, we find

$$k\cot\delta_0(k) = -\frac{1}{a} + \frac{1}{2}r_e k^2$$

The s-wave scattering cross section is then

$$\sigma = \frac{4\pi}{k^2}\sin^2\delta_0(k) = \frac{4\pi}{k^2 + \{\frac{1}{2}r_e k^2 - 1/a\}^2}$$

which reduces to Eq. (B-27) when $k \to 0$.

B-4 Scattering from a Complex Potential

When a particle is scattered from a target, part of the kinetic energy may be transformed into excitation energy of the projectile, the target nucleus, or both. At the same time, some of the nucleons from one may be transferred to the other. If enough energy is available in the collision, secondary particles may also be created. All such processes are inelastic in the sense that the exit channel of the reaction is different from the entrance channel. In general, a reaction consists of both elastic and inelastic scattering and the interaction potential is complex. The solution of the Schrödinger equation in such a case may still be represented by Eq. (B-8); however, the phase shifts can now be complex quantities as well.

In order to treat a broader class of scattering problems, we shall write the asymptotic form of the modified radial equation $u_\ell(r)$ for partial wave ℓ in terms of an incoming wave $\mathcal{I}_\ell(r)$ and an outgoing wave $\mathcal{O}_\ell(r)$,

$$u_\ell(r) \xrightarrow[r\to\infty]{} \mathcal{I}_\ell(r) - \eta_\ell\mathcal{O}_\ell(r) \tag{B-38}$$

in the place of Eq. (B-10). Here η_ℓ, the inelasticity parameter, is a way to measure the contribution of inelastic scattering, as we shall see later. [The definition of η_ℓ here is a more general one than that in Eq. (3-79), where η_ℓ is a real number, equivalent to the absolute value of η_ℓ here.] Each of the factors in Eq. (B-38) has a counterpart in (B-10),

$$\eta_\ell \sim e^{2i\delta_\ell} \qquad \mathcal{I}_\ell(r) \sim e^{-i(kr-\frac{1}{2}\ell\pi)} \qquad \mathcal{O}_\ell(r) \sim e^{i(kr-\frac{1}{2}\ell\pi)} \tag{B-39}$$

The elastic scattering cross section given in Eq. (B-19) may now be expressed as

$$\sigma^{\text{el}} = \frac{\pi}{k^2}\sum_\ell (2\ell+1)|1-\eta_\ell|^2$$

In addition, there are new terms contributing to the reaction that are not present in scattering by a real potential.

One way to see the difference between scattering by a real and a complex potential is to examine the intensities of the incoming and outgoing waves for partial wave ℓ. Using the last form of Eq. (B-10), we obtain the difference as

$$1 - |\eta_\ell|^2 = 1 - |e^{2\imath\delta_\ell}|^2$$

If the phase shift δ_ℓ is real, the difference vanishes and only elastic scattering can take place. For a complex phase shift, the difference does not vanish in general, as some of the incident flux is transferred to channels other than the incident one. This part of the scattering is represented by the "reaction" cross section

$$\sigma^{\text{re}} = \frac{\pi}{k^2}\sum_\ell (2\ell+1)(1 - |\eta_\ell|^2) \tag{B-40}$$

The total cross section is then the sum of those due to elastic scattering as well as the reaction,

$$\begin{aligned}\sigma^{\text{tot}} &= \sigma^{\text{el}} + \sigma^{\text{re}} \\ &= \frac{\pi}{k^2}\sum_\ell (2\ell+1)(|1-\eta_\ell|^2 + 1 - |\eta_\ell|^2) \\ &= \frac{2\pi}{k^2}\sum_\ell (2\ell+1)(1 - \Re\eta_\ell)\end{aligned} \tag{B-41}$$

We may compare this result with the scattering amplitude $f(\theta)$ at $\theta = 0$. From Eq. (B-16), we have

$$f(\theta=0) = \frac{1}{2ik}\sum_{\ell=0}^{\infty}(2\ell+1)(e^{2\imath\delta_\ell} - 1) = \frac{1}{2ik}\sum_{\ell=0}^{\infty}(2\ell+1)(\eta_\ell - 1)$$

where we have made use of the value

$$Y_{\ell 0}(\theta=0) = \sqrt{\frac{2\ell+1}{4\pi}}$$

Comparing this result with the final form of Eq. (B-41), we obtain the relation

$$\sigma^{\text{tot}} = \frac{4\pi}{k}\Im f(0) \tag{B-42}$$

known as the *optical theorem.*

Reaction channel. To discuss inelastic scattering involving nuclear particles in more detail, we need to define the concept of a *reaction channel.* It is used to describe a particular quantum-mechanical state of the system either before or after the scattering event. We shall examine here only two-body scattering, although the formalism itself can be generalized to include reactions involving three or more particles in the final state. The labels required to specify a reaction channel consist of three distinctive parts: those describing the internal degrees of freedom of the projectile or the scattered particle, those describing the corresponding quantities for the target or the residual nucleus, and those describing the relative motion between the two. For simplicity we shall use a single letter, c, the channel quantum number, to represent the complete set of labels,

$$c \equiv \{j_p\alpha_p, j_t\alpha_t; \gamma\mu : \ell m\}$$

where ℓ is the relative angular momentum and m is its projection on the quantization axis. The wave function of the projectile (or scattered particle) is represented by $\phi_{j_p\alpha_p}$, where j_p is the spin and α_p represents all the other quantum numbers required to specify the state for the projectile (or the scattered particle). The wave function of the target (or the residual) nucleus is given by $\psi_{j_t\alpha_t}$, where j_t is the spin and α_t represents all the other labels.

Since there are three different angular momenta involved here, it is useful to couple two of them together first. For this purpose, we shall define a function,

$$\Phi_{\gamma\mu} = (\phi_{j_p\alpha_p} \times \psi_{j_t\alpha_t})_{\gamma\mu}$$

the product of the wave functions of the projectile (or the scattered particle) and the target (or the residual) nucleus with their angular momenta coupled together to (γ, μ). It is convenient to treat the relative orbital angular momentum ℓ separately from the spins of the particles, as it is not usually observed directly in a measurement. The identification of one of the two particles involved in the scattering as the projectile and the other one as the target nucleus before the event, and one of the particles as the scattered particle and the other one as the residual nucleus after the event, is an artificial one without much significance in the center-of-mass system we are using here. To simplify the notation, we have omitted references to isospin.

Scattering solution. Instead of Eq. (B-39), we shall define the incoming and outgoing waves in the following way:

$$\begin{aligned} \mathcal{I}_c(\boldsymbol{r}) &= \frac{1}{r\sqrt{v_c}} i^\ell Y_{\ell m}(\theta, \phi) e^{-i(kr - \frac{1}{2}\ell\pi)} \Phi_{\gamma\mu} \\ \mathcal{O}_c(\boldsymbol{r}) &= \frac{1}{r\sqrt{v_c}} i^\ell Y_{\ell m}(\theta, \phi) e^{+i(kr - \frac{1}{2}\ell\pi)} \Phi_{\gamma\mu} \end{aligned} \tag{B-43}$$

where v_c is the center-of-mass velocity in channel c and is used to normalize the wave function in terms of probability current density, as we saw in Eq. (B-4). Consider first the simple case of a definite incoming channel c. The scattering wave function for this incident channel and all possible outgoing channels may be written as

$$\Psi_c(\boldsymbol{r}) = \mathcal{I}_c(\boldsymbol{r}) - \sum_{c'} S_{c'c} \mathcal{O}_{c'}(\boldsymbol{r}) \tag{B-44}$$

where $S_{c'c}$ is the matrix element relating the scattering amplitude from incident channel c to exit channel c'.

In general, the scattering process is described by the s-matrix (also referred to, on occasion, as the reaction matrix or the collision matrix). The matrix element

$$S_{c'c} = \langle \Psi_{c'}^{\text{out}}(\boldsymbol{r}) | \boldsymbol{S} | \Psi_c^{\text{in}}(\boldsymbol{r}) \rangle$$

is taken between wave functions in the incident channel c and outgoing channel c'. The superscripts on the wave functions are to remind us that the solution in channel c' must be obtained using the appropriate boundary condition for the outgoing wave and that in channel c for the incoming wave. We shall return to the topic of the s-matrix in the final section of this Appendix.

The general solution of the Schrödinger equation (B-3) outside the range of scattering potential V is a linear combination of those given in Eq. (B-44),

$$\Psi(\boldsymbol{r}) = \sum_c C_c\Big\{\mathcal{I}_c(\boldsymbol{r}) - \sum_{c'} S_{c'c}\mathcal{O}_{c'}(\boldsymbol{r})\Big\} \tag{B-45}$$

where the coefficients C_c depend on the initial conditions given by the particular arrangement of the incident beam and the target.

The asymptotic form of the incident wave function, with the projectile described by $\phi_{j_p\alpha_p}$, the target nucleus described by $\psi_{j_t\alpha_t}$, and the two particles approaching each other along the z-axis with relative wave function described by a plane wave (or a Coulomb wave if both particles carry charge), is given by

$$\begin{aligned}\Psi_{\rm inc}(\boldsymbol{r}) &= \frac{1}{\sqrt{v}}e^{ikz}\Phi_{\gamma\mu} \\ &\xrightarrow[r\to\infty]{} \sqrt{\frac{4\pi}{v}}\sum_\ell \sqrt{(2\ell+1)}\,\frac{i^\ell}{2ikr}\big\{e^{-i(kr-\frac{1}{2}\ell\pi)} - e^{i(kr-\frac{1}{2}\ell\pi)}\big\}Y_{\ell 0}(\theta)\Phi_{\gamma\mu} \\ &= \frac{i\sqrt{\pi}}{k}\sum_\ell \sqrt{(2\ell+1)}\,\big\{\mathcal{I}_{c(\ell,m=0)} - \mathcal{O}_{c(\ell,m=0)}\big\}\end{aligned} \tag{B-46}$$

in analogy with Eq. (B-12). For clarity, in addition to channel quantum number c, we have also given some of the implied labels explicitly in parentheses as part of the subscripts. The complete scattering wave function of Eq. (B-45) must contain a term describing an incident beam identical to that given in Eq. (B-46). Hence Eq. (B-45) may be written in the form

$$\begin{aligned}\Psi(\boldsymbol{r}) &\xrightarrow[r\to\infty]{} \frac{i\sqrt{\pi}}{k}\sum_\ell \sqrt{(2\ell+1)}\,\Big\{\mathcal{I}_{c(\ell,m=0)} - \sum_{c'} S_{c'c(\ell,m=0)}\mathcal{O}_{c'}\Big\} \\ &= \frac{i\sqrt{\pi}}{k}\sum_\ell \sqrt{(2\ell+1)}\Big\{\mathcal{I}_{c(\ell,m=0)} - \mathcal{O}_{c(\ell,m=0)} + \mathcal{O}_{c(\ell,m=0)} - \sum_{c'} S_{c'c(\ell,m=0)}\mathcal{O}_{c'}\Big\} \\ &= \Psi_{\rm inc}(\boldsymbol{r}) + \frac{i\sqrt{\pi}}{k}\sum_\ell \sqrt{(2\ell+1)}\Big\{\mathcal{O}_{c(\ell,m=0)} - \sum_{c'} S_{c'c(\ell,m=0)}\mathcal{O}_{c'}\Big\}\end{aligned}$$

We shall now work out the differential scattering cross section from this expression.

Cross section. Since the incident probability current density is normalized to unity because of Eq. (B-43), the differential scattering cross section is given by

$$\left(\frac{d\sigma}{d\Omega}\right)_{\gamma\mu\alpha;\gamma'\mu'\beta} = \frac{\pi}{k^2}\Big|\sum_{\ell\ell'} \sqrt{(2\ell+1)}\,S_{c'(\ell'\gamma'\mu'\beta)\,c(\ell,m=0,\gamma\mu\alpha)}Y_{\ell 0}(\theta)\Big|^2$$

where we have integrated over all the internal variables in the initial state, described by the product wave function $\Phi_{\gamma\mu}(j_p\alpha_p; j_t\alpha_t)$, and in the final state, described by the product wave function $\Phi_{\gamma'\mu'}(j_s\beta_s; j_r\beta_r)$. The expression is basically the same as Eq. (B-17) except that elements of the s-matrix between incident and final scattering states are used to replace the phase shifts. The summation over ℓ', the orbital angular momentum in the outgoing channel, is required since in a scattering experiment only the states of the scattered particle and the residual nucleus are observed and their relative angular momentum ℓ' is not usually identified. On integrating over the angles, we obtain the scattering cross section as

$$\sigma_{\gamma\mu\alpha;\gamma'\mu'\beta} = \frac{\pi}{k^2}\sum_{\ell}(2\ell+1)\left|S_{c'(\ell\gamma'\mu'\beta)\,c(\ell,m=0,\gamma\mu\alpha)}\right|^2 \tag{B-47}$$

in the same way as was done to arrive at Eq. (B-19). The reaction cross section is represented by terms with exit channels with $\beta \neq \alpha$.

For elastic scattering, the amplitude is given by the expression

$$T_{c'(\ell'm'\gamma'\mu'\beta)\,c(\ell,m=0,\gamma\mu\alpha)} = \delta_{\ell\ell'}\delta_{m'0}\delta_{\gamma\gamma'}\delta_{\mu\mu'}\delta_{\alpha\beta} - S_{c(\ell'm'\gamma'\mu'\beta)\,c(\ell,m=0,\gamma\mu\alpha)}$$

which, in its more general form, is known as the t-matrix. The elastic scattering cross section is then

$$\begin{aligned}\sigma^{\mathrm{el}}_{\gamma\mu\alpha;\gamma\mu\alpha} &= \frac{\pi}{k^2}\sum_{\ell}(2\ell+1)\left|1 - S_{c(\ell\gamma\mu\alpha)\,c(\ell,m=0,\gamma\mu\alpha)}\right|^2 \\ &= \frac{\pi}{k^2}\sum_{\ell}(2\ell+1)\Big\{1 - 2\Re S_{\gamma\mu\alpha;\gamma\mu\alpha(m=0)} + \sum_{\ell'}\left|S_{c(\ell\gamma\mu\alpha)\,c(\ell,m=0,\gamma\mu\alpha)}\right|^2\Big\}\end{aligned} \tag{B-48}$$

We can recover from this the relation given by Eq. (B-41) for total scattering cross section by adding to Eq. (B-48) the contribution from the reaction cross section contained in Eq. (B-47) and summing over all possible exit channels,

$$\sigma^{\mathrm{tot}}_{\gamma\mu\alpha;\gamma\mu\alpha} = \frac{\pi}{k^2}\sum_{\ell}(2\ell+1)\Big\{1 - 2\Re S_{c(\ell\gamma\mu\alpha)\,c(\ell,m=0,\gamma\mu\alpha)} + \sum_{\ell'\gamma'\mu'\beta}\left|S_{c'(\ell'\gamma'\mu'\beta)\,c(\ell,m=0,\gamma\mu\alpha)}\right|^2\Big\}$$

Because of the unitary property of the s-matrix,

$$\sum_{\ell'\gamma'\mu'\beta}\left|S_{c'(\ell'\gamma'\mu'\beta)\,c(\ell\gamma\mu\alpha)}\right|^2 = 1$$

where the summation is taken over all the possible channels, we have the result

$$\sigma^{\mathrm{tot}}_{\gamma\mu\alpha;\gamma\mu\alpha} = \frac{2\pi}{k^2}\sum_{\ell}(2\ell+1)\Big\{1 - \Re S_{c(\ell\gamma\mu\alpha)\,c(\ell\gamma\mu\alpha)}\Big\}$$

From this we obtain again the optical theorem in the same way as was done in deriving Eq. (B-42) from (B-41).

B-5 Coulomb Scattering

The discussions in §B-2 and §B-3 apply only to short-range potentials. For nuclear scattering this is quite adequate except for the electric charge carried by the participants. The Coulomb potential between two nuclei with charges Z_1e and Z_2e is given by

$$V_c(r) = \left[\frac{1}{4\pi\epsilon_0}\right]\frac{Z_1Z_2e^2}{r} = \alpha\hbar c\frac{Z_1Z_2}{r}$$

where the factor inside the square brackets converts the expression from cgs to SI units. Since the range of this potential is infinite, the techniques employed in §B-2 to find the scattering solution no longer apply. This is not a problem, as exact solutions are available (see, e.g., Messiah [104], Morse and Feshbach [106], and Blatt and Weisskopf [32]). A short summary of the results is given here.

For scattering involving only Coulomb potential, the Schrödinger equation can be written as

$$\left\{\nabla^2 + k^2 - \frac{2\gamma k}{r}\right\}\psi_c(\boldsymbol{r}) = 0 \tag{B-49}$$

where

$$k^2 = \frac{2\mu E}{\hbar^2} \qquad \gamma = \frac{Z_1 Z_2 \alpha \mu c}{\hbar k}$$

The regular solution of Eq. (B-49) has the form

$$\psi(\boldsymbol{r}) = e^{ikz} f(r - z)$$

where

$$kz = kr\cos\theta = \boldsymbol{k}\cdot\boldsymbol{r}$$

The function $f(\zeta)$ satisfies the differential equation,

$$\left\{\zeta\frac{d^2}{d\zeta^2} + (1-\zeta)\frac{d}{d\zeta} + i\gamma\right\} f(\zeta) = 0$$

with

$$\zeta = ik(r - z)$$

It is a type of Laplace equation,

$$\left\{u\frac{d^2}{du^2} + (\beta - u)\frac{d}{du} - \alpha\right\} f(u) = 0$$

with solution involving the confluent hypergeometric series

$$F(\alpha|\beta|u) = 1 + \frac{\alpha}{\beta}\frac{u}{1!} + \frac{\alpha(\alpha+1)}{\beta(\beta+1)}\frac{u^2}{2!} + \cdots$$

The normalized Coulomb wave function is then

$$\psi_c(\boldsymbol{r}) = e^{-\frac{1}{2}\pi\gamma}\Gamma(1+i\gamma)e^{ikz}F(-i\gamma|1|ik(r-z))$$

The definition of the gamma function $\Gamma(1+i\gamma)$ and its properties may be found in Abramowitz and Stegun [2].

At the origin, $F(\alpha|\beta|u) = 1$ and only the normalization factor remains,

$$\psi_c(0) = e^{-\frac{1}{2}\pi\gamma}\Gamma(1+i\gamma)$$

Using the identity that

$$|\Gamma(1+i\gamma)|^2 = \frac{\pi\gamma}{\sinh\pi\gamma}$$

we obtain the result

$$|\psi_c(0)|^2 = \frac{2\pi\gamma}{e^{2\pi\gamma}-1} \tag{B-50}$$

This gives the Fermi function $F(Z, E_e)$ of Eq. (5-67) for nuclear β-decay in the limit that the charge distribution in the daughter nucleus can be considered to be concentrated at a point located at the origin.

For scattering, we are more concerned with the asymptotic behavior of the wave function. As in Eq. (B-5), we need the values at large distances away from the origin and expressed as a sum of incident wave $\psi_i(\boldsymbol{r})$ and scattered wave $\psi_s(\boldsymbol{r})$,

$$\psi_c(\boldsymbol{r}) = \psi_i(\boldsymbol{r}) + \psi_s(\boldsymbol{r})$$

For $|r - z| \to \infty$, we have the result

$$\psi_i(\boldsymbol{r}) \longrightarrow e^{i\{kz+\gamma \ln k(r-z)\}}\{1 + \frac{\gamma^2}{ik(r-z)} + \cdots\}$$

$$\psi_s(\boldsymbol{r}) \longrightarrow \frac{1}{r}e^{i\{kr-\gamma \ln 2kr\}} f^c(\theta) + O(r^{-2})$$

The Coulomb scattering amplitude $f^c(\theta)$ is given by

$$f^c(\theta) = -\frac{\gamma}{2k\sin^2\frac{1}{2}\theta} e^{i\{\gamma \ln(\sin^2\frac{1}{2}\theta)+2\delta_0^c\}}$$

where

$$\delta_0^c = \arg\Gamma(1+i\gamma)$$

is the Coulomb phase shift for $\ell = 0$. Using this result, we obtain the Rutherford scattering formula

$$\left(\frac{d\sigma}{d\Omega}\right)_{\text{Ruth.}} = \left\{\frac{Z_1 Z_2 \alpha \hbar c}{4E\sin^2(\theta/2)}\right\}^2$$

This is the same expression as Eq. (4-7) except, here, the kinetic energy is represented by the symbol E to conform with the general practice in nonrelativistic scattering, rather than T in Eq. (4-7), where we need to make a distinction from the total relativistic energy.

We can also make a partial wave expansion for the solution to Eq. (B-49) in a way similar to that given in Eq. (B-8). Let

$$\psi_c(\boldsymbol{r}) = \sum_\ell \sqrt{4\pi(2\ell+1)}\,\frac{i^\ell}{kr}\,u_\ell^c(r) Y_{\ell 0}(\theta)$$

The modified Coulomb radial wave function $u_\ell^c(r)$ satisfies the radial equation

$$\left\{\frac{d^2}{d\rho^2} + 1 - \frac{2\gamma}{\rho} - \frac{\ell(\ell+1)}{\rho^2}\right\} u_\ell^c(r) = 0$$

where $\rho = kr$. The solution of this equation may also be expressed as a sum of $F_\ell(\gamma, \rho)$ and $G_\ell(\gamma, \rho)$, the regular and irregular Coulomb wave functions (see, e.g., Abramowitz and Stegun [2]),

$$u_\ell^c(\rho) = C_1 F_\ell(\gamma, \rho) + C_2 G_\ell(\gamma, \rho)$$

However, for scattering problems, it is more convenient to use

$$u_\ell^c(r) = e^{i\delta_\ell^c} F_\ell(\gamma, \rho) \tag{B-51}$$

where

$$\delta_\ell^c = \arg \Gamma(\ell + 1 + i\gamma)$$

is the Coulomb phase shift for partial wave ℓ.

Asymptotically, the Coulomb wave function has the properties

$$F_\ell(\gamma, \rho) \xrightarrow[r\to\infty]{} \sin \xi_\ell \qquad\qquad G_\ell(\gamma, \rho) \xrightarrow[r\to\infty]{} \cos \xi_\ell$$

where

$$\xi_\ell = \rho - \gamma \ln 2\rho - \tfrac{1}{2}\ell\pi + \delta_\ell^c$$

Applying this result to the right-hand side of Eq. (B-51), we can write the asymptotic form of the modified radial wave function in a manner similar to the final form of Eq. (B-10),

$$u_\ell^c(r) \xrightarrow[r\to\infty]{} \frac{i^{\ell+1}}{2kr} \left\{ e^{-i(kr - \gamma \ln 2kr))} - e^{2i\delta_\ell^c} e^{i(kr - \gamma \ln 2kr - \ell\pi))} \right\}$$

From this, we obtain the Coulomb scattering amplitude in terms of the phase shifts

$$f^c(\theta) = \frac{1}{2ik} \sum_\ell \sqrt{4\pi(2\ell+1)}\, (e^{2i\delta_\ell^c} - 1) Y_{\ell 0}(\theta)$$

similar to that given in Eq. (B-16).

B-6 Formal Solution to the Scattering Equation

There are two reasons to have a short discussion here on the formal solution to the scattering equation. The first is to define some of the terminology commonly used in scattering and related problems. The second is to make a connection with methods used in standard references on nuclear scattering.

We shall write the time-independent Hamiltonian as

$$H = H_0 + V \tag{B-52}$$

Normally H_0 consists of the kinetic energy operator only,

$$H_0 = -\frac{\hbar^2}{2\mu} \nabla^2 \tag{B-53}$$

as in Eq. (B-2). However, we may also choose to include in H_0 a part of the interaction, such as that due to Coulomb force or the optical model potential, as we did in §8-4. The potential V in Eq. (B-52), then, represents the *residual interaction*, the remainder of V that is not already included in H_0. For our purpose here, we shall further assume that any long-range part of the potential is included in H_0.

The eigenfunction of the scattering equation is the solution of the equation

$$(H_0 - E)\psi_{\boldsymbol{k}}^{\pm}(\boldsymbol{r}) = -V\psi_{\boldsymbol{k}}^{\pm}(\boldsymbol{r}) \tag{B-54}$$

where the superscript + on $\psi_{\boldsymbol{k}}(\boldsymbol{r})$ indicates that the solution satisfies *outgoing* boundary conditions and the superscript − refers to *incoming* boundary conditions. Our concern will be mainly with the former. The subscript $\boldsymbol{k}$, with magnitude $k = \sqrt{2\mu E}/\hbar$, displays the explicit dependence of the solution on energy.

The solution of the homogeneous equation

$$(H_0 - E)\phi_{\boldsymbol{k}}(\boldsymbol{r}) = 0 \tag{B-55}$$

forms a complete set satisfying the orthogonality condition

$$\int \phi_{\boldsymbol{k}'}^{*}(\boldsymbol{r})\phi_{\boldsymbol{k}}(\boldsymbol{r})\,d\boldsymbol{r} = \delta(\boldsymbol{k} - \boldsymbol{k}')$$

and having the closure property

$$\int \phi_{\boldsymbol{k}}^{*}(\boldsymbol{r}')\phi_{\boldsymbol{k}}(\boldsymbol{r})\,d\boldsymbol{k} = \delta(\boldsymbol{r} - \boldsymbol{r}')$$

For the simple case of Eq. (B-53) for H_0, we have plane waves, $\phi_{\boldsymbol{k}}(\boldsymbol{r}) \sim \exp(i\boldsymbol{k}\cdot\boldsymbol{r})$, as the solution for Eq. (B-55). On the other hand if, for example, the Coulomb potential is included as a part of H_0, we have the Coulomb wave functions as the solution instead.

Green's function. Using the method of Green's function, the solution of the scattering equation may be expressed in terms of an integral equation

$$\psi_{\boldsymbol{k}}^{+}(\boldsymbol{r}) = \phi_{\boldsymbol{k}}(\boldsymbol{r}) + \frac{2\mu}{\hbar^2}\int G^{+}(\boldsymbol{r},\boldsymbol{r}')V(\boldsymbol{r}')\psi_{\boldsymbol{k}}^{+}(\boldsymbol{r}')\,d\boldsymbol{r}' \tag{B-56}$$

The first term is the solution to the homogeneous equation of Eq. (B-55). The Green's function $G^{+}(\boldsymbol{r},\boldsymbol{r}')$ in the second term satisfies the equation

$$(H_0 - E)G^{+}(\boldsymbol{r},\boldsymbol{r}') = -\frac{\hbar^2}{2\mu}\delta(\boldsymbol{r} - \boldsymbol{r}') \tag{B-57}$$

with outgoing boundary conditions. In the simple case that H_0 involves only the kinetic energy, as given in Eq. (B-53),

$$G^{+}(\boldsymbol{r},\boldsymbol{r}') = -\frac{1}{4\pi}\frac{e^{ik|\boldsymbol{r}-\boldsymbol{r}'|}}{|\boldsymbol{r}-\boldsymbol{r}'|} \tag{B-58}$$

We shall use this simple form of the Green's function exclusively for the examples below.

It is easy to check that $\psi_{\boldsymbol{k}}^{+}(\boldsymbol{r})$ given in Eq. (B-56) is a solution to (B-54). On applying $H_0 - E$ to both sides of Eq. (B-56), we obtain the result

$$(H_0 - E)\psi_{\boldsymbol{k}}^{+}(\boldsymbol{r}) = (H_0 - E)\phi_{\boldsymbol{k}}(\boldsymbol{r}) + \frac{2\mu}{\hbar^2}(H_0 - E)\int G^{+}(\boldsymbol{r},\boldsymbol{r}')V(\boldsymbol{r}')\psi_{\boldsymbol{k}}^{+}(\boldsymbol{r}')\,d\boldsymbol{r}'$$

The first term on the right-hand side vanishes because of Eq. (B-55). For the second term, since $H_0 - E$ operates only on variable $\boldsymbol{r}$ and not on $\boldsymbol{r}'$, we may bring the operator inside the integral without changing the final result. Furthermore, since $\boldsymbol{r}$ appears only in $G(\boldsymbol{r},\boldsymbol{r}')$, we obtain, using Eq. (B-57), the result

$$(H_0 - E)\psi_{\boldsymbol{k}}^+(\boldsymbol{r}) = -\int \delta(\boldsymbol{r}-\boldsymbol{r}')V(\boldsymbol{r}')\psi_{\boldsymbol{k}}^+(\boldsymbol{r}')\,d\boldsymbol{r}' = -V(\boldsymbol{r})\psi_{\boldsymbol{k}}^+(\boldsymbol{r})$$

the same equality given in Eq. (B-54).

Scattering amplitude. It is easy to see how the scattering amplitude may be obtained from Eq. (B-56) using the explicit form of the Green's function given in Eq. (B-58). Let $\hat{\boldsymbol{r}} = \boldsymbol{r}/|r|$ be a unit vector along direction $\boldsymbol{r}$. In the asymptotic region,

$$|\boldsymbol{r}-\boldsymbol{r}'| \approx r - \hat{\boldsymbol{r}}\cdot\boldsymbol{r}'$$

since the integral over $\boldsymbol{r}'$ is effective only in the region of small r' where the short-range potential $V(\boldsymbol{r}')$ is nonvanishing. As a result, we may approximate the Green's function of Eq. (B-58) as

$$G^+(\boldsymbol{r},\boldsymbol{r}') \xrightarrow[r\to\infty]{} -\frac{1}{4\pi}\frac{e^{ikr}}{r}e^{-ik\hat{\boldsymbol{r}}\cdot\boldsymbol{r}'} = -\frac{1}{4\pi}\frac{e^{ikr}}{r}\phi_{\boldsymbol{k}'}^*(\boldsymbol{r}')$$

where we have taken $\boldsymbol{k}'$ to be along the direction of $\hat{\boldsymbol{r}}$. Equation (B-56) is now reduced to

$$\psi_{\boldsymbol{k}}^+(\boldsymbol{r}) = \phi_{\boldsymbol{k}}(\boldsymbol{r}) - \frac{e^{ikr}}{r}\frac{\mu}{2\pi\hbar^2}\int \phi_{\boldsymbol{k}'}^*(\boldsymbol{r}')V(\boldsymbol{r}')\psi_{\boldsymbol{k}}^+(\boldsymbol{r}')\,d\boldsymbol{r}' \qquad \text{(B-59)}$$

Comparing this result with Eq. (B-5), the scattering amplitude is identified as

$$f(\theta) = -\frac{\mu}{2\pi\hbar^2}\int \phi_{\boldsymbol{k}'}^*(\boldsymbol{r}')V(\boldsymbol{r}')\psi_{\boldsymbol{k}}^+(\boldsymbol{r}')\,d\boldsymbol{r}' = -\frac{\mu}{2\pi\hbar^2}\langle\phi_{\boldsymbol{k}'}|V|\psi_{\boldsymbol{k}}^+\rangle \qquad \text{(B-60)}$$

The result here is an exact one (in the asymptotic region) and is different from that of the first Born approximation given in Eq. (8-22), as $\psi_{\boldsymbol{k}}^+$, the solution of the scattering equation Eq. (B-54), appears in $f(\theta)$ in the place of $\phi_{\boldsymbol{k}}$. The differential scattering cross section is then

$$\frac{d\sigma}{d\Omega} = |f(\theta)|^2 = \frac{\mu^2}{4\pi^2\hbar^4}|\langle\phi_{\boldsymbol{k}'}|V|\psi_{\boldsymbol{k}}^+\rangle|^2$$

The usefulness of this expression is limited, as it requires a knowledge of $\psi_{\boldsymbol{k}}^+(\boldsymbol{r}')$, the complete solution to the scattering problem.

The result given by Eq. (B-59) is an integral equation, or "formal," solution of the scattering equation, as $\psi_{\boldsymbol{k}}^+$ itself appears on the right-hand side as well. Its value lies mainly in analytical works, such as a Born series expansion of the scattering wave function and scattering amplitude. To simplify the notation, we shall write Eq. (B-56) in the following way:

$$\psi_{\boldsymbol{k}}^+ = \phi_{\boldsymbol{k}} + G^+V\psi_{\boldsymbol{k}}^+ \qquad \text{(B-61)}$$

where, instead of $G^+(\boldsymbol{r},\boldsymbol{r}')$, we have used G^+, an operator for the Green's function defined by the relation

$$G^+(\boldsymbol{r},\boldsymbol{r}') = \langle\boldsymbol{r}|G^+|\boldsymbol{r}'\rangle$$

In terms of H_0 and E, the Green's function operator G^+ may be expressed as

$$G^+ = \lim_{\epsilon \to 0} \frac{1}{E - H_0 + i\epsilon} \tag{B-62}$$

where the factor $+i\epsilon$, with ϵ as some small positive quantity, is required to ensure that the operator corresponds to the outgoing boundary condition. The derivation of Eq. (B-62) may be found in quantum mechanics texts such as Merzbacher [103], Messiah [104], and Schiff [125].

Lippmann-Schwinger equation. It is easy to see that Eq. (B-62) is correct by substituting it into Eq. (B-61). The result

$$\psi_k^+ = \phi_k + \frac{1}{E - H_0 + i\epsilon} V \psi_k^+$$

is one way to write the Lippmann-Schwinger equation. The equation may be reduced to a more familiar form by operating from the left with $E - H_0 + i\epsilon$ and taking the limit $\epsilon \to 0$,

$$(E - H_0)\psi_k^+ = (E - H_0)\phi_k + V\psi_k^+$$

The first term on the right-hand side vanishes because of Eq. (B-55) and the rest of the equation is identical to Eq. (B-54).

If we replace ψ_k^+ on the right-hand side of Eq. (B-61) by its value in the same equation and repeat the process, we obtain an infinite series expansion of ψ_k^+ in terms of ϕ_k,

$$\begin{aligned} \psi_k^+ &= \phi_k + G^+V(\phi_k + G^+V\psi_k^+) \\ &= \phi_k + G^+V\phi_k + G^+VG^+V(\phi_k + G^+V\psi_k^+) \\ &= \Big(1 + \sum_{n=1}^{\infty}(G^+V)^n\Big)\phi_k \end{aligned} \tag{B-63}$$

This gives us a Born series expansion of the scattering amplitude if we substitute the expansion for ψ_k^+ into Eq. (B-60).

t-matrix. We have seen earlier that the scattering amplitude $(-\mu/2\pi\hbar^2)\langle\phi_{k'}|V|\psi_k^+\rangle$ given by Eq. (B-60) is not useful directly for calculating cross sections because of its dependence on ψ_k^+. For many purposes it is more convenient to define a transition matrix, or t-matrix, satisfying the relation

$$\langle\phi_{k'}|t|\phi_k\rangle = \langle\phi_{k'}|V|\psi_k^+\rangle \tag{B-64}$$

In terms of the t-matrix, the scattering amplitude is a function of matrix elements involving only ϕ_k, the solution of the homogeneous equation given in Eq. (B-55). Again, this is useful mainly for formal work, as the t-matrix itself cannot be written down unless we solve the scattering problem first. For the simple case of H_0 consisting of the kinetic energy operator only, the elements of the t-matrix involve only plane wave states.

Using the series expansion of ψ_k^+ given in Eq. (B-63), we can write the elements of the t-matrix as

$$\langle\phi_{k'}|t|\phi_k\rangle = \langle\phi_{k'}|V(1+\sum_{n=1}^{\infty}(G^+V)^n)|\phi_k\rangle$$

Since the equality holds for arbitrary ϕ_k and $\phi_{k'}$, we obtain a relation between the operators involved,

$$t = V(1+\sum_{n=1}^{\infty}(G^+V)^n)$$

This can be put in a more compact form. Since the summation is taken up to infinity, we can take one product of G^+ with V out of the summation and rewrite the equation in the form

$$t = V + VG^+V + VG^+V\sum_{n=1}^{\infty}(G^+V)^n = V + VG^+\Big\{V + V\sum_{n=1}^{\infty}(G^+V)^n\Big\}$$

The quantity inside the curly brackets is nothing but the transition operator t itself, and we obtain the result

$$t = V + VG^+t$$

a form that is convenient as the starting point of many other derivations.

s-matrix. The s-matrix may be expressed in terms of the t-matrix using the relation

$$\langle\phi_p|\boldsymbol{S}|\phi_q\rangle = \delta_{pq} - 2\pi i\delta(E_p - E_q)\,\langle\phi_p|\boldsymbol{t}|\phi_q\rangle$$

The definition of the s-matrix is usually introduced through the time development operator $U(t,t_0)$ in the interaction representation of quantum mechanics (see, e.g., Sakurai [121] and Schiff [125]).

For most elementary applications, the time dependence of a quantum-mechanical state is expressed in the Schrödinger representation. Here, the operators are time independent; all the time dependence resides with the wave functions $\Psi_s(t)$. Using Eq. (B-1), we obtain the result

$$i\hbar\frac{\partial}{\partial t}\Psi_s(t) = H\Psi_s(t) \qquad \text{(B-65)}$$

where the subscript s emphasizes that the wave function is in the Schrödinger representation. To simplify the notation, we have suppressed all arguments other than time. Alternatively, one can work in the Heisenberg representation where, in contrast, the wave function is time independent and all time dependence is built into the operators.

In the *interaction representation*, the time dependence of a system is partly in the operator and partly in the wave function. The Hamiltonian is divided into two parts

$$H = H_0 + H_I$$

Wave functions $\Psi(t)$ and operators $\hat{O}(t)$ in this representation are related to those in the Schrödinger representation through the transformations

$$\Psi(t) = e^{iH_0t/\hbar}\Psi_s(t) \qquad \text{(B-66)}$$

$$\hat{O}(t) = e^{iH_0t/\hbar}\hat{O}_s e^{-iH_0t/\hbar} \qquad \text{(B-67)}$$

As a result, the time development of a state in the interaction representation is given by the equation

$$i\hbar\frac{\partial}{\partial t}\Psi(t) = H_I(t)\Psi(t)$$

as can be seen by substituting the inverse of Eq. (B-66) into (B-65). For many purposes, such an approach can be simpler than working in the Schrödinger representation, especially if H_I is only a small part of the complete Hamiltonian.

We can now define the time development operator $\boldsymbol{U}(t_0, t)$ that takes a state from time t_0 to time t in the interaction representation

$$\Psi(t) = \boldsymbol{U}(t, t_0)\Psi(t_0)$$

On substituting this definition in to Eqs. (B-66) and (B-67), we obtain an equation for $\boldsymbol{U}(t_0, t)$,

$$i\hbar\frac{\partial}{\partial t}\boldsymbol{U}(t, t_0) = H_I(t)\boldsymbol{U}(t, t_0)$$

The solution of this equation may be given as an integral equation,

$$\boldsymbol{U}(t, t_0) = 1 - i\hbar\int_{t_0}^{t} H_I(t)\boldsymbol{U}(t, t_0)\,dt$$

The s-matrix operator is defined by the following relation:

$$\boldsymbol{S} = \lim_{\substack{t\to+\infty \\ t'\to-\infty}} \boldsymbol{U}(t, t')$$

It is easy to see that the matrix elements of operator $\boldsymbol{S}$ between specific initial and final states are proportional to the scattering amplitude, as both quantities are related to the probability of finding a system in the final state at $t = +\infty$ if it started out from an initial state at $t = -\infty$.

In terms of phase shifts, the element of the s-matrix for partial wave ℓ is given by

$$\langle\ell|\boldsymbol{S}|\ell\rangle \sim e^{2i\delta_\ell}$$

The analogous relation for the t-matrix element is

$$\langle\ell|\boldsymbol{t}|\ell\rangle \sim e^{i\delta_\ell}\sin\delta_\ell$$

The advantage of using the s-matrix for scattering problems is its unitarity and other symmetry properties that are convenient in more advanced treatments.

Bibliography

[1] S. Åberg, H. Flocard, and W. Nazarewicz. Nuclear shapes in mean field theory. *Ann. Rev. Nucl. Part. Sci.*, 40:439, 1990.

[2] M. Abramowitz and I.A. Stegun, editors. *Handbook of Mathematical Functions.* Dover, New York, 1965.

[3] M.C. Abreu et al. Production of $\rho + \omega$ and ϕ in p–W and S–U collisions at 200 GeV/nucleon. *Phys. Lett. B*, 368:239, 1996.

[4] F. Ajzenberg-Selove. Energy levels of light nuclei $A = 18 - 20$. *Nucl. Phys. A*, 475:1, 1987.

[5] S.V Akulinichev et al. Lepton-nucleus deep-inelastic scattering. *Phys. Rev. Lett.*, 55:2239, 1985.

[6] K. Alder et al. Study of nuclear structure by electromagnetic excitations with accelerated ions. *Rev. Mod. Phys.*, 28:432, 1956.

[7] K. Alder et al. Errata: Study of nuclear structure by electromagnetic excitations with accelerated ions. *Rev. Mod. Phys.*, 30:353, 1958.

[8] A. Aprahamian et al. First observation of a near-harmonic vibrational nucleus. *Phys. Rev. Lett.*, 59:535, 1987.

[9] A. Arima and F. Iachello. The interacting boson model. *Advances in Nuclear Physics*, 13:139, 1984.

[10] R.A. Arndt, J.S. Hyslop III, and L.D. Roper. Nucleon-nucleon partial-wave analysis to 1100 MeV. *Phys. Rev. D*, 35:128, 1987.

[11] D. Ashery and J.P. Schiffer. Pion absorption in nuclei. *Ann. Rev Nucl. Part. Sci.*, 36:207, 1986.

[12] J.J. Aubert et al. The ratio of the nucleon structure function F_2^N for ion and deuterium. *Phys. Lett. B*, 123:275, 1983.

[13] J. Audouze and S. Vauclair. *An Introduction to Nuclear Astrophysics.* Reidel Publishing Co., Dordrecht, Holland, 1980.

[14] F. Azgui et al. Feeding times of high-spin states in ^{152,154}Dy. *Nucl. Phys. A*, 439:573, 1985.

[15] R. Azuma et al. Constraints on the low-energy $E1$ cross section of $^{12}C(\alpha, \gamma)^{16}O$ from the β-delayed α spectrum of ^{16}N. *Phys. Rev. C*, 50:1194, 1994.

[16] H.W. Baer et al. Isobaric-analogue-state transitions in pion charge-exchange reactions above the $\Delta(1232)$ resonance. In R.J. Peterson and D.D. Strottman, editors, *Pion-Nucleus Physics*, volume 163 of *AIP Conf. Proc.*, page 67. American Institute of Physics, New York, 1987.

[17] C. Baglin et al. J/ψ and muon-pair cross-sections in proton-nucleus and nucleus-nucleus collisions at 200 GeV/nucleon. *Phys. Lett. B*, 270:105, 1991.

[18] C. Baglin et al. Ψ' and J/Ψ production in p–W, p–U and S–U interactions at 200 GeV/nucleon. *Phys. Lett. B*, 345:617, 1995.

[19] J.N. Bahcall. *Neutrino Astrophysics*. Cambridge University Press, Cambridge, 1989.

[20] J.N. Bahcall. The solar neutrino problem. *Sci. Am.*, 262:54, May 1990.

[21] J.N. Bahcall and B.R. Holstein. Solar neutrino from the decay of ^{8}B. *Phys. Rev. C*, 33:2121, 1986.

[22] R.M. Barnett et al. Review of particle physics. *Phys. Rev. D*, 54:1, 1996.

[23] C.J. Batty, G.T.A. Squier, and G.K. Turner. Forward pion-nucleus elastic scattering amplitudes. *Nucl. Phys. B*, 67:492, 1973.

[24] W. Bauhoff, H.V. von Geramb, and G. Pálla. Nonlocal and local equivalent miscroscopic optical potentials. *Phys. Rev. C*, 27:2466, 1983.

[25] J. Benecke et al. Rapidity-gap separation and study of single-diffraction dissociation in pp collisions at 12 and 24 GeV/c. *Nucl. Phys. B*, 76:29, 1974.

[26] W. Bertozzi et al. Contributions of neutrons to elastic electron scattering from nuclei. *Phys. Lett. B*, 41:408, 1972.

[27] F.E. Bertrand. Giant multipole resonances—perspective after 10 years. *Nucl. Phys. A*, 354:129c, 1981.

[28] F.E. Bertrand. Giant resonances—why protons. In J.M. Greben, editor, *Studying Nuclei with Medium Energy Protons,* TRIUMF Report TR-83-3, page 181. TRIUMF, Vancouver, 1983.

[29] G. Bertsch et al. Interactions for inelastic scattering derived from realistic potentials. *Nucl. Phys. A*, 284:399, 1977.

[30] H.A. Bethe. Nuclear physics: B nuclear dynamics, theoretical. *Rev. Mod. Phys.*, 9:69, 1937.

[31] H.A. Bethe. What holds the nucleus together. *Sci. Am.*, 189:58, September 1953.

[32] J.M. Blatt and V.F. Weisskopf. *Theoretical Nuclear Physics.* Wiley, New York, 1952.

[33] R.J. Blin-Stoyle. *Fundamental Interactions and the Nucleus.* North-Holland, Amsterdam, 1973.

[34] A. Bohr and B.R. Mottelson. *Nuclear Structure*, volume I. Benjamin, New York, 1969.

[35] A. Bohr and B.R. Mottelson. *Nuclear Structure*, volume II. Benjamin, New York, 1975.

[36] L. Brekke and J.L. Rosner. Baryon magnetic moments in the quark model—a status report. *Comments Nucl. Part. Phys.*, 18:83, 1988.

[37] D.M. Brink and G.R. Satchler. *Angular Momentum,* second edition. Clarendon Press, Oxford, 1968.

[38] D.A. Bromley. Nuclear molecules. *Sci. Am.*, 239:58, December 1978.

[39] B.A. Brown, W. Chung, and B.H. Wildenthal. Empirical renormalization of the one-body Gamow-Teller β-decay matrix elements in the $1s$-$0d$ shell. *Phys. Rev. Lett.*, 40:1631, 1978.

[40] M.B. Burbank and others. The ^{20}Ne(d,n)^{21}Na reaction. *Nucl. Phys. A*, 119:184, 1968.

[41] J. Bystricky, F. Lehar, and P. Winternitz. Formalism of nucleon-nucleon elastic scattering experiments. *J. Phys.* (Paris), 39:1, 1978.

[42] A.G.W. Cameron. Elemental and nuclidic abundances in the solar system. In C.A. Barnes, D.D. Clayton, and D.N. Schramm, editors, *Essays in Nuclear Astrophysics*, page 23. Cambridge University Press, London, 1982.

[43] L.S. Celenza and C.M. Shakin. *Relativistic Nuclear Physics.* World Scientific, Singapore, 1986.

[44] D.D. Clayton. *Principles of Stellar Evolution and Nucleosynthesis.* McGraw-Hill, New York, 1968.

[45] S. Cohen and D. Kurath. Effective interaction in the $1p$-shell. *Nucl. Phys.*, 73:1, 1965.

[46] C. Cohen-Tannoudji, B. Diu, and F. Laloë. *Quantum Mechanics,* English translation by S.R. Hemley, N. Ostrowsky, and D. Ostrowsky. Wiley, New York, 1977.

[47] M. Danos, V. Gillet, and M. Cauvin. *Methods in Relativistic Nuclear Physics.* North-Holland, Amsterdam, 1984.

[48] R. Davis. Attempt to detect the antineutrinos from a nuclear reactor by the ^{37}Cl$(\nu, e^-)^{37}$Ar reaction. *Phys. Rev.*, 97:766, 1955.

[49] A. de Shalit and H. Feshbach. *Theoretical Nuclear Physics.* Wiley, New York, 1974.

[50] A. de Shalit and I. Talmi. *Nuclear Shell Theory.* Academic Press, New York, 1963.

[51] H. de Vries, C.W. de Jager, and C. de Vries. Nuclear charge-density-distribution parameters from elastic electron scattering. *Atom. Data Nucl. Data Tables*, 36:495, 1987.

[52] S. Devons and I. Duerdoth. Muonic atoms. In M. Baranger and E. Vogt, editors, *Advances in Nuclear Physics*, volume 2, page 295. Plenum Press, New York, 1969.

[53] W. Dilg et al. Level density parameters for the back-shifted Fermi gas model in the mass range $40 < A < 250$. *Nucl. Phys. A*, 217:269, 1973.

[54] J. Dudek. Nuclear superdeformation at high spins. In A. Faessler, editor, *Progress in Particle and Nuclear Physics*, volume 28, page 131. Pergamon Press, Oxford, 1992.

[55] J.M. Eisenberg and W. Greiner. *Excitation Mechanisms of the Nucleus.* North-Holland, Amsterdam, 1970.

[56] H. Ejiri and M.J.A. de Voigt. *Gamma-Ray and Electron Spectroscopy in Nuclear Physics.* Clarendon Press, Oxford, 1989.

[57] S.R. Elliot, H.F. Hahn, and R.R. Moe. Limit on neutrinoless double-beta decay with majoron emission in ^{82}Se. *Phys. Rev. Lett.*, 59:1649, 1987.

[58] T. Ericson and W. Weise. *Pions and Nuclei.* Clarendon Press, Oxford, 1988.

[59] R.D Evans. *The Atomic Nucleus.* McGraw-Hill, New York, 1955.

[60] U. Fano. *Tables for the Analysis of Beta Spectra*, volume 13 of *Applied Mathematics Series.* National Bureau of Standards, Washington, DC, 1951.

[61] E. Feenberg and G. Trigg. The interpretation of comparative half-lives in the Fermi theory of beta-decay. *Rev. Mod. Phys.*, 22:399, 1950.

[62] A.L. Fetter and J.D. Walecka. *Quantum Theory of Many-Particle Systems.* McGraw-Hill, New York, 1971.

[63] R.P. Feynman and A.R. Hibbs. *Quantum Mechanics and Path Integrals.* McGraw-Hill, New York, 1965.

[64] W.A. Fowler and F. Hoyle. Nucleosynthesis in massive stars and supernovae. *Astrophys. J. Suppl. Ser.*, 91:201, 1964.

[65] M.A. Franey and W.G. Love. Nucleon-nucleon *t*-matrix interaction for scattering at intermediate energies. *Phys. Rev. C*, 31:488, 1985.

[66] D. Frekers et al. Elastic and inelastic scattering of 362 MeV polarized protons from ^{40}Ca. *Phys. Rev. C*, 35:2236, 1987.

[67] B. Frois et al. High momentum-transfer electron scattering from ^{208}Pb. *Phys. Rev. Lett.*, 38:152, 1977.

[68] R.F. Frosch et al. Electron scattering studies of calcium and titanium isotopes. *Phys. Rev.*, 174:1380, 1968.

[69] Y. Fukuda et al. Solar neutrino data covering solar cycle 22. *Phys. Rev. Lett.*, 77:1683, 1996.

[70] B. Gabioud et al. n-n effective range from the photon spectrum of the reaction $\pi^- d \to \gamma nn$. *Phys. Lett. B*, 103:9, 1981.

[71] A. Galonsky. Evidence for Gamow-Teller strength in broad bumps in (p, n) and (^{3}He,t) spectra. In C.D. Goodman et al., editors, *The (p, n) Reaction and the Nucleon-Nucleon Force*, page 191. Plenum Press, New York, 1980.

[72] D. Garreta et al. Scattering of antiprotons from carbon at 46.8 MeV. *Phys. Lett. B*, 135:266, 1984.

[73] D.W. Glasgow et al. The Los Alamos Laboratory neutron-neutron scattering program. In P.G. Young et al., editors, *Nuclear Data for Basic and Applied Science*. Gordon & Breach, New York, 1985.

[74] D. Gogny. Self-consistent pairing calculations. In G. Ripka and M. Porneuf, editors, *Nuclear Self-Consistent Fields*, page 333. North-Holland, Amsterdam, 1975.

[75] M. Goldhaber, L. Grodzins, and A.W. Sunyan. Helicity of neutrinos. *Phys. Rev.*, 109:1015, 1958.

[76] H. Goldstein. *Classical Mechanics,* second edition. Addison-Wesley, Reading, MA, 1980.

[77] P.A.M. Gram. Systemics of inclusive double charge exchange. In R.J. Peterson and D.D. Strottman, editors, *Pion-Nucleus Physics*, volume 163 of *AIP Conf. Proc.*, page 79. American Institute of Physics, New York, 1987.

[78] G.L. Greene et al. New determination of the deuteron binding energy and the neutron mass. *Phys. Rev. Lett.*, 56:819, 1986.

[79] Cross Section Evaluation Group. ENDF/B-VI Summary documentation. In P.F. Rose, editor, *Report BNL-NCS 17541*. Nuclear Data Center, Brookhaven National Laboratory, Upton, NY, 1991.

[80] F. Halzen and D. Martin. *Quarks and Leptons*. Wiley, New York, 1984.

[81] J.W. Harris and B. Müller. The search for the quark-gluon plasma. *Ann. Rev. Nucl. Part. Sci.*, 46:71, 1996.

[82] W. Hauser and H. Feshbach. The inelastic scattering of neutrons. *Phys. Rev.*, 87:366, 1952.

[83] K. Hikasa et al. Review of particle properties. *Phys. Rev. D*, 45:S1, 1992.

[84] S. Hofmann et al. The new element 112. *Z. Phys. A*, 354:229, 1996.

[85] R. Hofstadter and R. Herman. Electric and magnetic structure of the proton and neutron. *Phys. Rev. Lett.*, 6:293, 1961.

[86] J. Hufner and M. Thies. Pion-nucleus scattering and absorption as a solution of the Boltzmann equation. *Phys. Rev. C*, 20:273, 1979.

[87] C.H.Q. Ingram. Pion-nucleus scattering around the 3-3 resonance. In E.V. Hungerford III, editor, *Meson-Nuclear Physics—1979*, volume 54 of *AIP Conf. Proc.*, page 455. American Institute of Physics, New York, 1979.

[88] J. Jänecke and P.J. Masson. Mass predictions from the Garvey-Kelson mass relations. *Atom. Data Nucl. Data Tables*, 39:265, 1988.

[89] H.-Q. Jin et al. Identification and quadrupole-moment measurement of a superdeformed band in ^{84}Zr. *Phys. Rev. Lett.*, 75:1471, 1995.

[90] G. Jones. $NN \to \pi d$ and $NN \to NN\pi$—a review of experimental results. In R.D. Bent, editor, *Pion Production and Absorption in Nuclei—1981*, volume 79 of *AIP Conf. Proc.*, page 15. American Institute of Physics, New York, 1981.

[91] A.J. Kirwan et al. Mean-lifetime measurements within the superdeformed second minimum in ^{132}Ce. *Phys. Rev. Lett.*, 58:467, 1987.

[92] C.M. Ko. Role of deformation in deep ineastic heavy ion colliosions. *Phys. Lett. B*, 81:299, 1979.

[93] A.S. Kronfeld and P.B. Mackenzie. Progress in quantum chromodynamics with lattice gauge theory. *Ann. Rev. Nucl. Part. Sci.*, 43:793, 1993.

[94] L.D. Landau and E.M. Lifshitz. *Statistical Physics*, second edition. Pergamon Press, Oxford, 1969.

[95] C.M. Lederer and V.S. Shirley, editors. *Table of Isotopes*, seventh edition. Wiley, New York, 1978.

[96] T.D. Lee. *Particle Physics and Introduction to Field Theory*. Harwood, Chur, Switzerland, 1981.

[97] T.M. Liss and P.L. Tipton. The discovery of the top quark. *Sci. Am.*, 277:54, September 1997.

[98] R.M. Littauer, H.F. Schopper, and R.R. Wilson. Structure of the proton and neutron. *Phys. Rev. Lett.*, 7:144, 1961.

[99] V.A. Lubimov et al. An estimate of the ν_e mass from the β-spectrum of tritium in the valine molecule. *Phys. Lett. B*, 94:266, 1980.

[100] R Machleidt. The meson theory of nuclear forces and nuclear matter. In M.B. Johnson and A. Picklesimer, editors, *Relativistic Dynamics and Quark-Nuclear Physics*. Wiley, New York, 1986.

[101] R. Machleidt, K. Holinde, and Ch. Elster. The Bonn meson-exchange model for the nucleon-nucleon interaction. *Phys. Rep.*, 149:1, 1987.

[102] M.G. Mayer and J.H.D. Jensen. *Elementary Theory of Nuclear Shell Structure*. Wiley, New York, 1955.

[103] E. Merzbacher. *Quantum Mechanics,* second edition. Wiley, New York, 1961.

[104] A. Messiah. *Quantum Mechanics*. North-Holland, Amsterdam, 1966.

[105] M. Morita. *Beta Decay and Muon Capture*. Benjamin, Reading, MA, 1973.

[106] P.M. Morse and H. Feshbach. *Methods of Theoretical Physics*. McGraw-Hill, New York, 1953.

[107] W.M. Morse et al. π^+p, K^+p, and pp topological cross sections and inclusive interactions at 100 GeV using a hybrid bubble-chamber-spark-chamber system and a tagged beam. *Phys. Rev. D*, 15:66, 1977.

[108] W.D. Myers and W.J. Swiatecki. The macroscopic approach to nuclear masses and deformations. *Ann. Rev. Nucl. Part. Sci.*, 32:309, 1982.

[109] D.E. Nagle, Johnson M.B., and D. Measday. Pion physics at the meson factories. *Phys. Today*, page 56, April 1987.

[110] J.W. Negele. Structure of finite nuclei in the local-density approximation. *Phys. Rev. C*, 1:1260, 1970.

[111] P.J. Nolan et al. A shape change along the yrast line in ^{132}Ce? *J. Phys. G*, 11:L17, 1985.

[112] H.P. Noyes. The nucleon-nucleon effective range expansion parameters. *Ann. Rev. Nucl. Sci.*, 22:465, 1972.

[113] B.M. Nyako et al. Observation of superdeformation in ^{152}Dy. *Phys. Rev. Lett.*, 52:507, 1984.

[114] S. Okubo and R.E. Marshak. Velocity dependence of the two-nucleon interaction. *Ann. Phys.*, 4:166, 1958.

[115] D.H. Perkins. *Introduction to High Energy Physics,* third edition. Addison-Wesley, Menlo Park, CA, 1987.

[116] P.C. Peterson et al. Measurement of the Σ^0-Λ transition magnetic moment. *Phys. Rev. Lett.*, 57:949, 1986.

[117] W. Reuter, F.K. Merle, and H. Miska. Nuclear charge distribution and rms radius of ^{12}C from absolute elastic electron scattering measurements. *Phys. Rev. C*, 26:806, 1982.

[118] M. Rho and G.E. Brown. The role of chiral invariance in nuclei. *Comments Nucl. Part. Phys.*, 10:201, 1981.

[119] P. Ring and P. Schuck. *The Nuclear Many-Body Problem.* Springer-Verlag, New York, 1980.

[120] C.E. Rolfs and W.S. Rodney. *Cauldrons in the Cosmos.* University of Chicago Press, Chicago, 1988.

[121] J.J. Sakurai. *Advanced Quantum Mechanics.* Addison-Wesley, Reading, MA, 1967.

[122] D. Santos et al. Observation of excited superdeformed bands in ^{132}Ce and evidence for identical bands in the mass 130 region. *Phys. Rev. Lett.*, 74:1708, 1995.

[123] G.R. Satchler. *Direct Nuclear Reactions.* Oxford University Press, Oxford, 1983.

[124] H. Savajols et al. Lifetime measurement of superdeformed bands in $^{148-149}Gd$ and ^{152}Dy: evidence for structure-dependent elongations. *Phys. Rev. Lett.*, 76:4480, 1996.

[125] L.I. Schiff. *Quantum Mechanics,* third edition. McGraw-Hill, New York, 1968.

[126] P. Schwandt et al. Analyzing power of proton-nucleus elastic scattering between 80 and 180 MeV. *Phys. Rev. C*, 26:55, 1982.

[127] I. Sick and J.S. McCarthy. Elastic electron scattering from ^{12}C and ^{16}O. *Nucl. Phys. A*, 150:631, 1970.

[128] B. Singh et al. Review of β decay log ft values. *Nucl. Data Sheets,* (to be published).

[129] A. Sirlin. Remarks concerning the $O(Z\alpha^2)$ corrections to Fermi decays, conserved vector-current predictions, and universality. *Phys. Rev. D*, 35:3423, 1987.

[130] T. Sloan, G. Smadja, and R. Voss. The quark structure of the nucleon from the CERN muon experiments. *Phys. Rep.*, 162:45, 1988.

[131] J. Stachel and G.R. Young. Relativistic heavy ion physics at CERN and BNL. *Ann. Rev. Nucl. Part. Sci.*, 42:537, 1992.

[132] H.P. Stapp, T.J. Ypsilantis, and N. Metropolis. Phase-shift analysis of 310-MeV proton-proton scattering experiments. *Phys. Rev.*, 105:302, 1957.

[133] K. Stricker, J.A. Carr, and H. McManus. Pionic atoms and low energy elastic scattering. *Phys. Rev. C*, 22:2043, 1980.

[134] V.M. Strutinsky. Influence of nucleon shells on the energy of a nucleus. *Soviet J. Nucl. Phys.*, 3:449, 1966.

[135] W. Thomè et al. Charged particle multiplicity distributions in *pp* collisions at ISR energies. *Nucl. Phys. B*, 129:365, 1977.

[136] D.R. Tilley, H.R. Weller, and C.M. Cheves. Energy levels of light nuclei $A = 16 - 17$. *Nucl. Phys. A*, 564:1, 1993.

[137] W. Tobocman. *Theory of Direct Nuclear Reactions.* Oxford University Press, London, 1961.

[138] R. Vinh Mau. The Paris nucleon-nucleon potential. In M. Rho and D.H. Wilkinson, editors, *Mesons in Nuclei.* North-Holland, Amsterdam, 1979.

[139] H.V. von Geramb. The effective interaction in nuclei. In J.M. Greben, editor, *Studying Nuclei with Medium Energy Protons,* TRIUMF Report TR-83-3, page 1. TRIUMF, Vancouver, 1983.

[140] H.V. von Geramb, F.A. Brieva, and J.R. Rook. Effective nuclear matter interactions applied to finite nuclei. In H.V. von Geramb, editor, *Microscopic Optical Potentials*, volume 89 of *Lecture Notes in Physics*, page 104. Springer-Verlag, Berlin, 1979.

[141] G. Wallerstein et al. Synthesis of the elements in stars: forty years of progress. *Rev. Mod. Phys.*, 69:995, 1997.

[142] D. Ward et al. Rotational bands in ^{238}U. *Nucl. Phys. A*, 600:88, 1996.

[143] T.A. Weaver and S.E. Woosley. Nucleosynthesis in massive stars and the $^{12}C(\alpha,\gamma)^{16}O$ reaction rate. *Phys. Rep.*, 227:65, 1993.

[144] S. Weinberg. *The First Three Minutes.* Basic Books, New York, 1993.

[145] T.R. Werner and J. Dudek. Shape coexistence effects of super- and hyperdeformed configurations in rotating nuclei with $50 \leq Z \leq 74$. *Atom. Data Nucl. Data Tables*, 50:179, 1992.

[146] T.R. Werner and J. Dudek. Shape coexistence effects of super- and hyperdeformed configurations in deformed nuclei II: nuclei with $42 \leq Z \leq 56$ and $74 \leq Z \leq 92$. *Atom. Data Nucl. Data Tables*, 59:1, 1995.

[147] B.H. Wildenthal and W. Chung. Implications of experimental magnetic moment values in light nuclei for the presence and characteristics of mesonic exchange currents. In M. Rho and D.H. Wilkinson, editors, *Mesons in Nuclei*, page 721. North-Holland, Amsterdam, 1979.

[148] K. Wildermuth and Y.C. Tang. *A Unified Theory of the Nucleus.* Vieweg, Braunschweig, Germany, 1977.

[149] D. Wilkinson. Fundamental symmetries. *J. Phys. Soc. Jpn.*, 55 (Suppl.):347, 1986.

[150] C.Y. Wong. *Introduction to High-Energy Heavy-Ion Collisions.* World Scientific, Singapore, 1994.

[151] S.S.M. Wong. *Nuclear Statistical Spectroscopy.* Oxford University Press, New York, 1986.

[152] S.S.M. Wong. *Introductory Nuclear Physics.* Prentice-Hall, Englewood Cliffs, NJ, 1990.

[153] S.S.M. Wong. *Computational Methods in Physics and Engineering,* second edition. World Scientific, Singapore, 1997.

[154] C.S. Wu et al. Experimental test of parity conservation in beta decay. *Phys. Rev.*, 105:1413, 1957.

Index